Lecture Notes in Electrical Engineering

Volume 102

For further volumes:
http://www.springer.com/series/7818

Jame J. Park · Hai Jin · Xiaofei Liao
Ran Zheng
Editors

Proceedings of the International Conference on Human-centric Computing 2011 and Embedded and Multimedia Computing 2011

HumanCom & EMC 2011

Prof. Jame J. Park
SeoulTech Computer Science
and Engineering
Seoul University of Science
and Technology
Gongreung 2-dong 172
Seoul 139-743
South Korea
e-mail: parkjonghyuk1@hotmail.com

Prof. Hai Jin
Computer Science and Engineering
Huazhong University of Science
and Technology
Luoyu Road 1037
430074 Wuhan Hubei
People's Republic of China
e-mail: jinhust@gmail.com

Assoc. Prof. Xiaofei Liao
Computer Science and Technology
Huazhong University of Science
and Technology
Luoyu Road 1037
430074 Wuhan Hubei
People's Republic of China
e-mail: xfliao@mail.hust.edu.cn

Assoc. Prof. Ran Zheng
Computer Science and Technology
Huazhong University of Science
and Technology
Luoyu Road 1037
430074 Wuhan Hubei
People's Republic of China
e-mail: zhraner@hust.edu.cn

ISSN 1876-1100
ISBN 978-94-007-3713-6
DOI 10.1007/978-94-007-2105-0
Springer Dordrecht Heidelberg London New York

e-ISSN 1876-1119
ISBN 978-94-007-2105-0 (eBook)

Softcover reprint of the hardcover 1st edition 2011

Cover design: eStudio Calamar, Berlin/Figueres

Printed on acid-free paper

Springer is part of Springer Science+Business Media (www.springer.com)

Foreword

The 4th International Conference on Human-centric Computing (HumanCom'11) and The 6th International Conference on Embedded and Multimedia Computing (EMC'11), HumanCom'11-EMC'11, aim at providing an exciting platform and forum for researchers and developers from academia and industry to present their latest research in the fields of human-centric computing and EM (Embedded and Multimedia) computing.

This year HumanCom'11-EMC'11 received 145 submissions from authors in 12 countries. All papers were reviewed by HumanCom'11 and EMC'11 program committees, whose members are from USA, UK, Germany, France, Korea, China, and etc. Each paper received at least 3 reviews. Based on these reviews, the program co-chairs accepted papers into two categories:

Regular papers: 34 papers passed the most stringent selection. Each paper is up to 12 pages in the conference proceedings. The acceptance rate for select papers is 23.4%.

Short papers: 10 additional papers are of sufficient publishable valuable. Each paper is up to 8 pages.

For HumanCom'11-EMC'11, we have invited three distinguished speakers for keynote speeches:

- Graph-based Structure Search, by Xuemin Lin (University of New South Wales, Australia and Eastern China Normal University, China)
- Location Significance Ranking from Quality Enhanced Trajectory Data, by Xiaofang Zhou (University of Queensland, Australia)
- Ubiquitous Neural Interface, by Bin Hu (Lanzhou University, China and Birmingham City University, UK)

Two workshops were held in conjunction with HumanCom'11-EMC'11:

- The 2011 International Annual Workshop on Parallel and Distributed Computing (PDC 2011)
- The 2nd International Workshop on Engineering and Numerical Computation (EMC 2011)

The proceedings of the workshops are also included in this volume.

We would like to thank all the authors of submitted papers for their works and their interests in the conferences. We would like to express our sincere appreciation for all members of the program committees. It was from these reviews we have identified a set of submissions that were clear, relevant, and described high quality work in related research fields. It was through these reviews, authors received objective and often detailed feedback from a diverse group of experts. In addition, we would like to thank the steering committee members Prof. Stephen S. Yau, Laurence T. Yang and Hamid R. Arabnia for their invaluable advice and guidance. Our special thanks also go to all workshop organizers for their professional expertise and excellence in organizing the attractive workshops. The conference proceedings would not have been possible without the supports of these individuals and organizations.

At closing, it is our hope that all of these efforts have helped to improve and promote human-centric computing and EM computing researches in China, other Asian countries, United States and beyond.

May 2011

Jame J. Park
Hai Jin
Xiaofei Liao
Ran Zheng

HumanCom 2011 Organizing Committee

Steering Committee

Stephen S. Yau	Arizona State University, USA
James J. (Jong Hyuk) Park	Seoul National University of Science and Technology, Korea
Laurence T. Yang	St Francis Xavier University, Canada
Hamid R. Arabnia	The University of Georgia, USA

General Chair

Hai Jin	Huazhong University of Science and Technology, China

Program Chair

Xiaofei Liao	Huazhong University of Science and Technology, China

Workshop Co-Chairs

Qun Jin	Waseda University, Japan
Deqing Zou	Huazhong University of Science and Technology, China

Publication Chairs

Sang-Soo Yeo	Mokwon University, Korea

Publicity Chairs

Ching-Hsien (Robert) Hsu	Chung Hua University, Taiwan
Wenbin Jiang	Huazhong University of Science and Technology, China

Local Arrangement Committee

Minghui Zheng	Hubei University for Nationalities, China
Yingshu Liu	Huazhong University of Science and Technology, China

Web and System Management Committee

Shizheng Jiang	St Francis Xavier University, Canada
Jingxiang Zeng	Huazhong University of Science and Technology, China

Registration and Finance Chair

Na Zhao	Huazhong University of Science and Technology, China

Program Committee

Computer-Assisted Learning, Cognition and Semantic Systems

Rodrigo Martinez-Bejar	Universidad de Murcia, Spain
Shu Chen	Cheng Southern Taiwan University, Taiwan
Shu-Ching Chen	Florida International University, USA
Wenjian Jiang	Orange Labs Beijing, China
Y.M. Hwang	National Cheng-Kung University, Taiwan

Human–Computer Interaction and Social Computing

Ashkan Sami	Shiraz University, Iran
Chia-Hung Yeh	National Sun Yat-sen University, Taiwan
Guandong Xu	Victoria University, Australia
Keqiu Li	Dalian University of Technology, China
Lin Liu	University of South Australia, Australia
Mehul Bhatt	University of Bremen, Germany
Ruay-Shiung Chang	National Dong Hwa University, Taiwan
Ryszard Tadeusiewicz	AGH University of Science and Technology, Poland

Privacy, Security and Trust Management

Claudio Ardagna	University of Milan, Italy
Dieter Gollmann	The Hamburg University of Technology, Germany
Indrakshi Ray	Colorado State University, USA
Jordi Forne	Technical University of Catalonia, Spain
Margaret Tan	Nanyang Technological University, Singapore
Masakazu	Soshi Hiroshima City University, Japan
Tanveer Zia	Charles Sturt University, Australia
Zhenglu Yang	University of Tokyo, Japan

Ubiquitous Computing, Mobile Systems and Applications

Chih-Lin Hu	National Central University, Taiwan
Dennis Pfisterer	University of Luebeck, Germany
Dragan Lvetic	University of Novi Sad, Republic of Serbia
Farhad Arbab	Centrum Wiskunde and Informatica, Netherlands
Han-Chieh Chao	National Ilan University, Taiwan
Jenq-Muh Hsu	National Chiayi University, Taiwan

Jen-Wei Hsieh	National Taiwan University of Technology, Taiwan
Jun Wu	National Pingtung Institute of Commerce, Taiwan
Kuei-Ping Shih	Tamkang University, Taiwan
Naveen Chilamkurti	La Trobe University, Australia
Paolo G. Bottoni	Sapienza University of Rome, Italy
Paolo Trunfio	University of Calabria, Italy
Qishi Wu	University of Memphis, USA
Stefan Fischer	University of Luebeck, Germany
Susumu Date	Osaka University, Japan
Yeh-Ching Chung	National Tsing Hua University, China
Yo-Ping Huang	National Taipei University of Technology, Taiwan

Virtualization Technologies for Desktop Applications

Baoliu Ye	Nanjing University, China
Chuliang Weng	Shanghai Jiaotong University, China
Guangtao Xue	Shanghai Jiao Tong University, China
Guojun Wang	Central South University, China
Juan Tourino	University of A Coruna, Spain
Song Wu	Huazhong University of Science and Technology, China
Yingwei Luo	Peking University, China

EMC 2011 Organizing Committee

Steering Chair

James J. (Jong Hyuk) Park	Seoul National University of Science and Technology, Korea

General Chair

Hai Jin	Huazhong University of Science and Technology, China

Program Co-Chairs

Ran Zheng	Huazhong University of Science and Technology, China
Tao Gu	University of Southern Denmark, Denmark

Workshop Chair

Deqing Zou	Huazhong University of Science and Technology, China

Publication Chair

Sang-Soo Yeo	Mokwon University, Korea

Steering Committee

Borko Furht	Florida Atlantic University, USA
Edwin Sha	University of Texas at Dallas, USA
Jorg Henkel	Karlsruhe Institute of Technology, Germany
Laurence T. Yang	St. Francis Xavier University, Canada
Morris Chang	Iowa State University, USA
Sethuraman Panchanathan	Arizona State University, USA
Vincenzo Loia	University of Salerno, Italy

Publicity Chairs

Ching-Hsien (Robert) Hsu	Chung Hua University, Taiwan
Wenbin Jiang	Huazhong University of Science and Technology, China

Local Arrangement Chairs

Minghui Zheng	Hubei University for Nationalities, China (Chair)
Yingshu Liu	Huazhong University of Science and Technology, China

Web and System Management Chairs

Shizheng Jiang	St Francis Xavier University, Canada
Jingxiang Zeng	Huazhong University of Science and Technology, China

Registration and Finance Chair

Na Zhao	Huazhong University of Science and Technology, China

Program Committee

Cyber-Physical Systems and Real-Time Systems

Bjorn Andersson	CISTER/IPP Hurray Research Unit, Portugal
Borzoo Bonakdarpour	University of Waterloo, Canada
Damir Isovic	Malardalen University, Sweden
Feng Xia	Dalian University of Technology, China
Gang Zeng	Nagoya University, Japan
Gerhard Fohler	University of Kaiserslautern, Germany
Lei Zhang	Institute of Computing Technology Chinese Academic of Sciences, China
Linh T.X. Phan	University of Pennsylvania, USA
Meikang Qiu	University of New Orlean, USA
Moris Behnam	Malardalen University, Sweden
Nicolas Navet	Institut National des Sciences Appliquees de Rennes, France
Qingxu Deng	Northeastern University, USA
Shengquan Wang	University of Michigan – Dearborn, USA
Tei-Wei Kuo	National Taiwan University, Taiwan
Wei Zhang	Virginia Commonwealth University, USA
Zhiwen Yu	Northwestern Polytechnical University, China
Zonghua Gu	Zhejiang University, China

Distributed Multimedia Systems

Bill Grosky	University of Michigan, USA
Cheng-Hsin Hsu	Deutsche Telekom, USA
Dewan Ahmed	University of Ottawa, Canada
Guntur Ravindra	National University of Singapore, Singapore
Laszlo Boszormenyi	Klagenfurt University, Austria
Mei-Ling Shyu	University of Miami, USA
Nabil J. Sarhan	Wayne State University, USA
Oscar Au	The Hong Kong University of Science and Technology, Hong Kong
Pascal Lorenz	University of Haute Alsace, France
Ralf Klamma	RWTH Aachen University, Aachen, Germany
Razib Iqbal	University of Ottawa, Canada
Savvas Chatzichristofis	Democritus University of Thrace, Greece
Sheng-Wei (Kuan-Ta) Chen	Academia Sinica, Taiwan

Embedded Systems, Software and Application

Chih-Chi Cheng	Massachusetts Institute of Technology, USA
Ching-Hsien Hsu	Chung Hua University, Taiwan
Chih-Wei Liu	National Chiao Tung University, Taiwan
Eugene John	University of Texas at San Antonio, USA
Jogesh Muppala	Hong Kong University of Science and Technology, Hong Kong
John O'Donnell	University of Glasgow, UK
Luis Gomes	Universidade Nova de Lisboa, Portugal
Nicolas Navet	Institut national de recherche en informatique et en automatique, France
Olivier DeForge	Institut National des Sciences Appliquees de Rennes, France
Seon Wook Kim	Korea University, Korea
Tien-Fu Chen	National Chung Cheng University, Taiwan
Ya-Shu Chen	National Taiwan University of Science and Technology, Taiwan
Yuan-Hao Chang	National Taipei University of Technology, Taiwan

Multimedia Computing and Intelligent Services

Akihiro Sugimoto	National Institute of Informatics, Japan
Andrew D. Bagdanov	Universitat Autonoma de Barcelona, Spain
Chris Poppe	Ghent University, Belgium
Datchakorn Tancharoen	Panyapiwat Institute of Technology, Thailand
Debzani Deb	Indiana University of PA, USA
Jingdong Wang	Microsoft Research Asia, China
Lyndon Kennedy	Yahoo! Research, USA
Naoko Nitta	Osaka University, Japan
Qi Tian	University of Texas at San Antonio, USA
Xiaoyi Jiang	University of Munster, Germany
Xiaoyong Wei	Sichuan University, China
Yadong Mu	National University of Singapore, Singapore
Yiannis Kompatsiaris	Informatics and Telematics Institute, Greece
Yiyu Cai	Nanyang Technological University, Singapore

Multimedia Software Engineering

Angela Guercio	Kent State University, USA
Brent Lagesse	Oak Ridge National Laboratory, USA
Charles Shoniregun	Infonomics Society, UK
Dimitrios Katsaros	University of Thessaly, Greece
Farag Waleed	Indiana University of PA, USA
Galyna Akmayeva	Infonomics Society, UK

PDC 2011 Executive Committee

General Chair

Daoxu Chen	Nanjing University, China
Zhiying Wang	National University of Defense Technology, China

Program Chair

Guihai Chen	Nanjing University, China
Xiaofei Liao	Huazhong University of Science and Technology, China

ENC 2011 Executive Committee

Steering Committee Chairs

Deqing Zou	Huazhong University of Science and Technology, China
Yunfa Li	Hangzhou Dianzi University, China

Workshop Chairs

Chuliang Weng	Shanghai Jiao Tong University, China
Song Wu	Huazhong University of Science and Technology, China
Wanqing Li	Hangzhou Dianzi University, China

Contents

Part IV HumanCom 2011 Session 3: Network or Distributed Algorithms, Applications

Part V HumanCom 2011 Session 4: Privacy, Security and Trust Management

Part VI HumanCom 2011 Session 5: Ubiquitous Computing, Mobile Systems and Applications

Part VII HumanCom 2011 Session 6: Virtualization Technologies for Desktop Applications

Part VIII EMC 2011 Session 1: Cyber-Physical Systems and Real-Time Systems

Part XI EMC 2011 Session 4: Multimedia Computing and Intelligent Services

Part XII PDC 2011

Part I
Keynote Speech

Chapter 1
Graph-Based Structure Search

Xuemin Lin

Abstract Recent years have witnessed an emergence of graph-based applications that strongly demand efficient processing of graph-based queries and searches. In many real applications such as chem- and bio-informatics, road networks, social networks, etc., efficiently and effectively conducting structure search on graphs is a key. The problem of structure search on graphs and its variants are NP-complete. The presence of a large number of graphs makes the problem computationally more intractable. In this talk, I will introducc recent techniques in graph structure search with the focus on our recent work published in SIGMOD, VLDB, ICDE, etc. My talk will cover the problems of substructure search, superstructure search, and substructure similarity search.

1.1 Biography

Xuemin Lin is a Professor in the school of Computer Science and Engineering at the University of New South Wales and a concurrent Professor in the school of Software at Eastern China Normal University. He currently heads the database research group in the School of Computer Science and Engineering at UNSW. Xuemin got his Ph.D. in Computer Science from the University of Queensland (Australia) in 1992 and his B.Sc. in Applied Mathematics from Fudan University (China) in 1984. During 1984–1988, he studied for Ph.D. in Applied Mathematics at Fudan University. His current research interests lie in data streams, graph

X. Lin (✉)
University of New South Wales, Sydney, Australia
e-mail: lxue@cse.unsw.edu.au

X. Lin
Eastern China Normal University, Shanghai, China

J. J. Park et al. (eds.), *Proceedings of the International Conference on Human-centric Computing 2011 and Embedded and Multimedia Computing 2011*, Lecture Notes in Electrical Engineering 102,
DOI: 10.1007/978-94-007-2105-0_1,

databases and information networks, keyword search, probabilistic queries, spatial and temporal databases, and data qualities. He published over 50 papers in both theoretical computer science and database systems; many of them are published in major conferences and journals, including SIGMOD, VLDB, KDD, ICDE, WWW, TODS, TKDE, VLDBJ, etc. He co-authored several best (student) papers, including ICDE2007 (best student paper award). He served as the Program Committee Chair for Computing: The 4th Australasian Theory Symposium (CATS'98), 6th International Computing and Combintorics Conference (COCOON2000), 6th Asia Pacific Web Conference (APweb04), the joint conference of 9th Asia Web Conference and 8th Web-Age Information Management Conference (APweb/WAIM2007), and 19th and 20th Australasian Database Conference (ADC08, ADC09). He is a conference co-chair of 14th International Conference on Database Systems for Advanced Applications (DASFAA09). He has been also involved as a PC member in a number of conferences, including SIGMOD, VLDB, ICDE. Currently, he is an associate editor of ACM Transactions on Database Systems.

Chapter 2
Location Significance Ranking from Quality Enhanced Trajectory Data

Xiaofang Zhou

Abstract The proliferation of GPS-enabled mobile devices has contributed to accumulation of large-scale data on trajectories of moving objects, and presented an unprecedented opportunity to discover and share new knowledge, such as location significance. Existing technologies lack the ability to provide meaningful rankings on locations respective to user communities due to lack of consideration for some fundamental characteristics of trajectory data such as uncertainty, lack of semantics and lack of context. These problems can be addressed by building solutions on quality enhanced trajectory data and achieving ranking of physical locations in the geographical space similar to what was achieved by page ranking in cyberspace. In this talk we will look at emerging issues, recent research results and new opportunities in this multidisciplinary research area involving trajectory data processing, data quality management and information ranking.

2.1 Biography

Prof. Xiaofang Zhou is a Professor of Computer Science at the University of Queensland, and the Head of the Data and Knowledge Engineering Research Division at UQ. He is the Convenor and Director of ARC Research Network in Enterprise Information Infrastructure (a major national research collaboration initiative in Australia), and the founding chair of ACM SIGSPATIAL Australian Chapter. Xiaofang received his B.Sc. and M.Sc. degrees in Computer Science from Nanjing University, China, and Ph.D. in Computer Science from the University of Queensland. Before joining UQ in 1999, he worked as a researcher in

X. Zhou (✉)
University of Queensland, Brisbane, Australia
e-mail: zxf@uq.edu.au

J. J. Park et al. (eds.), *Proceedings of the International Conference on Human-centric Computing 2011 and Embedded and Multimedia Computing 2011*, Lecture Notes in Electrical Engineering 102,
DOI: 10.1007/978-94-007-2105-0_2, © Springer Science+Business Media B.V. 2011

Commonwealth Scientific and Industrial Research Organisation (CSIRO), leading its Spatial Information Systems group. His research focus is to find effective and efficient solutions for managing, integrating and analyzing very large amount of complex data for business, scientific and personal applications. He has been working in the area of spatial and multimedia databases, data quality, high performance query processing, Web information systems and bioinformatics, co-authored over 200 research papers with many published in top journals and conferences such as SIGMOD, VLDB, ICDE, ACM Multimedia, The VLDB Journal, ACM Transactions and IEEE Transactions. He was the Program Committee Chair of IFIP Conference on Visual Databases (VDB 2002), Australasian Database Conferences (ADC 2002 and 2003), International Conference on Web Information Systems Engineering (WISE 2004), Asia Pacific Web Conferences (APWeb 2003 and 2006), International Conference on Databases Systems for Advanced Applications (DASFAA 2009 and DASFAA 2012), the 22nd International Conference on Database and Expert Systems Applications (DEXA 2011) and the 29th International Conference on Data Engineering (ICDE 2013). He has been on the program committees of over 150 international conferences, including SIGMOD (2007–2009), VLDB (2008, 2009), ICDE (2005–2007, 2009–2011), WWW (2006, 2008, 2009) and ACM Multimedia (2011). Currently he is an Associate Editor of The VLDB Journal, IEEE Transactions on Knowledge and Data Engineering, World Wide Web Journal and Springer's Web Information System Engineering book series. In the past he was an Associate Editor of Information Processing Letters and an Area Editor of Encyclopedia of Database Systems. Xiaofang is an Adjunct Professor of Renmin University of China appointed under the Chinese National Qianren Scheme.

Chapter 3
Ubiquitous Neural Interface

Bin Hu

Abstract With innovative developments in Computer Science, e.g. Pervasive Computing, Affective/Cognitive Computing, and their amazing applications, the Human Computer Interaction (HCI) has been impacted by some emerging research and technologies, such as, Brain–Computer Interface (BCI), Electrocenphlogram Interface (EEGI), Neural Human–Computer Interface (NHCI/NI), and Cognitive Interaction etc. I believe that Cognitive/Neural Interface, no matter what it should be called, may perhaps be one of the most promising and life altering technologies in existence today. The implementation of these technologies could launch the world into a new era, especially, in the ways of assisted living, Health Care, E-commerce, Security (Biometrics), Learning, like Stephen Hawking said: "We must develop as quickly as possible technologies that make possible a direct connection between brain and computer, so that artificial brains contribute to human intelligence rather than opposing it".

3.1 Biography

Dr. Bin Hu, Professor, Dean, at the School of Information Science and Engineering, Lanzhou University, China. Professorship at ETH, Zurich; Readership, the leader of the Intelligent Contextual Computing Group, at Birmingham City University, UK. My research fields are Affective Computing, Neuro-Interface, Context Aware Computing, Pervasive Computing and has published about 100 papers in peer

B. Hu (✉)
Lanzhou University, Lanzhou, China
e-mail: bh@lzu.edu.cn

B. Hu
Birmingham City University, Birmingham, UK

J. J. Park et al. (eds.), *Proceedings of the International Conference on Human-centric Computing 2011 and Embedded and Multimedia Computing 2011*, Lecture Notes in Electrical Engineering 102,
DOI: 10.1007/978-94-007-2105-0_3,

reviewed journals, conferences, and book chapters. The works have been funded as Principle Investigator by quite a few famous International Funds, e.g. EU FP7; EPSRC/HEFCE, UK; NSFC, China and industry. I have served more than 30 international conferences as a chair/pc member and offered about 20 talks in high ranking conferences or universities, also served as editor/guest editor in about ten peer reviewed journals in Computer Science and membership of the European Commission Health Care section.

Part II
HumanCom 2011 Session 1: Computer-Assisted Learning, Cognition and Semantic Systems

Chapter 4
Dynamic Recommendation in Collaborative Filtering Systems: A PSO Based Framework

Jing Yao and Bing Li

Abstract In collaborative filtering (CF) recommender systems, a user's favorites usually can be captured while he rating or tagging a set of items in system, then a personalized recommendation can be given based on this user's favorites. As the CF system growing, the user information it hosts may increase fast and updates frequently, which makes accurate and fast recommending in such systems become more difficult. In this article, a particle swarm optimization based recommending framework is introduced, which enhances the ability of traditional CF system to adapt dynamic updated user information in practice with steady and efficient performance. The experiments show the proposed framework is suitable for dynamic recommendation in CF system.

Keywords Recommendation · PSO · Dynamic · CF

J. Yao · B. Li (✉)
State Key Laboratory of Software Engineering, Wuhan University, 430072 Wuhan, China
e-mail: sklse@vip.qq.com

J. Yao
e-mail: 910061765@qq.com

B. Li
School of Computer, Wuhan University, 430072 Wuhan, China

J. Yao
Information Management School, Hubei University of Economics, 430072 Wuhan, China

J. J. Park et al. (eds.), *Proceedings of the International Conference on Human-centric Computing 2011 and Embedded and Multimedia Computing 2011*, Lecture Notes in Electrical Engineering 102,
DOI: 10.1007/978-94-007-2105-0_4,

4.1 Introduction

As information grows rapidly on web, how to organize the data, provide more useful information for users and make the recommending process be more efficient become the challenges of many web systems.

A collaborative filtering (CF) recommender system usually gives user personalized recommendations based on users' history information, while it faces some ignorable problems as the system scaling up. The common used CF algorithms for recommending usually have to check the history records of every user. As the number of users growing up, the traversal checking approach will inevitably lead to intolerant latency. Moreover, as time goes by, users' history information left in CF system will update constantly and frequently, this may severely affect the accuracy of user-based recommendation and make the recommending process has to be done repeatedly for a specific user. Since that, how to make the CF recommending process to fit the dynamic environment, with acceptable efficiency and accuracy is demanded. Therefore, a particle swarm optimization (PSO) based framework is suggested as the overall solution.

Our proposed framework is aim to enhance the ability of traditional CF system to adapt dynamic updated history information of users, enable system to scale up with guaranteed efficiency and accuracy of recommendation by combining PSO algorithm and a proposed user relationship network. Moreover, adapting PSO algorithm in our framework can be a new contribution to the community of interest.

As a computational intelligence algorithm, PSO usually works fine when searching the best fitness location in a continuous two-dimensional searching space with satisfactory efficiency especially when searching space is large. Since that, PSO and other computational intelligence algorithm have great potential to solve recommending problem of CF system with large dataset. While in a CF system, another obvious problem is that the searching space composed by users' information is discrete and not continuously exists, which makes the PSO approach cannot be adapted directly. Therefore, a user relationship network is needed to organize the relationship between different user's data for further PSO calculation.

The proposed framework is tested with three groups of experiments. Two of them are based on real MovieLens data sets besides a set of simulations.

This paper is organized as follows. In Sect. 4.2, related researches have been described. In Sect. 4.3, the overall of the recommendation approach is represented. Section 4.4 shows the experiment in detail. Section 4.5 concludes this paper with future works.

4.2 Related Works

This section covers related works and their nature connection to this research.

4.2.1 Recommendation Systems

Recommendation systems use the opinions of a group of customers to help individuals in that group more effectively identify content of interest from a potentially overwhelming set of choices [1]. The most common used techniques for recommendation system are CF and content-based filtering (CBF) by now. CF approaches need history information provided by users. Recommendations are created for target user u from the items preferred by similar users (user-based) or from items that received similar ratings to the items that u likes (item-based) [2]. In contrast, CBF techniques rely on item descriptions and generate recommendations from items that are similar to those the target user has liked in the past [3]. Many researchers prefer combining CF and CBF techniques to get more accurate and better coverage recommendation, such as [4] and [5]. An important limitation of CF systems is their lack of scalability: as the number of users and items increases, this approach can lead to unacceptable latency for providing recommendations during user interaction [6]. Several works with more sophisticated method has been done to minimize such limitation. Singular value decomposition (SVD) technology proposed in [6] tries to reduce the dimension of data matrix of a tagging system for efficiency of recommendation. PSO recommender system [7] and wasp: a multi-agent system for multiple recommendations problem [8] follows the computational intelligence way but lack of considering dynamic updated data. Our framework focuses on providing dynamic recommendation and increasing the scalability of CF system in a simple way.

4.2.2 Particle Swarm Optimization

PSO is a heuristic search technique firstly developed by Kennedy and Eberhart [9] in 1995 which simulated the movements of a flock of birds or a school of fish aiming to find food. PSO is common used to solve optimization problems through evolution technique by now. In PSO, a swarm consists of a group of particles which represents a population of candidate solutions. Each particle within the swarm has initial velocity and position. At each round, particles move around in the search-space according to simple mathematical formulae. The movements of the particles are guided by the best found positions in the search-space which are updated as better positions are found by the particles.

4.3 The Proposed Framework

4.3.1 The Overall Architecture

Our proposed framework contains four major components: active user analyzer, auto maintainer, PSO matcher and recommender. The internal relationships are shown as Fig. 4.1.

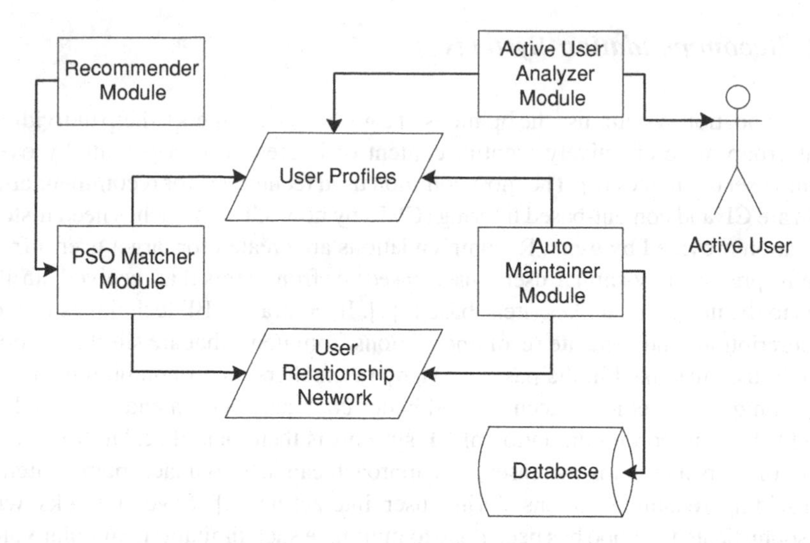

Fig. 4.1 The architecture of proposed framework

In this paper, a user who will be given recommendation is named an active user. The active user analyzer is responsible for analyzing and retrieving information from an active user. If the active user has a profile in database, then the active user analyzer just make some necessary update to keep user's information up to date. Otherwise, a new profile of the active user will be created. In this way, the user profiles within database will be dynamic updated, new users will be taken into account for future recommendation. The format of a user's profile is described in Sect. 4.3.2.

The auto maintainer is an individual thread which is responsible for storing users' profile into permanent storage and maintaining a profile based user relationship network which is mentioned in Sect. 4.3.3. The auto maintainer works asynchronous with other components, which guarantees the user relationship network been prepared at spare time, and cuts down the time cost when a recommending is launched.

The PSO matcher tries to find out the best matched profile within user relationship network correspond to a given active user's profile, then the result will be passed to the recommender component for further operation.

As the number of user profiles will increase as time goes on, the exhaustive method for matching will lead to intolerable delay, while other methods include PSO with limited iteration times may save much time for the same task.

The PSO matcher bases on PSO, with the help of prepared relationship network, recommending in a large-scale dynamic updated CF system will be possible.

The process for recommendation is as follows:

1. Auto maintainer maintains the user relationship network in a separate thread, which shifts huge amount of works to be done at the spare time.
2. When a new recommendation process launched.

Table 4.1 A user's profile

Items	Rating
1	5
2	2
3	4
…	…
N	4

- The active user's information is collected by active user analyzer.
- Based on the user relationship network and the information of the active user, the problem domain is narrowed, a subset of the user relationship network is marked.
- A PSO search process for a specific active user is launched by PSO matcher. The best match user is found finally.
- According to the matched user, recommender component gives the recommendation consists of items that the matched user have but the active user do not.

4.3.2 Profile Generation

In several works, recommendations are often made based on user's profile like occupation, age and gender. There is no doubt a user's background will affect his/her chooses when navigating the system, but some issues are not ignorable, such as a user may not provide any personal information in advance or people do not like to share personal information. Besides these issues, an important problem is people's mood, interests, and favorites will change along the time. To reflect the effect of dynamic changing, our approach adopts user history behaviors as user's profile instead of common used background information. According to the system like MovieLens with rating mechanism, a user's profile can be represented as Table 4.1:

As time goes by, users' profiles can be updated dynamically as they rate more items or changing items' rating value. In this way, recommendation made based on users' profiles will be more accurate.

4.3.3 Relationship Network

As mentioned above, a common met problem in recommending process of a CF system is that the relationship between different users is difficult to be judged. This can lead to the solution of the user profile matching limited to the time consuming exhaustive method.

Suppose user X and Y have very close favorites on some items, then X and Y may have same opinion on a new item. Also, if there is a user Z has quite different

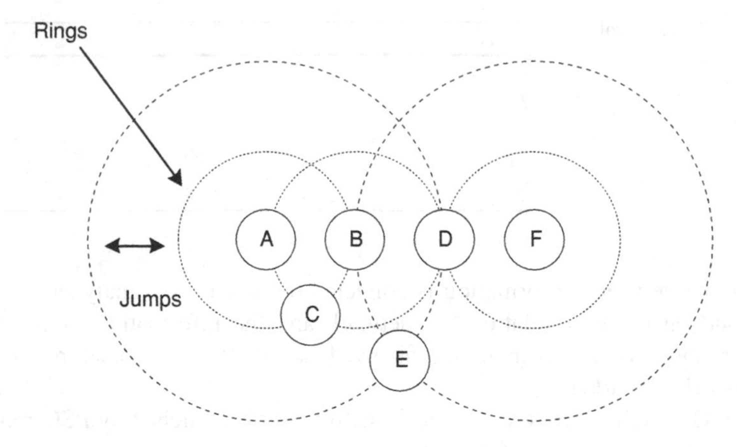

Fig. 4.2 An example of user relationship network

tastes from X's, and then there is high probability that Y and Z's favorites may be quite different as well. Based on this hypothesis, a user relationship network can be made.

To generate the user relationship network, we make following definitions for convenience:

- Neighbors: for a given user X, any other users have similar profile to him are his direct Neighbors. If user X and user Z both are direct neighbors of user Y, but Z is not X's direct Neighbors, then Z is X's indirect Neighbors.
- Jumps: jumps represent the rough distance between different set of users. If users are direct neighbors, then they have 1 Jump distance. If users are indirect Neighbors, they have N user between them, then they have N + 1 Jump distance.
- Rings: a users neighbors are grouped into different Rings, marked from Ring 1 to N, according to different Jumps.

Figure 4.2 represents a more detailed example to illustrate the user relationship network. Suppose the searching space includes the profiles of user A, B, C, D, E and F. User A has direct Neighbors B and C, the distance between A and B is 1 Jump, B and C are on the Ring 1 of A. User B has direct Neighbors A, C and D. Similarly, user F has direct Neighbor D and indirect Neighbors B and E. Table 4.2 represents the user relationship network between user A, F and other users.

Say T is the user to be matched, if there is a similarity between A and T, then the position of T in Fig. 4.2 is supposed to be near A, within the Ring 1 of A, otherwise T should be outside. If there is no similarity between A, B and T, then the position of T should be close to E and F.

To measure more precise difference within one Jump distance, Euclidean algorithm can do great help. Finally, Table 4.3 which describes the relative distance about A, F to other users can be generated.

Table 4.2 The relationship network of user A and F

Distance	User A	User B
1Jump	B, C	D
2Jump	D, E	B, E
3Jump	F	A, C

Table 4.3 The distance table of user A and F

Distance	User A	User F
Closest to farthest	C	D
	B	E
	E	B
	D	C
	F	A

4.3.4 PSO Algorithm

After problem domain been found, the most direct way to find the best match user is to calculate the similarity between active user and each other user within problem domain by using similarity measurement such as Euclidean distance score, Pearson correlation score and Jaccard coefficient. To cut down the time cost, PSO is picked as our major algorithm for matching, Pearson algorithm is picked for fitness value calculation.

As mentioned above, PSO algorithm can help finding the best position in searching space in limited iteration. A set of particles move through the whole searching space based on the fitness values which are calculated by using each particle's position and the target.

The standard PSO function steps include:

- initialize a swarm's size, position and speed
- evaluated fitness of each particle in the swarm
- compare the fitness of each particle to pbest, if better replace the pbest
- compare the fitness of each particle to gbest, if better replace the gbest
- update position and speed of each particle
- if terminate condition does not meet, loop back to step 2

The standard update position function includes:

$$\text{vid} = \text{w} \cdot \text{vid} + \text{c1} \cdot \text{r1} \cdot (\text{pbest} - \text{xid}) + \text{c2} \cdot \text{r2} \cdot (\text{gbest} - \text{xid}) \tag{4.1}$$

$$\text{xid} = \text{xid} + \text{vid} \tag{4.2}$$

Where

- pbest is the personal best position of particle i
- gbest is the global best position of all the particles
- xid is the current position of particle i

- vid is the velocity of particle i
- w is a random inertia weight set to 0.5
- c1 and c2 are constants usually set to 1.494 [10]
- r1 and r2 are random number between 0 and 1

4.3.5 Pearson Algorithm

The Pearson algorithm (PA) is based on the k Nearest Neighbor algorithm. The correlation coefficient is calculated as the equation shown below.

$$P(X,Y)=\frac{\sum XY-\frac{\sum X\sum Y}{N}}{\sqrt{\left(\sum X^2-\frac{(\sum X)^2}{N}\right)\left(\sum Y^2-\frac{(\sum Y)^2}{N}\right)}} \quad (4.3)$$

4.4 Experiments and Evaluation

In this section, a set of experiments have been carried out. Two experiments are based on real MovieLens data set while a group of simulation with random number of users and items also have been done and evaluated.

4.4.1 Data Set and Experiment Parameters

The first real experimental data set is MovieLens data set which contains 100,000 ratings from 943 users on 1,682 movies. The data set is reorganized into profile structure which represents a user's favorites. The user profiles then randomly divided into two parts, 80% of training set and 20% of test set respectively. The profiles within test set are tested against training set, if the most similar profile pair was found, then users with the most similar favorites within data set are clear, and the recommendation can be given. After one user's profile has been tested, it will be add into training set immediately, which keeps the training set dynamic updated.

As each user within MovieLens data set rated at least 20 movies, while under real circumstance, users may rate items lease than 20 items. Since that, a group of simulation has been done. All simulations are based on randomly generated data set which contains total 1,000 users and 3,000 items; each user randomly rates 1–40 items with random rating from 1 to 5. The data set is divided into two parts, 80% of training set and 20% of test set respectively as well, and the training set is configured to update dynamically.

The last experiment based on real MovieLens data set which contains 100,000 ratings from 943 users on 1,682 movies. As users may rate items from time to

Table 4.4 Experiment 1

Round	R(avg)
1	0.000252
2	0.000388
3	0.000329

Table 4.5 Experiment 2 with random data set 1

Round	R(avg)
1	6.617E-06
2	3.308E-06
3	2.2056E-06

time, then the recommendation should be dynamic as well. In this experiment, 0.995 of user profiles are random picked as training set, while other 0.005 user ratings are marked with random time, each user rates one item at a time. The user profiles of test set are dynamically changed as time goes on.

The parameters used in experiments are kept same. PSO parameters are set to W = 0.5, C1 = C2 = 1.494 [10], swarm size = 30, maximum iterations for each run = 80. Among all the experiments, Euclidean algorithm is used for generating user relationship network, Pearson algorithm is used for fitness value calculation. For each data set, experiments have been done three round.

4.4.2 Evaluation Criterion and Experiment Results

All experiments are followed the same evaluation criterion.

For each round of user profile matching, the given user to be matched is tested against all the other users in the training set by using Pearson algorithm. Then the generated similarities are queued into a list Q represents the results from the best to the worst.

Then the result i generated by PSO approach can be compared to items in Q, the distance between i and the best item is marked as Di and the differential ratio against the size of Q is marked as Ri. If size of test set is N, then the average difference R(avg) of all active users can be calculated as:

$$R(\mathrm{avg}) = \frac{\sum_{i=1}^{n} Ri}{N} \tag{4.4}$$

Experiment Result: Tables 4.4, 4.5, 4.6, 4.7, 4.8

Analysis of Results:

The three different experiments show that the proposed framework works fine under all different conditions. In all experiments the average of R is lower than 0.0006 which means in most cases our matching process can find the top

Table 4.6 Experiment 2 with random data set 2

Round	R(avg)
1	1.977E-05
2	1.296E-05
3	1.082E-05

Table 4.7 Experiment 2 with random data set 3

Round	R(avg)
1	6.260E-06
2	6.441E-06
3	8.681E-06

Table 4.8 Experiment 3

Round	R(avg)
1	0.000582
2	0.000569
3	0.000568

one matchable item within the problem domain. The overall performance of the framework keeps steady, and the gap between the results of different experiments may caused by discrete degree among the users' information. The time cost of recommending for a specific user mainly depends on the PSO process, more precisely depends on the configuration of PSO. Since PSO process does not check every user's history information, and has limited iteration, the time costs will not grow up linearly and this will improve users' interactive experience.

4.5 Conclusion and Future Work

In this paper, a PSO based framework for recommendation which support dynamic updated user's information has been proposed and tested. The experiments also show the relationship between different users in a CF system can be measured by proposed user relationship network. We are currently running a long time test on the framework to measure the performance and the overall efficiency. The improvements can be made in the process of user relationship network generation, the PSO algorithm can be modified to deal with more complex situation as well.

Acknowledgments This work is supported by the National Basic Research 973 Program of China under grant No. 2007CB310801, the National Natural Science Foundation of China under grant No. 60873083, 60803025, 60970017, 60903034, 61003073 and 60703018, the Natural Science Foundation of Hubei Province for Distinguished Young Scholars under grant No. 2008CDB351, the Young and Middle-aged Elitists' Scientific and Technological Innovation Team Project of the Institutions of Higher Education in Hubei Province under grant No. T200902 and the Natural Science Foundation of Hubei Province under grant No. 2010CDB05601.

References

1. Resnick P, Varian HR (1997) Recommender Systems. Commun ACM 40:56–58
2. Resnick P et al (1994) GroupLens: an open architecture for collaborative filtering of netnews. In: ACM conference on computer-supported cooperative work (CSCW 94), ACM Press, pp 175–186
3. Pazzani MJ, Billsus D (2007) Content-based recommendation systems. In: Brusilovsky P, Kobsa A, Nejdl W (eds.) The adaptive web. LNCS, vol 4321. Springer, Heidelberg, pp 325–341
4. Faqing WU, Liang HE et al. (2008) A collaborative filtering algorithm based on users partial similarity. In: 10th international conference on control, automation, robotics and vision, Hanoi, Vietnam, pp 17–20
5. Chen Dongtao, XuDehua (2009) A collaborative filtering recommendation based on user profile weight and time weight. In: International conference on computational intelligence and software engineering, pp 1–4
6. Barragáns-Martínez AB, Rey-López M et al (2010) Exploiting social tagging in a web 2.0 recommender system. IEEE internet computing, IEEE computer society, Los Alamitos
7. Ujjin S, Bentley PJ (2003) Particle swarm optimization recommender system. In: Proceedings of the 2003 IEEE swarm intelligence symposium, pp 124–131
8. Dehuri S, Cho SB, Ghosh A (2008) Wasp: a multi-agent system for multiple recommendations problem. In: 4th international conference on next generation web services practices, pp 159–166
9. Kennedy J, Eberhart R (1995) Particle swarm optimization. In: Proceedings of the IEEE international conference on neural networks, vol 4, Piscataway, pp 1942–1948
10. Eberhart RC, Shi Y (2001) Particle swarm optimization: developments, applications and resources. In: Proceedings of the 2001 congress on evolutionary computation, vol 1, pp 81–86

Chapter 5
Automatic Chinese Topic Term Spelling Correction in Online Pinyin Input

Sha Sha, Liu Jun, Zheng Qinghua and Zhang Wei

Abstract Usually people use the input software to type Chinese language on a computer. The software takes a three-step approach: (1) receive the English keyboard input; (2) convert it to Chinese words; (3) output the words. Traditional Chinese spelling correction algorithms focus on the errors in the output Chinese, but ignore the errors introduced in the original keyboard input. These algorithms do not work well because the errors in the output are usually not the type of the typographical errors, which these algorithms are good at. In this paper, we propose a novel Chinese spelling correction model directly targeting at the original keyboard input. We integrate this model to an online Chinese input method, to improve the spelling suggestion feature. Experiments using real-word data show that this model helps the spelling suggestion achieve a 93.3% accuracy.

Keywords Topic term · Correction · Complementary · Pinyin · Principle of locality

S. Sha (✉) · L. Jun · Z. Qinghua
MOE KLINNS Lab and SPKLSTN Lab, Department of Computer Science and Technology, Xi'an Jiaotong University, 710049 Xi'an, China
e-mail: summershasha@gmail.com

L. Jun
e-mail: liukeen@xjtu.edu.cn

Z. Qinghua
e-mail: qhzheng@xjtu.edu.cn

Z. Wei
Microsoft Corporation, China Redmond, WA 98052 USA
e-mail: wzhan@microsoft.com

J. J. Park et al. (eds.), *Proceedings of the International Conference on Human-centric Computing 2011 and Embedded and Multimedia Computing 2011*, Lecture Notes in Electrical Engineering 102,
DOI: 10.1007/978-94-007-2105-0_5,

5.1 Introduction

More and more Chinese people use online education websites to learn knowledge. These websites usually provide some search options to cover the web site documents. Similar to the general web search engine, they take the queries that users type in, and return ranked relevant documents. The queries are usually in Chinese, and are often the concepts of certain knowledge domains. In this paper, we call them the "topic terms". These misspelled words often lead to poor search results. Some misspellings are simple typographic errors such as typing "光纤" (spelled in Pinyin as guang xian) as "光前" (spelled in Pinyin as guang qian). The cognitive error is more complicated such as typing "单模光纤" (mono-mode optical fiber) as "光纤" (optical fiber). Often a spelling correction feature is provided as part of the search option to help the users.

Chinese characters are encoded so they cannot be inputted into computer by typing the keyboard directly. Usually a Chinese language input software must be used, which takes a three-step approach: accept the English keyboard input; convert the English input to a set of possible Chinese words; and accept user's choice to print the chosen Chinese word or character to the computer. Pinyin input method is the most popular type. Although some of the spelling errors happen in the third stage, more often they are introduced by the user in the first stage. When user's original input is wrong, the input software takes it, and produces the wrong output.

The English spelling correction research has a long history. It has been widely studied in the natural language processing and information retrieval communities. However, the differences between English and Chinese make these research results less applicable to Chinese text [1]. Recent research on Chinese spelling correction focus on how to correct the errors in the final chosen Chinese on computers, but ignore the errors introduced in the original keyboard input. Their performances on the real world scenarios are not promising.

In this paper we propose a novel Chinese spelling correction algorithm, which corrects a user's original keyboard input, to further improve the accuracy of Chinese topic term spelling correction. The contributions of our work are: (1) Algorithm improves the Chinese Pinyin input method by rebuilding the user English keyboard inputs; (2) organize the topic terms in our lexicon in a two-layer structure with Chinese Pinyin and tone annotation. (3) Introduce the principle of locality in learning to the correction candidates ranking.

The rest of this paper is structured as follows. Section 5.2 reviews the related research. Section 5.3 studies and classifies the spelling errors in the Chinese topic terms input, and present the idea of rebuilding the original user inputs for the Pinyin-based input method. Section 5.4 describes our algorithm and the workflow. Experimental results are evaluated in Sect. 5.5. Section 5.6 concludes the paper.

5.2 Related Work

English text auto-correction was derived from the spelling auto-correction and text recognition in the 1960s. This work can be divided into three parts [2]: (1) Non-word error detection, which was studied during early 1970s to early 1980s; (2) Isolated-word error correction, on which the research started from 1960s to present; (3) Context-dependent word correction research, which began in 1980s.

The spelling correction is studied to correct the errors in the queries [3]. Statistical models have been widely used. Brill and Moore proposed generic string-to-string edit operations [4], Toutanova and Moore's model utilized the phonetic information [5]. Both works need training data, while the latter also requires a pronunciation lexicon. Then methods for automatic learning of these models came out by using the query logs [6, 7]. To further improve the performance of former methods, the extended information from web search results are used to address real-word spelling correction problem [3].

Since the English words are naturally separated by spaces, while Chinese are not. This nature of Chinese makes the correction much more difficult than that of English. The research on the Chinese text auto-correction started from early 1990s. It includes two facets: text auto-detection and correction. The Chinese text error detection methods can be concluded into three classes: (1) Utilize the text features such as context, morphological features and word inflection [8, 9]; (2) Transition probability based analysis on connecting relationship between adjacent words [10]; (3) Linguistics rules and knowledge based method [11]. Along with the popularity of web search, Chinese spelling correction research based on N-gram model is also conducted [12].

No matter what methods are involved in existent research, their operation targets are all the final chosen Chinese appeared on the computer's screen, but not the user's original keyboard inputs.

5.3 Chinese Term Spelling Errors in Pinyin Input Method

5.3.1 *Factors in Chinese Term Spelling Errors*

English and Chinese have very different factors in spelling errors, because they have distinct language features. These differences are: (1) English is phonetic, while Chinese is ideographic. (2) English uses spaces to separate words, while Chinese does not. That is why we need to split Chinese paragraphs into smaller units before correction. (3) There are morphological changes on English words, but there is no simple correspondence rule in Chinese. So the semantic analysis on Chinese is much difficult than that on English. (4) English is inputted into computer letter by letter. Chinese is encoded. So the Chinese character printed on computer screen must be one in Chinese encoded character set. Chinese (text) errors are character-misuse errors, in that the wrong character has same (similar) pronunciation or plesiotype as the correct one. (5) English character set is composed by 26 letters and punctuation, while that of Chinese has more than 91,251 characters [13].

5.3.2 The Features of Chinese Terms and Error Classification

Large numbers of log studies show that web queries are relatively short, and the average length is about 1.85 Chinese characters [14] or 2.3 English words [15]. So there is little context information can be used. Moreover, in an e-learning application, these query strings are usually Chinese topic terms, so the spelling correction results should not only be correctly-spelled words, but also the Chinese domain terms. When it comes to the Chinese input methods, two keyboard access models are widely adopted: Pinyin input method and character grapheme coding input method [16]. This paper focuses on the Pinyin input methods.

According to the features of Chinese input methods and queries in online search, we divide Chinese term spelling errors into three categories:

Homophonic errors. This type of errors usually occurs when user uses the Pinyin input method. The scenario is that a wrong Chinese character is chosen because it has the same or similar pronunciation or Pinyin spelling as the right character. For example: 光纤 (guang xian) → 光线 (guang xian), 先 (xian) → 西安 (xi'an), etc.

Plesiotype error. This error happens when using the character grapheme coding input method, such as Zhengma input, Wubi input, etc. These input edcode based on Chinese character's grapheme. Each character is divided into several radicals, and each radical is stated by a letter [17]. Two characters have identical or similar graphemes are often mixed. such as中毒 (Zhengma: jv) → 土壤 (Zhengma: jv).

Non-words error. This can be explained in three cases: (1) Mis-spelling of a Chinese character's Pinyin spelling, e.g. hit a wrong key on the keyboard, forget a key, etc. For example, 光纤 (guangxian) →黄线 (huangxian), (2) Wrong Chinese word is chosen caused by user's cognitive level. For instance: 数据链路→ 数据 (character missed), (3) User chooses the wrong word from the wordlist provided by input software. Such as: 光纤 (gx) → 光学 (gx).

From the classification of the errors, we can find that the spelling error types have a strong correlation with the types of the input software. The statistics of web data shows that the Pinyin input software downloads counts for about 90% of the total Chinese input software downloads, while that of the character grapheme coding input software is just about 8% [18]. This implies that the number of the Pinyin software users is far more than that of other input software users. So our study will focus on the homophonic errors and non-word errors.

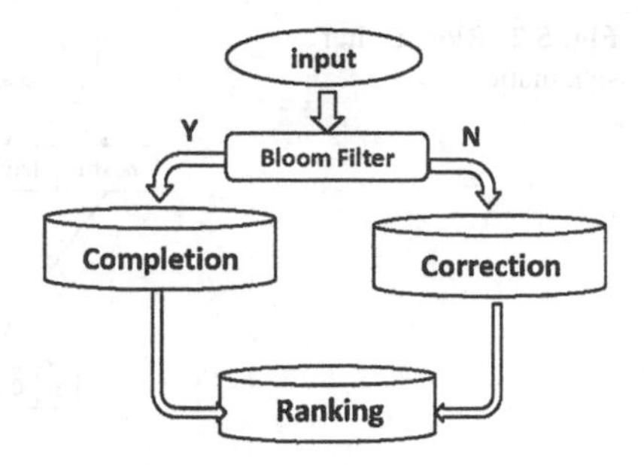

Fig. 5.1 Chinese term correction workflow

5.4 Chinese Topic Term Errors Correction in the Online Pinyin Input Method

We propose a novel method to correct the spelling errors occurred when users use the online Pinyin input method. The workflow of our method is presented in Fig. 5.1

When user inputs a query in Chinese characters, this input will firstly be checked by a Bloom filter [19]. The Bloom filter, conceived by Burton Howard Bloom in 1970 [20, 21], is a space-efficient probabilistic data structure which is used to test whether an element is a member of a set. We define a legal Chinese term as a Chinese topic term in a given domain.

If this input does not belong to the given valid word repository, we assign it to the correction module, because we believe it is incorrectly spelled. Correction module searches in a valid word repository for candidates which may be the legal Chinese terms that the user wants to input.

If this input belongs to the given valid word repository, it will be send to completion module. This module provides related Chinese terms for the user to select. Each related Chinese term contains the user's input string. For example, when a user input a Chinese string "光纤", the candidate Chinese terms will be "单模光纤" and "多模光纤". This completion is helpful when user's cognitive level is limited, and the candidates provide clues for user to refine his query.

At last, all the candidates are organized as a ranked list. The top ranked candidate should be the "closest" terms to the user's original intention. The details of each step in our proposed model are described here.

5.4.1 Error Detection

Since the total number of Chinese topic terms in a given domain is limited and relatively small, we choose the Bloom filter as an error detective tool. All the terms in a certain knowledge field compose the data set.

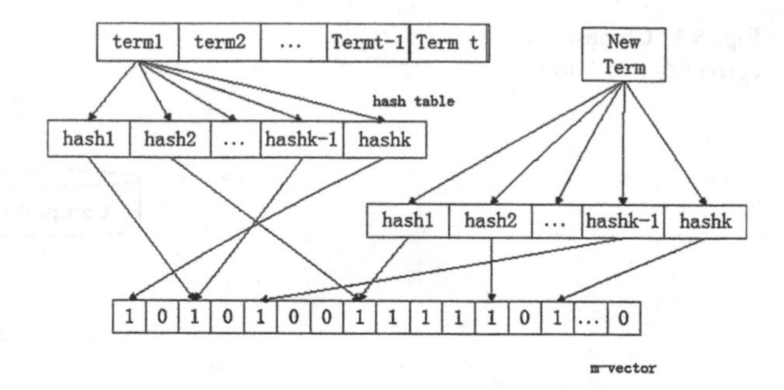

Fig. 5.2 Bloom filter schematic

Figure 5.2 shows the schematic of Bloom filter. Bloom filter is a bit array used for testing whether an element is in a set quickly. When it is empty, all bits set to 0. Meanwhile a hash table containing k different hash functions must be defined, each of which maps an element to one of the array positions with a uniform random distribution.

Building a Bloom filter for a data set is to add all elements in the set to the empty filter. Adding an element is to feed it to the hash table to get k array positions, and set the bits at these positions to 1. After adding all elements of the set, Bloom filter for this set is done. To test whether an element is in the set, feed it to the same hash table to get k array positions. If any of the bits at these positions are 0, the element is not in the set; if it were, all the mapping bits would have been set to 1 when it was inserted. If all are 1, then either the element is in the set, or the bits have been set to 1 during the insertion of other elements. So when an element which is not in the set feeds to the hash table and get k array positions, the bits at these positions may all be 1. This phenomenon is called false positive, which is due to hash collision.

Because the smallest false positive rate of a Bloom filter is [19]:

$$\left(1/2\right)^k \approx 0.1682^{m/n} \tag{5.1}$$

and:

$$k \approx 0.7\frac{m}{n} \tag{5.2}$$

k is the number of hash function used for building the filter; m is the length of the filter; n is the amount of words in the word repository. In our experiment, n is 979, and the given smallest false positive rate is about 0.5%, so m should bigger than $979 \times 16 = 15{,}664$. To minimize the hash collision, we assign a prime value 15,667 to m, which is bigger than 15,664 and the closest with 15,664.

To ensure the smallest false rate, k should be 11 or 12 in response to (5.6). Testing on the relationship between the number of hash used and the false positive rate of the filter demonstrates that when there is little hash functions, the false positive rate will decline sharply with the increase of hash functions used. But its decline ratio grows down quickly when the number of hash functions grows larger. Moreover, the test result shows that the false positive rate fluctuates around 0.5% with little

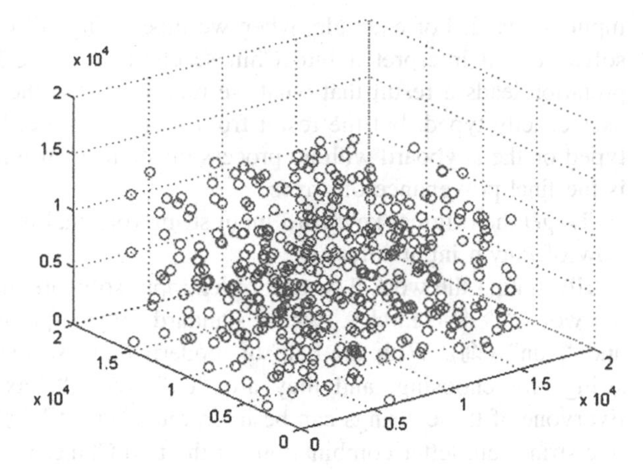

Fig. 5.3 Associated hash efficiency of Mod, DJB, and PJW

diversification when hash functions are more than three. Less hashes means less calculation and higher efficiency. Since the practice false rate is close to the given value when hash number is three, we decided to use this three hashes configuration.

There are 21 popular hash functions, which are FNV1, Rolling hash, Bernstein's hash, Pearson's hash, CRC hash, Universal hash, Zobrist hash, FVN, Thomas Wang's algorithm, RS hash, JS, PJW, ELF, BKDR, SDBM, DJB, DEK, AP, BP, and Mod [22]. The three hash functions chosen from them should be independent of each other. In our research, we conducted experiments for the hash functions' independent verification. First, three hash functions are selected randomly, and are combined as a test sample; Next, Bloom filters are built with these samples, and the positive false rate of each filter are examined. Finally, the best sample with the smallest false positive rate is selected.

In our experiment, the smallest false positive rate is about 0.6%, which is close to the given one (0.5%). The three hashes in the sample S_s are: Mod, PJW and DJB. These three are used to build the final Bloom filter in error detection.

The hash results of Mod, PJW and DJB on the 979 Chinese terms are all of even lines, which mean all terms are well-proportioned hashed, and these three hash functions perform well on the 979 Chinese terms individually. Figure 5.3 shows the associated hash result of the three functions. In the figure, the three hash functions make up a three dimensional space, and the 'o' denotes the location where a term is hashed by these three hash functions.

We can find that all terms are homogeneous mapped with the three hashes from the Fig. 5.3, which proves that the associated hash efficiency of Mod, PJW and DJB is high.

5.4.2 Input Reversion

When using the Pinyin input method in the online search box, we type an English string on the keyboard. This string will be interpreted as Chinese characters by the input software. For example, when we type “yingyu” on the keyboard, the input software will interpret it into Chinese characters “英语” (English). This interpretation leads a result that what we received from the search box is not what a user exactly typed, but the result from input software. However, the letter string typed on the keyboard with no process of input software, we call it original input, is the final provenance of errors.

To get the original keyboard input string (original input), we reversed the work flow of Pinyin input software.

Illustrating the work flow of morden input software, there is an example. Given the word “查询 (search)”, whose standard Pinyin spelling strings are “cha” (查) and “xun” (询). With the help of modern input software, there are totally five strings for choosing, and they are “cx”,“chx”,“chax”,“cxu”, and “chaxun”. Everyone of these strings can be interpreted into “查询”. We can find that these five strings are letter combinations of the two Chinese characters’ standard Pinyin spelling strings.

To reverse this process, when we get the interpretation result (Chinese word), we reverse each character into its standard Pinyin spelling string(s) firstly, and then make letter combination of these strings. The combination results are possible inputs for this Chinese word. For example, when we get a Chinese word “查询”, firstly, we reverse the characters “查” and “询” into their standard Pinyin spelling strings, and the results are two letter strings: “cha” (for “查”) and “xun” (for “寻”); secondly, we make letter combination with “cha” and “xun”; finally, we get five possible inputs: “cx”, “chx”, “chax”, “cxu”, and “chaxun”.

The following spelling correction and suggestion are both based on this reversion. For preparation, we reverse all terms in the legal valid word repository. From the reversion results, we get a legal Pinyin string set, in which each string corresponding to at least one legal Chinese word. A Chinese term is corresponding to a Pinyin string means that: this Pinyin string is in its reversion results list, or this Pinyin string is a sub-string of one result in its reversion Pinyin string list.

5.4.3 Error Correction Method

As to homophonic errors, the original keyboard input string is legal, so we just need to match this string in the legal string set, and get all the corresponding Chinese terms as the spelling correction candidates.

The correction of non-word error is more complex. Since the original keyboard input is A-to-Z letter string that is similar to English words, its error causes have similar features with that of the English words as well. We correct the original input by adopting the editing method [23, 24], which is based on the statistical analysis on spelling errors.

Edit. The operation of insertion, deletion, alteration and transposition on letter string.

Edit distance. The least operation times of edit.

Statistical results show that the wrong string is the edit result of the right string. And the operations of edit are reversible. So we could get the right string, which is what the user wants to type in reality, by reversing-edit those candidate strings, which are reversed from the received string in the search box.

There is a statement in several literatures about spell error correction that the probability of edit distance between the right and wrong spelling to be 1 is 80–95% [23, 24]. This coverage of errors in e-learning is enough, so we just anti-edit those possible strings once at most. And the four operations are:

Deletion. delete each letter of the input to form new strings. For an input whose length is n, this operation will produce n new strings.

Alteration. replace each letter of the input with another 25 letters to form new strings. For an input whose length is n, this operation will produce $25 \times n$ new strings.

Transposition. reverse the position of every two adjacent letters in the input. For an input whose length is n, this operation will produce $n - 1$ strings.

Insertion. insert the 26 letters after each letter of the input string, and one letter per time. For an input whose length is n, this operation will produce $26 \times n$ new strings.

When we operate once on the wrong string with these four operations, the right string were supposed to be in the editing results. So all the strings in edit results, which can be matched in the legal string set, may be what the user tends to input originally, and whose corresponding Chinese terms could all be regarded as correction candidates.

5.4.4 Completion Method

The purpose of this method is to suggest longer terms related with the input. That means the input is a sub-string of suggested terms.

5.4.5 Lexicon Organization

For both correction and complication modules, finding the corresponding Chinese terms for a given letter sting efficiently is required. To do this, we organize terms into clusters. Each cluster has a Pinyin string as tag, and the terms belong to this cluster are the corresponding terms to the tag string.

Then these clusters are inserted in a Trie tree, and the tag of each cluster is the key.

The process of finding the term for a given Pinyin string and tone string are:

Use the string *str* to be retrieved as a key to retrieve the Trie tree *trieTree*.

When we process the last letter of *str*, and come to the node $node_i$ of *trieTree*:

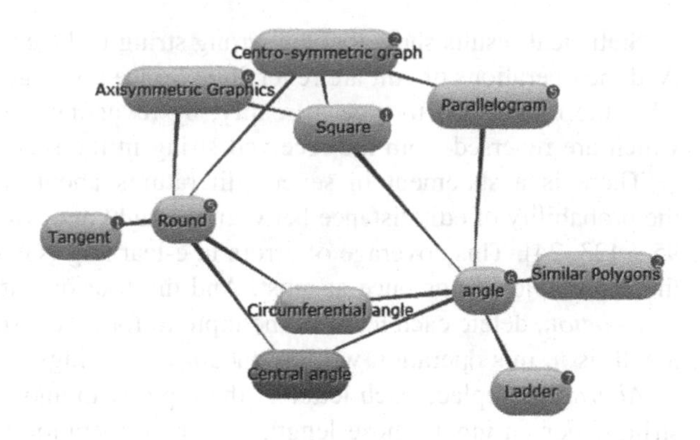

Fig. 5.4 Fragment of the knowledge map of "Plane Geometry"

1. If $node_i$ has term lists, return this list;
2. If $node_i$ has no term lists, return all the term lists of $node_i$'s child nodes and end the whole retrieve process.

When we come to the node $node_i$, and cannot find the next node with the rest part of *str*, return the term lists of both $node_i$ and all child node of $node_i$, and end the whole retrieve process.

This is fuzzy matching process, which is similar with the longest prefix matching.

5.4.6 User Behavior Analysis in Semantic Network

Recently, semantic networks are borrowed for presenting the relationships between Chinese topic terms. Figure 5.4 shows a fragment of it. Each node in the graph is a topic term. Lines between nodes are kinds of semantic relationships. Since these relationships are directed, this network is a directed graph. Semantic distance can be defined as the length of the shortest path from one node to another.

After this system is applied to practice for about one month, we got about 2,711,109 query logs of about 300 users' information seeking activities. From this logs, we found that the distance between two continuous clicked nodes is one in 83.1%, two in 6.3%, and lager than two in 10.6%. That means knowledge learning is localized, which means the two Chinese terms clicked in series are semantic close to one another. So the knowledge term node closed with the last clicked one is more likely to be required.

5.4.7 Ranking Method

Based on previous analysis, the rank of a candidate term is determined by two main factors: the string inputted currently and the term last learned. Citing the sort theory of probability [23, 24], it demands to calculate:

Table 5.1 Accuracy testing results

Item	Remark
Test set	15,737
Average length of candidates list	6
Top accuracy	81.7%
The first two accuracy	87.7%
The first six accuracy	93.3%
Failed suggestion item	756
Failed percentage	4.8%

$$\text{argmax}_{t_c} p(t_c|i) \cdot p(t_c|t_p) \tag{5.3}$$

$p(t_c|t_p)$: The probability that candidate term t_c will be learned when the user learned t_p in last retrieval. It is obtained by searching logs statistics.

$p(t_c|i)$: The probability that a user means to input candidate string t_c when he inputs string i. Since the input strings are diverse and cannot be controlled, $p(t_c|i)$'s calculation is impossible. Introducing Bayesian theory [25], $p(t_c|i)$ is able to be replaced by (5.5):

$$p(t_c|i) = p(p(i|t_c)p(t_c))/p(i) \tag{5.4}$$

In (5.5), three probabilities are referred: $p(i)$—the probability of string i is inputted; $p(t_c)$—the probability of string t_c will be learned which is also statistical results mining from user logs; $p(i|t_c)$—the probability of string t_c being misspelled into string i. Although $p(i)$ is not predictable since we do not know what the i will be, its value is same for all the string i. That means $p(i)$ won't affect the order of candidates. Then:

$$\text{argmax}_{t_c} p(t_c|i) \cdot p(t_c|t_p) = \text{argmax}_{t_c} p(i|t_c)p(t_c)p(t_c|t_p) \tag{5.5}$$

To calculate $p(i|t_c)$, we employ a hypothesis: There are t_{c1} and t_{c2} in candidates list, if the distance from t_{c1} to i is bigger than that from t_{c2} to i, then $p(t_{c1}|i) \gg p(t_{c2}|i)$. So, we can divide the whole candidates list into several groups by their distance to the input string i, and sort these groups by distance. Then in each group, candidates can be further ranked with (7):

$$p = p(t_c)p(t_c|t_p) \tag{5.6}$$

5.5 Experimental Result

Test set in Table 5.1 is derived from user behavior records in real on line study website (Yotta [26]). We attached a monitor on the input box. When a user input something in the box, the monitor will be waked up. If the user chooses a word from the correction suggestion list, the monitor will record where the chosen word is in the list, and how long the suggestion list is. If the user doesn't choose any item from our suggestion list, we will compare user's input with the item in our suggestion list. If there is an item same with the user's input, we will record where this item's order number, and the total length of the suggestion list. Or else, we will give "1" as the chosen suggestion order number, and the suggestion length will be recorded as well. Each monitor result is a number tuple $< No, len >$. *No*: chosen suggestion's order number; *len*: suggestion list's length. If the first number of a tuple is below 0, that means our suggestion is failed. From the monitor data in one month, we got 15,737 items composing our test set.

Accuracy. The probability that the term meeting user's original intention is listed on the top of candidates list.

Top accuracy. the probability that the top term in candidates list is right the one what a user tends to input in reality; *the first two accuracy*: the probability that the term which meets the user's intension is ranked on the top or the second in the list; *the first six accuracy* is similar with the first two accuracy. Since the average length of correction suggestion list is 6, we take "the first six accuracy" as the average accuracy of this correction method. The accuracy results tabled above states a good performance of our system, and the average accuracy is about 93.3%.

Failed suggestion items. the amount of tuples in which the first number is "1".

Failed percentage. the percentage that failed suggestion items are in the test set, and that is 756/15,737 = 4.8%.

5.6 Conclusion and Future Work

This paper develops the method of Chinese term auto-correction in an e-learning environment. Borrowing ideas from existing text error classification, combining with the unique features of Chinese term and its input process, we categorize conventional Chinese term errors into two groups: homophonic error and non-word error. Against with the cause of errors, the correction module is designed. While suggestion module aims at replenish the final candidates list in the consideration of locality principle of learning. When these candidates being presented to the users, their ranks in the list reflect the influence of both what a user has input and what the user has learn previously. The average accuracy rate is about 93.3%, and it proves that our method is effective.

In our paper, we focus on the errors caused by Pinyin input methods. Since Wubi input is also wildly used, the next work will put emphasis on the error analysis associate with this input method.

Acknowledgment The research was supported in part by the National Science Foundation of China under Grant Nos.60825202, 60921003, 60803079, 61070072; the National Science and Technology Major Project (2010ZX01045-001-005); the Program for New Century Excellent Talents in University of China under Grant No.NECT-08-0433; the Doctoral Fund of Ministry of

Education of China under Grant No. 20090201110060; Cheung Kong Scholar's Program; Key Projects in the National Science & Technology Pillar Program during the 11th 5-Year Plan Period Grant No. 2009BAH51B02; IBM CRL Research Program—Research on BlueSky Storage for Cloud Computing Platform.

References

1. Jinshan M (2004) Detecting chinese text errors based on trigram and dependency parsing
2. Kukich K (1992) Technique for automatically correcting words in text. ACM Comput Surv 24(4):377–439
3. Chen Q, Li M, Zhou M (2007) Improving query spelling correction using web search results. In: Proceedings of EMNLP-CoNLL, pp 181–189
4. Brill E, Moore RC (2000) An improved error model for noisy channel spelling correction. In: Proceedings of the 38th annual meeting of the ACL, pp 286–293
5. Toutanova K, Moore R (2002) Pronunciation modeling for improved spelling correction. In: Proceedings of the 40th annual meeting of ACL, pp 144–151
6. Ahmad F, Kondrak G (2005) Learning a spelling error model from search query logs. In: Proceedings of EMNLP, pp 955–962
7. Reynaert M (2004) Text induced spelling correction. In: Proceedings of COLING, pp 834–840
8. Zhang L, Zhou M, Huang C et al (2000) Automatic chinese text error correction approach based on fast approximate chinese word-mathing algorithm. Microsoft research china paper, collection, pp 231–235
9. Zhang L, Zhou M, Huang C et al (2000) Automatic detecting/correcting errors in Chinese text by an approximate word-matching algorithm, In: Annual meeting of the ACL proceedings of the 38th annual meeting on association for computational linguistics, pp 248–254
10. Edward, Riseman DM (1971) Contextual word recognation using binary dirams. IEEE on computers 20(4):397–403
11. Li J, Wang X, Wang P, Wang S (2001) The research of multi-feature chinese text proofreading algorithms. Comput Eng Sci (3):93
12. Zhipeng C, Yuqin L, Liu H et al (2009) Chinese spelling correction in search engines based on N-gram model. 4(3)
13. Quantity amount of Chinese character. http://www.sdtaishan.gov.cn/sites/liaocheng/shenxian/articles/J00000/1/1630615.aspx
14. Yu H, Yi Y, Zhang M, Ru L, Ma S (2007) Research in search engine user behavior based on log analysis. J Chin Inf Process 21(1):109–114
15. Silverstein C, Henzinger M, Marais H, Moricz M (1998) Analysis of avery large Alta Vista query log. Digital system research center, Technical report p 014
16. Chinese input methods Wikipedia (2010) http://zh.wikipedia.org/zh/中文输入法 Accessed 5 Oct 2010
17. Snowling, MJ, Hulme C (2005) The science of reading: a handbook. In: Blackwell handbooks of developmental psychology, vol 17. Wiley-Blackwell, pp 320–322. ISBN 1405114886
18. Ranking of Chinese input methods downloaded (2008.02). http://download.zol.com.cn/download_order/ime_order.html
19. Agarwal S, Bloom filter: designing a spellchecker. http://ipowerinfinity.wordpress.com
20. Bloom BH (1970) Space/time trade-offs in hash coding with allowable errors. Commun ACM 13(7):422–426
21. Agarwal S (2006) Approximating the number of differences between remote sets. IEEE information theory workshop, Punta del Este, Uruguay, Trachtenberg Ari 217
22. Hash algorithm. http://www.java3z.com/cwbwebhome/article/article5/51002.html

23. Damerau F (1964) A technique for computer detection and correction of spelling errors. Commun ACM 7:171–176
24. Norvig P How to Write a Spelling Corrector. http://norvig.com/spell-correct.html
25. Bayes' theorem. http://en.wikipedia.org/wiki/Bayes'_theorem
26. Zheng Q, Yanan Q, Liu J (2010) Yotta: a knowledge map centric e-learning 1. In: System, IEEE 7th international conference on e-business engineering, pp 42–49

Chapter 6
An Ontology-Based Approach for Semantic Web Service Searching

Shuijing Shen, Bing Li, Bo Shao and Dingwei Li

Abstract Web service is released, combined and invoked through the web. With the continuous growth of web services, finding the web service to meet user requirement has become a challenging problem. The current web service discovery technology such as UDDI which provides keyword-based search is unable to fulfill the user's demand of semantic search. In view of that web services have some functional and non-functional information described by natural language, and the sparsity of user requirement description, we put forward a semantic web service discovery method depending on a concept network which is built on the domain ontology and local ontology in registration model. This method fully uses the registered user's knowledge of web service to build conceptual network and processes the user's search text to mine the true intentions, thus enabling more effective service discovery.

Keywords Ontology · Semantic search · Web service · DVM

S. Shen (✉) · B. Li · B. Shao · D. Li
State Key Laboratory of Software Engineering, Wuhan University, 430072 Wuhan, China
e-mail: Shenshuijing06@163.com

B. Li
School of Computer Science and Complex Network Research Center, Wuhan University, 430205 Wuhan, China
e-mail: bingli@whu.edu.cn

J. J. Park et al. (eds.), *Proceedings of the International Conference on Human-centric Computing 2011 and Embedded and Multimedia Computing 2011*, Lecture Notes in Electrical Engineering 102,
DOI: 10.1007/978-94-007-2105-0_6,

6.1 Introduction

The web has developed from storing simple pictures and texts to providing a wide range of services, such as flight information, weather forecasts and a range of B2B applications [1]. As one of the main ways to provide services now, the application of web services has been developed rapidly. In the case of highly dynamic and context dependent, the service's automatic discovery is the step of effective service integration and coordination in multi-knowledge distributed environment. Because of the limitations of service search based on UDDI, we require a service discovery mechanism combined with semantic information to bridge the gap between web2.0 and the semantic web.

With the development of all kinds of web service description languages, such as OWL, the semantic web has been developed rapidly [1].Ontology provides a vocabulary to describe a concept sharing in a certain domain, and services labeled with the ontology have become part of the implementation of Semantic Web. Although many implementations of semantic search are based on labeling the web services by the ontology now, the requirements of general users are unprepared, and have not been labeled by the corresponding ontology. Therefore, although the requirement represents certain knowledge, it has not been formally described. This produces the gap between the web service described by formal ontology and user requirement described by unstructured or semi-structure text information. In this paper, we use swarm intelligence methods, based on the WordNet ontology and existing techniques of text processing, to label the unstructured plain text messages inputted by users with the concept in corresponding ontology to establish the semantic link between user requirement and registration services to realize the semantic search for web services.

6.2 Related Work

There have been a lot of researches on semantic search for web services. McIlraith et al. [1] proposed a web service search method based on similarity of ontology, it used a basic ontology which contained all the basic concepts as reference ontology, and achieved web service discovery by looking for the similar set of the reference ontology. In the first phase of ontology matching, the descriptions of web services play a key role, and the method uses traditional natural language processing technology to process the description. Ding et al. [2] proposed the Woogle system that achieves a more fine-grained matching method. It finds the similar operations mainly through clustering the method's parameters to realize web service discovery.

In summary, most semantic web service search methods use ontology and natural language processing technology. Each model approach will be suitable for some certain data sets. In order to find a suitable semantic search method for the web service, we propose a search method based on WSSR register model in this

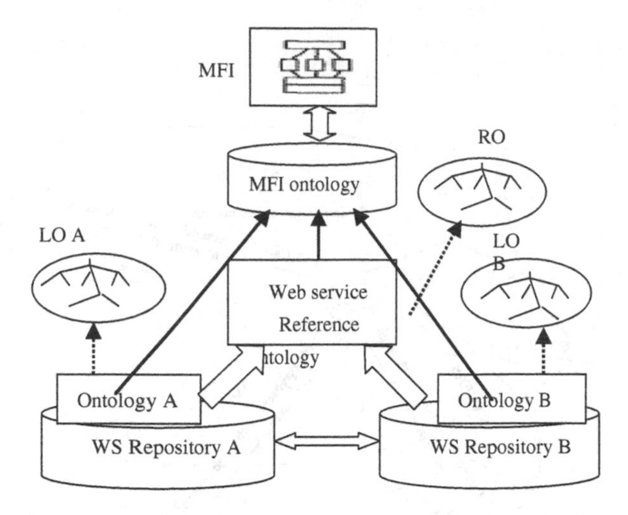

Fig. 6.1 Web service registry model based on MFI

paper. First, classify the input text using natural language processing technology to find the domain. Then, calculate the semantic distance between the classified input text and description texts of web services labeled by one domain ontology concept. Finally, search the web service through the concept network.

6.3 Service Query System

6.3.1 Service Register Model

Now, web services have already supported interoperability in syntax, but because of the different service register models, such as ebXml, UDDI, etc., semantic interoperability has been unable to be perfect. Therefore, in this paper, we use web service registry model which is based on MFI supporting semantic interoperability to achieve semantic search of web service. The model includes the registration of web services and ontology. And according to ISO SC32 19763-3:MFI4OR [3] (Metamodel Framework for Interoperability: Metamodel for Ontology Registration), the WSRR (Web Service Registry & Repository) system registers a domain reference ontology (RO) and some local ontologies (LOs) evolved from reference ontology, and the overall structure is shown in Fig. 6.1 [4].

When registering web services in WSSR, it requires one to label the IOP of the service by domain reference ontology or local ontology making web services attached with semantic information. For semantic search subsystem, the registration process is more or less equivalent to supervised learning process; the registered web services would be instances to establish relationships between the ontology concepts and semi-structured text data in the background of WSSR system. Meanwhile, when registering the user-defined ontology, it can add the

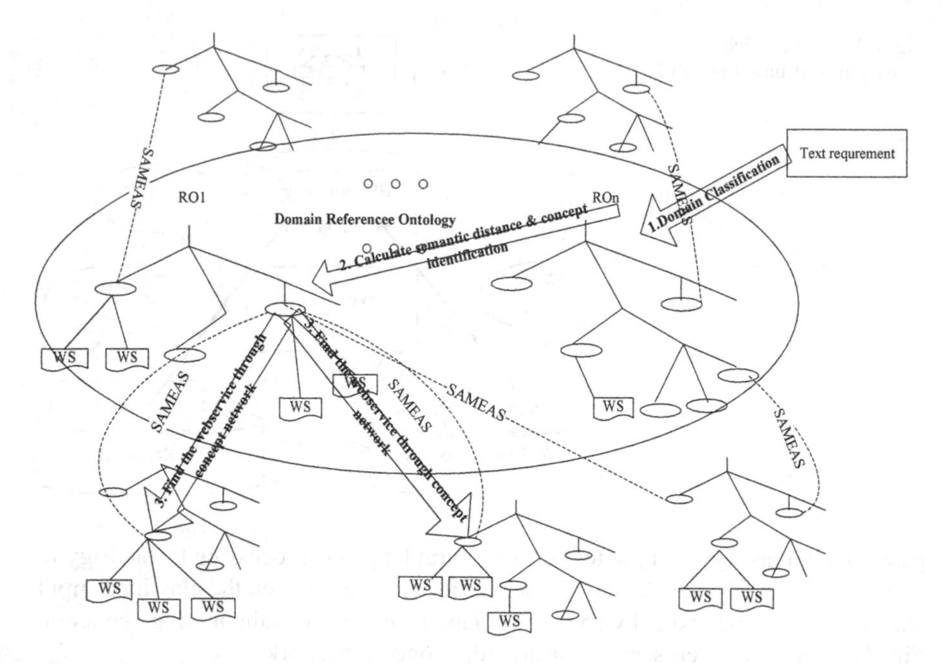

Fig. 6.2 Web service registry model based on MFI

SAMEAS relationship between certain concepts of the LO and the corresponding RO, thus forming the concept network in register library to achieve semantic service search across ontologies.

6.3.2 Service Search

WSRR system offers two search modes, general search and advanced search. General search provides an input box as the user interface.

The process of general search is shown in the Fig. 6.2.

- Identify domain through the analysis of the input text.
- Calculate semantic distance between the input text and the domain ontology (RO) concepts to obtain the concept nodes that meet the requirement.
- Check out the web services that associate with the corresponding concepts through SAMEAS relationship between ontology.

The main difference between general search and advanced search is the determination of the concept nodes. The rest process of advanced search is similar to the third step in general search, only adding the definition of matching level [5].

Text Classification, Identify Domain. In regard to the identification of domain, we can determine the input texts' domain type by the training set. Because of the significant differences between domains, lots of methods have high accuracy. So we focus on speed and self-learning ability of the method in the stage of domain identification. The WSRR system applies the n-gram language model which calculates the posterior probabilities of the input requirement document then obtain the domain with largest values.

Data sources. With the increase in the number of services, Many software services communities on the Internet such as Yahoo Pipes (Yahoo Pipe. http://pipes.yahoo.com/pipes/), Programmable Web (programmableWeb. http://www.programmableweb.com/) have become an important third-party platform as a transaction venue for service developers and consumers. So in this paper the training set data is extracted from the third-party platforms, Seekda (seekda. http://webservices.seekda.com/) and ProgrammableWeb.

Calculate the probability that belongs to a domain according to n-gram language model [6]. According to n-gram language model and the estimation method Jelinek and Mercer (1980) proposed in 1980, its calculate formula is [6]:

$$\Pr(wi|w_1^{i-1}) = \sum_j \lambda j(w_1^{i-1})\Pr^{(j)}(wi|w_1^{i-1}) \tag{6.1}$$

In this formula, w_1^{i-1} represents the historical information of the text, represents the current information, $\Pr^{(j)}(wi|w_1^{i-1})$ represents the conditional probability that the current information determined by the historical data of the text in j-gram language model. In the n-model, the historical information keeps n bits of information at most. And in the experiment, we apply the Witten-Bell's method to calculate the coefficient.

Compute semantic distance. After the first step, we could have known the domain which the input text may belong to, and then obtain the reference ontology of the domain to calculate the semantic distance between the input text and the conceptual nodes in the ontology to find the ideal matching concept finally. We will change the classical DVM [7] (document vector model) to SDVM (skew document vector model) which can obtain the synonym relations easily.

We use the WordNet to calculate the semantic distance. WordNet [8] is a large lexical database in English, developed under the direction of George A. Miller at Princeton University. WordNet includes huge number of concepts and relationships to provide an expanded knowledge repository for creating the model of common sense reasoning. WordNet establishes mutual relations between words according to the semantic. Nouns and verbs are organized by synonyms phrases; superordinate and subordinate relationship between the words is formed between synonyms phrases and words [9], and eventually forms a tree structure.

According to Thanh Dao's algorithm of calculating semantic distance of concepts based on WordNet, the definition of word similarity is as follows.

Definition 1 (word similarity): Define the semantic similarity between two words wd1, wd2 as Simword(wd1,wd2).

$$\mathrm{Sim}_{\mathrm{word}}(\mathrm{wd1}, \mathrm{wd2}) = \left(\frac{1}{1 + \mathrm{length}}\right) \tag{6.2}$$

In this formula:

$$\text{length} = \min\{d\text{syn1}, \text{syn2}|\text{syn1} \in \text{wd1}'s \text{ synonymset}, \text{syn2} \in \text{wd2}'s \text{ synonymset}, \text{d} : \text{shortest length}\}$$

In service registration phase, it requires registrants to provide service description as original data, and formalize the input words through the Porter stemming algorithm (Porter Stemming Algorithm. http://tartarus.org/~martin/PorterStemmer/). As the structure of adverbs and adjectives are not organized in accordance with is a relationship in WordNet, it is unable to calculate the semantic distance between them [10]. So when inserted into the database, these words will be filtered out.

Definition 2 (Document similarity): Define the document similarity between two documents doc1, doc2 as Simdoc(doc1, doc2). doc1, doc2 represents a document respectively, the structure of a document is:

Document=(<Word1, num1>, <Word2, num2>, <Word3, num3>, ...<Wordn, numn>);

Before introducing the document similarity algorithm, we introduce the formula for calculating inner product and the angle of two vectors under any linear space.

Assume that *e1*, *e2*,..., *en* is base of the *n*-dimension european space Vn, for any two vectors of Vn $x = (x1, x2, \ldots, xn)^T$, $y = (y1, y2, \ldots, yn)^T$, according to the law of inner product [11], we can obtain, $(x,y) = z^T Ay^T$, Ay^T, $A \in R^{n \times n}$, representing the measurement matrix of the base.

$$A = (a_{ij}) = \begin{pmatrix} (e_1, e_1) & \cdots & (e_1, e_n) \\ \vdots & \ddots & \vdots \\ (e_n, e_1) & \cdots & (e_n, e_n) \end{pmatrix} \tag{6.3}$$

6.3.2.1 Algorithm 1 (Algorithm of calculating the similarity on SDVM):

Input: User input document, the document in ontology concept

Output: semantic similarity between two documents

1. Extract different words as dimension of the space from user input document and the document in concept ontology
2. Calculate similarity between any concept, and establish measurement matrix A
3. If the similarity between the two dimensions is 1, then merge the two dimensions
4. If the similarity between the two dimensions is less than Threshold ε, then set the component of measurement matrix as 1

Table 6.1 Domains in WSRR system

Domain	Count
Logistics	12
Weather	12
Hotel	8
Email	46
Finance	106
Bioinfo	8
Image	60
Tool	73
Authentication	4

5. Else set the value of the element in measurement matrix as its similarity
6. According to measurement matrix A, adjust the user input document
7. for each (each dimension i in input document)
8. for each (each dimension j in conceptual document)
9. if($A_{i,j} > \eta$ and the value of input document is 0 in this dimension)
10. modify the value of input document in this dimension to 1

Using inner product formula for n-dimensional oblique coordinate system to calculate the semantic distance between user input document vector and the concept ontology vector.

Algorithm 1 is based on document vector: The model is degenerated into the document vector model when we do not merge the two dimensions whose similarity is 1, and setting ε to a very small value. Line 6 to line 10 are to adjust user's input text to add more semantic information according to the following heuristic rules: humans will add their own "domain knowledge" when comprehend something. Calculation of metric matrix A is the bottleneck of the algorithm, it is mainly caused by loading WordNet and calculating the shortest path. So in WSRR system, the inverted indexes of the distance between words are added to the database to improve query speed.

Search the web services. After the two steps above, we will get the following results: the similar domain of the input text, the concept in the reference ontology close to the input text and the web services that are labeled by the proper concepts in RO and some concepts established by the SAMEAS relationship between RO and LOs. Finally, we sort the results according to the weight of words in each web service.

6.4 Case Analysis

In this section, the method mentioned earlier will be tested by examples to verify the validity of the method. In WSRR system, they are divided into 9 domains; Table 6.1 shows the number of web services with description information under the corresponding field.

Table 6.2 Domain classification results

Domain	Score	Log2(P(Domain¦input))
Logistics	3.046785844949356	112.73107626312618
Weather	4.1056216334551285	151.90800043783975
Hotel	3.444778015884542	127.45678658772805
Email	4.409435246506814	163.14910412075213
Finance	4.182092522461988	154.73742333109357
Bioinfo	3.406898423976135	126.055241687117
Image	4.380387165631729	162.07432512837397
Tool	3.7850020747681965	140.04507676642328
Authentication	4.589039886498082	169.794475800429

Table 6.3 Domain classification results

Domain	Score	Log2(P(Domain¦input))
VehicleTrack	0.5116	0
TrackByGPS	0.8226	0.4522
DispatchVehicle	0.6529	0
WareHouse	0.6104	0

Users use general search, and enter "service to track the cars by GPS" in the text box, the system then prints out details about domain identification and concept matching in the background. In the experiment, the result is as Table 6.2.

The system gets the domain of Logistics according to the above result, and then loads the ontology of the domain and matches the concepts of ontology. The ontology concept tree of Logistics includes 54 concepts.

Since ontology tree is large, we take the "VehicleTrack", "TrackByGPS", and "DispatchVehicle" concepts as an example. Matching results are as Table 6.3.

From the results, the n-gram gives the wrong result, and the SDVM and DVM are better than n-gram model here. Although the DVM give the correct result in the example, when we want to get the top n results, it always gave the unsatisfactory results for almost all equals to zero. From a statistical point of view, the SDVM gives better results. Therefore, the system applying SDVM will choose the web services that are labeled by concept "TrackByGPS" and the corresponding SAMEAS concept "TrackedBaggageByGPS" in my_logistics1.owl of ontology to users. The idea of SDVM can be applied to solve some semantic issues. What' s more, many web services can be found after they were labeled with the corresponding concept no matter whether they have description information.

6.5 Conclusion

This paper presents a method to achieve the semantic search of web services through the ontology concept network based on WordNet and also compares the method with DVM and n-gram model. This method is designed based on the informality of user requirement and the sparsity of requirement description, fully mining the limited relationship within web services, providing a new method different from keyword match for the semantic web application to achieve semantic web service search. The semantic search method in this paper is based on the ontology annotation in register model and the RO and LO ontology reduces the number of semantic matching and improves the speed of semantic search effectively.

Acknowledgments This work is supported by National Basic Research Program (973) of China under grant No. 2007CB310801, National High Technology Research and Development Program (863) of China under grant No. 2006AA04Z156, National Natural Science Foundation of China under grant No.60873083, 60803025, 60703009, 60303018, 60970017 and 60903034. Natural Science Foundation of Hubei Province for Distinguished Young Scholars under grant No. 2008CDB351.

References

1. McIlraith S, Son T, Zeng H (2001) Semantic Web services. IEEE Intell Sys 16(2):46–53
2. Ding X et al (2004) Similarity search for web services. In: Proceedings of the 30th VLDB conference, pp 372–383
3. Keqing H (2004) Information technology framework for metamodel interoperability. Part-3: metamodel framework for ontology registry. ISO/IEC JTC1, Tech Rep: ISO/IEC 19763-3
4. Zeng C, Keqing H, Zhitao T (2008) Towards improving web service registry and repository model through ontology-based semantic interoperability, GCC2008, pp 747–752
5. Paolucci M, Kawmura T, Payne T, Sycara K (2002) Semantic matching of web services capabilities. In: Proceedings of the international semantic web conference (ISWC), pp 333–347
6. Brown PF, Della Pietra VJ, PV deSouza, Mercer RL (1990/1992). "Class-based n-gram models of natural language." In: Proceedings of the IBM Natural Language ITI. Paris, France, March 1990, pp 283–298. Also in Comput Linguist 18(4): 467–479
7. Salton G, Wong A, Yang C (1975) A vectorspace model for automatic indexing. Commun ACM 18(11): 613–620. Also reprinted in SparckJones and Willett (1997), pp 273–280
8. Miller A, (1990) WordNet: an on-line lexical resource. J. Lexicography, 3(4)
9. 李杉, 李兵, 潘伟丰,侯婷婷. 一种mashup服务描述本体的自动构建方法,小型微型计算机系统, 2010
10. Ted Pedersen. WordNet::similarity–measuring the relatedness of concepts.
11. 方保鎔. 《矩阵论》. 清华大学出版社. 2004

Chapter 7
Applying Data Visualization to Improve the e-Health Decision Making

Md. Asikur Rahman and Sherrie Komiak

Abstract e-Health is the combined use of electronic communication and information technology in the health sector. Number of patients and diseases are increasing everyday because of the population growth. An e-health database or a decision making in e-health field may involve huge amount of data. Dealing with huge amount data is a complex task. To deal with such challenge, we can take the advantage of advance technology like information visualization techniques. This paper presents a snapshot of how we will apply different visualization techniques such as zooming, focusing, multiple coordinated views, and details on demand etc. on the e-Health data to visualize the medical and health information in very interactive way. The proposed application can help the doctors and medical students with their decision making. Doctors or medical students will be able to visualize the information based on their needs which will improve the decision accuracy and decision time as well.

Keywords e-Health · Information visualization · Decision support system · Cognitive fit theory

---

Md. Asikur Rahman (✉)
Computer Science, Memorial University of Newfoundland (MUN), St. John's,
NL A1B 3X5, Canada
e-mail: asikur.rahman@mun.ca

S. Komiak
Business Administration, Memorial University of Newfoundland (MUN), St. John's,
NL A1B 3X5, Canada
e-mail: skomiak@mun.ca

J. J. Park et al. (eds.), *Proceedings of the International Conference on Human-centric Computing 2011 and Embedded and Multimedia Computing 2011*, Lecture Notes in Electrical Engineering 102,
DOI: 10.1007/978-94-007-2105-0_7,

7.1 Introduction

> Implementing a computerized record system... could save 60,000 lives, prevent 500,000 serious medication errors, and save $9.7 billion each year.
>
> —Leapfrog [7]

> Existing technology can transform health care.... If all Americans' electronic health records were connected in secure computer networks... providers would have complete records for their patients, so they would no longer have to re-order tests.
>
> —Gingrich and Kennedy [1]

Advances in the information Systems can have a significant effect on health sector. The World Health Organization defines e-health as: "the cost-effective and secure use of information and communications technologies in support of health and health-related fields, including health-care services, health surveillance, health literature, and health education, knowledge and research". Key application areas of e-Health include electronic medical records, telemedicine and telecare services, consumer health informatics, and Internet-based technologies and services.

The transformational power of information technology in altering the nature of competition in an industry and creating value for both firms and consumers has long been acknowledged in diverse industry sectors such as airlines, financial services, and retailing. The common characteristic among industries that have experienced such transformations is that they are information intensive—that is, a significant proportion of their value-creation activities occur through the storage, processing, and analysis of data. Now-a-days e-Health systems gain popularity because of the technological advances. By increasing the population growth, the number of patients and diseases are increasing as well which involved huge amount of data.

Sometimes it might be required to analyze heuristic information for doctors or medical students which can improve their learning and decision quality. For example, medical students may need to learn from previous cases about how to treat a disease; doctors or medical students may need to do some comparative analysis to diagnose a disease or research on new medicine to treat a disease. In these cases they need to view historical data.

Dealing with these huge amounts of data is a real challenge. In the past doctors and medical students can retrieve information by using traditional queries. The information (e.g., blood test results) is normally presented in tabular format. However, if the retrieved result shows too much information then deal with the tabular representation is really tough. Because in that case we have to see the whole table by using scrolling up and down, and then have to compare results with other tables which is a time consuming task and it will also have odd effect on the human perceptual process.

To overcome this problem we can take the advantage from the advance data visualization technologies. In this study we have proposed a system whose purpose is to deal with the medical data using various visualization techniques which will overcome the problem of traditional system and also improve the performance of decision making of the doctors or medical students. The main contribution of this paper is the design of a system which will work as decision support system in which doctors or medical students will be able to visualize different medical information of the patients based on different parameters which will improve the performance of the decision quality.

The remainder of this paper is organized as follows: in Sect. 7.2, literature review is briefly described. In Sect. 7.3 the research model is proposed. In Sect. 7.4, a system with advanced data visualization techniques is designed. Finally, conclusion and future work are presented.

7.2 Literature Review

The term e-Health has been in use since the year 2000. This area is full of opportunities for research. Some works have been done in this field. Electronic health records (EHRs) contain a wealth of information. Categorical event data such as complaints, diagnoses, treatments, etc., are important, and play important roles in health providers' decision making. However, past research efforts have been focused on numerical data and single-record visualization techniques. Discovering patterns of categorical events across multiple records are supported in limited ways in [6].

Augmented and Mixed Reality technology provides to the medical field the possibility for seamless visualization of text-based physiological data and various graphical 3D data onto the patient's body. This allows improvements in diagnosis and treatment of patients. Key issues in developing such applications are the tracking methodology, the display technology and most of all ensuring good usability. There have been several research groups who extended the state of the art in employing these technologies in the medical domain and addressing issues of human–computer interaction, interaction design, and usability engineering [5]. Most of the works have been done in personal health record (PHR). A review of the implementation challenges of PHR is provided by Halamka, Mandl and Tang [2]. Particular evaluation of PHR functions, adoption and attitudes of healthcare providers and patients towards PHRs, PHR related privacy and security, and PHR architecture are discussed by Kaelber et al. [3].

7.3 The Research Model

Our research model draws its theoretical foundation from cognitive fit theory [9] which has been widely used by many research domains for problem solving and it could be applied to many dimensions of fit. Cost-benefit theory is the overall framework for the cognitive fit theory. According to cost-benefit theory, decision makers' trade-off the effort required to make a decision vis-à-vis the accuracy of the outcome. The error and effort required to make a decision and may therefore induce decision makers to change strategy is influenced by the factors: (1) task and (2) context. According to Payne [4], task variables are those "associated with the general structural characteristics of the decision problem". Context variables are those related to the actual values of the objects under consideration. Task variables have been shown to influence effort. Context variables influence accuracy. To apply cost-benefit theory to decision making using graphs and tables, we first need to determine the characteristics of both the problem representations and the tasks they support. We view problem representation and decision-making task as independent.

Let us assume we are considering graphs and tables derived from equivalent data, so that all information in one is inferable from the other, with a different type of information predominating in each. Graphs are spatial problem representations because they present spatially related information; that is, they emphasize relationships in the data. They do not present discrete data values directly. The data in a graph are accessed using perceptual processes. On the other hand, tables are symbolic problem representations because they present symbolic information; that is, they emphasize discrete data values. The data in a table are accessed using analytical processes. As we already know that, the theory of cognitive fit applies cost-benefit theory to decision making using graphs and tables in two ways. First, the simple form of the theory, which addresses information acquisition and well-defined evaluation, is a special case of cost-benefit theory. Second, the more traditional view of cost-benefit theory involving strategy shift applies to decision making on more complex tasks where a number of appropriate strategies may be available.

When the types of information emphasized in the decision-making elements (problem representation and problem-solving task) match, the problem solver is able to use processes (and therefore formulate a mental representation) that also emphasize the same type of information. Consequently, the processes the problem solver uses to act both on the problem representation and on the task will match. The resultant, consistent mental representation will facilitate the decision-making process. Hence, cognitive fit leads to an effective (i.e., accurate or precise) and efficient (i.e., fast) problem solution.

Using the cognitive fit theory and cost-benefit theory, our research model describes that matching the problem representation directly to the task has significant effects on problem solving performance (Fig. 7.1).

In our research model we have used clustering technique and multiple coordinated views (MCVs) as the independent variable. From the theoretical foundation it is clear that perceptual processes require less effort than analytical processes which will make the decision time faster. Perceptual processes involve graphical representation. Again analytical processes involve symbolic representation. Tables are symbolic problem representations which will improve the accuracy of the information. When problem representation and problem-solving task are defined

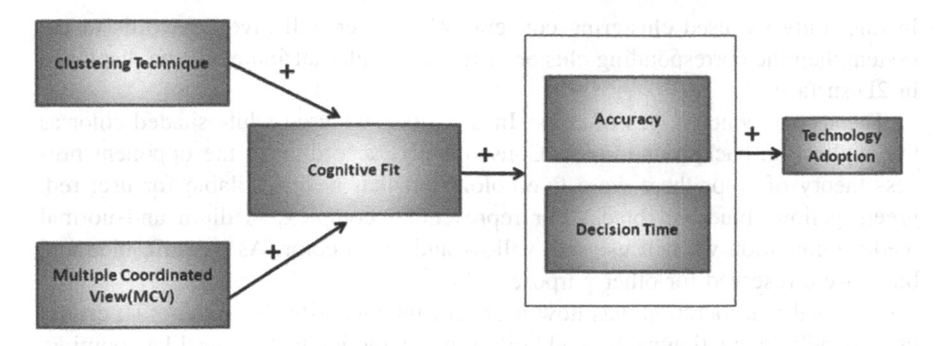

Fig. 7.1 Cognitive fit model in problem solving

properly then it will positively effect on the mental representation for task solution, which will lead problem solving performance. Therefore in this study we proposed our system in such a way so that it will take both the advantage of graphical representation and tabular representation which will improve the performance in terms of decision time and accuracy.

7.4 System Design

Base on the theoretical foundation and research model we can generate some hypotheses which will discover whether these hypotheses are true and assess the usefulness of the proposed system.

Using the proposed visualization based system, we are thinking that a user will be able to complete the specified task faster than with a traditional general system. Because when the processes the problem solver uses to act both on the problem representation and on the task will match. The resultant, consistent mental representation will facilitate the decision-making process. In our proposed research model we are using clustering technique and MCVs for problem representation and task completion. Therefore the resultant mental representation will facilitate the decision-making process.

7.4.1 Data Collection

We have proposed visualization based decision support system in this study. The purpose of designing this system is to see what the value is of providing various visualization techniques such as multiple coordinated view (MCV), zooming, focusing, details-on-demand etc.

Various usability issues were examined when designing this application. The first concern was how we could display the result of the searched attribute.

In this study we used clustering concept. When user will give keywords to the system then the corresponding cluster of that particular attribute will be displayed in 2D-surface.

The second concern was of color. In this study we used white shaded color as the application background color. Consequently, according to the opponent process theory of color there were five colors left that were available for use: red, green, yellow, blue and black. For representing complex, medium and normal medical condition we will use red, yellow and green color. As a result, blue and black were reserved for other purpose.

The final consideration was how user will interact with the system. To ensure user friendly interaction we will add different interaction techniques like zooming, focusing, and multiple coordinated views and so on.

7.4.2 Visualization Based System

This section describes the main characteristics and functionalities of the proposed decision support system.

7.4.2.1 Visualization Techniques

In the proposed visualization system there will be an option where user will give the search keyword based on their need. After giving the search keyword the retrieved result will be displayed on the screen (see Fig. 7.2). For example, user wants to see that what was the treatment for a particular disease like "Migraine". Each table contains multiple attributes. Here Medication Table contains the disease name, treatment, condition, important findings, and symptoms. Personal_info Table, Family_history table, Allergies Table also contains multiple attributes. When a keyword is given for a particular attribute then the corresponding information will be displayed for the corresponding table in MCVs (see Fig. 7.2). Then the user will be able to analyze data simultaneously in all views. Here each point on medication table represents the search result for "Migraine". Where red points indicate severe condition, yellow points indicate medium condition and green points indicate normal condition of the diseases. Therefore, when any keyword is given to the system then based on that keyword corresponding result will be displayed as points and other tables will display results based on the first attribute of each table.

7.4.2.2 Coordination

One thing is important in this system. We have used three colors for three conditions of diseases. In this case how user will be able to know that which point

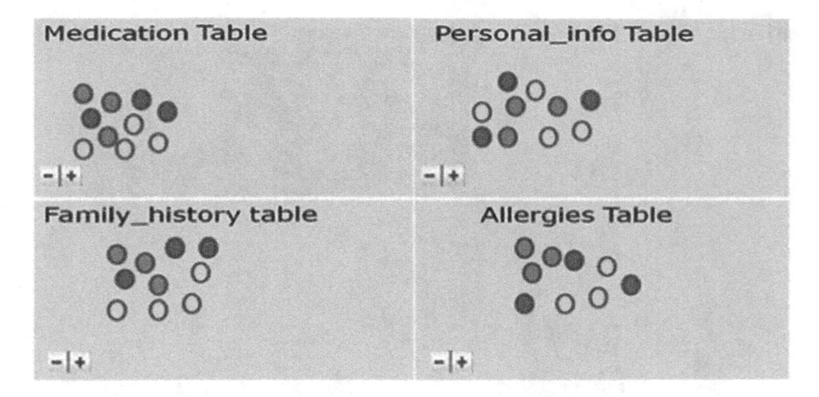

Fig. 7.2 Searched result in MCVs

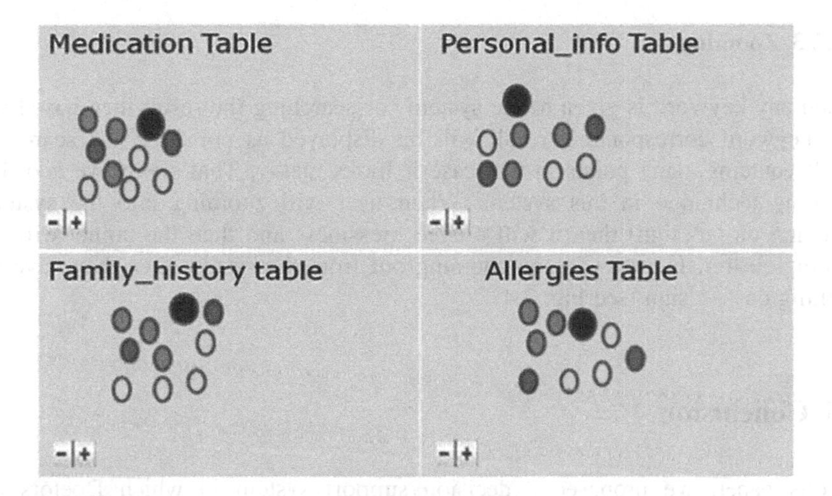

Fig. 7.3 Coordination technique for a particular point

corresponds to which other points in different table. To identify this thing another level of interaction is needed and that is hovering the mouse. When users hover mouse on any points then that point will be highlighted as blue and the shape will become a bit larger. By which user will be able to pre-attentively identify all the related values in the multiple coordinated views which also will lead faster perception process (see Fig. 7.3).

Details-on-demand are set for all visualization techniques. The user selects the attributes that will be presented as additional information about each data item using the tabular box (see Fig. 7.4). In this case user needs another level of interaction. For viewing the details result of any point, users have to click on that particular point and then details information will be displayed in a tabular box.

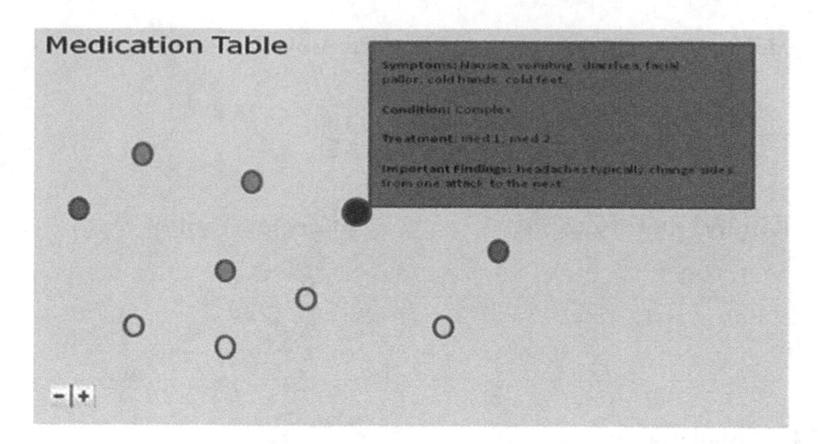

Fig. 7.4 Details information of a particular point

7.4.2.3 Zooming

When any keyword is given to the system for searching the result then based on that keyword corresponding result will be displayed as points. If the searched result contains many points in that case it looks messy. That's why we add the zooming technique in this system. When user will zooming into the system (clicking on '+' sign) then it will reduce messiness and then the points will be looking clearer. If user wants to zooming out from the system then they have to clicking on '–' sign (see Fig. 7.4).

7.5 Conclusion

In this paper, we proposed a decision support system in which Doctors or medical students (more specifically medical students) will be able to visualize different medical information of the patients based on different parameters and which will improve the performance of the decision quality. From theoretical foundation we have known that, graphical representation will make the decision time faster and tabular representations will improve the accuracy of the information. In our proposed system initially we displayed the searched result as a cluster form in 2D-surface which is correspondent with the graphical representation. Again for seeing the details information of any particular point we add extra level of interaction to the corresponding point. So that when user will click on the particular point then the details information will be displayed in tabular format. Therefore, in our proposed system both the decision time and accuracy will be achieved based on the well known theoretical foundation.

In short, the traditional system for visualizing patient's information is the history of progression, and the proposed visualization based system using various visualization techniques is a step in advancement of information technology.

7.6 Future Work

The scope of user interaction is currently lesser than it is expected in this proposed version. The future work will be to modify the proposed system to incorporate more interaction techniques such as filtering, panning etc., so that the user can find their desired information in the easiest possible way. Again we have a plan to collect real data from some reputed e-Health system provider. Then we will add various features to improve the system based on the customer needs. And we are also planning to run user evaluation so that we can compare our proposed system with the traditional general system.

References

1. Gingrich N, Kennedy P (2004) Operating in a vacuum, p 23. New York Times, NY
2. Halamka JD, Mandl KD, Tang PC (2008) Early experiences with personal health records. J Am Med Inf Assoc 15(1):1–7
3. Kaelber DC, Jha AK, Bates DW (2008) A research agenda for personal health records (PHRs). J Am Med Inf Assoc 156(6):729–736
4. Payne J (1982) Contingent decision behavior. Psychol Bull 92(2):382–402
5. Reinhold B, Johannes C, Steve W (2007) Some usability issues of augmented and mixed reality for e-Health applications in the medical domain. In: Proceedings of the third human–computer interaction, pp. 255–266
6. Shneiderman B (2010) Electronic health records search and visualization. Available at http://www.cs.umd.edu/hcil/lifelines2/
7. The Leapfrog Group Fact Sheet (2004) Available at http://www.leapfroggroup.org/FactSheets/LF_FactSheet.pdf
8. Vessey I, Galletta D (1991) Cognitive fit: an empirical study of information acquisition. Inf Sys Res 2(1):63–84
9. Vessey I (1991) Cognitive fit: a theory-based analysis of the graphs versus tables literature. Decis Sci 22(2):219–240

Chapter 8
A Psychological Decision Making Model Based Personal Fashion Style Recommendation System

Jiyun Li, Xiaodong Zhong and Yilei Li

Abstract In this paper, we give out the design and implementation of a personal fashion style recommendation system. The user style preference modeling was based on not only the macroscopic Interactive Genetic Algorithm based preference fitting but also the microscopic psychological factor adjustment. By integrating the psychological decision making model into user preference modeling, the system intends to approach a human like decision making in terms of the decision context. Experiments on the subjects from both fashion and non fashion professionals show the effectiveness of the model.

Keywords Psychological model · Decision making · Personalization · Fashion style

8.1 Introduction

Since its beginning in the early 1980s till now computer aided fashion design has made remarkable progress. Various efforts have been made to make the system more intelligent. One stream of these efforts is to make the system learn the user's style preference thus automatically generates or recommend personalized fashion

J. Li (✉) · X. Zhong · Y. Li
School of Computer Science and Technology,
Donghua University, Shanghai, China
e-mail: jyli@dhu.edu.cn

X. Zhong
e-mail: zxd_rambler@sina.com.cn

Y. Li
e-mail: liyilei@dhu.edu.cn

J. J. Park et al. (eds.), *Proceedings of the International Conference on Human-centric Computing 2011 and Embedded and Multimedia Computing 2011*, Lecture Notes in Electrical Engineering 102,
DOI: 10.1007/978-94-007-2105-0_8,

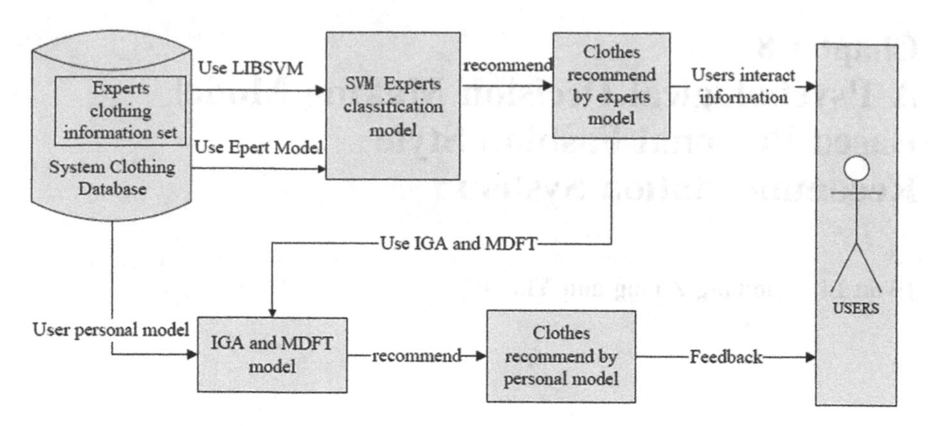

Fig. 8.1 System architecture of PBFRS

style to him. How to build the preferential model of the customer according to user's operation in the system is the major concern of fashion style personalization. Interactive Genetic Algorithm (IGA) [1–3] is the most popular method adopted by these researchers. IGA did capture the user's subjective feeling about the design by eliciting the user's evaluation about the garments on the one hand, while on the other hand, being a global optimization algorithm, IGA do the global search always with no differentiation between the originally large choice data set and the final small choice alternatives. There is a paradigm conflict between this kind of choice decision making and the one that belongs to real human being. Human beings actually follow different process during they make choices from large amount of choice candidates. According to psychology literature on human decision making, there is generally a quick eliminate by aspects process and a relatively slow deliberate process among the small set of alternatives. Several irrational effects occurred during the latter process [4, 5]. This makes the final choice usually not explainable by the rational model. Concerned about this aspect, we proposed a model incorporates the psychological decision making model into IGA, and developed the Psychological decision making model based personal Fashion style Recommendation System, PBFRS for short.

The rest of the paper is organized as follows, Sect. 8.2 presents the system architecture of PBFRS, Sect. 8.3 details the design and implementation of the key modules in PBFRS, Sect. 8.4 gives out the implementation of the system, and Sect. 8.5 draws the conclusion and gives out some further research directions.

8.2 System Architecture of PBFRS

In PBFRS, the fashion style was divided into eight categories. On the left side of Fig. 8.1 is the system clothing database. In PBFRS, each garment was made of four parts; the parts and their parameters are stored in the system database.

The garments recommended to the user are generated by searching in the database for different parts according to the user's style preference

We can see that there are three models in between users and system clothing database. The SVM model i.e. support vector machine was to establish expert evaluation classification. This information was used as the default recommendation from the system before a user has any interactions with the system; IGA was used build the preferential model of users to tackle the personalization part of fashion style recommendation; By incorporating the psychological decision making model, Multi-alternative decision making (MDFT) model to be specific, into the final preferential set of IGA the system successfully tackle the classic irrational effects in multi-alternative decision making, especially the contextual effects, like similarity effect, attraction effect and compromise effect.

8.3 Key Modules in PBFRS

8.3.1 SVM Experts Classification Mode

In order to use SVM to do the fashion style classification, first we have to decide the feature number of a fashion style. The features that mainly affect a clothing fashion style includes Silhouette, Waist design, length, Collar type, Sleeve type, Pocket type and Facing design [6].

RBF kernel function was adopted in this multi-classification problem. When parameters $c = 1.573$, $g = 0.1623$, the classification accuracy reaches 82.22%. Considering of the subjective aspects of fashion style and also the results only work as a default value before user's interaction, this accuracy is acceptable.

8.3.2 IGA and MDFT Personal Preferential Model

The default fashion styles recommended by SVM only stands for the experts' taste. It cannot fully meet the requirements and style taste of the user. We adopt IGA and multi-alternative decision field theory (MDFT) [7] to build the preferential model of each user. In this model, IGA works as the extensive candidate searching process while MDFT works as the deliberating process on the small number of final alternatives given by IGA.

Fashion designers establish their fashion style by splitting, combining, aggregating and aligning of fashion style elements, including points, lines, faces etc. On the parts level, the style elements are collar, piece, sleeve, separation line and ornament according to style from strong to weak. To be specific, the degrees of effect are 50, 30, 15, and 5%, accordingly. Thus the scheme of IGA is as shown in Fig. 8.2.

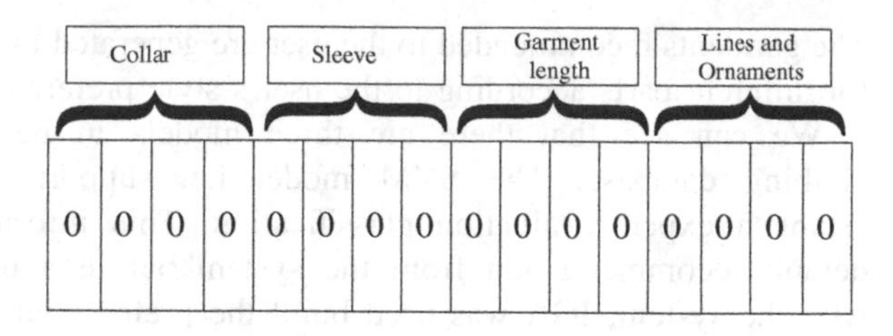

Fig. 8.2 Coding scheme of IGA

- Design of fitness function of IGA:
  Fitness function design is the key part of IGA. Here fitness stands for the preferential degree of a user to a given garment. It is decided by the pattern of the garment, evaluation score from the user and the degree of effect of the part to the whole garment style. So we have

$$C\_\text{fitness} = \sum R_i$$

where C_fitness stands for the fitness of the whole garment, and R_i stand for the fitness of different part of the garment.

$$R_i = \frac{\sum (\text{Point}_j * \text{Part}_i)}{\sum j}$$

where Point_j stands for the score given by the user to part j, Part_i ($\text{Part}_i \in \{50\%, 30\%, 15\%, 5\%\}$) is the degree of style effect of part i. Σj is the total amount of garments in the choice set.
After 200 generation's evolution, we got the fittest four garments as the user's preferential choice set:

$$\text{Choice set} = \begin{pmatrix} A_{11} & A_{12} & A_{13} & A_{14} \\ A_{21} & A_{22} & A_{23} & A_{24} \\ A_{31} & A_{32} & A_{33} & A_{34} \\ A_{41} & A_{42} & A_{43} & A_{44} \end{pmatrix} \begin{pmatrix} \text{Fitness\_1} \\ \text{Fitness\_2} \\ \text{Fitness\_3} \\ \text{Fitness\_4} \end{pmatrix}$$

where Fitness_i is the fitness value of ith garment.

- MDFT fitness micro adjustment:
  MDFT is a computational personal decision making model in psychology. MDFT provides for a mathematical foundation leading to a dynamic, stochastic theory of decision behavior in an uncertain environment. It is the expansion of DFT which was proposed by Jerome Busemeyer and Jim Townsend [8] in 1993. MDFT formally described the preference changing process of the user as a stochastic process as in (8.1)

$$P(t+h) = SP(t) + V(t+h) \tag{8.1}$$

where,
P(t) is the preferences matrix of the alternatives at time *t*;
S is the inhibition matrix. It is decided by the psychological distance of the two compared alternatives.

While

$$V(t) = CMW(t)$$

where, *C* is the contrast matrix, represents the process used to compare options; *M* stands for the attributes' matrix in subjective deliberating space; in PBFRS they are EV_{SC} (simple vs. complicate) and EV_{FT} (fashion vs. traditional) for each garment. They are derived from A_{ij} of the garment according to Kansei engineering technology [9], semantic mapping to be specific;
W(t) is the weight vector. It represents the amount of attention allocated to each attribute at each moment. It can be achieved by parameters' fitting using the initial evaluation given by the user. It is a better representation of the user's preference than the pure IGA searching.

- Find the most similar garments for recommendation:
  In order to find the most similar garments for recommendation, we have to first map the M value in subjective deliberating space back into IGA searching space, that means, from EVSC and EVFT to A_{ij}, Then traversing the garments in Database, find the garment with the highest similarity value with the garments in the model recommend to the user.
  As a garment in the database is coded into a four-element tuple B (B_1, B_2, B_3, B_4), the style similarity between B and A_j is defined in (8.2)

$$\text{similarity}_j = \text{Fitness\_}i \times \sum_{i=1}^{4} \mathrm{d}(\mathrm{A}_{ij}, \mathrm{B}_i) \times \text{degree}_i \tag{8.2}$$

  where d(,) represents the Euclidian distance of style between A_j and B. degree_i stands for the degree of effect of part *i* to the whole garment style. The preference of the user to garment B is defined as:

$$\text{Preference}_B = \max(\text{similarity}_j)$$

  Here, *i* stands for *i*th part and *j* stands for the *j*th garment. $i, j = 1, 2, 3, 4$.
- Training strategy:
  The effectiveness of the output of the system was evaluated by the following process:
  For a given period of time the user can re-score the recommended garments, and got the average of these scores Res_avg, then compare it with the average of the initial garment generation Exp_avg, which was also given by the same user.
  If test_avg = Res_avg−Exp_avg > test_num, then we think that the recommendation was successful. test_num is given by the system. Otherwise,

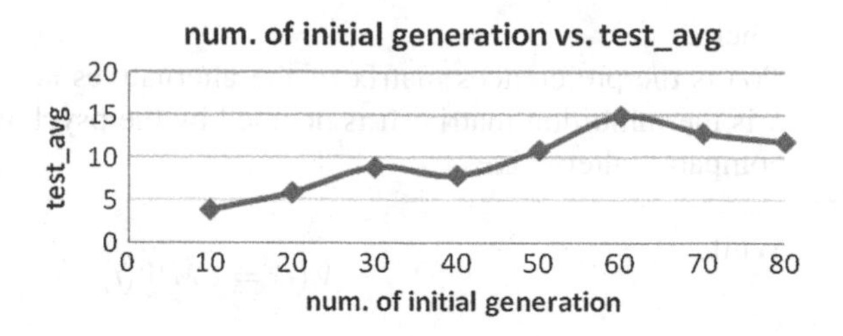

Fig. 8.3 test_avg with different number of garments in the initial generation

PBFRS takes the recommended garment group as the initial generation, start another round of IGA. Figure 8.3 shows the relationship between different initial generation number and among10 users.

From Fig. 8.3 we can see that we achieve the highest test_avg when the number of garments in the initial generation of IGA.

8.4 Implementation of PBFRS

Figure 8.4 is the flow chart of the system.

After registration or login, the user is first shown the default fashion style recommended by the system by SVM model. Then he will be asked to score the garments recommended. With this information, IGA and MDFT personal preferential model start working. Through training, the model can gradually learn the user's style preference and make the recommendation more and more acceptable to the user. Figure 8.5 is the screenshot of one of the recommendation given by the system.

8.5 Conclusions

In this paper, we give out the design and implementation of a personal fashion style recommendation system PBFRS. The major difference between PBFRS and other systems is that its preferential model of the user is built upon not only the rational IGA optimization, but also the psychological decision making model MDFT to handle the irrational part of human beings in multi-alternative decision making, especially the contextual effects. The architecture and key modules of the system and its implementation were presented in the paper. We also invite more than 10 subjects including both fashion professionals and non professionals to use the system. We do find the three contextual effects reported in psychological decision making literatures in the choice results made by the subjects in the sample choice set. The results also show that more than 85% of the subjects got a positive

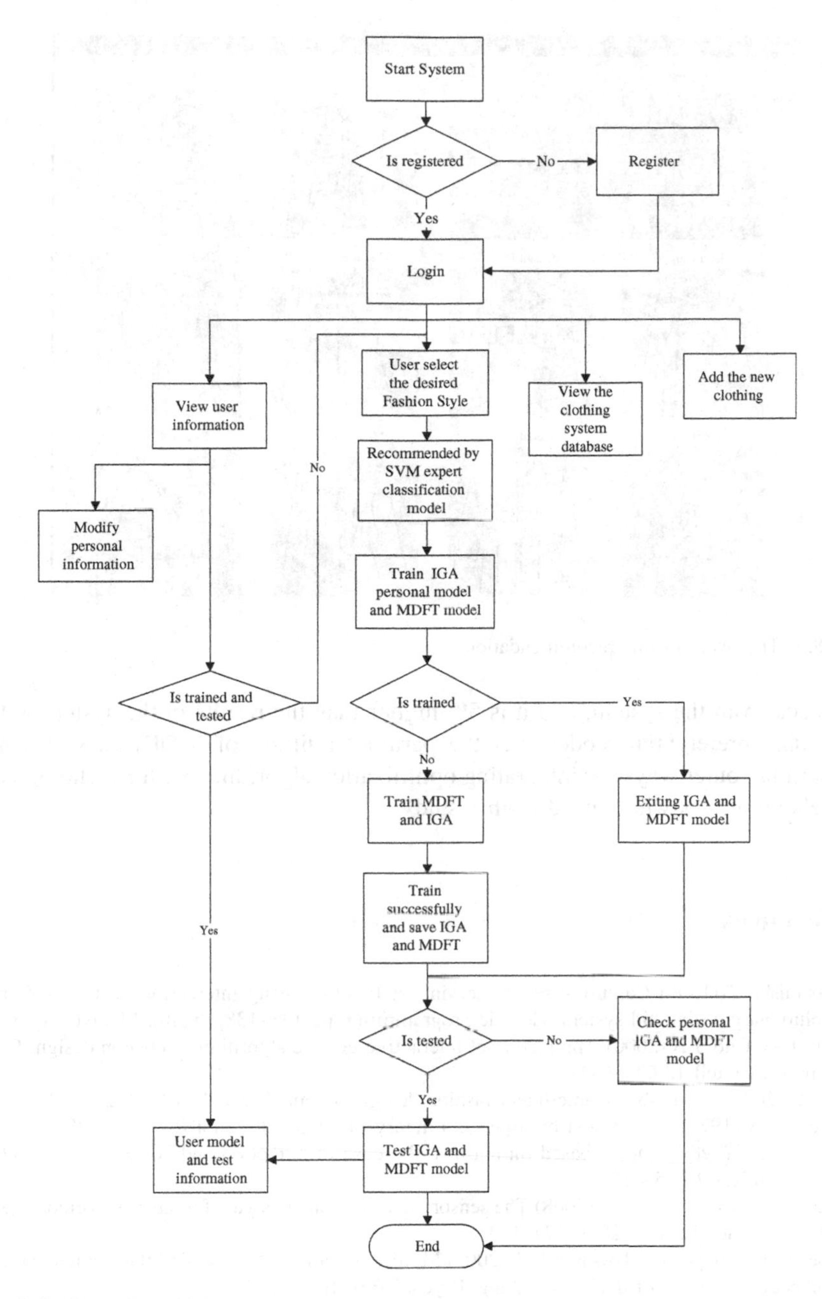

Fig. 8.4 Flow chart of the system

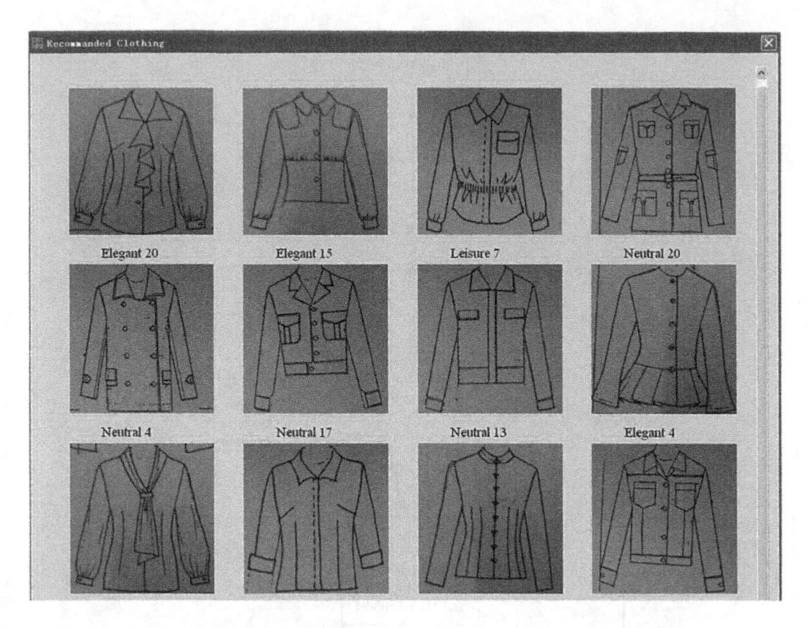

Fig. 8.5 The result of one recommendation

feedback from the system, and it is 5% higher than the results in the system with only IGA preferential model. Yet the parameter fitting of MDFT is still time consuming, other ways for integrating optimization algorithms with psychological models are issues which need further study.

References

1. Nakanishi Y (1996) Capturing preference into a function using interactions with a manual evolutionary design aid system. Genetic programming, pp 133–138, Stanford University, CA
2. Kim H-S, Cho S-B (2000) Application of interactive genetic algorithm to fashion design. Eng Appl Artif Intell 13:635–644
3. Li JY (2006) A web-based intelligent fashion design system. J Donghua Univ 23(1):36–41
4. Tversky A (1972) Elimination by aspects: a theory of choice. Psychol Rev 79(4):281–299
5. Simonson I (1989) Choice based on reasons: the case of attraction and compromise effects. J Consum Res 16:158–174
6. Wang Y, Chen Y, Chen Zg (2008) The sensory research on the style of women's overcoats [J]. Int J Cloth Sci Technol 20(3):174–183
7. Roe R, Busemeyer JR, Townsend JT (2001) Multi-alternative decision filed theory: a dynamic connectionist model of decision making. Psychol Rev 108:370–392
8. Busemeyer JR, Townsend JT (1993) Decision field theory: a dynamic-cognitive approach to decision making in an uncertain environment. Psychol Rev 100(3):432–459
9. Simon TW, Schutte JE, Axelsson JRC, Misuo N (2004) Concept, methods and tools in sensory engineering. Theor Issues Ergon Sci 5:222

Chapter 9
A Multi-Modal Augmented Reality Based Virtual Assembly System

Changliang Wu and Heng Wang

Abstract A multi-modal and augmented reality based virtual assembly system is described. It uses computer vision technology, speech and vision gesture to enhance user interaction experience. Augmented reality reserves virtual reality interactions and combines reality with assemble scene to improve assembly feelings.

Keywords Augmented reality · Virtual assembly · Multimodal

9.1 Introduction

Traditional assembly is based on real model. It's time consuming and requires special equipment. Virtual assembly is based on computer system to view, manipulate and assemble model. There are representative system such as VADE system of Washington State University [1], VTS system of University of Maryland [2], VDVAS of Zhejiang university [3], etc. Augmented reality technology enhances user's perception of environment. There are several augmented reality based assembly systems. Schwald [4] developed a training system to help users learn assembly. STARMATE study group of IST (information societies technology) [5] developed a system support complex assembly based on real environment. Klinker [6] developed a augmented system for training staffs of maintaining nuclear power plant. Operators use head mounted display to see real model tagged with name, and to be guided by the instructions through the assemble process. For interaction in 3D

C. Wu (✉) · H. Wang
Graphics and Interactive Technology Lab, The Key Lab of Machine perception and intelligent (MOE), Peking University, Beijing, China
e-mail: wucl@graphics.pku.edu.cn

H. Wang
e-mail: hw@pku.edu.cn

J. J. Park et al. (eds.), *Proceedings of the International Conference on Human-centric Computing 2011 and Embedded and Multimedia Computing 2011*, Lecture Notes in Electrical Engineering 102,
DOI: 10.1007/978-94-007-2105-0_9,

environment unlike 2D user interface as GUI/WIMP, multi-modal input and output should be integrated, including speech, gestures.

This paper focus on the user interaction technology in augmented reality based virtual assembly system with comprehensive study of augmented reality, computer vision tracking and registering, as well as constraints guided assembly locating and multi-modal interaction. We provide a convince way for user to manipulate objects with realistic feelings in virtual assemble system.

9.2 Augmented Reality Environment

It's an important part of the system to fuse a virtual object into real world. A virtual model must keep consistent with real world in space relationship, logic and structure relations. Those mainly include two parts, mark tracking and camera calibration.

For camera calibration we use Zhang's algorithm [8], to obtain camera internal parameters and distortion by using images captured from different directions. The algorithm includes two models: (1) pinhole model; (2) distortion model. The three items of the formula below are distortion of Radial distortion, tangent distortion and thin prism distortion, with k, p, s are coefficients for radial distortion, tangent distortion and thin prism distortion separately.

$$\delta_x(X,Y) = k_1X\left(X^2 + Y^2\right) + \left(p_1\left(3X^2 + Y^2\right) + 2p_2XY\right) + s_1\left(X^2 + Y^2\right) \quad (9.1)$$

$$\delta_y(X,Y) = k_2X\left(X^2 + Y^2\right) + \left(p_2\left(X^2 + 3Y^2\right) + 2p_1XY\right) + s_2\left(X^2 + Y^2\right) \quad (9.2)$$

We get transform matrix from identifying marks to do connected component analysis of the captured images. All rectangular areas are candidates. We match candidate areas with mark template images. Suppose input image pattern is x, template images are c, $S(x) = \frac{x * c}{\| x \| * \| c \|}$ is measure scale. x is closer to c, if S(x) is loser to 1. Using multiple marks can reduce errors. Different mark template images have different codes. It can be easily identified. All marks are in a plane, forming an array in a rectangular area with certain distance. Every mark's positions in 3D coordinate are known. Using every 3D position (X, Y, Z) and 2D image position (x, y) pairs we can get transform matrix of camera. We then use the average value to reduce errors.

9.3 Computer Vision Based 3D Tracking and Locating

The system uses infrared light pen as an interaction tool. Two CCD cameras with infrared filters can easily track the infrared light pen. We use two camera based computer vision technology to recover the 3D position of the pen. There is a point

Fig. 9.1 3D position recovery

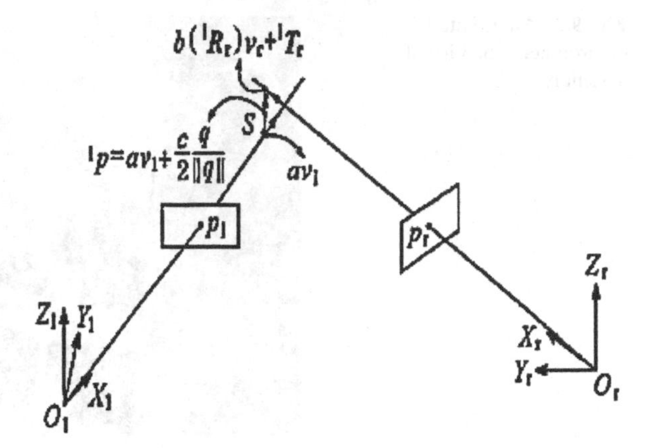

P located in the space where both two cameras can capture. The positions in two captured images are:

$$\begin{cases} p_l = (x_l,\ y_l) \\ p_r = (x_r, y_r) \end{cases} \tag{9.3}$$

We use camera's internal parameter to transform image positions to 3D position in camera's coordinates.

$$p = \begin{bmatrix} (x - C_x)s_x \\ (y - C_y)s_y \end{bmatrix} = \begin{bmatrix} p_x \\ p_y \end{bmatrix} \tag{9.4}$$

$[C_x, C_y]$ is the center of the image, $[s_x, s_y]$ is scalar factor. Define two vectors: $v_1 = \begin{bmatrix} p_{1x} \\ p_{1y} \\ f_1 \end{bmatrix}, v_2 = \begin{bmatrix} p_{rx} \\ p_{ry} \\ f_r \end{bmatrix}$, f_1 and f_2 are the focal length of two cameras. Ideally these two vectors cross on the location of the target. But practically because of errors and distortion of camera, v_1 and v_2 are hardly cross. This paper chooses the middle point of the vertical segment as the target P. As Fig. 9.1 shows, vector v_2 after rotation and translation becomes $b({}^{l}R_r)v_2 + {}^{l}T_r$, ${}^{l}R_r$, ${}^{l}T_r$ are rotation matrix and translation matrix of the right camera relative to left camera. Because segment S is vertical to both vectors v_1 and v_2, the outer product $q = v_1 \times ({}^{l}R_r v_2)$, the normalized form is $\frac{q}{\| q \|}$.

One endpoint of S is av_1, another is $b({}^{l}R_r)v_2 + {}^{l}T_r$, we got:

$$av_1 + c\frac{q}{\| q \|} = b\left({}^{1}R_r\right)v_r + {}^{1}T_r \Rightarrow av_1 - b\left({}^{1}R_r\right)v_r + c\frac{q}{\| q \|} = {}^{1}T_r \tag{9.5}$$

c is scalar factor, [a b c] are unknown parameters. The coordinate of P in left camera coordinates is ${}^{1}p = av_1 + \frac{c}{2}\frac{q}{\| q \|}$. The coordinate of P in world coordinates

Fig. 9.2 Augmented environment for virtual assembly

is ${}^{w}p = {}^{w}M_{l}\,{}^{l}p$; ${}^{w}M_{l}$ is the transform matrix of left camera coordinates and world coordinates, which is obtained through cameras' external parameters.

Therefore we can get 3D location of the infrared light pen through two cameras. We can use the pen to select, translate and rotate virtual assembly models. It's an important interaction mode of the system.

As shown in Fig. 9.2, we can select and manipulate virtual assembly models with the infrared light pen. The blue rectangles are virtual buttons, we can define different manipulate type when the buttons are pushed or not. When the first button color is blue means the button is not pushed, we can translate the selected object; when we pushed the button with the pen, the color change to red, means we can rotate the selected object. We can define more buttons to add more system commands.

9.4 Multi-Modal Interaction

The system combines speech recognition, gesture recognition and other input modals in augmented reality environment with a new interface of multi-modal interaction.

We use Microsoft's speech recognition engine SAPI to recognize user's speech commands, such as operation help, simple operations, etc. which are implemented as speech commands. Vision gesture is an assistant interaction. The system recognizes some static gestures and dynamic gestures. The static gestures' algorithm includes two parts, firstly, making image segmentation according to skin color; secondly, recognizing gestures by matching template gesture images using statistics method. Figure 9.3 shows an example the system recognize the gesture of "palm close". Our dynamic gesture recognition method first tracks hands by using mean-shift algorithm, then recognizes the hands' movement by using neural network machine learning algorithm. We can define some simple manipulations such as translation, rotation, etc. through vision gesture. For example, move left means translate object left, close the palm means stop the movement, etc.

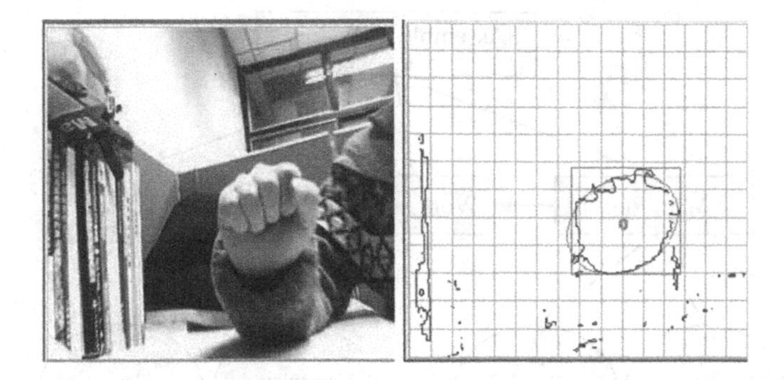

Fig. 9.3 Gesture recognition example

A multimodal input fuser takes input from various recognizers such as the speech recognizer and gesture recognizer, fuses the results into a complete semantic frame for the application. The system accepts semantic frames and provides user feedback through multimodal output. We use the Johnston's algorithm [11] in which the input modes, after full recognition, are semantically represented by underspecified typed feature structures. The multimodal fusion is conducted through a unification process. Only compatible speech and gesture results are unified and the most probable unified type feature structure is taken as the action by the system.

9.5 Assembly Process Control

Assembly target is represented as a tree structure. The root node is the entire model, the non-leaf nodes are the sub model. Assembly parts are the leaf nodes. In the virtual environment, precise locating is a difficult task. Using constraints to help precise locating model is an important method. Here we use a method of using pre-defined workflow script to constrain the movement of the assembly parts models. It achieved precise locating and operation guiding. In each work step we defined manipulated object ID, manipulate type (translate, rotate, Spiral), movement parameters, etc. and slave object ID, manipulate type, movement parameters, etc. Furthermore, the system monitors user's operation mistakes according to constraints, and analyze the operation log to help users intelligently (Fig. 9.4).

9.6 System Design and Application

We developed a plane engine assembly system. The platform is windows XP operation system, using VC++ 2005 IDE and OGRE graphics rendering engine. The system architecture is shown in Fig. 9.6. The system implements the plane

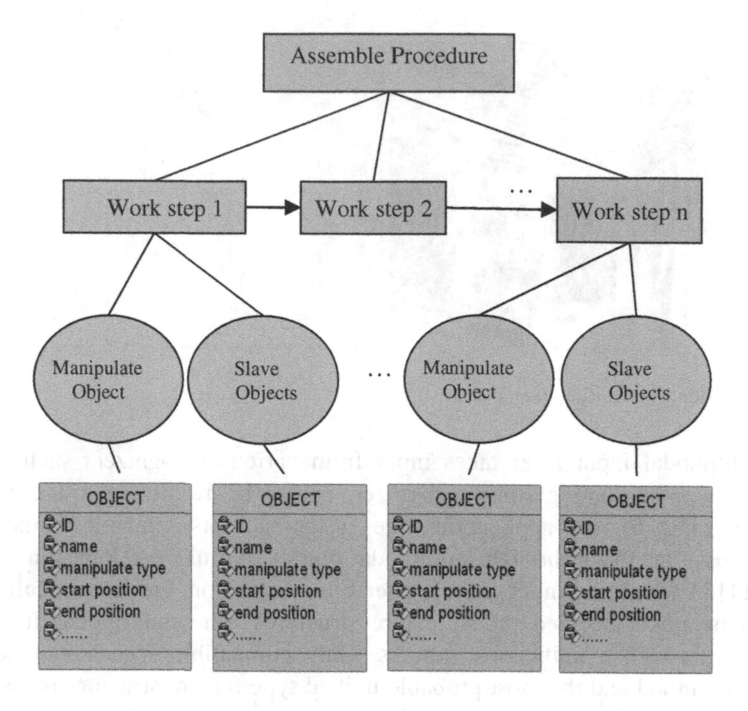

Fig. 9.4 Assembly process control

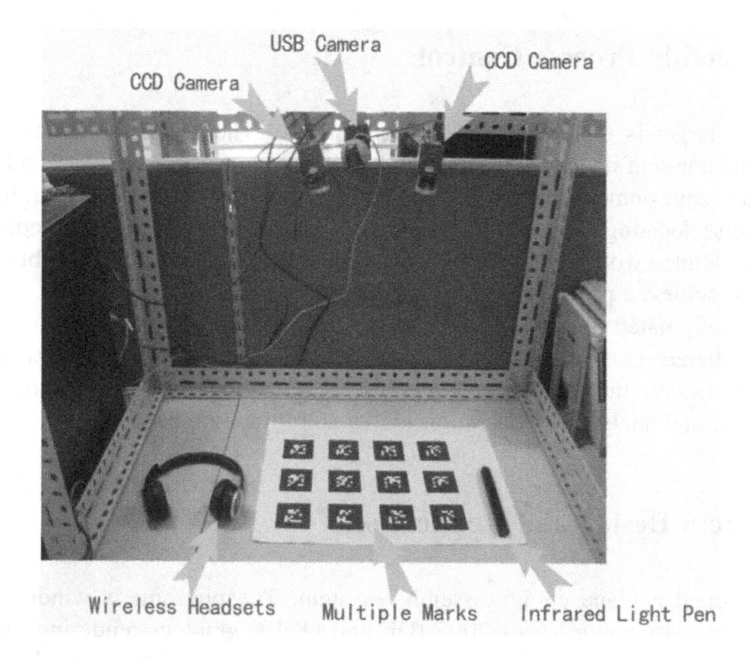

Fig 9.5 Experiment environment

Fig 9.6 System architecture

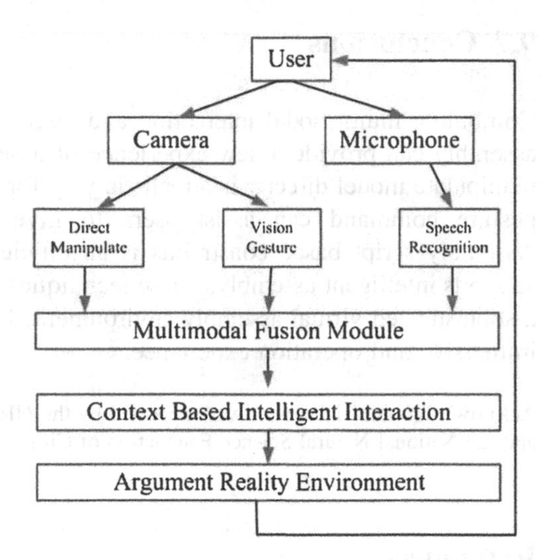

engine assembly in augmented reality environment. The hardware includes one Logitech USB camera, two black and white CCD cameras, two Infrared filter lenses, an Infrared light pen and wireless headsets. Figure 9.5 shows the experiment environment. The plane engine CAD model is transformed to mesh file which is used in OGRE. Then according to assembly relations we define the workflow script XML file. OGRE graphics rendering engine takes charge of model rendering, scene management. We use OpenCV lib to process captured images and camera calibration, etc.

The system mainly includes multi-modal input module, virtual reality fusion module, three-dimensional locating, assembly process control module, etc. The multi-modal input module processes input information streams from different modal, such as mouse, keyboard, speech, vision gestures, etc. It combines different inputs to a unified command. Virtual reality fusion module add virtual plane engine model into real scene. Operators can use an infrared light pen to operate models directly. The system constrains the movement of the assembly parts to precise location, according to the workflow script. Furthermore according to the constraints the system makes record of the mistakes of the operator, then help operators to review their performance. Users also can give command through speech or gestures.

We have evaluated the system with a scenario of plane engine disassembly. The disassembly model consists of 30 subassemblies and contains more than 2,000,000 triangles. The procedure includes more than 20 work steps. The experiment shows augmented reality environment provides a more immersive disassembly feeling. The multiple interactive ways, including direct manipulating, speech commands and gestures, give users a bigger operation space and more freedom of movement, though with the weakness of less accuracy.

9.7 Conclusions

Combining multi-modal interaction and augmented reality technology in virtual assembly can provide a new experience of assembly. Using infrared light pen to manipulate model directly is a realistic way for interaction. Speech command and gesture command can assist users to give system command conveniently. Assembly script based constraints is an efficient way to precise assemble, also supports intelligent assembly. These techniques are brought together to implement a sophisticated virtual assembly environment. It improves users' virtual assembly immersive and operation experience.

Acknowledgment This work was supported by the NHTRDP 863 (Grant No. 2009AA012105) and the National Natural Science Foundation of China (Grant No. U0735004).

References

1. Jayaram S, Jayaram U, Wang Y et al (1999) VADE : a virtual assembly design environment. IEEE Comput Graph Appl 19(6):44–50
2. Brough JE, Schwartz M, Gupta SK et al (2007) Towards development of a virtual environment-based training system for mechanical assembly operations. Virtual Real 11(4):189–206
3. Huagen W, Shuming G, Qunsheng P (2002) VDVAS : an integrated virtual design and virtual assembly environment. J Image Graphics 7(A):27–35
4. Schwald B, Laval B (2003) An augmented reality system for training and assistance to maintenance in the industrial context. J WSCG 11:425–432
5. Schwald B, Figue J, Chauvineau E et al (2001) STARMATE: using augmented reality technology for computer guided maintenance of complex mechanical elements. In: Proceedings of the e-business and e-work conference, Venice
6. Klinker G, Creighton O, Dutoit AH et al (2001) Augmented maintenance of power plants: a prototyping case study of a mobile AR system. In: Proceedings of IEEE and ACM international symposium on augmented reality, New York
7. Lu P, Chen Y, Dong C (2009) Stereo vision based 3D game interface. The sixth international symposium on multispectral image processing and pattern recognition
8. Zhang Z (2000) A flexible new technique for camera calibration. IEEE Trans Pattern Anal Mach Intell 22(11):1330–1334
9. Wan H, Gao S, Peng Q et al (2004) MIVAS: a multi-modal immersive virtual assembly system. In: ASME conference Proceedings
10. Gupta SK, Anand DK, Brough JE, Kavetsky RA, Schwartz M, Thakur A (2008) A survey of the virtual environments-based assembly training applications. Virtual manufacturing workshop
11. Johnston M (1998) Unification-based multimodal parsing. In: proceedings of the seventeenth international conference on computational linguistics (COLING '98)

Chapter 10
A Novel Memory-Based Storage Optimization Approach for RDF Data Supporting SPARQL Query

Feng Zhao, Delong Wu, Pingpeng Yuan and Hai Jin

Abstract Due to sparse of RDF data, RDF storage approaches using triple table or binary file rarely show high storage usage and high query performance. To achieve the goal of decreasing storage space and improving the efficiency and generality of query on RDF data, a memory-based storage optimization approach supporting SPARQL query is proposed in this paper. For storage efficiency, strings are transferred to 32-bit integer identifiers, RDFS/OWL is used to organize the RDF data storage model, vertical partition method and other ways are utilized to split and organize the RDF triples. For query generality, the storage model also partially supports SPARQL query. Furthermore, by getting the statistics of the underlying storage data, we construct a SPARQL query optimized module. Experiments show that our query optimization module greatly improves the performance of the SPARQL query. Compared with Jena Memory and RDF-3X, our storage method has higher performance.

Keywords RDF · Memory-based storage · Storage optimization · SPARQL

F. Zhao · D. Wu · P. Yuan · H. Jin (✉)
Service Computing Technology and System Lab, Cluster and Grid Computing Lab, Huazhong University of Science and Technology, 430074 Wuhan, China
e-mail: hjin@mail.hust.edu.cn

F. Zhao
e-mail: zhaof@mail.hust.edu.cn

J. J. Park et al. (eds.), *Proceedings of the International Conference on Human-centric Computing 2011 and Embedded and Multimedia Computing 2011*, Lecture Notes in Electrical Engineering 102,
DOI: 10.1007/978-94-007-2105-0_10, © Springer Science+Business Media B.V. 2011

10.1 Introduction

Semantic web aims to generate more machine understandable data, which is an effort by W3C to enable integration and sharing of data across different applications and organizations [1]. *Resource Description Framework* (RDF) [2], widely used in semantic web, represents data as statements about resources using a graph connecting resource nodes and their property values with labeled arcs representing properties. RDF data can be transferred into a set of statements, also called triples, representing in the form <*subject*, *predicate*, *object*>. Currently, approaches to store RDF data can be divided into two categories: disk-based and memory-based.

Disk-based storage for RDF data, such as *Jena* [3], *Sesame* [4], and *3store* [5], is to store triples in a relational database with a three-column table, one triple for a record in a table. Under this way, indexes are commonly built to improve query performance. The main demerit of disk-based storage is the low performance. To alleviate these deficiencies, a vertical partitioning approach is presented to make RDF storage efficiency and speed up RDF query [6], that a triples table is rewritten into *n* two column tables, where *n* is the number of unique properties in the data. Since almost all queries are properties bound, this approach can gain more storage efficiency and higher performance. However, it does not use schema information to optimize the storage model.

In recent years, memory-based approach to store RDF data is another trend. *Jena Memory* loads all the triples into memory and uses hash function to index subject, predicate object which truly can gain high performance but at the same time it consumes a lot of memory [7]. *BRAHMS* stores RDF/OWL data in main memory which separates the RDF/OWL data into schema and instance [8]. This storage system still uses triple table as its primary model and makes several indexes on the model to speed up semantic association query. However, it does not support SPARQL, which is the recommended language by W3C for RDF query [9], and becomes increasingly important as a RDF query language.

Our work is mainly inspired by memory-based storage approach. The goal of our research is to decrease storage space and improve the efficiency and generality of query. In this paper, vertical partition and main memory are utilized to construct a memory model to store RDFS/OWL data. Considering there is a lot of useful information in the schema which can be used to construct an optimized storage model [10], we also divide the RDFS/OWL data into schema and instance as *BRAHMS* [8].

The rest of this paper is organized as follows. Sect. 10.2 surveys of the related work on RDF storage and technology supporting SPARQL query. Section 10.3 describes the design of system architecture of main memory-based storage. Section 10.4 presents our storage optimization approach. In Sect. 10.5, implementations of supporting interface on SPARQL and query optimizer based on the storage model are proposed. Section 10.6 gives the experimental results and the final is the conclusion.

10.2 Related Work

Currently, lots of RDF storage systems are available. The physical storage layers underlying those systems generally rely on RDBMS, file-system or main-memory.

The most typical RDF data storage representative systems relying on RDBMS are *Jena SDB*, *Sesame*, *3store*, and *Oracle* [3–5, 11]. These solutions use giant triples table to store statements. Specific RDF query, such as SPARQL or RDQL [12], must be converted to SQL in the higher level of RDF layers, and be sent to the RDBMS which will optimize and execute the SQL query over the triple store. In order to gain storage efficiency and high query performance, some other RDF storage systems are implemented using disk files as the storage back-end rather than RDBMS, such as *RDF-3X*, *Kowari*, *Jena TDB* [13–15]. *RDF-3X* makes use of six indexes (SPO, SOP, OSP, OPS, PSO, POS) using B+ tree and compresses on each index to make storage efficiency, but still occupies amount of disk space. *Kowari* uses AVL tree to store the triples. Since the storages are based on disk, more optimization can be made on the storage model [13]. Normally, disk-based approach always gains a better performance than RDBMS triple's storage approach [14].

With the development of more and cheaper memory, building a memory-based RDF data storage system becomes a new trend. If the data structures are implemented in memory, the data can be operated in real-time that saving disk I/O overhead. A well-structured RDF storage system designed in main memory can greatly improves the performance of query, such as *SwiftOWLIM* [15] and *BRAHMS* [8]. These two systems employ the hash tables and indexes. *BRAHMS* is a read-only RDF/OWL storage system focusing on fast semantic association discovery, which uses graph algorithms liking depth-first-search (DFS) and breadth-first-search (BFS).

For storage optimization, vertical partition approach for semantic web data management is introduced to store RDF data [6], which uses a fully decomposed storage model. This approach has many advantages, including well support for NULL and multi-valued properties, less storage space usage compared with giant triple table based method. If vertical partition method implements in column-oriented databases [6, 16], less storage consumption and higher performance can be gained. Another storage optimization technology is schema and instance. When storing RDF schema and instance data in RDBMS, a schema-aware database schema layout can be adopted rather than schema oblivious approach: a big table to store all the triples. In schema-aware approach, RDF schema data and RDF instance data are separated, and both are partitioned into many tables, usually by the class or property types of the resource [17]. Experiments show that the schema-aware approach outperforms the schema-oblivious approach on query processing [18].

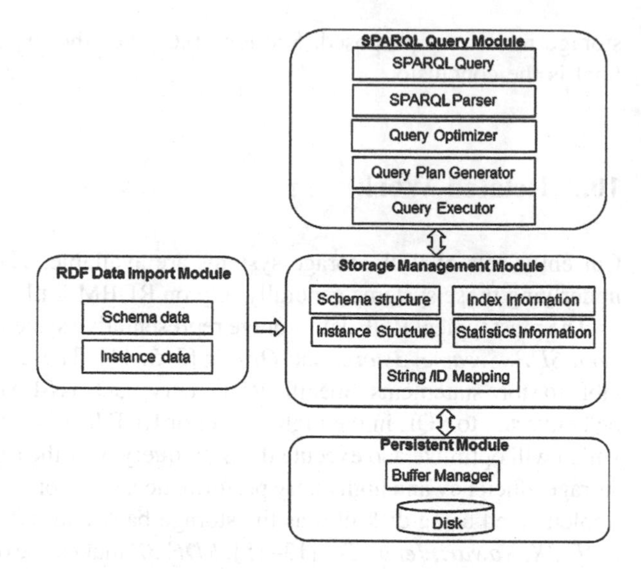

Fig. 10.1 Architecture of memory-based RDF storage model

10.3 Memory-Based Storage Model for RDF Data

In order to manage RDF data efficiently and get higher performance of SPARQL queries, some principles are followed when designing memory-based RDF storage supporting SPARQL query system:

- Store RDF data in main memory.
- Use integer identifiers to encoding RDF data.
- Use schema to optimize the storage structure.
- Fast join support for SPARQL query.
- Build optimized module to optimize SPARQL query.

The architecture of our memory-based RDF storage model is depicted in Fig. 10.1, which is separated into four modules:

- RDF Data Import Module: This module transfers various specifications' RDF data into triples form so that the *Storage Management Module* can use the triples. In this module, we use Redland Raptor parser [19] as the RDF parser. Since our model separates RDF data into schema space and instance space, so this module must use the schema triples that generate by RDF parsers to construct the primitive storage module, and then classifies the instance triples into corresponding storage structure.
- Storage Management Module: String/ID mapping is used for fast translating the triple strings form into triple integer identifiers form or vice versa. When *RDF Data Import Module* imports the schema triples, it translates the strings form triple into integer identifiers form and divides the schema triple into different types, more details in Sect. 10.4. The instance triples are classified into different tables according to their predicate types (the predicate belongs to which schema

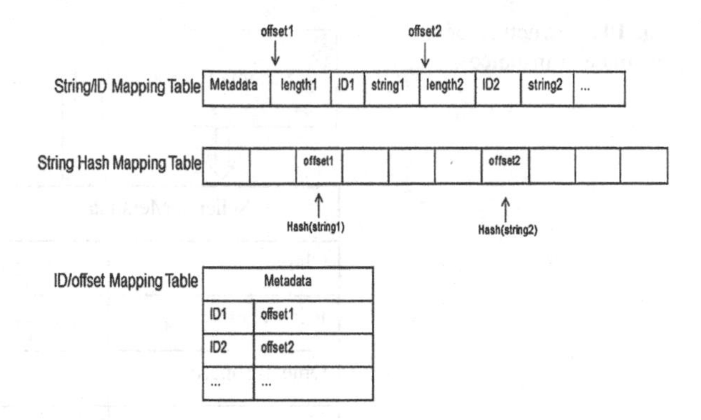

Fig. 10.2 String/ID mapping

vocabulary). For fast query, we also make indexes on the structure, and store the statistics metadata about the underlying tables for query optimization.

- Persistent Module: Loading RDF raw files into the storage model may take a long time, so we build this module for the subsequent load operation. It maps the memory buffers or memory structures in main memory to files in the disk. Experiments show that the time loading memory buffers or structures from disk files can be ignored.
- SPARQL Query Module: This module is responsible for receiving SPARQL query, parsing the query, and translating the query patterns into the physical storage operations.

10.4 Storage Optimization

10.4.1 String/ID Mapping

The URIs and literals of primitive triples tend to be very long strings. For storage optimization, a String/ID mapping module is proposed to map the strings into integer identifiers, as shown in Fig. 10.2. There are three tables for fast finding string by integer identifiers or vice versa. The *String/ID Mapping Table* stores the ID and corresponding string as a record, this table can be accessed by the address offset. To make finding string by ID quickly, an *ID/offset Table* is built, in this table the order of ID are sequenced, so it is very fast when using ID to find string. The *String Hash Mapping Table* uses a linear hash function to hash the string and get the hash code as the address in this table, and stores the offset in the address or linear probing, so it makes finding ID by string also very fast. All tables in this module are implemented using memory buffers and can be flushed into disk files.

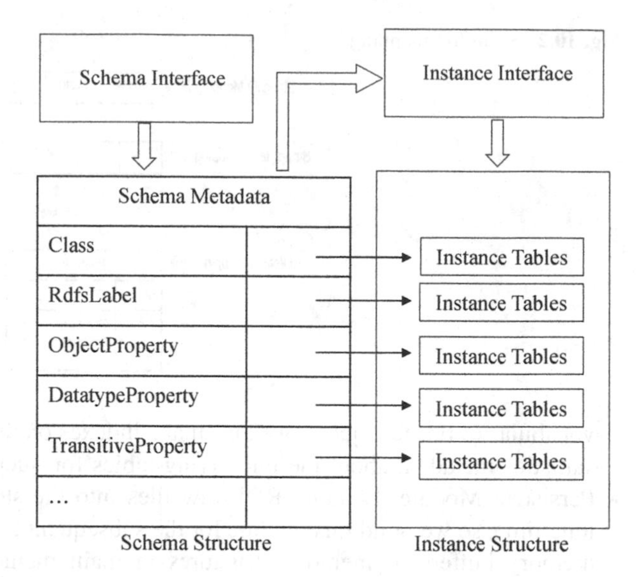

Fig. 10.3 Structure of schema and instance

10.4.2 Schema and Instance Data Storage

Our model partitions instance RDF data into different triple sets by their characteristics and stores then into multiple tables, so there are two steps to load RDF data. Firstly, vocabularies are collected from schema space using the scheme file defining the vocabularies used in instance data. For example, we extract the LUBM schema file [20], and get Class, such as *ObjectProperty*, *DatatypeProperty* and *TransitiveProperty*, vocabularies that defined in RDFS/OWL. For example, we get *Article* and *Book* as the Class, *affiliateOf* as the *ObjectProperty*, *age* as the *DatatypeProperty* and *subOrganizationOf* as the *TransitiveProperty*. A structure seen in Fig. 10.3 is built when finish loading all the schema triples. Since we transfer the triple strings into integer identifiers using the String/ID mapping module, we save the vocabularies' integer identifiers in the Schema Metadata, for example the Class vocabulary *Article* which integer identifier is 1, and *ObjectProperty* vocabulary *affiliateOf* which integer identifier is 5.

Secondly, the instance triples are loaded into corresponding instance tables according to their class types or property types. For class type instance triples, we classify them into different classes. We have a lot of *Class Tables*, and we sort all the *Class Instance Tables* by integer identifiers. We store other properties just like the vertical partition method, but instead we sort the first column, group the second column and sort by integer identifiers. This somehow likes *Hexastore* [21] method, but we only build the corresponding properties into PSO and POS. For example, we only build the properties belong to *ObjectProperty* into PSO and POS which always contain multi-valued, but for *RdfsLabel* property and those properties that belong to *DatatypeProperty*, we only build PSO to store the triple because these vocabularies' range is literal. So we can store RDF data more compactly and less redundantly.

10.4.3 Statistics Collection

The order of SPARQL query can have greatly influence to the query performance, so a query optimized module needs to be built. In order to establish a query optimized module, some underlying statistics data must be stored.

Just like *Jena ARQ* and *RDF-3X*, our model also uses selectivity estimation to estimate query cost. As we use schema triples to construct the instance storage structure, partition the instance triples according to their class types or property types, and sort the corresponding columns by integer identifiers, so we can easily get the accurate result set size for each query pattern. For example, the query mode *<?x p1 o1>* in a SPARQL query, we can quickly get the accurate result set size of this pattern through using *p1* and *o1* to search the POS table, and for the query mode *<?x p1 ?y>*, as we use group-method to store the integer identifiers, so we can keep the result set size of *p1* only by calculating the triples store in this property, so the store format is *<p1, count>*. Generally the properties of a RDF data set are not many, so the space of these statistics information can be negligible.

10.5 SPARQL Query Execution

10.5.1 SPARQL Conversion

Liking *RDF-3X*, we translate the SPARQL into another form that can be operated by underlying storage model or optimized by our query optimization module. *RDF-3X* is open source, so we use *RDF-3X* SPARQL query parser to parse SPARQL query. Through using the parser, we can get projection variables, triple patterns, and some keywords like DISTINT, FILTER, and LIMIT. Since our model converts the strings to integer identifiers (or ids) to store, we use our String/ ID module to translate the literals or URIs of triple patterns into ids. The variables also are converted to ids that contain Boolean flags. Figure 10.4 depicts the SPARQL query conversion process. We use *P* to represent triple node pattern in SPARQL query, seen in Fig. 10.4(1).

If a query consists of multiple triple node patterns, we must join the results of the individual patterns. We use a join edge to represent a join between two triple node patterns, e.g. we transform the SPARQL query in Fig. 10.4(1) into another form in Fig. 10.4(2). From the join graph, we can see that some joins are redundant, so we must make a join query plan to prune the redundant joins to execute the query graph.

10.5.2 SPARQL Query Optimization

As discussed before, we have collected the statistics of each table in the storage model. It is easy to get the accurate result set size of each triple node pattern.

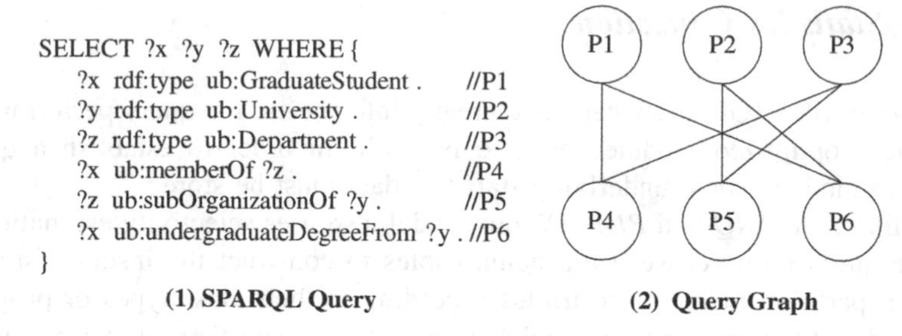
SELECT ?x ?y ?z WHERE {
?x rdf:type ub:GraduateStudent . //P1
?y rdf:type ub:University . //P2
?z rdf:type ub:Department . //P3
?x ub:memberOf ?z . //P4
?z ub:subOrganizationOf ?y . //P5
?x ub:undergraduateDegreeFrom ?y . //P6
}
(1) SPARQL Query
P1
P2
P3
P4
P5
P6
(2) Query Graph

Fig. 10.4 SPARQL query conversion process. (1) SPARQL query, (2) Query graph

Similar to *Jena ARQ*, we use greedy method to generate an optimized query plan, but our selectivity estimation is more accurate. We modify the greedy method in [22] to make it more suitable for our query model and use Triple Node Pattern Selectivity and Join Edge Selectivity as follow. We use *TripleNodeSel* to represent the triple node patterns selectivity, it equals the result set size of the triple node pattern which can easily get from statistics module. The Join Edge Selectivity is expressed as following: *JoinEdgeSel* = *FirstTripleNodeSel* * *SecondTripleNodeSel*. The *FirstTripleNodeSel* is the selectivity of the first triple node pattern's result set size, and *SecondTripleNodeSel* is the other triple node pattern's result set size. Multiplying the two results we get the join edge selectivity, denoted as *JoinEdgeSel*.

Steps using greedy method and selectivity to generate an optimized query plan are following: (1) If the query graph only contains one triple node pattern; binding the output projection and add it to the query plan. (2) If the query graph contains two or more triple node patterns, calculates Triple Node Selectivity and the Join Edge Selectivity that defined above. (3) Select the join edges' minimum selectivity and order the triple node patterns of this join edge by their Triple Node Selectivity. Add the minimum selectivity triple node into the query plan, and then add the second triple node belong to the same edge, bind the output projection. (4) Select a minimum selectivity edge of the edges which one triple node pattern has been added to the query plan and the other triple node pattern does not. (5) Loop step 4 until all triple node patterns are add to the query plan.

10.5.3 Query Execution

Using the query optimized module, we generate a left deep join tree query plan. Figure 10.5 depicts a possible execution flow of a SPARQL query in Fig. 10.4(1). We execute the join tree from bottom, since the tables of our model are sorted by integer identifiers, so we use merge join get the results. As shown in Fig. 10.5, the storage structures corresponding to the first two triple node patterns (*P1* and *P2*) are sorted, so we only need to perform merge join and put the intermediate results into the an *intermediate table structure* that implemented in our model which can

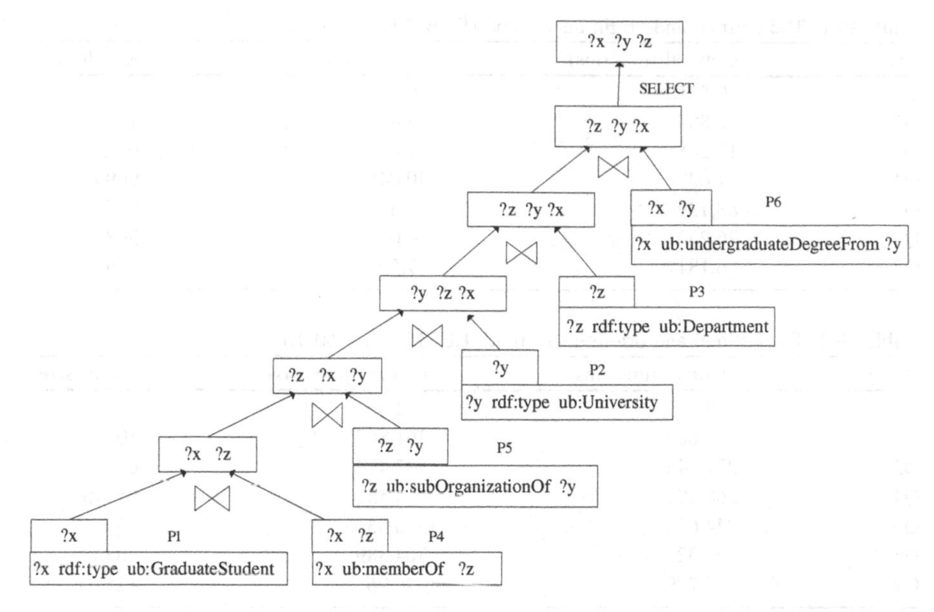

Fig. 10.5 A SPARQL query execution flow

append the corresponding record according to the binding variables and be sorted by variable dimension. But for second join, we firstly sort the input *intermediate table structure* by the join variable, and then perform merge join. At the last join, we use String/ID mapping module to translate the integer identifiers into the strings. The output result can be stored in a *string intermediate table* which stores the strings. At last we do the projection to get the final result sets.

10.6 Experimental Results

Our model is implemented in C++ and can be used as an embedded RDF storage system or a query server to support remote query request. We have tested our query optimized module, which proves that for some query it can gain a significant enhancement in query speed. We compare our storage system with *Jena Memory* and *RDF-3X* using *LUBM* benchmark query [20], the experimental results show that our storage system has a large performance advantage. The experiments are performed on a server with eight 2.50 GHz Intel Xeon CPUs and 16 GB physical memory, 270 GB hard disk, which runs RedHat Enterprise Linux 4.

Jena Memory [7] and *RDF-3X* [11] are chosen to test with our model. In the test, *Jena Memory* version is Jena-2.5.7; *RDF-3X* version is rdf3x-0.3.4. Both our model and *RDF-3X* are compiled using g++ 3.4.6, with optimization parameter-O2. We use *LUBM Data Generator* [20] to generate two datasets LUBM Univ (10,0) and LUBM Univ (60,0), and use Raptor Parser to parse the dataset to *N*-Triples format

Table 10.1 The optimal and origin query time (LUBM Univ (10,0))

Query	Optimal time (ms)	Original time (ms)	Result size
Q1	0.525	0.77	10
Q2	2.885	33.852	10
Q3	12.208	11.629	0
Q4	34.023	40.045	9099
Q5	88.139	127.449	21
Q6	76.233	94.072	368
Q7	6.181	7.59	125

Table 10.2 The optimal and origin query time (LUBM Univ (60,0))

Query	Optimal time (ms)	Original time (ms)	Result size
Q1	0.469	2.051	10
Q2	17.602	161.867	10
Q3	237.943	237.473	0
Q4	268.29	314.552	65046
Q5	774.623	1380.53	159
Q6	645.328	704.389	2658
Q7	34.279	46.96	125

Table 10.3 The optimal and origin query time (LUBM Univ (10,0))

Query	Our model	Jena Memory	RDF-3X	Result size
Q1	0.14	4.0	1.415	10
Q2	0.479	8.0	4.391	10
Q3	11.936	626	19.161	0
Q4	1171.15	169	5055.1	9099
Q5	72.728	351	1641.73	21
Q6	183.798	989	766.441	368
Q7	4.12	10	4.505	125

since *RDF-3X* can only parse *N*-Triples. *LUBM* Univ (10,0) occupies about 195 MB disk space in *N*-Triples format and LUBM Univ (60,0) occupies about 1.36 GB disk space in the same format. As *Jena* is implemented in Java, we set the running JVM parameter with-Xmx1000M when using LUBM Univ (10,0) as dataset and-Xmx2000M when using LUBM Univ (60,0) as dataset.

For *Storage Optimization Module*, we use *LUBM* Univ (10,0) and LUBM Univ (60,0) dataset to test the query optimized module. Seven SPARQL queries are selected to test (see APPENDIX A). All the times in Tables 10.1 and 10.2 are collected from the query input into query module until the last join is finish, but not including the query result set print time. All the query times are in millisecond. The original time is the query time that does not use query optimized module, and optimal time is the query time that use query optimized module. The experiments show that the query optimized module has improved the query response time in most queries except Q3. Since Q3 has no result set, the query module can statically detect that, so the optimal time will be longer.

Table 10.4 The optimal and origin query time (LUBM Univ (60,0))

Query	Our model	Jena Memory	RDF-3X	Result size
Q1	0.408	5.0	1.445	10
Q2	11.241	8.0	4.044	10
Q3	88.496	11477.0	128.746	0
Q4	30079.6	147499.0	47844.1	65046
Q5	996.517	21859.0	12908.1	159
Q6	5702.65	34864.0	11265.3	2658
Q7	25.619	195	4.999	125

Moreover, we do comparison with *Jena Memory* and *RDF-3X*. Tables 10.3 and 10.4 show the query response time. But in Table 10.3, we can see that *Jena Memory* has a higher performance in Q4, and in Table 10.4, *RDF-3X* has a higher performance in Q7. In most cases, our model has a higher performance than *Jena Memory* and *RDF-3X*.

10.7 Conclusions

To make semantic web widely used, it is required to develop an efficient RDF data management tools. But current triple table approach scales poorly mainly because it uses a big table to store all the triples, proper schema is not used to make storage optimized. We present a new RDF storage and query system which stores the whole data in main memory and uses schema data to organize the storage structure and optimized the structure by their RDFS/OWL class or property types. We build a query optimizer inside the SPARQL query module. The experiment of query optimized module shows that the SPARQL optimizer of our model can achieve a higher performance. The comparison with *Jena Memory* and *RDF-3X* shows that our storage system has a higher performance in most cases. In the future, we plan to support more SPARQL operators and inferences about the RDF data.

Acknowledgments This paper is supported by the Key Project in the National Science and Technology Pillar Program of China under grant No.2008BAH29B00.

Appendix A: SPARQL Queries to Query LUBM Dataset

*Q*1: SELECT ?x WHERE {
?x rdf:type ub:ResearchGroup.
?x ub:subOrganizationOf http://www.Department0.University0.edu.
}
*Q*2: SELECT ?x WHERE {

?x rdf:type ub:FullProfessor.
?x ub:name ?y1.
?x ub:emailAddress ?y2.
?x ub:worksFor http://www.Department0.University0.edu.
}
*Q*3: SELECT ?x ?y ?z WHERE {
?x rdf:type ub:UndergraduateStudent.
?y rdf:type ub:University.
?z rdf:type ub:Department.
?x ub:memberOf ?z.
?z ub:subOrganizationOf ?y.
?x ub:undergraduateDegreeFrom ?y.
}
*Q*4: SELECT x WHERE {
?x ub:name ?y.
?x rdf:type ub:Course.
}
*Q*5: SELECT ?x ?y ?z WHERE {
?x ub:memberOf ?z.
?x rdf:type ub:GraduateStudent.
?z ub:subOrganizationOf ?y.
?x ub:undergraduateDegreeFrom ?y.
?y rdf:type ub:University.
?z rdf:type ub:Department.
}
*Q*6: SELECT ?x ?y ?z WHERE {
?y ub:teacherOf ?z.
?x rdf:type ub:UndergraduateStudent.
?x ub:advisor ?y.
?z rdf:type ub:Course.
?x ub:takesCourse ?z.
?y rdf:type ub:FullProfessor.
}
*Q*7: SELECT ?x ?y WHERE {
?x ub:worksFor ?y.
?x rdf:type ub:FullProfessor.
?y ub:subOrganizationOf http://www.University0.edu.
?y rdf:type ub:Department.
}

References

1. Berners-Lee T, Hendler J, Lassila O (2001) The semantic web. Sci Am 5:34–43
2. Resource description framework, http://www.w3.org/RDF/
3. Jena: a semantic web framework for Java, http://jena.sourceforge.net/
4. Sesame: RDF schema querying and storage, http://www.openrdf.org/
5. Harris S, Lamb N, Shadbolt N (2009) 4store: the design and implementation of a clustered RDF store. In: Proceedings of 5th international workshop on scalable semantic web knowledge base systems, Washington, pp 94–109
6. Abadi DJ, Marcus A, Data B (2007) Scalable semantic web data management using vertical partitioning. In: Proceedings of 33rd international conference on very large data bases. ACM Press, Austria, pp 411–422
7. Jing Y, Jeong D, Baik DK (2009) SPARQL graph pattern rewriting for OWL-DL inference queries. Knowl Inf Syst 5:243–262
8. Janik M, Kochut K (2005) BRAHMS: a workbench RDF store and high performance memory system for semantic association discovery. In: Maciej J, Krys K (eds) ISWC 2005. LNCS, vol 3729. Springer, Heidelberg, pp 431–445
9. SPARQL query language for RDF, http://www.w3.org/TR/rdf-sparql-query/
10. Chong EI, Das S, Eadon G, Srinivasan J (2005) An efficient SQL-based RDF querying scheme. In: Proceedings of 31st international conference on very large data bases. ACM Press, Trondheim, pp 1216–1227
11. Thomas N, Gerhard W (2008) RDF-3X: a RISC-style engine for RDF. In: Proceedings of 34th international conference on very large data bases. ACM Press, Auckland, pp 647–659
12. RDQL: a query language for RDF, http://www.w3.org/Submission/RDQL/
13. Valencia GR, Garcia SF, Castellanos ND, Fernandez-Breis JT (2011) OWLPath: an OWL ontology-guided query editor. IEEE Trans Sys Man Cyber—Pt A: Sys Hum 41:121–136
14. TDB: a SPARQL database for Jena, http://openjena.org/TDB/
15. SwiftOWLIM Semantic repository, http://www.ontotext.com/owlim/
16. Abadi DJ (2007) Column stores for wide and sparse data. In: Proceedings of 3rd Biennial conference on innovative data systems research. ACM Press, Asilomar, pp 292–297
17. Lu J, Ma L, Zhang L, Brunner JS, Wang C, Yue P, Yu Y (2007) SOR: a practical system for OWL ontology storage, reasoning and search. In: Proceedings of 33rd international conference on very large data bases. ACM Press, Austria, pp 1402–1405
18. Rohloff K, Dean M, Emmons I, Ryder D, Sumner J (2007) An evaluation of triple-store technologies for large data stores. In: Meersman R, Tari Z (eds) OMT 2007Ws, Part II. LNCS, vol 4806. Springer, Heidelberg, pp 1105–1114
19. Raptor RDF Syntax Library, http://librdf.org/raptor/
20. Lehigh University Benchmark, http://swat.cse.lehigh.edu/projects/lubm/
21. Weiss C, Karras P, Bernstein A (2008) Hexastore: sextuple indexing for semantic web data management. In: Proceedings of 34th International Conference on very large data bases. ACM Press, Auckland, pp 1008–1019
22. Stocker M, Seaborne A, Bernstein A, Kiefer C, Reynolds D (2008) SPARQL basic graph pattern optimization using selectivity estimation. In: Proceedings of 17th international conference on World Wide Web. ACM Press, Beijing, pp 595–604

Chapter 11
An Extended Supervised Term Weighting Method for Text Categorization

Bifan Wei, Boqin Feng, Feijuan He and Xiaoyu Fu

Abstract When Support Vector Machines (SVM) are exploited for automatic text categorization, text representation and term weighting have a significant impact on the performance of text classification. Conventional supervised weighting methods only focus on the frequency characteristics of feature terms, without consideration of semantic characteristics of them. Inspired by supervised weighting method, semantic distance between terms and categories is introduced into term weights calculation. The first step is modeling each category with two vectors of feature terms, which are called category core terms, and acquiring these terms by machine learning methods. Second, the semantic distance between feature terms and category core terms is calculated based on semantic database. Third, the global weight factor is replaced by the sematic distance to calculate the weight of every term. Based on the standard benchmark Reuters-21578, this kind of term weighting schemas can generally produce satisfied results of classification using SVMlight as classifier with default parameters.

Keywords Term weighting · Semantic distance · Support vector machines · Text categorization

B. Wei (✉) · B. Feng
The Electronic and Information Engineering School, Xi'an Jiaotong University, Xi'an, China
e-mail: weibifan@sohu.com

B. Feng
e-mail: bqfeng@mail.xjtu.edu

F. He
Internet Education School, Xi'an Jiaotong University, Xi'an, China
e-mail: hfj@mail.xjtu.edu.cn

X. Fu
MOE KLINNS Lab and SPKLSTN Lab, Department of Computer Science and Technology, Xi'an Jiaotong University, Xi'an, China
e-mail: yykfxy@yahoo.com.cn

J. J. Park et al. (eds.), *Proceedings of the International Conference on Human-centric Computing 2011 and Embedded and Multimedia Computing 2011*, Lecture Notes in Electrical Engineering 102,
DOI: 10.1007/978-94-007-2105-0_11,

11.1 Introduction

There are two basic phases during automatic text categorization (TC): term selection (also called feature selection) and term weighting. The phase of term selection is selecting some important terms from original terms set, often with parameter constrains, for example, aggressivity defined by the number of original terms set divided by the number of selected terms set. Term weighting assigns a proper value for every selected term generated from term selection, and this value indicates how important a term is for the performance of TC. Term weighting can be used not only for TC and text clustering, but also for query matching and other information retrieval tasks.

The two phases of term selection and term weighting can be replaced by one integrated phase. The weights of term can be computed before feature selection and some terms with higher weight will be selected for classifier. In this way, term selection can be performed by term weighting.

In recent years, it is a hot topic in TC to build text classifier using Support Vector Machine (SVM) technology. In general, this kind of approaches is better than other basic classification algorithms using only one kind of classification algorithms [1]. For TC, term weighting schemas or term-frequency transformations have a larger impact on the performance of SVM than the kernel itself [2]. Dimensionality reduction is not the central consideration in TC while using SVM, because SVM do well with a great number of feature terms of text.

The calculation process of the term weighting is divided into three parts, namely, the local weighting factor, the global weighting factor and normalization. These three parts are also known as the feature frequency components, text collection components and normalization components, respectively [3]. For Supervised Term Weighting (STW) [4], the global weighting factor is a statistic related to category and it also is called global statistical factor, which is calculated by some algorithms such as χ^2-test (χ^2, Chi-square), Odd Ratio (OR), Expected Mutual Information (MI) or Information Gain (IG), Pointwise Mutual Information (PMI), Relative Frequency (RF), and so on.

Conventional dominant calculation methods of term weighting are all based on term frequency or term occurrence probability. However, human beings classify the documents not only depending on word frequency (such as repentance of some words), but also depending on the semantic of words. In this paper, words and terms may be exchanged to represent similar meaning. Based on this observation, one new weight schema based on STW was proposed to evaluate the semantic role of words during the calculating term weights for TC.

The rest of this paper is structured as follows: Section 11.2 introduces the popular traditional term weighting methods and the state-of-the-art supervised term weighting methods. Section 11.3 explains in detail the process of the new supervised term weighting method. In Sect. 11.4 we experiment these the new idea on Reuters-21578, the standard benchmark of TC research. Experiments have been performed with one classifier learning method (SVM), six term weighting algorithms (MI, and RF and so on). Section 11.5 concludes this paper with future work.

11.2 Related Work

The term weighting methods are divided into two categories according to whether the method involves the prior information to determine term weights. The first is supervised term weighting method (STW) which uses this known membership information and the second is unsupervised term weighting (UTW) method, which does not use this information. That is to say, the UTW method does not make use of the information on the category membership of training documents.

The typical UTW methods stem from information retrieval related technologies, such as binary, term frequency (TF), TF.IDF and its various variants, and so on. Salton and Buckley [3] summarized the term weighting methods for different information retrieval tasks, based on SMART system. They presented different term weighting recommendations varied with IR tasks, such as document term weighting and query term weighting. They concluded that the text represented by appropriate weighted single terms can produce better results than that represented by other complex indexing methods.

Debole and Sebastiani [4] firstly applied the results of term selection to term weighting, and proposed the STW method. They pointed out that the algorithms of STW are more efficient compared with those of the UTW and the performance of TC based on STW is not inferior to that of the UTW. They also pointed out that the STW method based on Gain Ratio could produce very good results in general and given an improvement of 11% in macro-averaging.

Man et al. [5, 6] compared some typical algorithms of STW and UTW, and proposed a new algorithm of STW, TF.RF. They pointed out these STW methods have mixed performance from the controlled experimental results and the new algorithm, TF.RF, has a consistently better performance than other term weighting methods [7]. The global statistical factor of TF.RF is represented in the following equation.

$$\mathrm{RF}(t_m, c_n) = \log\left(2 + \frac{b}{c}\right) \tag{11.1}$$

In (11.1), b represents the number of documents that contain term t_m and belong to category c_n, and c represents the number of documents that contain term t_m but do not belong to the category c_n.

Altincay and Erenel [8] evaluated six widely used STW schemas of TC, including TF.RF, TF.MI, TF.Prob, TF.χ^2 and so on. They gave a reasonable explanation for the relative performance of different term weighting schemes by the ways how they use ratio and difference of term occurrence probabilities in generating the term weights. They also gave some directions of different algorithms to improve the performance of TC. They concluded that the selection of category dependent form of weights in terms of β and Δ is a promising research direction, where β representing ratio of terms and Δ representing difference of term occurrence probabilities.

The previous term weighting schemas including STW and UTW are all based on the frequency of terms, and the frequency may be the term occurrence in one document, for example Term Frequency (TF); the number of documents containing the term in data set, for example Inverse Document Frequency (IDF); or the ratio of some kinds of terms used in TF.RF method.

11.3 The New Method of Term Weighting

In the STW, the global statistical factor represents the information on the membership of training documents to categories during calculating term weights. Inspired by STW, an Extended Supervised Term Weighting (ESTW) method is proposed in this paper. The basic idea behind ESTW is using the semantic distance between feature terms and categories as global weighting factor, namely, replacing global statistical factor of STW with the semantic distance.

Semantic similarity calculation can be based on different models or text corpora, and different algorithms were proposed to reflect the relationship between words or concepts from different aspects. Semantic distance of same pair of words varies with models. Semantic similarity calculation based on the lexical database WordNet [9] has been unanimously approved in related research fields . In this paper, therefore, one WordNet-based algorithm was adopted to acquire the semantic distance between feature terms and categories.

11.3.1 Acquirement of Category Core Terms

Human beings do not directly depend on the word frequency in document for text classification. Based on this observation and inspired by Formal Concept Analysis (FCA), this paper assumes that every category has some representative objects whose attributes represent the attributes of itself. Similar to the extension of formal concept, these objects are the extension of categories. For TC, the terms selected for representing categories can be used as the extension of categories and are called category core terms (CCT). In this article we focus on binary TC, in which a document either belongs to a category or do not belongs to this category, called positive case or negative case respectively. So the CCT are divided into positive CCT and negative CCT. For example, the positive CCT of category “earn” are “vs, cts, shr, net, qtr, said, revs, note, loss, profit” and the negative CCT are “minister, agriculture, secretary, statistics, council, crop, corn, soviet, exporters, coffee”. Every category is represented by two vectors of selected terms.

Selection of CCT is similar to the selection of feature terms and theses two processes apply some same mathematical formulas to acquire certain terms from a great deal of original terms. However, there are two major differences between the two processes. Firstly, The number of results is significantly different. There are only dozens of CCT for every category, while the number of results of term selection is often hundreds or even tens of thousands. For example, the number of CCT in our experiment is 10 for each category, while the number of term selection is 5,000 for every category in general. Secondly, CCT are divided into two parts: positive CCT and negative CCT. The former represent the typical attributes of positive examples, and the latter characterize negative examples. In contrast to two types of terms for each category, term selection only select one kind of terms for every category.

The CCT can be acquired by manual selection of domain experts, and can be obtained by machine learning methods. Manual selection is high cost and has poor portability. The methods based on Machine learning are always borrowed from term selection algorithms, such as χ^2-test, PMI, IG, Odds Ratio and so on, as shown below.

$$X^2(t_m, c_n) = \frac{T[P(t_m, c_n)P(\bar{t}_m, \bar{c}_n) - P(t_m, \bar{c}_n)P(\bar{t}_m, c_n)]^2}{P(t_m)P(\bar{t}_m)P(c_n)P(\bar{c}_n)} \tag{11.2}$$

$$\mathrm{PMI}(t_m, c_n) = \log \frac{P(t_m, c_n)}{P(t_m)P(c_n)} \tag{11.3}$$

$$\mathrm{IG}(t_m, c_n) = \sum_{t \in \{t_m, \bar{t}_m)\}} \sum_{c \in \{c_n, \bar{c}_n)\}} \log \frac{P(t, c)}{P(t)P(c)} \tag{11.4}$$

$$\mathrm{OR}(t_m, c_n) = \frac{P(t_m|c_n)(1 - P(t_m|\bar{c}_n))}{(1 - P(t_m|c_n))P(t_m|\bar{c}_n)} \tag{11.5}$$

$$\mathrm{RF}(t_m, c_n) = \log\left(2 + \frac{b}{c}\right) = \log\left(2 + \frac{P(t_m, c_n)}{P(t_m, \bar{c}_n)}\right) \tag{11.1}$$

In these equations, probabilities are interpreted on an event space of documents (e.g. $P(t_m, c_n)$ indicates the probability that, for a random document x, term t_m occurs in x and x belongs to category c_n), and are estimated by maximum likelihood. In (11.2), the symbol T represents the total numbers of all training samples.

We conducted different combinations of (11.2–11.5) to generate positive CCT and negative CCT based on the category "acq" of Reuters-21578, using (11.8) for calculating term weights. When using only one formula such as (11.2–11.5) for selecting positive CCT and negative CCT, we found the classification performance based on the χ^2 is better than the performance based on other equations. When different equations were used for producing positive CCT and negative CCT, it was found that the combination of positive CCT using χ^2 method and negative CCT using PMI method can produce better performance of TC.

In the following sections, for every category the terms with top 10 largest values are selected for positive CCT by means of χ^2 method and the terms with least 10 smallest values are pick out for negative CCT by PMI method.

11.3.2 Semantic Distance Between Feature Terms and Categories

Semantic Weighting Factor (SWF) is the semantic distance between feature terms and categories. The value of SWF is acquired by calculating weighted distance between feature terms and terms of CCT. The calculation process is divided into two steps. The first step is calculating the semantic similarities between feature terms and every term of CCT. The second step is calculating the weighted distance between feature terms and categories.

Firstly, the semantic similarities between one feature term and every term of CCT is calculated by the method based on WordNet. The approaches of semantic similarity calculation include the algorithms based on path and the algorithms based on information content. Lin presented an information theoretic definition of similarity, which was called Lin method below [10]. Jiang and Conrath [11] presented a new approach for measuring semantic similarities/distances between words or concepts, and the approach combines a lexical taxonomy structure with corpus statistical information. This approach was called JCn method.

The function for calculating semantic similarities between words is denoted by SIM (Term1, Term2). The implementation of the function can be based on Lin method, JCn method or others [12].

Because of limits of implementation, we only compared the Lin method and JCn method base on the category acq of Reuters-21578, using (11.8) for calculating term weights. The classification performance using Lin method was better than the performance using JCn method and then the Lin method was used in the following sections.

In the following equations, the positive and negative CCT are denoted by Positive Core Term (PCT) and Negative Core Term (NCT), respectively, where $i = 1, \ldots,$ indicating different feature terms, where $m = 1, \ldots,$ representing different categories, $n = 1, \ldots, N$, representing different core terms. N is defined as 10 in following experiments.

The first step is calculating the weighted distance between feature terms and positive CCT and negative CCT respectively.

$$\mathrm{PD}_m(\mathrm{Term}_i) = \frac{1}{\mathrm{PN}_{mi}} \sum_{n=1}^{N} \mathrm{SIM}(\mathrm{PCT}_{mn}, \mathrm{Term}_i) \tag{11.6}$$

where PN_{mi} representing how many times the $\mathrm{SIM}(\mathrm{PCT}_{mn}, \mathrm{Term}_i)$ is not zero. The number of PN_{mi} varies with Term_i and PCT_{mn}. The PD is the abbreviation of "Positive Distance".

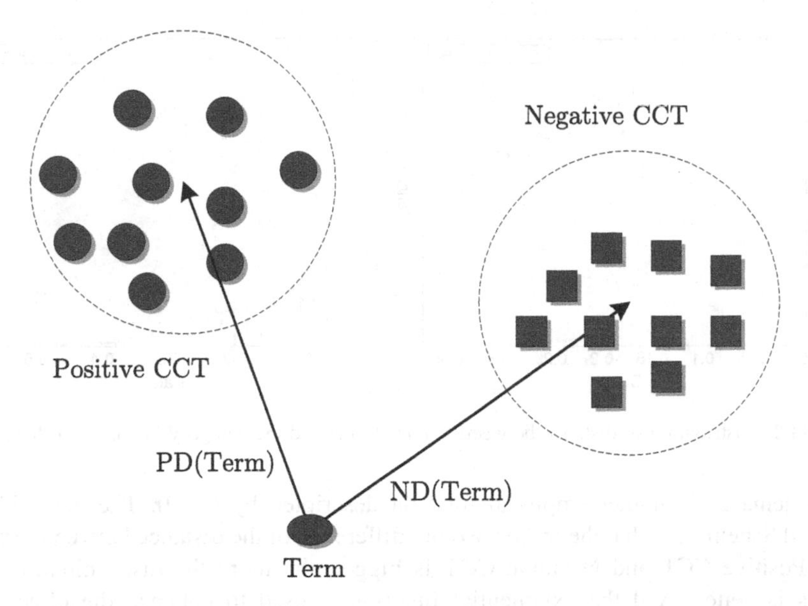

Fig. 11.1 The illustration of PD (Term) and ND (Term)

$$\text{ND}_m(\text{Term}_i) = \frac{1}{\text{NN}_{mi}} \sum_{n=1}^{N} \text{SIM}(\text{NCT}_{mn}, \text{Term}_i) \tag{11.7}$$

where NN_{mi} representing how many times the $\text{SIM}(\text{NCT}_{mn}, \text{Term}_i)$ is not zero. The number of NN_{mi} varies with Term_i and NCT_{mn}. The ND is the abbreviation of "Negative Distance".

For each category, PD (Term) and ND (Term) represent the semantic distance between terms and Positive CCT, and the distance between terms and Negative CCT respectively, illustrated in Fig. 11.1. In Fig. 11.1, the circles and squares represent the terms of Positive CCT and Negative CCT respectively.

The second step is obtaining the weighted semantic distance between feature terms and categories. This paper proposed two schemas to calculate the value of SWF.

Schema 1 is one general form of (11.1) and this method is called Semantic Distance 1 (SD_1) defined by (11.8). The basic idea behind Schema 1 is that if one term has similar distance between Positive CCT and Negative CCT, the term is neutral for TC and should have small SD value.

$$\text{SD}_{1m}(\text{Term}_i) = \log\left(K_1 + K_2 \times \frac{\text{PD}_m(\text{Term}_i)}{\text{ND}_m(\text{Term}_i)}\right) \tag{11.8}$$

where K_1 and K_2 is parameters. Parameter K_1 and K_2 are used to scale the importance of ratio of $\text{PD}_m(\text{Term}_i)$ and $\text{ND}_m(\text{Term}_i)$. For example, if K_1 is big and K_2 is small then the ratio has weak impact on the Semantic Distance. The range of K_1 and K_2 is [0.1, 2] in general. Where $m = 1, \ldots, M$, representing different categories.

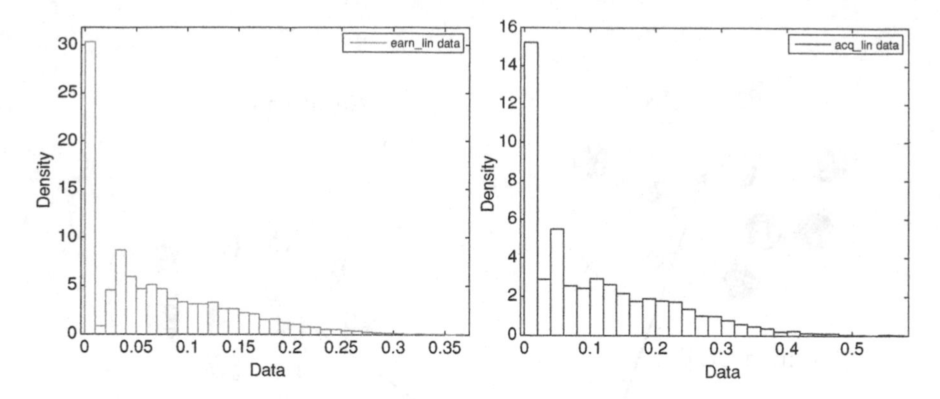

Fig. 11.2 Distribution of distance between feature terms and the category "earn" and "acq"

Schema 2 is another empirical formula descripted by (11.9). The basic idea behind Schema 2 is that the more absolute difference of the distance between terms and Positive CCT and Negative CCT is bigger, the more the discrimination of terms is better. And the exponential function is used to enhance the effect of absolute difference because the absolute difference is always small.

$$\mathrm{SD}_{2m}(\mathrm{Term}_i) = K_1 + e^{K_2 \times |\mathrm{PD}_m(\mathrm{Term}_i) - \mathrm{ND}_m(\mathrm{Term}_i)|} \tag{11.9}$$

where K_1 and K_2 are parameters. Parameter K_1 and K_2 are used to scale the importance of absolute difference of $\mathrm{PD}_m(\mathrm{Term}_i)$ and $\mathrm{ND}_m(\mathrm{Term}_i)$. For example, if K_1 is big and K_2 is small, this means the absolute difference has little impact on the results of the right part of the equation. The range of K_1 and K_2 is [0.1, 2] in general. Where $m = 1, \ldots, M$, representing different categories.

Using standard Reuters-21578 benchmark, when (11.9) is used to calculate the semantic distance between feature terms and categories, it was observed that the semantic distance distribution of different categories is similar to each other. Figure 11.2 illustrates the distribution of distance between feature terms and category "earn" and "acq".

In Fig. 11.2, the x-axis and y-axis indicate the semantic similarities and the density, respectively. From Fig. 11.1, it can also be concluded that for a number of feature terms, the distance between them and categories is zero, in particular, amounting to about 30% between them and category "earn". The semantic distance mainly fall in the range of [0, 0.5]. After calculating, the average semantic distance between feature terms and categories is about 0.1. That is to say, every feature terms has weak relation with categories and the distribution indicates that each feature term alone has limited effect on the performance of TC.

Similar to STW equations, ESTW equation is divided into three parts. The first part is term frequency factor, the second part is SWF and the third part is normalization. We call our method as TF.SD, defined as following.

$$\mathrm{TF.SD} = \mathrm{TF} \times \mathrm{SD} \tag{11.10}$$

where TF representing term frequency, and SD representing SD1 or SD2 defined by (11.8) or (11.9). So TF.SD1 represents the ESTW schema using (11.8). Similarly, TF.SD2 represents the ESTW schema using (11.9). Cosine normalization is used in this paper. Cosine normalization is used in this paper.

11.4 Experimental Analysis

11.4.1 Evaluation Metrics

The $F1$ measure and Macro $F1$ measure are two common evaluation metrics for TC and in this paper these two metrics are used to compare different term weighting schemas.

$$F1 = \frac{2 \times \text{Precision} \times \text{Recall}}{\text{Precision} + \text{Recall}} \tag{11.11}$$

$$\text{Macro } F1 = \frac{1}{M}\sum_{i=1}^{M} F1_i \tag{11.12}$$

where $F1_i$ representing $F1$ measure of different categories, and M representing the number of all the categories.

11.4.2 The Preprocessing of Reuters-21578

The Reuters collection include a set of newswire stories classified under categories and it was used widely by most of the experimental work in TC. In this paper, the top 10 classes of the ModApte split of Reuters-21578 is selected for test benchmark. The χ^2 algorithm was used to perform term selection and selected 5,000 feature terms for every category.

The performance of SVM classifier always varies with how the documents are preprocessed. The following steps in detail show the preprocessing of documents.

1. Preparing ModApte split based on the readme file of Reuters-21578. For training sample, the value of "TOPICS" attribute is "YES" and the value of "LEWISSPLIT" attribute is "TRAIN". For testing sample, the value of "TOPICS" attribute is "YES" and the value of "LEWISSPLIT" attribute is "TEST".
2. Getting rid of stop words.
3. Erasing the words whose length is less than three letters and the words which contain numbers.

4. The stemming is omitted because it is difficult to calculate the semantic similarities between the stem of words. In fact, the stemming has little effect on the performance of SVM classifier.
5. In the ModApte split of Reuters-21578, there are many "noise document", whose body only contain some "Blah", such as the document with newID = 689, newID = 15185 and so on. These documents originally were created to test the tolerance of classification algorithms. If the "noise document" is dropped, the evaluation is deficient in some degree. So in this paper, these documents were retained for testing the robust of term weighting schemas.
6. Removing the documents that have no category label. The 1,833 documents with no category label are removed from the 9,603 training set and the 280 documents with no category label are removed from the 3299 testing set.

After all these steps, These are 7,770 documents in training set, and 3,019 documents in testing set, and 10,075 terms in the all. In this paper, the classification of Reuters-21578 benchmark was tested by SVMlight with default parameters [13].

11.4.3 Results and Analyses

The first experiment was executed to verify the effectiveness of schema 1 based on the top 10 classes of the ModApte split. The following Table 11.1 shows the F1 and Macro $F1$ results of classification by using SVMlight with different parameters of K_1 and K_2, represented by $\mathrm{SD1}(K_1, K_2)$.

In Table 11.1, the TF.χ^2 was used as baseline schema. From this table, it can be concluded that the Macro $F1$ of the top 10 classes of Reuters-21578 have little change with different parameters of K_1 and K_2. The overall performance of TF.SD1 is better than that of baseline.

The second experiment was carried out to test the performance of schema 2 based on the same top 10 classes of the ModApte split. Table 11.2 shows the $F1$ and Macro $F1$ results of classification by using SVMlight with different parameters of K_1 and K_2, represented by $\mathrm{SD2}(K_1, K_2)$.

From Table 11.2, it can be concluded that the Macro F1 of the top 10 classes of the ModApte split have more change with different parameters of K_1 and K_2 compared with schema 1. The overall performance of TF.SD2 is better than that of baseline and the TF.SD1.

The following Table 11.3 shows the $F1$ and Macro $F1$ results of classification by using the semantic distance with (11.8) and (11.9) based on the top 10 classes of the ModApte split. Table 11.3 also depicts the results by using STW methods including TF.RF, TF.χ^2, TF.IG and TF.OR.

The bold numbers are the highest value of very row. From Table 11.3 the following conclusions can be made.

Table 11.1 Performance comparison of different parameters of schema 1

Category	TF.χ^2	TF.SD1 (0.5, 2)	TF.SD1 (0.5, 1)	TF.SD1 (0.1, 1)	TF.SD1 (1, 1)	TF.SD1 (2, 1)
Earn	96.75	98.49	98.17	98.17	98.49	98.58
Acq	92.70	96.2	96.00	95.84	96.20	96.05
Money-fx	73.53	78.86	78.74	74.50	78.86	78.86
Crude	73.75	88.89	87.10	88.06	89.18	89.87
Grain	91.58	92.31	93.84	93.15	92.31	91.93
Trade	80.33	79.65	78.45	78.45	80.17	79.49
Interest	63.16	74.46	75.32	75.21	74.46	75.54
Wheat	90.32	84.67	84.06	85.92	84.67	82.71
Ship	78.26	83.23	84.15	86.23	83.95	83.02
Corn	93.22	88.46	90.57	88.46	87.38	86.87
Macro *F*1	83.36	86.52	86.64	86.40	86.57	86.29

Table 11.2 Performance comparison of different parameters of schema 2

Category	TF.χ^2	TF.SD2 (1, 1)	TF.SD2 (1, 2)	TF.SD2 (1, 3)	TF.SD2 (2, 1)	TF.SD2 (2, 2)
Earn	96.75	98.44	98.44	98.11	98.34	98.44
Acq	92.70	96.33	96.61	96.49	96.62	96.55
Money-fx	73.53	81.23	81.56	80.90	80.56	81.23
Crude	73.75	90.16	89.41	89.92	89.12	89.64
Grain	91.58	91.04	92.63	91.17	92.63	92.20
Trade	80.33	77.92	78.60	79.15	78.60	78.07
Interest	63.16	75.32	75.86	75.00	75.32	75.86
Wheat	90.32	83.33	83.71	81.82	81.82	82.44
Ship	78.26	83.02	83.75	83.02	83.75	83.75
Corn	93.22	87.87	88.00	86.00	88.00	88.87
Macro *F*1	83.36	86.47	86.86	86.16	86.48	86.71

Table 11.3 Performance comparison of different term weighting schemas

Category	TF.SD1 (0.5, 1)	TF.SD2 (1, 2)	TF.RF	TF.χ^2	TF.IG	TF.OR
Earn	98.17	98.44	98.31	96.75	96.65	96.71
Acq	96.00	96.61	96.62	92.70	92.56	92.47
Money-fx	78.74	82.56	81.44	73.53	75.58	75.51
Crude	87.10	89.41	90.26	73.75	83.01	76.01
Grain	93.84	92.63	92.73	91.58	89.73	90.73
Trade	78.45	78.60	79.13	80.33	79.69	79.34
Interest	75.32	75.86	75.86	63.16	66.04	65.04
Wheat	84.06	82.71	84.89	90.32	89.61	88.61
Ship	84.15	83.75	81.76	78.26	78.53	79.53
Corn	90.57	88.00	88.68	93.22	94.74	93.74
Macro *F*1	86.64	86.86	86.97	83.36	84.61	83.77

1. There is no algorithms performing coherently best than other algorithms with all the 10 classes.
2. The TF.SD1, TF.SD2 and TF.RF respectively achieved the best results in three categories.
3. The values of Macro F1 of The TF.SD1, TF.SD2 and TF.RF are very close and is higher than other STW methods by about 2%.

11.5 Conclusion and Feature Work

In this paper, we proposed one new approach to term weighting based on semantic similarity. The first step is model category with two feature vectors. The second step is calculating the semantic distance between feature terms and categories by means of defining the semantic distance between feature terms and the CCT of every category. The third step is replace the global statistical factor of STW with the semantic distance to obtain the term weights.

We tested the new approach based on the standard top 10 classes of Reuters-21578 benchmark and concluded that this approach generally was better than other common STW such as χ^2, IG and OR, and is similar to the TF.RF weighting schema.

In our experiment, the WordNet is one close semantic database and there are a great number of unknown words during the computation of term weighting. If the open semantic database such as Wikipedia is used, the performance should be better than that of current experiment.

Acknowledgments The research was supported in part by the National Science Foundation of China under Grant Nos. 60825202, 60921003, 60803079, 61070072; the National Science and Technology Major Project (2010ZX01045-001-005); the Program for New Century Excellent Talents in University of China under Grant No. NECT-08-0433; the Doctoral Fund of Ministry of Education of China under Grant No. 20090201110060; Cheung Kong Scholars Program; Key Projects in the National Science & Technology Pillar Program during the Eleventh Five-Year Plan Period Grant No. 2009BAH51B02; IBM CRL Research Program; Research on BlueSky Storage for Cloud Computing Platform.

References

1. Sebastiani F (2002) Machine learning in automated text categorization. ACM Comput Surv 34(1):1–47
2. Leopold E, Kindermann JR (2002) Text categorization with support vector machines. How to represent texts in input space? Mach Learn 46:423–444
3. Salton G, Buckley C (1988) Term-weighting approaches in automatic text retrieval. Inform Process Manag 24:513–523
4. Debole F, Sebastiani F (2004) Supervised term weighting for automated text categorization. In: Text mining and its applications, vol 138, pp 81–97

5. Man L, Chew-Lim T, Hwee-Boon L (2007) Proposing a new term weighting scheme for text categorization. Proceedings of the 21st national conference on artificial intelligence, vol 1, AAAI Press, Boston, Massachusetts, pp 763–768
6. Man L, Sam-Yuan S, Hwee-Boon L, Chew-Lim T (2005) A comparative study on term weighting schemes for text categorization. Neural networks, IJCNN '05. In: Proceedings 2005 IEEE international joint conference on, vol 1, pp 546–551, vol 541
7. Man L, Tan CL, Su J, Lu Y (2009) Supervised and traditional term weighting methods for automatic text categorization. IEEE Trans Pattern Anal Mach Intell 31:721–735
8. Altlnay H, Erenel Z (2010) Analytical evaluation of term weighting schemes for text categorization. Pattern Recog Lett 31:1310–1323
9. Miller GA (1995) WordNet: a lexical database for English. Commun ACM 38:39–41
10. Lin D (1998) An information-theoretic definition of similarity. In: Proceedings of 15th international conference on machine learning. Morgan Kaufmann, San Francisco, CA, pp 296–304
11. Jiang JJ, Conrath DW (1997) Semantic similarity based on corpus statistics and lexical taxonomy. In: International conference research on computational linguistics (ROCLING X), pp 19–33
12. Pedersen T, Patwardhan S, Michelizzi J (2004) WordNet::Similarity: measuring the relatedness of concepts. Demonstration papers at HLT-NAACL 2004, Association for computational linguistics, Boston, Massachusetts, pp 38–41
13. Joachims T (1998) Text categorization with support vector machines: learning with many relevant features. In: Ndellec C, Rouveirol C (eds) Machine learning: ECML-98, vol 1398. Springer, Heidelberg, pp 137–142

Part III
HumanCom 2011 Session 2: Human–Computer Interaction and Social Computing

Chapter 12
Building Ontology for Mashup Services Based on Wikipedia

Kui Xiao and Bing Li

Abstract Tagging as a useful way to organize online resources has attracted many attentions in the last few years. And many ontology building approaches are proposed using such tags. While tags usually associated with concepts in some databases, such as WordNet and online ontologies. However, these databases are stable, static and lack of consistence. In this paper, we build an ontology for a collection of mashup services using their affiliated tags by referring to the entries of Wikipedia. Core tags are filtered out and mapped to the corresponding Wikipedia entries (i.e., URIs). An experiment is given as an illustration.

Keywords Ontology · Mashup · Service · Tag · Wikipedia

12.1 Introduction

Tags are widely used in well-known Web2.0 applications and popular medium to annotate and share content. Some tags are provided by the creator of content, such as the programmable web, the tags of mashup services and APIs on the platform are provided by their creators. And another category of tags come from the

K. Xiao · B. Li (✉)
State Key Laboratory of Software Engineering, Wuhan University,
430072 Wuhan, China
e-mail: bingli@whu.edu.cn

K. Xiao
e-mail: xiaokui1024@gmail.com

K. Xiao
School of Computer and Software, Wuhan Vocational College of Software Engineering,
430205 Wuhan, China

J. J. Park et al. (eds.), *Proceedings of the International Conference on Human-centric Computing 2011 and Embedded and Multimedia Computing 2011*, Lecture Notes in Electrical Engineering 102,
DOI: 10.1007/978-94-007-2105-0_12,

common users who are interested in the resources, namely folksonomies. For example, the Delicious website describes bookmarks with folksonomies. All the tags on that platform are created by average users.

According to tags, internet resources could be divided into different categories. Every resource is described by several tags, so that people can find out a group of resources with a specific tag. However, tags are always flat list of terms, and users may get little help from the terms in the process of resource searching. For instance, tags on the Programmable platform are stable and static terms. Users could not get the exact services with several terms.

To help the searching process, applications need to get the meaning of users. In other words, systems should be aware of which kind of resources users want. It is necessary to employ intelligent searching technologies to improve the process of resource retrieval. Generally speaking, researchers are favor of describing resources with ontologies to improve the resource retrieval process.

In recent years, a lot of researches focus on building ontology with tags. As we know, it is difficult to build an ontology for only one resource, because there are few tags for each resources on these platforms. Consequently, people often build an ontology for a group of resources, and there are many tags that can be used as concepts of the ontology. Of course, the resources in the group must be similar to each other, for example, they have a same topic.

The document is structured as follows: First, we describe related work. Then, we present the approach on how to select related tags, after that we introduce the method of building ontology by turning related tags into meaningful concepts with the supports of the Wikipedia website. Next, the Experiment is given to show the process from the service tagspace to a simple ontology. Finally, we present the conclusions.

12.2 Related Work

Tag-related challenges have been studied by the semantic web research community for a long time. Research has focused on comprehending the inherent characteristics of tags and exploring their emergent semantics. Additionally, there are several main databases that could provide semantic web entities for common tags, such as Wikipedia, WordNet, other online ontologies, etc. All of these databases have been employed to enrich tags by the research community. However, each of them provides semantic information with respective characteristic.

Wikipedia provides entries on a vast number of named entities and very specialized concepts. The database provides a large coverage knowledge resource developed by a large community, which is very attractive for information extraction applications [1]. Hepp et al. [2] show that standard Wiki technology can be easily used as an ontology development environment for named classes, reducing entry barriers for the participation of users in the creation and maintenance of lightweight ontologies, proves that the URIs of Wikipedia entries are surprisingly reliable identifiers for ontology concepts, and demonstrates the applicability of our approach in a use case.

WordNet is maintained by a small group of people, and there is a lot of research work that is based on the database. Liniado et al. [3] investigate a new approach, based on the idea of integrating an ontology in the navigation interface of a folksonomy, and it describes an application that filters delicious keywords through the WordNet hierarchy of concepts, to enrich the possibilities of navigation. While, the WordNet database may not be updated in time.

Online ontologies are also import data source for research work. For example, dAquin et al. [4] present the design of Watson, a tool and an infrastructure that automatically collects, analyses and indexes ontologies and semantic data available online in order to provide efficient access to this huge amount of knowledge content for semantic web users and applications.

Finally, we use Wikipedia, instead of other semantic databases, to turn tags into concepts. Unlike WordNet and online ontologies, the Wikipedia databases are updated by average users all the time, so the resources always stay in newest version.

12.3 Approach

Mashup service platforms have some common features as follows: each service is described by several tags; services are divided into different group according to their tags; the tags are created by the service providers. From what has been discussed above, we may safely come to the conclusion that the tags that are attached to mashup services provide littlc help to the process of service retrieval.

In this section we propose an approach to build ontologies for every mashup category with the service tags, so that the process of service retrieval could become more intelligent. First of all, we select the core tags from all the tags of a mash-up service category. Every category could be viewed as a domain. Secondly, core tags of each domain are connected with Wikipedia entries. It is clear that every Wikipedia entry has an URI and could be used as an entity of RDF documents. In our current approach, we support only plain RDF and completely leave out any kind of hierarchical order.

12.3.1 Tag Selection

As far as we know, every mashup service is described by several tags. It goes without saying that some tags have noting with the domain of the mashup service. These tags may be names of some city, product, or platform and so on. A mashup service sometimes is related to a city, or might be installed on a specific product such as iphone, or need to call the APIs of some famous platforms such as Delicious. It is obvious that these tags do not describe the features of the service domain.

So we must put great emphasis on selecting core tags of service domain. There is no doubt that the tags that are employed by many mashup services in a domain must characterize this domain. These tags are viewed as core tags of the domain. It is important to count the number of all the tags in the domain. By taking account of the frequencies of all the tags in the domain, we may find out the core tags of a domain.

12.3.2 Ontology Building

To create RDF documents, we have to transform the core tags of the domain into RDF elements. Generally speaking, RDF elements take the form of URI. Of course, some objects of RDF statement are datatype properties. Wikipedia is a great source for building RDF documents. Entries of the Wikipedia database are viewed as ontology concepts; each of them has a URI. What is more important is that a tag is often defined as a Wikipedia entry and then is transformed into an entity.

Since there is no explicit knowledge representation model in the background, a Wikipedia entry can be anything; it is not clear whether it refers to an instance, a concept, or a property. By social convention, Wikipedia contains mostly entries that are proper nouns and does not include relationships and properties. We solve this issue by omitting this distinction between instance and concept, which is no significant problem in pure RDF.

After selection, the core tags of domain are led up to Wikipedia entries so that tags are translated into ontology concepts. Every concept is identified by an URI; as a result, concept can be used as entity of RDF document. It must be pointed out that some tags, that have more than one entry, must be handled manually.

In addition, some core tags are discarded beforehand. Currently, we just focus on noun tags. Although a few of tags, such as "social", emerge frequently, we still do not handle them. It is same with the tags which have no entry for mapping.

Another problem is we cannot find properties and relationships as entries in Wikipedia. We address this problem by using properties and relationships defined in popular existing ontologies, namely Dublin Core elements [5, 6] together with Wikipedia entries.

12.4 Experiments and Results

It is considered that the programmable web is a popular platform of mashup services and APIs. In our experiments, we collected the description tags of 55 mashup services in "microblogging" category. The proportion was about 1 percent

Table 12.1 Core tags of "microblogging" domain

Tags	Number in the collection	Number in programmable web	Proportion
Microblogging	55	220	0.250
Social	19	508	0.037
Messaging	9	286	0.031
Photo	13	693	0.019
RSS	2	107	0.019
Blog	2	109	0.018
Events	2	152	0.013
News	4	326	0.012
Video	7	583	0.012
Visualization	2	194	0.010

of all the mashup services in the Programmable platform. Next, we count the number of the tags in our collection, and the number of these tags in all the programmable web. Then we calculate the proportion of the former to the latter for every tag in our collection.

In order to acquire the core tags of the "microblogging" category, we have to make use of thresholds as follows:

- The number of tag in all the programmable web must be bigger than 100.
- The proportion of tag that calculated above must be bigger than 0.01.

Table 12.1 shows the core tags of the 55 mashup services collection. Evidently, all these tags closely related to the "microblogging" category.

After that, we created a RDF document with the core tags about the "microblogging" category. There is no doubt that the tag "microblogging" is chief concept of the domain. All other tags, namely concepts, should link to the chief concept. Firstly, we found out the Wikipedia entry about the concept "microblogging", and then sought the text about other concepts in the entry page. On the contrary, we next found out the Wikipedia entries about other concepts, and then sought the text about the term "microblogging".

There are some concepts that we had to omit. "social" is an adjective, while we just focused on noun in current work. In addition, "photo", "RSS", and "visualization" had no connection with "microblogging" in the Wikipedia entries. So the remaining tags would be used to create RDF documents. The entry of each tag has an URI, and could be viewed as an entity. The properties between entities came from Dublin Core elements. The details of the RDF documents are showed in Fig. 12.1:

- Microblogging is a type of Blog.
- Microblogging is the publisher of Messaging.
- Microblogging describes Events.
- Microblogging is the source of News and Video.

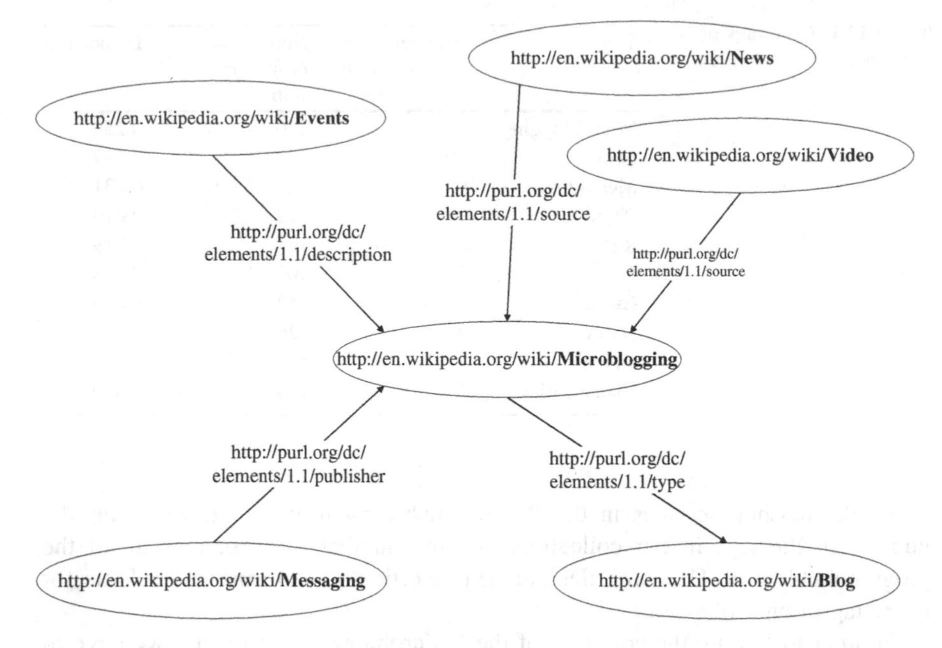

Fig. 12.1 RDF graph of "microblogging" domain

It is necessary to clarify that the experiment results are only some preliminary results. In the next, more domain services and tags will be introduced to our experiments to valid the method. Furthermore, ontologies built with other knowledge database, such as WordNet, will be introduced to compare with the Wikipedia ontologies.

12.5 Conclusions

We have shown that Wikipedia could be used as a basic database to build ontology for mashup services. Core tags are selected beforehand; corresponding Wikipedia entries and URIs are found out as ontology elements later. The results of the approach are RDF documents. In our case, the properties of RDF derived from Dublin Core elements. Next step, we will try to include more subsumption hierarchy or axioms. In addition, textual description is also a source of tags. Making use of textual description about various resources is very significant.

Acknowledgments This work is supported by National Basic Research Program (973) of China under grant No.2007CB310801, National High Technology Research and Development Program (863) of China under grant No.2006AA04Z156, National Natural Science Foundation of China under grant No.60873083, 60803025, 60703009, 60303018, 60970017 and 60903034. Natural Science Foundation of Hubei Province for Distinguished Young Scholars under grant No. 2008CDB351.

References

1. Ahn D, Jijkoun V, Mishne G, M̈uller K, de Rijke M, Schlobach S (2004) Using wikipedia at the TREC QA track. In: Proceedings of TREC-13
2. Hepp M, Bachlechner D, Siorpaes K (2006) Harvesting wiki consensus—using wikipedia entries as ontology elements. In: Third European semantic web conference, Budva, Montenegro
3. Laniado D, Eynard D, Colombetti M (2007) Using WordNet to turn a folksonomy into a hierarchy of concepts. In: Proceedings of fourth SWAP
4. dAquin M, Sabou M, Dzbor M, Baldassarre C, Gridinoc L, Angeletou S, Motta E (2007) Watson: a gateway for the semantic web. In: Fourth European semantic web conference, Innsbruck, Austria
5. Dublin core metadata initiative, "Dublin core metadata element set, version 1.1: reference description", http://dublincore.org/documents/dces/
6. Dublin core metadata initiative, "DCMI metadata terms", http://dublincore.org/documents/dcmi-terms/

Chapter 13
A Hybrid Model for Simulating Human Crowd

Muzhou Xiong, Hui Li, Hanning Wang, Wencong Zeng, Minggang Dou and Dan Chen

Abstract The characteristic of crowd movement is reflected in two aspects, i.e., crowd behaviour and individual behaviour. The existing models, either macroscopic or microscopic model can simulate only one of them. In order to simulate both in one model, we propose a hybrid model for simulating human crowd. In the proposed model, we use macroscopic model to simulate the factors from external environment, and microscopic for agent's behaviour. Agent can choose its movement speed and direction by its own desire, under the constraints from those external factors. In each simulation time step, the macroscopic and microscopic model are executed sequentially, and at the end of each time step, the information for macroscopic model will be updated from the simulation result by the microscopic one. Case study is also conducted for the proposed model. From the simulation result, it indicates that the proposed model is able to simulate the crowd and agent behaviour in dynamic environment.

This study was supported in part by the Fundamental Research Funds for the Central Universities, (China University of Geosciences (Wuhan), No. CUG100314 and No. CUGL100608), and National Natural Science Foundation of China (grant No. 60804036).

M. Xiong (✉) · H. Li · H. Wang · W. Zeng · M. Dou · D. Chen
School of Computer Science, China University of Geosciences,
430074 Wuhan, People's Republic of China
e-mail: mzxiong@gmail.com

J. J. Park et al. (eds.), *Proceedings of the International Conference on Human-centric Computing 2011 and Embedded and Multimedia Computing 2011*, Lecture Notes in Electrical Engineering 102,
DOI: 10.1007/978-94-007-2105-0_13, © Springer Science+Business Media B.V. 2011

13.1 Introduction

Crowd behaviour and its movement pattern is an ubiquitous and fascinating social phenomenon. It has become a mesmerizing question for psychologists, sociologists, computer scientist, and even for architecture designers and urban planners, on how the crowd behaviour and movement can be predicted and depicted. However, it is scarcely possible to get real statistics of crowd movement from real life, especially in the extraordinary situation within emergency or fatal event occur. Crowd simulation [10] then becomes an effective approach to study how crowd moves under diverse what—if scenarios, and has been widely applied to transportation systems [4], sporting and general spectator occasions [2], and fire escapes [9].

Currently, macroscopic and microscopic methods are the two main modelling methods for crowd simulation. The microscopic methods, such as the methods used in [5, 7, 8, 12], emphasize the issues of individual characteristics, including pedestrian's psychological and social behaviors, communication amongst pedestrians, and individual decision making decision making processes. With the ability to describe individual aspects, microscopic models can generate a fine-grain simulation result. However, with the increase of the scale of individuals in simulation, the computation complexity will definitely increase, and the simulation performance will not be tolerable in some extreme conditions. On the other hand, since microscopic methods only consider issues about individual itself, it can only reflect the individual behaviour and movement pattern. Global and local crowd information can scarcely be reflected in these models. Whereas, macroscopic methods (e.g., methods used in [3, 6]) focuses on the movement features of the whole crowd. It can then generate a simulation result reflecting the tendency and movement pattern about the whole crowd, which is what the microscopic methods lack. As for the aspect of simulation performance, macroscopic methods adopt partial differential equations (PDEs) for modelling crowd behaviour and movement, and the performance of these PDEs solution only depends on the environment size and the number and shape of cells. It is obvious that the scale of the simulating crowd has no impact on the performance of model execution. Hence, macroscopic methods usually owns a higher execution speed than microscopic method. However, macroscopic methods do not consider any individual issues, it is hardly to produce a fine-grain simulation result, and it can not reflect the individual diversity in simulation.

In order to exploit the advances both of microscopic (individual diversity and fine-grained simulation result) and macroscopic method (high execution efficiency and global crowd movement pattern), existing work has attempted to combine different models together and build a *hybrid model*. The existing methods fall into three categories. Methods in [1, 11] build their model by adopt microscopic and macroscopic model into different layers. For example, in [11] the macroscopic model is used at the crowd level to generate simple rules to govern the movement of individuals; while the microscopic model generates the collision avoidance for each individual in the crowd. These methods can decrease the time necessary for model execution. However, is cannot consider several issues from crowd or environment int the same model. Another type, like the one proposed in [13], incorporates both a complete macroscopic and microscopic model and executes them inter-changeably. However, the execution efficiency is achieved on the assumption that the crowd movement is mostly stable (or becomes stable eventually). The third type, as proposed in [14], adopts a complete microscopic and macroscopic model and execute them simultaneously by apply them into different parts of the environment. It aims to solve the problem that the macroscopic model may be not applicable some some scenario. In this method, the environment is divided into several non-intersecting areas. Microscopic model is applied into areas where macroscopic model is not applicable for the crowd in those areas, and microscopic model is applied into the remaining areas. In this method, model execution efficiency is improved by applying macroscopic method into more areas. However, the global crowd movement pattern can not be reflected in the areas simulated by microscopic model, since there is no global characteristic in the applied microscopic model. In contrast to these methods, we have previously proposed a multi-resolution modelling [13] approach, which attempts to combine both macroscopic and microscopic model. The approach of this method is to make the two models work iteratively: the macroscopic model governs the simulation when the crowd movement is stabilized; if there is an event which makes the crowd movement unstable, the simulator will switch to microscopic mode and choose the microscopic model to simulate the crowd movement. Finally, the simulation will adopt the macroscopic model when the crowd movement becomes stable again.

This paper proposes a hybrid method to exploit the advances by macroscopic and microscopic method. In this model, the macroscopic issues or environmental issues, like position information, density and crowd control factors, are simulated by macroscopic model. By contrast, microscopic model only concerns individual factors of agents in simulation. This individual factors impact how agent makes decision and movement in dynamic environment. Since every agent needs to perceive environment information and make a further decision and movement as response, the result by microscopic model can be used as input for microscopic model. In other words, the microscopic model receives the result by the macroscopic model, and each agent makes a further individual movement according to those macroscopic information. At the end of each simulation time step, the macroscopic model adjusts its result according to the simulation result by microscopic model and then begin the execution for the next simulation time step. In existing work for microscopic model, each agent needs to perceive the environment information and then produce a local view for an individual. This means every agent will execute a similar procedure to obtain local environmental information in each time step. Hence, the process can be simplified in the microscopic model for each agent. By contrast, the process of obtaining the environmental information is handled by the macroscopic model and will be only executed once. In term of this, the proposed model is able to get a higher execution efficiency than those exiting work. On the other hand, the environmental issues are modelled from the aspect of crowd level, it can easily reflects the crowd characteristic and integrated in the microscopic model. Actually, the existing work can hardly reflect such crowd characteristic.

The rest of the paper is organized as follows. The macroscopic model is depicted in Sect. 13.2. Section 13.3 gives the microscopic model for the hybrid modelling method. Case study of the hybrid model is conducted in Sect. 13.4. The paper is finally concluded in Sect. 13.5.

13.2 Macroscopic Model

The macroscopic model describes environmental and crowd movement, including environment description, position field, density distribution and description of guide information. In this section, it describes how the macroscopic model is established to evaluate these factors. The following section gives how these factors govern the agent decision making and behaviour in a dynamic simulation environment.

13.2.1 Environment Description

A typical environment for crowd simulation usually includes entrance, exit and obstacle. Entrance determines the locations from which the crowd comes, and exit represents the position towards which the crowd moves. As for obstacle, it determines which areas in the environment is not accessible for crowd movement. The environment specification will further affects the model establishment of position field, density distribution, guiding information.

The environment is divided into different cells with the same size. One of the following properties,including entrance, exit, obstacle and accessible area, is assigned to each cell. These information is composed of the *position field* defined in Sect. 13.2.2. Each cell is also assigned an average agent density, and all such densities compose the density distribution for the simulation environment. Detailed density distribution is described in Sect. 13.2.3. If guiding information is located at a cell, it is also assigned as an attached property for the cell (described in Sect. 13.2.4).

13.2.2 Position Field

The movement of agent should be governed by a selected path, and the agent's movement in simulation will follow it. The procedure of this work in crowd simulation is called *path planning*. Here we assume every agent comes from an entrance and goes towards an exit in simulation. So the simulation of agent's movement is always started from an entrance and ends at the arrival of the exit. There may exist several exits, so the path planning should also indicates how to choose an exit located at the end of the selected path. It is also assumed that each agent in the simulation always prefers a shorted path form entrance to exit. Hence the function of path planning in our model is to find the shortest path to exit for each agent in a specified location. In our hybrid model, the path planning is determined by the *Position Field*, where no other factor is considered. How other factors affect the path planning will be introduced later in this paper.

Position field defines a value for each cell in the simulation environment, indicating the shortest distance between the current location to exit. Values for all the cells compose the position field. Based on the position field, the model can calculate the shortest path for each cell, as well as the exit at the end of the selected path. Since the assumption for path planning attempts to find the shortest path and arrive an exit in the environment.

There are two special types for cells in simulation environment: one is the cell is part of an exit; and the other one is the cell is in an obstacle. The value assignment for these two types of cells is defined as follows: infinity is assigned to cells in obstacle, and 0 is assigned to cells in exit. Since it is assumed that agent always prefers to the shortest path to an exit, it should moves to the neighbored cell where the value difference of the position field between the next selected cell and the current one should decreases as fast as possible. Generally, if we use $\phi(x,y)$ indicates the value of position field at Point (x,y) and the movement direction of the agent at the point is described as $\boldsymbol{d}(x,y)$, the two functions should holds the relationship defined in (13.1),

$$\boldsymbol{d}(x,y) // -\nabla\phi(x,y) \tag{13.1}$$

where // indicates that the directions of the two vectors are parallel. In other words, the direction of agent movement should always follow the opposite direction of the gradient of the position field.

As described above, the path planning here only considers the factor of position. It can then be assumed that the speed of agent movement is constant around the environment. Suppose that the speed is represented as v_0. Then the time for agent movement per unit distance along the selected path can be represented as $1/v_0$, and here we represent the time as t, i.e., $t = 1/v_0$. In term of this, the shortest path from the current location to an exit indicates that the agent can arrive at the exit in the shortest time. So, the shortest path can be evaluated by time. In other word, the agent can move to the exit in the shortest time with speed v_0 from the current location (cell), along the shortest path. It is obvious that the time used per unit distance along the shortest path and the position field holds the following relationship:

$$t = |\nabla\phi(x,y)| \tag{13.2}$$

which is a partial differential equation known as *Eikonal Equation*.

Because t in (13.2) can be specified by any positive value, we can then solve the Eikonal equation by methods proposed in [15]. When the position field is determined, or the position field for each cell is clear, we can then determine the agent's moving direction by solving the (13.1). Definitely, if agent moves exactly along such direction, the movement trajectory must be the same to the shortest path mentioned above.

13.2.3 Density Distribution

Density distribution is depicted by another scalar field in this model. In each cell, there is a value recording the average pedestrian density in the cell. Density distribution for the whole environment can be stored in a matrix according to the environment division, which hence is considered as a scalar field. The main function of the field is to impact and direct pedestrian to move around the environment. The basic idea of such impact is that pedestrian tends to move to areas with lower density, but not to move to higher density area even those areas are nearer to the current selected goals. Details about how density impact pedestrian's behavior will be introduced in Sect. 13.3.

13.2.4 Description of Guide Information

In real life, sigh board can help pedestrians to find the desired goal efficiently. In the proposed macroscopic model, such information is tagged in specified cells. When it appears within agent's eyesight, it shows the shortest path to the goal for agent and the agent will be aware of it immediately. Actually, the indicated path should follow the direction according to the opposite direction of $\nabla\phi$ defined in Formula (13.2).

13.3 Microscopic Model

The macroscopic model is mainly used to describe the environment information to direct pedestrian how to behave and move around the environment. For the microscopic model, the basic idea is to use an agent to simulate an individual in the simulation system, and the simulation result is reflected by the position and velocity of each agent. The simulation is divided into time steps, and the macroscopic and microscopic model will be executed once in each time step. The simulation result by the macroscopic model is explained and executed by the microscopic model to direct agent to behave and move around the environment.

13.3.1 Agent Properties

An agent holds a property set, S, defined by Formula (13.3),

$$S = \{x, y, H, v_x, v_y, v_{\max}, \rho, \text{sight}, \theta_{\text{sight}}, w_{\text{dir}}, w_{\text{pos}}, w_\rho\} \tag{13.3}$$

where x and y represents the agent's current position; H reflects the agent's current healthy status; v_x and v_y holds current magnitudes of the agent's velocity along x and y coordinates, respectively; $v_{\max}$ holds the value of the possible maximum speed; ρ represents the density at point (x, y); sight means the agent's maximum distance of eye sight, and θ_{sight} means the maximum angle the agent can see from its current position; there are three weight factors shows how agent's behaviour is affected by external factors from either crowd or environment, i.e., w_{dir} (crowd movement), w_{pos} (position factor) and w_ρ (density factor).

As introduced above, the paper applies cell automata as the microscopic model, this means each cell of the environment can hold one and only one agent at, and an agent can only move from one cell to the neighboured one in one time step.

13.3.2 Agent Movement Rules

The agent movement is reflected by two factors, speed and moving path. The two factors is mainly determined by the agent's desire. For example, when an agent desires to get to a position, it will choose a speed it usually holds and a path which holds the shortest distance from the current location to the desired position. The speed and path selected by the agent's desired is named *preferred* speed and path in this paper. An agent can move with the *preferred* speed and path without any impact from external factors. However, an agent cannot always move around an vacant environment. In fact, an agent's movement is always impacted by external factors from environment and other agents (i.e., crowd). The values of these factors are recorded in macroscopic model result, and these values will impact how an agent moves in the same simulation time step.

Actually, an agent cannot choose a path directly pointing to the goal from its current location. The obstacle in the environment will force the agent to make a detour around the obstacle to achieve the goal. In order to navigate an agent around the environment and select an appropriate path, the microscopic model will use the position field from the macroscopic model, and direct agent how an agent moves to the neighboured cell. As analyzed in Sect. 13.2.2, an agent should moves along the direction parallel to the opposite direction of $\nabla\phi$. Here we use an attraction to describe how the neighboured cell to attract an agent to move towards the cell, define in (13.4),

$$P_k = \frac{\phi_{\max} - \phi_k}{\phi_{\max} - \phi_{\min}} \tag{13.4}$$

where k means which cell the agent currently is, and $\phi_{\max}$ represents the maximum value of ϕ amongst all cells in the environment (with the exception of infinity of ϕ), $\phi_{\min}$ is the minimum value of all cells (i.e., 0 at exists in the environment), and ϕ_k is the value of ϕ at cell k.

Another external factor impacting agent movement is density. On the one hand, density can determines speed of agent movement; on the other hand, density will also impact how agent select its movement path. It is known that agent speed drops as the density around it increases. In our model, we apply the speed-density relationship proposed in [6], shown in Formula (13.5).

$$f(\rho) = \begin{cases} A, & \rho \le \rho_t \\ A\sqrt{\frac{\rho_t}{\rho}}, & \rho_t < \rho \le \rho_c \\ A\sqrt{\frac{\rho_t \rho_c}{\rho_m - \rho_c}\frac{\sqrt{\rho_m - \rho}}{\rho}}, & \rho_c < \rho \le \rho_m \end{cases} \tag{13.5}$$

where ρ represents the crowd density around the agent, and the typical values of ρ_t, ρ_c, ρ_m, and A are set as 0.8 m^{-2}, 2.8 m^{-2}, 5.0 m^{-2}, and 1.4 ms^{-1} respectively in [6].

It can also be observed that pedestrian will make a detour to avoid entering into a crowded area. An attraction in our model is proposed to describe how cell with lower density attracts agent moving towards it, and the attraction is calculated by Formula (13.6),

$$Den_k = \frac{\rho_{\max} - \rho_k}{\rho_{\max} - \rho_{\min}} \tag{13.6}$$

where $\rho_{\max}$ and $\rho_{\min}$ represents the maximum and minimum density in the environment at current time, and ρ_k is the density at cell k. By this definition, agent will always moving towards area with lower density when achieving its goal.

As crowd density becomes higher, agent will chooses its moving direction according to the one holding by most of the agents around it. In each simulation time step, an agent can only move to the neighboured cells from its current cell, and the maximum number of its select is 8. Since these neighboured cells can be occupied by obstacle, the number of candidate moving direction decreases accordingly. Here we apply a formula [defined in (13.7)] to evaluate how neighboured candidate cell attracts agent moving towards it.

$$P_{\mathrm{dir}_k} = \frac{S_{\mathrm{dir}(k)}}{\sum_{i=1}^{m} S_{\mathrm{dir}(i)}} \tag{13.7}$$

where p_{dir_k} means the attraction from the neighboured cell k (the possible maximum value of k is 8); $S_{\mathrm{dir}(k)}$ means number of agent in eyesight of the agent moving along the direction from the agent's current cell to cell k; hence

$\sum_{i=1}^{m} S_{\mathrm{dir}(i)}$ means the sum of agent number moving along all the m candidate moving direction in the agent's eyesight, where the value of m is not larger than 8.

The selection of agent's moving direction is determined by three factors from external environment, i.e., current position, density and crowd movement. In this model, we use three attraction formulas to depict how these factors impact agent's movement, separately [defined in Formulas (13.4), (13.6) and (13.7)]. Formula (13.8) defines how they work together to direct agent to choose its moving direction.

$$V_k = w_{\mathrm{pos}_k} P_{\mathrm{pos}_k} + w_{\rho_k} \mathrm{Den}_k + w_{\mathrm{dir}_k} P_{\mathrm{dir}_k} \tag{13.8}$$

where V_k means the attraction from cell k to the specified agent. Three weight factors are introduced here, i.e., w_{pos_k}, w_{ρ_k} and w_{dir_k}, represent the impact degree of position, density and crowd movement, respectively. The three weight factors holds the following constrains defined in Formula (13.9), i.e., sum of them equals to 1.

$$w_{\mathrm{pos}_k} + w_{\rho_k} + w_{\mathrm{dir}_k} = 1 \tag{13.9}$$

There are several guide tags recorded in the macroscopic model, which are used to direct agents in the environment how to get the exit by the shortest path. In real life, pedestrian sometimes can not find an exit quickly, and these tags are assigned to help agent find an exit. If there is at least one such guide tag appears in an agent's eyesight, it can help the agent select the optimal path to achieve the exit. In this case, the weight factor by position should be dominant amongst the three factors, and the value of it is set as 0.7, and w_{dir_k} and w_{ρ_k} is set as 0.2 and 0.1, respectively. On the other hand, if there is no guide tag in agent's eyesight, the crowd movement should be dominant (the value is set as 0.7), and w_{ρ_k} and w_{pos_k} is set as 0.2 and 0.1, respectively.

13.3.3 Cell Selection

Agent movement regulation follows the constrains of general cell automata model, i.e., agent can only moves from current cell to the neighboured cells (with the maximum number 8). The attractions of these neighboured cells can be calculated by Formula (13.8), and an agent will always move towards the cell with the maximum value of attraction (i.e., the value of V_k). If there are more than one cell with the same value of attraction, the agent will choose one of them randomly.

If there is only one agent in the simulation environment, the agent can select cell without any constraints. As density increases, agents may choose the same cell for the next simulation time step, and competition is then generated. The competition amongst agents for the same cell is evaluated by the value defined in Formula (13.10), and agent with the largest value of competition$_i$ can enter the cell in the next simulation time step.

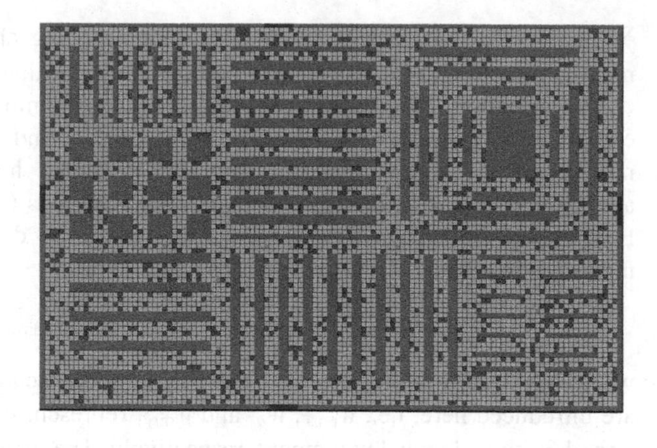

Fig. 13.1 Snapshot of simulation result for scenario 1

$$\text{competition}_i = \frac{H_i}{\text{cost}_i} \tag{13.10}$$

In the formula, i discriminates agent, H_i is the health degree, and cost_i describes the cost of the agent moving from current cell to the targeted cell. H_i reflects agent's characteristic of age and strength. An adult can hold a higher value of H_i than child, elderly man and the disabled. The cost of movement cost_i is set as follows: 1 is assigned if the current and targeted cell share an edge, and 1.414 if the two cells only share a vertex.

Finally, if an agent cannot select an cell to move, it should stop at the current cell and wait for the next chance for movement.

13.4 Case Study

In this section, two cases were studied by applying the proposed model. The environment is divided into cells by the size of 0.4×0.4 m, and simulation time step is set as $0.5\,s$.

The first scenario is shown in Fig. 13.1, where there are two exits in the environment (red colored areas located at left and right side). Obstacles is indicated by blue blocks and agents is represented by solid dots (blue for adult, dark green for child, slight green for elderly agent, and cyan for the disabled). These agent types are reflected by health degree in agent properties (defined in Formula (13.3). Figure 13.1 shows a snapshot of the simulation result. Figures 13.2 and 13.3 show the position field and density distribution respectively at time of Fig. 13.1. Since there are six areas defined as obstacle, it can be observed that the value of position field and density distribution is 0 at the corresponding positions in Figs. 13.2, 13.3

For the other simulation scenario, the crowd distribution is shown in Fig. 13.3, which represents the initial status for the crowd. Obstacles are represented by red

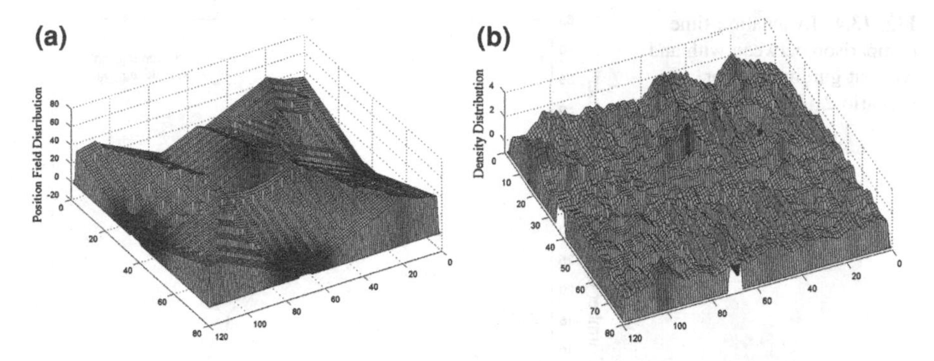

Fig. 13.2 Position field and density distribution around environment for scenario 1

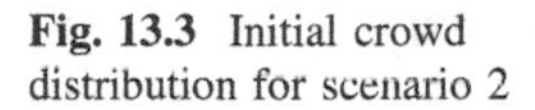

Fig. 13.3 Initial crowd distribution for scenario 2

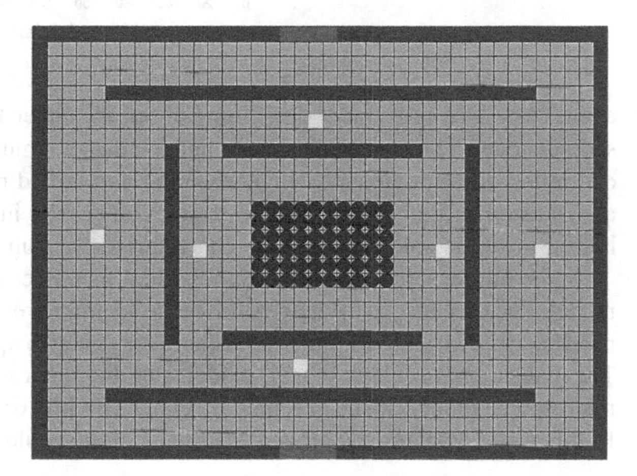

color, and guiding tags are tagged with yellow color. Figure 13.4 shows simulation time used for the crowd evacuation process, i.e., how much time the crowd takes for them to leave the environment. Two experiments are conducted for this case, one with guiding tag and the other without guiding tag. It can be observed that the simulation time for the one with guiding tag is about two times faster than the one without guiding tags. It indicates that guiding tag is helpful for agents to direct evacuation of agent with high efficiency, which is satisfied with the evacuation process in real life.

13.5 Conclusions and Future Work

This paper proposes a hybrid model, integrating macroscopic and microscopic model, to simulate human crowd in dynamic environment. The macroscopic model simulates crowd movement tendency. For the microscopic model, it directs how

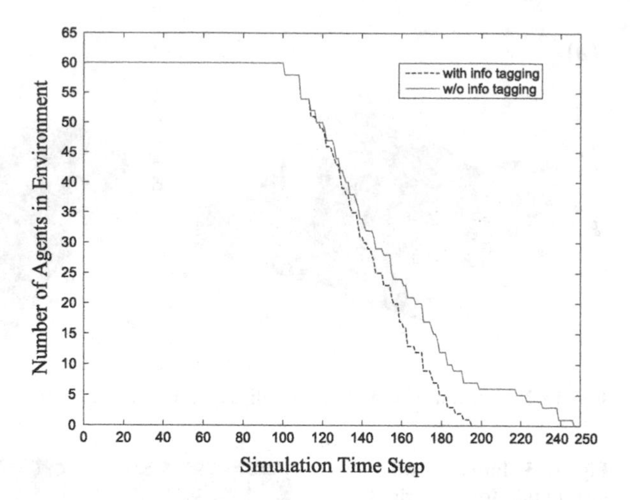

Fig. 13.4 Evacuation time comparison between with and without guiding tag for scenario 2

agent selects its movement direction and speed, under the constraints by macroscopic model. The two models share and exchange simulation results, so that they can reflect the current state of environment and crowd movement. Two cases are also studied in this paper, and the simulation results indicates that the proposed hybrid model is able to reflect the characteristic of human and crowd movement.

In the future, some external and individual factors can be further added into the model, such as fire and emotion. In order to improve the simulation efficiency, parallel and distributed simulation techniques can also applied into the simulation. The execution of macroscopic model can be processed by GPGPU, and the microscopic model can be executed by either GPGPU or high performance cluster. In this case, synchronization and load balancing should be further considered.

References

1. Banerjee B, Abukmail A, Kraemer L (2008) Advancing the layered approach to agentbased crowd simulation. In: Proceedings of 22nd international workshop on principles of advanced and distributed simulation (PADS 2008). Roma, Italy, pp 185–192
2. Bradley GE (1993) A proposed mathematical model for computer prediction of crowd movements and their associated risks. In: Smith RA, Dickie JF (eds) Proceedings of the international conference on engineering for crowd safety. Elsevier Publishing Company, London, UK, pp 303–311
3. Chenney S (2004) Flow tiles. In: Proceedings of the 2004 ACM SIGGRAPH/ Eurographics symposium on computer animation. San Diego, California, USA, pp 233–242
4. Daly P, McGrath F, Annesley T (1991) Pedestrian speed/flow relationships for underground stations. Traffic Eng Control 32(2):75–78
5. Helbing D, Farkas I, Vicsek T (2000) Simulating dynamical features of escape panic. Lett Nat 407:487–490
6. Hughes RL (2002) A continum theory for the flow of pedestrians. Transp Res Part B Methodol 36(6):507–535

7. Pan X, Han CS, Law KH (2005) A multi-agent based simulation framework for the study of human and social behavior in egress analysis. In: Proceedings of the international conference on computing in civil engineering. Law Cancun, Mexico
8. Pelechano N, O'Brien K, Silverman B, Badler N (2005) Crowd simulation incorporating agent psychological models, roles and communication. In: Proceedings of the first international workshop on crowd simulation. EPFL, Lausanne-Switzerland
9. Tanaka T (1991) A study for performance based design of means of escape in fire. In: Cox G, Langford B (eds) Proceedings of the 3rd international symposium on fire safety science. Elsevier, Amsterdam, pp 729–738
10. Thalmann D, Musse SR (2007) Crowd simulation. Springer, London
11. Treuille A, Cooper S, Popović Z (2006) Continuum crowds. In: Proceedings of ACM SIGGRAPH, pp 1160–1168
12. Xiong M, Lees M, Cai W, Zhou S, Low M (2010) Analysis of an efficient rule-based motion planning system for simulating human crowds. Vis Comput 26(5):367–383
13. Xiong M, Cai W, Zhou S, Low MYH, Tian F, Chen D, Ong DWS, Hamilton BD (2009) A case study of multi-resolution modeling for crowd simulation. In: Proceedings of the 2009 spring simulation multiconference (SpringSim 2009). San Diego, California, USA
14. Xiong M, Lees M, Cai W, zhou S, Low MYH (2009) Hybrid modelling of crowd simulation. Procedia Comput Sci 1(1):57–65
15. Zhang Y, Zhao H, Qian J (2006) High order fast sweeping methods for static Hamilton–Jacobi equations. J Sci Comput 29(1):25–26

Chapter 14
Automatic Inference of Interaction Occasion in Multiparty Meetings: Spontaneous or Reactive

Zhiwen Yu and Xingshe Zhou

Abstract In this paper, we propose an automatic multimodal approach for inferring interaction occasions (spontaneous or reactive) in multiparty meetings. A variety of features such as head gesture, attention from others, attention towards others, speech tone, speaking time, and lexical cue are integrated. A support vector machines classifier is used to classify interaction occasions based on these features. Our experimental results verified that the proposed approach was really effective, which successfully inferred the occasions of human interactions with a recognition rate by number of 0.870, an accuracy by time of 0.773, and a class average accuracy of 0.812. We also found that the reactive interactions are easier to recognize than the spontaneous ones, and the lexical cue is very important in detecting spontaneous interactions.

Keywords Multiparty meeting · Interaction occasion · Multimodal recognition

14.1 Introduction

Human interaction is one of the most important characteristics of group social dynamics in multiparty meetings. We are developing a smart meeting system for capturing human interactions and recognizing their types, such as proposing an idea, giving comments, expressing a positive opinion, and requesting information [1, 2]. The human interactions are recognized based on various audiovisual and

Z. Yu (✉) · X. Zhou
School of Computer Science, Northwestern Polytechnical University,
710072 Xi'an, Shaanxi, People's Republic of China
e-mail: zhiwenyu@nwpu.edu.cn

X. Zhou
e-mail: zhouxs@nwpu.edu.cn

J. J. Park et al. (eds.), *Proceedings of the International Conference on Human-centric Computing 2011 and Embedded and Multimedia Computing 2011*, Lecture Notes in Electrical Engineering 102,
DOI: 10.1007/978-94-007-2105-0_14,

high-level features, e.g., gestures, attention, speech tone, speaking time, interaction occasion, and information about the previous interaction. While most features are automatically extracted, the high-level feature of interaction occasion is manually labeled in our previous system.

Interaction occasion is a binary variable that indicates whether an interaction is spontaneous or reactive. In the former case, the interaction is initiated spontaneously by a person (e.g., proposing an idea or asking a question). The latter denotes an interaction triggered in response to another interaction, e.g., acknowledgment is always a reactive interaction.

To make our recognition system applicable in real-life settings, automatic mechanisms are required to infer interaction occasion based on the basic features, e.g., gaze and speaking time. During meeting capture, we really observed that there are some correlations between the interaction occasion and human behavior. For instance, when a person performs a spontaneous interaction, he or she usually looks through most participants in the meeting and speaks relatively long.

Detecting interaction occasions is also useful to segment a meeting into a sequence of sessions that can be used for discovering human interaction patterns in multiparty meetings [3]. In this paper, we propose an automatic multimodal approach for inferring interaction occasions in multiparty meetings. A variety of features such as head gesture (nodding, shaking or normal), attention from others (how many people looking at the target person during the interaction), attention towards others (how many people the speaker looks through during the interaction), speech tone (question or non-question), speaking time (how long the target person speaks), and lexical cue (whether there are any key words in common with the previous interaction) are integrated. A support vector machines (SVM) [4] classifier is used to classify interaction occasions based on these features. To the best of our knowledge, this is the first work to investigate the automatic detection of the occasions of human interaction in multiparty meetings.

The rest of this paper is organized as follows. In Sect. 14.2, we discuss related works. Section 14.3 introduces the automatic approach for interaction occasion inference, including feature extraction and the inference model. In Sect. 14.4, the experimental results are presented. Finally, we conclude the paper in Sect. 14.5.

14.2 Related Work

There have been a number of works done in the past decade in capturing and analyzing multiparty meetings. Most of the previous systems focus on low-level analysis of the captured video and audio data, such as person identification, speech recognition, summarization, attention detection, and activity recognition (see [5] for a full review).

Numerous studies have been specially conducted to extract higher-level semantics in meetings, such as agreement, emotion, group interest, statement intention, influence, and group functional role.

Germesin and Wilson [6] propose a system for the automatic detection of agreements in meetings based on various types of features including lexical, prosodic, and structural features. Hillard et al. [7] proposed a classifier for the recognition of an individual's agreement or disagreement utterances using lexical and prosodic cues.

Kumano et al. [8] propose to automatically discover the interpersonal emotions through facial expression recognition, e.g., how each person feels about the others, or who affectively influences the others the most. Jaimes et al. [9] proposed an approach for affective meeting video analysis.

Researchers in ICSI [10] and IDIAP [11] recognized group interests (i.e., hot spots) using audio–visual features. Yu et al. [12] use the degree of crossover of utterances and gaze changes as well as the number of nods and feedbacks as indicators for recognizing group interests.

The Discussion Ontology [13] was proposed for obtaining knowledge such as a statement's intention and the discussion flow in meetings. Our previous work [1, 2] analyzes seven types of human intentions or attitudes towards a topic in multiparty meetings such as proposing an idea, requesting information, commenting on a topic, acknowledgement, asking someone's opinion, expressing a positive opinion, and disagreeing with a proposal.

Rienks et al. [14] attempted to automatically detect participant's influence levels in meetings through speech-related features. The meeting mediator at MIT [15] detects dominance of participants in a meeting using sociometric feedback.

Banerjee and Rudnicky [16] adopted a decision tree classifier to recognize participant's roles (e.g., discussion participators, presenter, information provider, and information consumers) in meeting. Zancanaro et al. [17] applied SVM to detect group functional roles in face-to-face meetings, such as orienteer, information giver, recorder, and follower. Garg et al. [18] proposed an approach to recognize participant roles in meetings based on lexical information and social network analysis.

Although the existing works have analyzed multiparty meetings in various aspects of low-level structure features as well as higher-level semantics, we have yet to see an investigation to automatically recognize the occasions of human interaction in meetings as we do in this paper.

14.3 Interaction Occasion Recognition

14.3.1 Feature Extraction

The interaction occasion recognition is based on such features as head gestures, attention from others, attention towards others, speech tone, speaking time, and lexical cue. These features are extracted from raw audio–visual and motion data recorded with video cameras, microphones, and motion sensors [1].

Head gesture. Head gestures are very common in detecting human attitudes in meeting discussion (acknowledgement, agreement, or disagreement). They are also useful to determine the occasion of an interaction. For instance, reactive interactions are usually accompanied with nodding or shaking of the head, while spontaneous interactions are not.

We determine nodding through the vertical component of the face vector calculated from the position data. We first determine the maximum and minimum values of the vertical component of the face vector in a time window (here, we set one second). Next, we calculate θ_1 (the difference between the maximum and minimum values) and θ_2 (the difference between the current value and the minimum). Then the nodding score is calculated as: $\text{Score} = (\theta_1/11.5) \times (\theta_2/11.5)$. Here, 11.5 is empirically set as the normalization constant. If the calculated score is above a preset threshold (e.g., 0.80), we consider the head gesture to be nodding.

Head-shaking is detected through the horizontal component of the face vector. We first calculate a projection vector of the face vector on the horizontal plane. To distinguish head-shaking from normal changes in facial orientation, we count the switchback points of the projection vector's movement in a time window (e.g., 2 s). If the number of switchback points is larger than the preset threshold (e.g., 2), we consider that it is a shaking action.

Attention from others. Attention from others is an important determinant of interaction occasion. For example, when a user is performing a spontaneous interaction, such as proposing an idea, he is usually being looked at by most of the participants. The number of people looking at the target user during the interaction can be determined by their face orientation. We measure the angles between the reference vectors (from the target person's head to the other persons' heads) and the target user's real face vector (calculated from the position data). Face orientation is determined as the one whose vector makes the smallest angle.

Attention towards others. Attention towards others means how many people the speaker looks through during the interaction. It is observed that when a person carries out a spontaneous interaction, he or she usually looks through most participants in the meeting other than concentrating on a particular person. This feature is measured by the speaker's face orientation and its changes within the timeslot of the interaction.

Speech tone and speaking time. Speech tone indicates whether an utterance is a question or a non-question statement. Speaking time is another important indicator in detecting the occasion of an interaction. A spontaneous interaction, such as putting forward a proposal usually takes a relatively longer time, but it takes a short time to acknowledge (reactive interaction). Speech tone and speaking time are automatically determined using the Julius speech recognition engine [19], which segments the input sound data into speech durations by detecting silence intervals longer than 0.3 s. We classify segments as questions or non-questions using the pitch pattern of the speech based on Hidden Markov Models [20] trained with each person's speech data. The speaking time is derived from the duration of a segment.

Lexical cue. The lexical cue used here is a binary variable that indicates whether there are any key words in common with previous interaction. It is useful to identify reactive interactions as human usually repeats certain words when responding to the other one whom holds the topic. The key words of a speech segment are automatically extracted using the Julius speech recognition engine [19].

14.4 Inference Model

With the basic feature aggregated, we adopt the SVM classifier for interaction occasion recognition. Given the features of an instance, the SVM classifier sorts it into one of the two classes of interaction occasions: spontaneous and reactive. SVM has been proven to be powerful in classification problems, and often achieves higher accuracy than other pattern recognition techniques [21]. It is well known for its strong ability to separate hyper-planes of two or more dimensions.

Our system uses the LIBSVM library [22] for classifier implementation. Before learning and inference, we need to represent each data instance as a vector of real numbers. For the attributes with continuous values (e.g., speaking time, attention from others, and attention towards others), we use their original values in the vector. If an attribute has categorical values (e.g., head gesture, speech tone, and lexical cue), we use linear numbers to represent it. For instance, head gesture is a three-category attribute (nodding, shaking, normal), which is represented as 0, 1, and 2. The learning and testing data is prepared with the format that LIBSVM can process.

The meeting content is first segmented into a sequence of interactions. Sample interactions are selected and fed to the SVM as training data, while others are used as a testing set (details of training and test data are presented in Sect. 14.4).

14.5 Experimental Results

14.5.1 Data

For the experiments we use the meetings captured in our previous system [1]. The meeting data set is recorded by employing multiple sensors, such as video cameras, microphones, and motion sensors. From the four meetings in the data set, we use the job selection meeting in which no presentation slides were used in order to avoid gaze attraction by the presentation. In the meeting four participants were seated around a table and discussing job selection issues, i.e., the factors that they would consider in seeking a job, such as salary, working place, employer, position, interest, etc. The meeting lasted around 10 min. To obtain the ground truth data for the interaction occasion recognition, one master student annotated the meeting. We ultimately got a total of 213 interactions of which 35 are spontaneous and 178 are reactive.

14.5.2 Evaluation Metrics

We adopt three metrics to evaluate the effectiveness of our recognition approach: accuracy by number, accuracy by time, and class average accuracy.

Accuracy by number. Measuring the inference accuracy using recognition rate by number is a typical method. Accuracy by number (R_N) is defined as:

$$R_N = \frac{m}{N} \tag{14.1}$$

where N stands for the total number of interactions while m denotes the number of interactions our approach correctly infers their occasions.

Accuracy by time. The measure of accuracy by number mainly focuses on how many interactions whose occasions are correctly inferred while ignore the diversity of the interactions in terms of time. For example, the inference approach might work well in recognizing the occasions of short interactions, such as *acknowledgement*, but not in long interactions, such as *propose*, which will make the accuracy by number seem good, but a large period of a meeting is not correctly recognized. The accuracy by time measures the percentage of time that interaction occasions are correctly recognized for the duration of the labeled interactions. It is calculated as follows:

$$R_T = \frac{\sum_{k=1}^{m} T_k}{\sum_{i=1}^{N} T_i} \tag{14.2}$$

where R_T represents the accuracy by time, m and N have the same meaning as those in (14.1), T_k represents the duration of interaction k that is correctly inferred.

Class average accuracy. As the reactive interactions appear much more frequent than the spontaneous ones, our dataset is considered to be imbalanced. Although the recognition accuracy by number is acceptable, we need to examine the class average accuracy that is common in analyzing datasets with dominant classes [23]. It is calculated as:

$$R_C = \frac{\sum_{c=1}^{C} R_{N\,c}}{C} \tag{14.3}$$

where R_C represents the class average accuracy, C is the number of classes and R_{Nc} the recognition rate by number for class c.

14.5.3 Results

In this section, we present the results according to three experiments: overall performance, single feature exploration, and effect of different feature combinations. Within the 213 interactions in the job selection meeting, 69 of them were chosen as the learning set, and the other 144 were used for testing.

Table 14.1 Overall performance

Interaction occasion	Accuracy by number (R_N)	Accuracy by time (R_T)	Class average accuracy
Spontaneous	0.727	0.741	/
Reactive	0.897	0.788	/
Total	*0.870*	*0.773*	*0.812*

Table 14.2 Interaction occasion recognition accuracy using each feature alone

Feature	Accuracy by number (R_N)	Accuracy by time (R_T)	Class average accuracy (R_{N-R}+R_{N-S})/2
Head gesture	0.819	0.726	(1 + 0)/2 = 0.5
Attention from others	0.819	0.726	(1 + 0)/2 = 0.5
Attention to others	0.819	0.726	(1 + 0)/2 = 0.5
Speech tone	0.819	0.726	(1 + 0)/2 = 0.5
Speaking time	0.785	0.701	(0.932 + 0.115)/2 = 0.524
Lexical cue	0.819	0.726	(1 + 0)/2 = 0.5

R_{N-R} accuracy by number of reactive interactions, R_{N-S} accuracy by number of spontaneous interactions.

Experiment 1: *overall performance.* In this experiment, we use all the six features to infer interaction occasion. The experimental results are depicted in Table 14.1. The overall recognition rate (by number) is 0.870, the accuracy by time is 0.773, and the class average accuracy is 0.812. From both accuracies by number and by time, we can see that the reactive interactions are easier to recognize than the spontaneous ones. The accuracy by time of spontaneous interactions is a little higher than that by number, but the difference is not substantial. The accuracy by time of reactive interactions and all testing set is lower than that by number really indicates that our approach performs better in recognizing the occasions of short interactions.

Experiment 2: *single feature exploration.* In this experiment we examined whether we can use each of the six features in stand-alone to discriminate the interaction occasions. Table 14.2 shows the results. As we can see, it is very difficult to infer the occasion of interaction by using single feature. The accuracies by number for head gesture, attention, speech tone, and lexical cue are all 0.819. But they did not make sense, because all spontaneous interactions were wrongly classified into the category of reactive. Hence the class average accuracies are 0.5. Although the speaking time can distinguish spontaneous and reactive a little bit, the result is not acceptable.

Experiment 3: *effect of different feature combinations. To test the effect of different feature combinations on interaction occasion inference, different feature subsets were configured and fed into the SVM inference model. The feature set configuration and recognition results are presented in Table 14.3. There are a total of six different feature subsets in this experiment. Set 1 contains features*

Table 14.3 Interaction occasion recognition results with different feature subsets

Feature sets	Accuracy by number (R_N)	Accuracy by time (R_T)	Class average accuracy ($R_{N\text{-}R}+R_{N\text{-}S}$)/2
Set 1—features about attention: (f2, f3)	0.819	0.726	(1 + 0)/2 = 0.5
Set 2—features about head movement: (f1, f2, f3)	0.819	0.726	(1 + 0)/2 = 0.5
Set 3—features about speech: (f4, f5)	0.812	0.743	(0.932 + 0.182)/2 = 0.557
Set 4—features about head movement and speech: (f1, f2, f3, f4, f5)	0.768	0.712	(0.879 + 0.182)/2 = 0.531
Set 5—features about head movement and lexical cue: (f1, f2, f3, f6)	0.797	0.750	(0.810 + 0.727)/2 = 0.769
Set 6—features about speech and lexical cue: (f4, f5, f6)	0.855	0.768	(0.879 + 0.727)/2 = 0.803

f1 head gesture, *f2* attention from others, *f3* attention to others, *f4* speech tone, *f5* speaking time, *f6* lexical cue, $R_{N\text{–}R}$: accuracy by number of reactive interactions, $R_{N\text{–}S}$: accuracy by number of spontaneous interactions

about attention (attention from others and towards others), Set 2 includes features about head movement (i.e., head gesture, attention from others, and attention towards others), Set 3 includes speech related features (i.e., speech tone and speaking time), Set 4 is the combination of Sets 2 and 3, Set 5 contains features about head movement and lexical cue, and Set 6 includes features about speech and lexical cue. From the results shown in Table 14.3, first we can observe that the recognition performance of basic combination of attention features (Set 1), or head movement features (Set 2) is the same as that of using single feature. The basic combination of speech features (Set 3) is more effective, but the result is still not acceptable. To our surprise, adding head movement to speech features (Set 4) could not improve the inference performance as we expected, but degrade by 0.044 in accuracy by number and 0.031 in accuracy by time. So far we have no satisfactory explanation for this but we plan to investigate this further in detail in future study. The best result appears when using feature Set 6 (combining speech and lexical cue). All the recognition rate by number, accuracy by time, and class average accuracy are quite good. It works well in recognizing both spontaneous and reactive interactions. The result of Set 5 is also acceptable. Sets 5 and 6 are better than other feature sets in inferring spontaneous interactions. Their common feature is the lexical cue. Hence we can draw a conclusion that the lexical cue is very important in detecting spontaneous interactions.

14.6 Conclusion

In this paper, we presented the study of inferring a kind of semantics in multiparty meetings, interaction occasion–whether an interaction is spontaneous or reactive. The experimental results verified that our proposed approach is really effective.

This extracted higher-level semantic information is useful to recognize the type of human interaction and to segment a meeting into a sequence of sessions. For future work, we plan to incorporate the automatic recognition results in our interaction recognition and mining system [1, 3].

Acknowledgments This work was partially supported by the National Natural Science Foundation of China (No. 60903125), the Program for New Century Excellent Talents in University (No. NCET-09-0079), and the Natural Science Basic Research Plan in Shaanxi Province of China (No. 2010JM8033).

References

1. Yu ZW, Yu ZY, Aoyama H, Ozeki M, Nakamura Y (2010) Capture, recognition, and visualization of human semantic interactions in meetings. The 8th IEEE international conference on pervasive computing and communications (PerCom 2010), pp 107–115
2. Yu ZW, Yu ZY, Ko Y, Zhou X, Nakamura Y (2009) Inferring human interactions in meetings: a multimodal approach. The 6th international conference on ubiquitous intelligence and computing (UIC 2009), pp 14–24
3. Yu ZW, Yu ZY, Zhou X, Becker C, Nakamura Y (2011) Tree-based mining for discovering patterns of human interaction in meetings. IEEE Trans Knowl Data Eng
4. Vapnik VN (1995) The nature of statistical learning theory. Springer, Heidelberg
5. Yu Z, Nakamura Y (2010) Smart meeting systems: a survey of state-of-the-art and open issues. ACM Comput Surv 42(2):1–20
6. Germesin S, Wilson T (2009) Agreement detection in multiparty conversation. In: Proceedings of ICMI, pp 7–14
7. Hillard D, Ostendorf M, Shriberg, E (2003) Detection of agreement vs. disagreement in meetings: training with unlabeled data. In: Proceedings of HLT-NAACL, pp 34–36
8. Kumano S, Otsuka K, Mikami D, Yamato J (2009) Recognizing communicative facial expressions for discovering interpersonal emotions in group meetings. In: Proceedings of ICMI, pp 99–106
9. Jaimes A, Nagamine T, Liu J, Omura K, Sebe N (2005) Affective meeting video analysis. In: IEEE international Conference on multimedia and expo, pp 1412–1415
10. Wrede B, Shriberg E (2003) Spotting hotspots in meetings: human judgments and prosodic cues. In: Proceedings of Eurospeech, pp 2805–2808
11. Gatica-Perez D, McCowan I, Zhang D, Bengio S (2005) Detecting group interest-level in meetings. In: Proceedings Of IEEE ICASSP, pp 489–492
12. Yu ZW, Yu ZY, Ko Y, Zhou X, Nakamura Y (2010) Multimodal sensing, recognizing and browsing group social dynamics. Pers Ubiquit Comput 14(8):695–702
13. Tomobe H, Nagao, K (2006) Discussion ontology: knowledge discovery from human activities in meetings. In: Proceedings of JSAI, pp 33–41
14. Rienks R, Zhang D, Gatica-Perez D, Post W (2006) Detection and application of influence rankings in small group meetings. In: Proceedings of ICMI, pp 257–264
15. Kim T, Chang A, Holland L, Pentland A (2008) Meeting mediator: enhancing group collaboration using sociometric feedback. In: Proceedings of CSCW, pp 457–466
16. Banerjee S, Rudnicky AI (2004) Using simple speech based features to detect the state of a meeting and the roles of the meeting participants. In: Proceedings of the 8th international conference on spoken language processing (Interspeech)
17. Zancanaro M, Lepri B, Pianesi F (2006) Automatic detection of group functional roles in face to face interactions. In: Proceedings of ICMI, pp 28–34

18. Garg NP, Favre, S, Salamin H, Tur DH, Vinciarelli A (2008) Role recognition for meeting participants: an approach based on lexical information and social network analysis. In: Proceedings of ACM Multimedia, pp 693–696
19. Julius Speech Recognition Engine. http://julius.sourceforge.jp/en/
20. Rabiner L (1989) A tutorial on Hidden Markov Models and selected applications in speech recognition. Proc IEEE 77(2):257–286
21. Hsu CW, Chang CC, Lin CJ (2005) A practical guide to support vector classification. Technical report
22. Chang CC, Lin CJ (2001) LIBSVM: a library for support vector machines, Software available at http://www.csie.ntu.edu.tw/~cjlin/libsvm
23. Kasteren T, Noulas A, Englebienne G, Kröse B (2008) Accurate activity recognition in a home setting. In: UbiComp, pp 1–9

Chapter 15
Managing Interactions in the Collaborative 3D DocuSpace for Enterprise Applications

Tong Sun, Jonas Karlsson, Wei Peng, Patricia Wall and Zahra Langford

Abstract A novel system is proposed in this paper to manage the interactions in the context of document contents to enhance user collaboration experiences. By explicitly capturing, tracking and analyzing the document interactions in the collaborative 3D space, the system can provide instant feedback to influence users' behaviors and/or enable the adaptive contents. Furthermore, the proposed system can persistent these document interactions to content repository for selective and contextual "replay" or "review".

Keywords 3D virtual environment · Collaborative social context · Interaction management · User experiences

T. Sun (✉) · J. Karlsson · W. Peng · P. Wall · Z. Langford
Xerox Research Center Webster, 800 Phillips Road, Webster, NY 14580, USA
e-mail: Tong.Sun@xerox.com

J. Karlsson
e-mail: Jonas.Karlsson@xerox.com

W. Peng
e-mail: Wei.Peng@xerox.com

P. Wall
e-mail: PWall@xerox.com

Z. Langford
e-mail: Zahra.Langford@xerox.com

J. J. Park et al. (eds.), *Proceedings of the International Conference on Human-centric Computing 2011 and Embedded and Multimedia Computing 2011*, Lecture Notes in Electrical Engineering 102,
DOI: 10.1007/978-94-007-2105-0_15,

15.1 Introduction

There is an emerging trend to embrace collaborative Web 2.0 technologies (e.g. wiki, blogs, social networking) in document management systems to support enterprise applications, such as human resource management, customer relationship management, and workflow management. However, these Web 2.0 enabled collaborations are primarily centered on the evolution of the content (i.e. text, image, video, audio), while important aspects of human interaction (e.g. social presences and/or visual cues) [7, 12, 13, 16, 17] are neglected and not explicitly managed. Furthermore, there is very little support for the direct social interactions among participants in existing document management systems. As the 3D web matures, collaborative 3D virtual environments (e.g. second life [10], open Wonderland [15], Unity3D/Jibe [21], open simulator [14]) are emerging as non-game virtual world platforms to support enterprise collaborations. Largely derived from massive multi-player online games (MMOs), these 3D virtual worlds are built to inherently support rich social interactions. Many technology companies, including IBM, Cisco, Intel, and Nortel, are seriously investing in this area for transformed business opportunity and growth [3, 4]. We (Xerox—a technology leader in document services) believe the combination of the document content management capabilities with the collaborative 3D virtual worlds provides an exciting opportunity to revolutionize the user experiences around digital content, and a new business platform to create innovative xerox document services.

The *Collaborative 3D DocuSpace* [18] bridges a traditional document management system (such as xerox DocuShare [24] or Microsoft sharepoint) with a collaborative 3D virtual environment. In this paper, we propose a novel system to track and analyze interactions in the context of document contents (we call them *document interactions* or *interactions* for short) in this 3D space. These include interactions between avatars and document(s) (e.g. collaborative authoring, tagging, search, etc.), and interactions among avatars within a well-defined document context (e.g. conversations around a presentation slide, discussions in front of a report or book or picture, etc.). By properly tracking and analyzing these document interactions, the system can provide instant feedback to influence the users/avatars' behaviors and/or to enable the adaptive contents. By explicitly capturing the document interactions, the system can store them back into the content repository, and make them available for "replay" or "review." Furthermore, these persistent document interactions can also be served as additional "social" contexts to improve content search and navigability of the repository. An early prototype of the proposed system has been developed using the Open Simulator platform [14]. And an initial evaluation in which users access digital assets from a museum content repository for "story telling" purposes has shown positive and promising user experiences.

The primary elements of the proposed interaction management system include:

1. *A reference scheme* to determine the association between an interaction and its document context (e.g. what report or book this discussion is referencing,

which presentation slide this comment is specifically made for). Although there are many existing tools in 3D virtual worlds to capture the multi-modal interactions (e.g. instant message logs, voice recorder, avatar location sensor, camera position sensor, etc.), object references in the collaborative environment still remain a challenge and unresolved task [22].
2. *An interaction semantic scheme* that extends the repository's meta-data structure with the interaction context (e.g. when and where the interaction happens, who participates, what has been said or discussed, and how participants feel, etc.). In other words, these interaction-related semantic attributes provide a sense of social life [2] context to the contents that is beyond what existing document management systems or even Web 2.0 technologies can offer today. Through this semantic scheme, users can tell not only what is in the content and who created/updated it at what time, but also to whom the content has been presented and exposed and what has been said or expressed about the content. The document interactions can also be explicitly stored or persistent in the repository according to this semantic scheme and served as additional search or navigation contexts for repository users.
3. *A context sensitive scheme for interaction retrieval* that allows the persistent interactions to be selectively retrieved or replayed in a 3D DocuSpace based on user/avatar specific choices. This interaction retrieval or replay scheme is more than just text, audio or video replay. By using the programmatic flexibility in the 3D virtual worlds, the historical document interactions can be retrieved or re-played based on a particular document reference (e.g. slide title, picture name), a specific participant name (what this participant has done, said or viewed, and in what way), or the hot-spot (where most participants and interactions have taken place), etc.

15.2 Related Work

The related work can be categorized into two inter-related aspects: one is the fusion of document management system with 3D technologies (including interactive visualization and shared 3D virtual space), the other is the collaborative content-centric interaction management.

Prior studies have shown that 3D interfaces are more effective than their 2D counterparts for computer-aided design and modeling tasks. A comparative evaluation of a 2D and 3D version of a document management system is described in [5] to assess the usability benefits of 3D interactive system for document management. The results showed no significant difference between task performance in 2D and 3D, however there was a significant "subjective" preference for the interactive 3D interfaces. Later, a multi-user collaborative 3D Information place, called TimePlace [8], was designed to support social navigation based on spatial organization, interaction metaphor, and narrative settings. Rather than a pure 3D information visualization, TimePlace contains 2D information artifacts (i.e. content information, activity information, communicational information) as well as 3D realistic objects (e.g. the virtual environment settings, such as building, room, furniture, plant, etc.). However, it focused on design strategies for replicating a physical real-world representation in a shared 3D virtual world, but did not address the seamless integration of the 3D user experiences and collaborative interactions with content management system.

There are methods proposed to capture and analyze user-centric interactions in collaborative virtual environments for various purposes, such as interest management [6], group behavior discovery [9], and presence measurement [13]. With a focus on managing content-centric interaction to support ad-hoc collaborative processes in a digital library, the proposed hybrid multi-agent system in [20] does not directly apply in immersive 3D virtual worlds. In past years, many 3D enterprise applications (e.g. Tele-place [9], IBM/Lotus sametime 3D [11], the university libraries Island in second life [23]) started providing integration with a document system. However, these applications often jumble efforts to bring avatars or people together around "static" contents mainly for facilitating "socialization" in a virtual meeting room or learning place. The support for document content access from a repository in the 3D virtual worlds is very limited; few specifically manage the collaborative interactions around adaptive documents.

15.3 The Key Features

The collaborative 3D DocuSpace [18] provides a seamless integration from a repository collection to an initialized shared 3D space, including the environment settings (e.g. meeting room, conference auditorium, or story garden) and the document content layout. As avatars navigate through this content-centric 3D space, we need a mechanism to track and capture collaborative interactions.

15.3.1 Tracking and Capturing Document Interactions

By using the unique programmatic computation capability in virtual worlds, a set of tools are available to be deployed to track and monitor multi-modal document interactions:

- Avatar sensing tool: senses the avatar name, ID, and position over time. It captures the avatars who present in the space and where in the space each avatar travels or visits. Based on this, *a traffic heat map* (a graphical representation of the traffic data in a 3D space) can be constructed.
- Chat message logging tool: logs the chat messages or instant messages sent among avatars. It captures what has been said in text by participating avatar(s).

- Camera position tracking tool: tracks the avatar's camera position, which indicates where avatar is looking at regardless the avatar's physical location. It provides a kind of avatar attention indicator. Based on this, similarly *an attention heat map* (a graphical representation of which positions in a 3D space have been viewed by avatars) can be constructed as well.
- Audio recorder: records the voice communication among avatars. The recorded voice files can be further analyzed and disseminated based on voice recognition technologies, which will not be discussed in this proposal.
- Video recorder: records the video around its designated space. The recorded video files can be further analyzed and disseminated based on motion image recognition technologies, which will not be discussed in this proposal.
- Avatar animation detection tool: detects the avatar's gestures, expressions and other animations. In a virtual world, an avatar's mood (e.g. happy, sad, bored, excited) or opinion (e.g. agreed, disagreed) usually can be expressed via animation scripts (e.g. nod/shake head, smile/cry, jump up, raise hand, etc.) that manipulate the avatar's facial expression or body movement. Often, a small set of common gestures or expressions (e.g. nod/shake head, smile/cry, jump up, raise hand, etc.) can be invoked via a GUI message/command. A rule set is created beforehand between these common gestures or expressions and a set of pre-defined mood/opinion terms. For instance, if an avatar "shakes head," then this avatar "disagreed;" if an avatar "smiles," then this avatar is "happy." As soon as the avatar animation is detected, a message will be logged as a mood/opinion feedback according to these pre-determined rules. In other words, whenever an avatar XYZ's "shake head" animation is detected, a message of "avatar XYZ disagreed" will be logged. This tool helps to capture the social and visual cues of interactions that often lost in existing Web 2.0 collaboration technologies. Although it is possible to apply many sophisticated motion detection algorithms to determine the avatar's mood/opinion, it is not the focus of this paper.
- Document tagging and annotation tool: provides a UI text box or note-card drop box next to a document so that additional comments or tags can be input or submitted.

All these sensing, tracking and detection tools function within a pre-set range in the initialized 3D space. However, none of these tools addresses the problem of mutual references during the interactions. In other words, all captured interactions are lumped together as a data blob, and there is no way to differentiate what interactions applied to which documents, especially when multiple documents are presented. Without a proper mutual reference scheme, the document interactions would be out of context and therefore have less value. Also as pointed out by a previous work practice study [22], the problem of mutual reference remains unresolved and a challenge in existing virtual worlds.

In this paper, we present the following solutions to address the mutual reference of a document in the Collaborative 3D DocuSpace:

- Deploy one set of above tools in a pre-defined *3D cell* (i.e. a rectangular prism), whose size is the pre-set function range for each sensing/tracking/detecting tool. The interactions captured by these tools are associated with all document contents being laid out within this 3D cell, if not explicitly stated otherwise.

- Assign a unique chat messaging channel for specified document(s), so that all textual conversations captured in that channel are associated with the specified document(s).
- Require each avatar to explicitly mention, select or highlight the referenced document(s) via a GUI command before he/she speaks or types messages/comments/tags. Otherwise, the interactions are deemed to refer to all documents in a 3D DocuSpace.
- Use the physical proximity to determine the document references in the interactions, in which captured interactions are associated with the closest document(s) (in terms of physical distance in between the participants and the documents).

15.3.2 Instant Feedback and Adaptive Contents

One purpose of tracking and analyzing interactions in the context of content is to provide instant feedback to influence user/avatar behaviors and/or enable the adaptive content in the virtual world. The feedback includes the attention map, the comments from textual or audio messages, and the avatar's mood/opinion (inferred from their gestures/expressions). More importantly, this feedback can be specifically targeted for individual documents or a collection of content via our above reference scheme, instead of simply being stored as a data blob in most of the note taking and sharing applications (e.g. Microsoft oneNote).

For instance, when the presenter is aware of the decreasing attention density from the attention map (which means he/she is losing the virtual audience's attention), he/she might decide to adjust the speed of presentation, or upload new content into the space, or ask questions to stimulate more engaging interactions. In another case when the mood/opinion feedback is telling the presenter that most audience disagree with him/her, the presenter might decide to add more supporting contents into the virtual space. These adaptive contents are very attractive and powerful for many content-centric collaboration scenarios, such as in a virtual conference room or a virtual classroom. We seamlessly integrate the social experiences and document contents in the 3D DocuSpace, something that no existing document management systems and 3D virtual worlds can provide alone.

15.3.3 Capturing and Storing Document Interactions

Through the reference scheme described earlier, document interactions can be tracked and analyzed in the context of specific content(s). In this paper, we propose

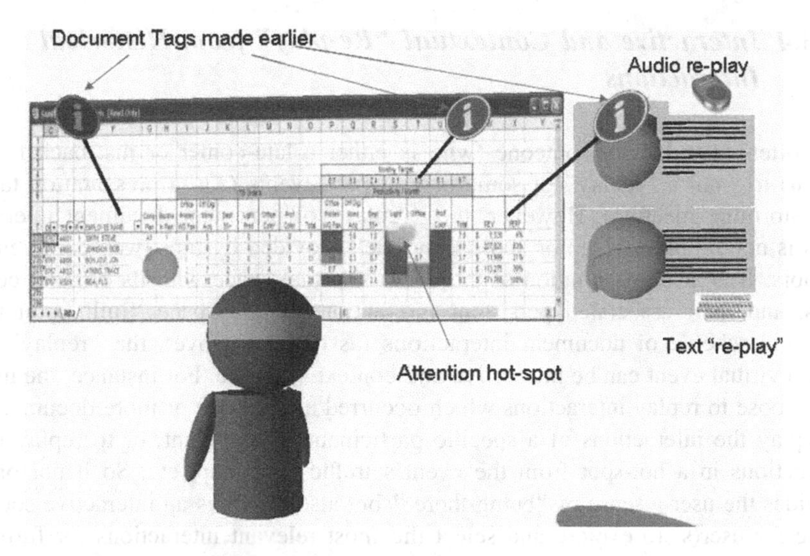

Fig. 15.1 Examples of multi-modal document interactions

a semantic scheme for representing and indexing document interactions and their relationship with contents. Tracked interactions by the above tools can be indexed and stored back to the repository just like any other collection or content object. Each persistent document interaction inherits the generic properties from repository's meta-data structure (i.e. a unique identity, a title, a summary or brief description, and links). Meanwhile, the following meta-data attributes are also added to the semantic scheme of persistent document interactions:

- Participating avatar names and IDs,
- A trail map for all participating avatars (using physical coordinates in the 3D virtual space),
- A traffic heat map (indicates the total number of participants or visitors),
- An attention heat map (indicates where the participants are viewing or paying attention),
- The link attributes to the interaction files (e.g. comments, discussions and feedbacks in textual, audio, video),
- The link attributes to the original contents (i.e. the context of contents based on the reference scheme described in previous section).

The persistent document interactions are stored, organized, and retrieved according to this extensive semantic meta-data scheme. Furthermore, the semantic scheme of document interactions can be served as additional social contexts for repository contents as well. It can enable a more efficient content search and navigation in the repository. Examples of multi-modal document interactions in the context of a specific content (e.g. an Excel sheet) are illustrated in Fig. 15.1.

15.3.4 Interactive and Contextual "Re-play" from Historical Interactions

It is often desirable for someone (who is either a late-comer or just back for a revisit) to want a "replay" of some part of past events (e.g. a presentation talk, brainstorming meeting). However, the "replay" of persistent document interactions is not simply a video or audio play-back provided by most web conference vendors. It is a re-instantiation of the event's virtual space with its original contents, and their associated persistent interactions in the space. Built upon the semantic scheme of document interactions (as defined above), the "replay" of such a virtual event can be interactive and context sensitive. For instance, the user can choose to replay interactions which occurred around one or more documents, to replay the interactions of a specific participant in the event, or to replay the interactions in a hot-spot from the event's traffic heat map, etc. So it not only provides the user a sense of "being there," but also provides an interactive social space for user(s) to explore and select the most relevant interactions for further review without a complete audio/video re-run. Therefore, this interactive and contextual interaction "replay" mechanism provides a unique yet powerful union of content(s) and their social experiences that could ultimately enhance user experiences for knowledge intensive applications, such as distance learning, document communications, field support services, etc.

The requester can also choose to "replay" the interactions in either a shared space or a private space. In a shared space, the requester will be able to "re-play" the interactions with other user/avatars in the same virtual space at the same time. However, a private space is a separate instance where the requester can watch the "re-play" alone. Figure 15.2 illustrates an example of interaction replay scenario, where a user "replays" the persistent interactions around a document in a private space. Depending on the mode of interactions, different "replay" stations in a virtual space can be activated. For instance, a message billboard or whiteboard or pop-up box displays the textual messages, comments, or tags; an audio speaker station plays the audio segments while a movie screen plays the video file.

15.3.5 Experience-Mediated Content Search and Navigation

As stated earlier, document interactions encompass the contents and social experiences. The persistent document interactions described above also serve as additional social contexts for the contents in their lifecycles. For instance, a presentation might have been presented in a series of project review meetings, customer training sessions, and external conferences during its lifecycle. The document interactions captured during these events by the proposed system can provide information beyond the slide contents, such as who the presenters are, to whom it has been presented, what feedback was gathered and, by whom.

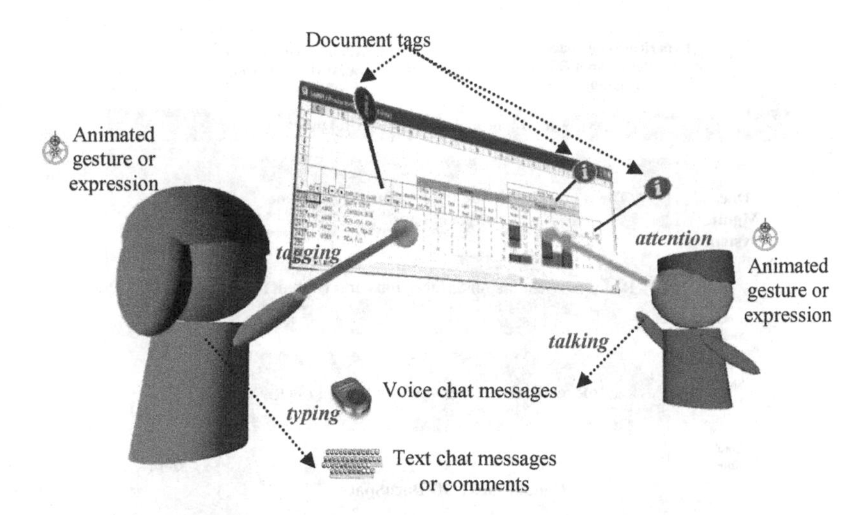

Fig. 15.2 An interaction re-play scenario

This information provides far more meaningful contexts that go beyond the evolution of content itself (e.g. author, keywords, versions, etc.). The proposed system could enable a paradigm shift for information retrieval from the content-driven to a combination of content and social experiences. This is true for the repository navigation scheme as well, since the contents can be organized or structured based on their social experiences. For instance, in the virtual technology conference scenario, a user can look for contents that have caught the most attention from participants, or retrieve the contents that have been visited by a specific event.

15.4 The Architecture of the Document Interactions Management System

The document interactions management system proposed in this paper comprises of the following architectural elements (as shown in orange boxes in Fig. 15.3):

1. A set of tools that track and monitor the multi-modal interactions in a 3D space, including an avatar sensing tool, chat message logging tool, avatar camera position tracking tool, voice recorder, video recorder, avatar animation detection tool, and document tagging tool. The list can go on as many new types of tools are becoming available in the open source or other software development communities.
2. The reference scheme that explicitly links the interactions with the contents, from which interactions are tracked and monitored in the context of contents. The reference scheme defines the following approaches to either allow participants explicitly reference the content(s) or identify the implicit references during the interactions:

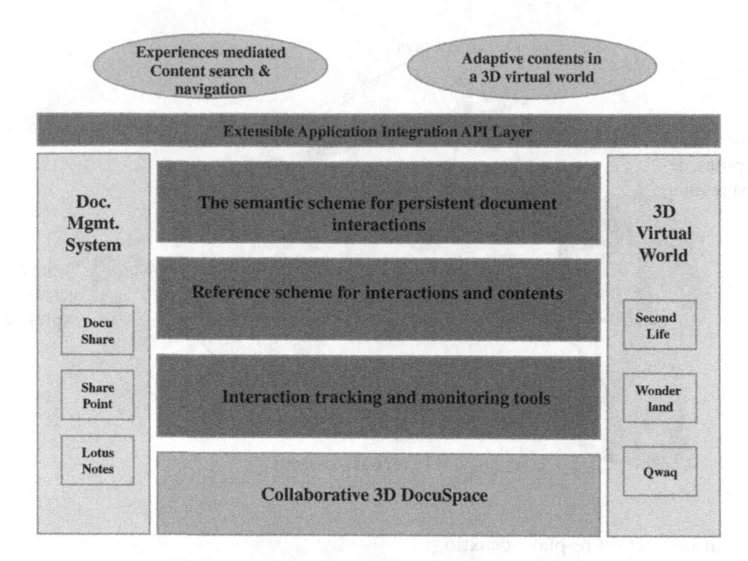

Fig. 15.3 The high-level architecture for the document interactions management system (the *boxes* in *orange* represent the components of the proposed system)

- Approaches required for providing explicit references via a specific GUI command or designated communication channel.
- Approaches for identifying the implicit references (a) via pre-setting a 3D cell or a defined range for tracking/monitoring tools, within which all interactions are applied to all contents in these spaces; (b)via the physical proximity of participants to the contents, from which all interactions in the 3D DocuSpace are applied to the contents that are close to them.

3. The interaction semantic scheme that extends the repository meta-data structure with interaction contexts captured in a 3D space, upon which the interactions can be persistent in the repository and served as extensive social experience contexts for content search and navigation. A schematic diagram is shown as in Fig. 15.4. The persistent document interaction is treated as a first level object in the repository system like any other root object, such as *Collection* or *Document*, but with specially extended meta-data attributes describing the multi-modal interactions captured in the 3D DocuSpace and their relationships with original contents.
4. The extensible APIs at the application layer that can be used to build value-added applications, such as "adaptive contents in 3D DocuSpace," and "experience-mediated content search and navigation in the repository" (as we described in previous sections). Other potential applications could include "adaptive 3D space according to the dynamic interactions," "experiential learning and knowledge transferring," etc.

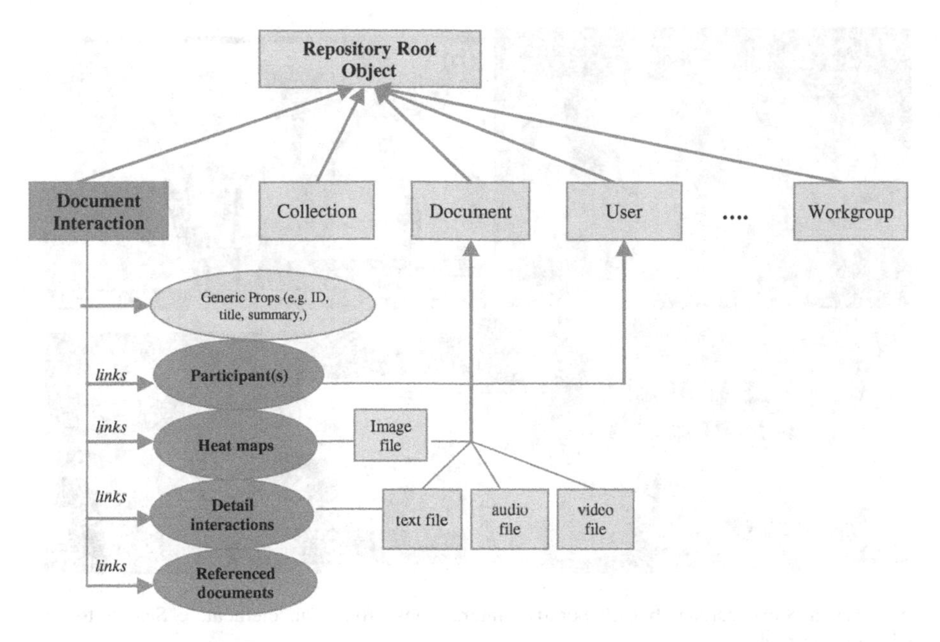

Fig. 15.4 The schematic diagram for persistent document interactions (the symbols in orange color represent the extensions made by the proposed system)

15.5 Prototype Implementation and Evaluation

An initial system prototype, the 3D virtual story garden application was developed using the open source platform opensim [14] in collaboration with The Henry Ford Museum. The goal was to enable users to present a subset of a digital collection as a story to others in the 3D DocuSpace. To narrow the initial scope of the system, we define stories as a sequence of elements from a digital collection, together with text captions, semantic tags, and longer text annotations. The digital elements can include images, multi-page documents and videos. The 3D virtual story gardens takes the stories from a content repository and instantiates them in a multi-user 3D virtual world. In a story garden, the sequence of images and documents of a story is laid out along a spiral path, which is traversed as the story is presented (see Fig. 15.5a). The story garden also contains natural outdoor features, such as grass, trees and flowers to complete the effect of being in a garden. Each story element is displayed as a billboard which has mechanisms to display and create tags and annotations. Figure 15.5b shows a group of avatars around a story element in the garden. The tracked interactions include content tagging/annotating, avatar movements, and text-chat conversations. Any avatar movements or text-chat conversations that take place within a defined 3D cell that is 6 horizontal virtual meters and 10 vertical virtual meters from the center of an element are recorded and associated with that element. These recorded interactions are indexed in the

Fig. 15.5 **a** Story garden. **b** Collaborative interactions around an element. **c** Search tower. **d** Interaction replay

repository and can be searched and retrieved for later re-play by clicking a "re-play" button on the element's billboard. In this case, the avatar is teleported to a shared replay area (see Fig. 15.5d) that shows the element as it appeared in the past, together with past interactions, such as the avatar movements (in orange-blue figure) and text-chat that took place around it. A search facility is provided in the form of a "search tower," shown in Fig. 15.5c. Users can search the museum repository by entering a search term in text-chat. Stories are represented by doors along the walls of the tower, extending to multiple floors if needed. By clicking on a door, the user is transported to the associated Story Garden space.

This 3D Story Garden application was evaluated by seven members of The Henry Ford Museum staff. The evaluation included two 60 min training sessions to provide instructions and experience with basic operations and navigation in the 3D environment. This was followed by a story presentation session, led by one of the museum participants. The story, titled 'Hacking the Model T' was about the customization of Ford Model T cars. The story was created from a collection of museum artifacts in story maker, a web-based prototype for exploring collections and building stories around collection elements [1]. Analysis of the text-chat log recorded during the story sharing session revealed a rich and varied set of interactions among the participants. In addition to interactions related to the story (e.g. guiding and confirming understanding of story contents), and questions and instructions related to the story garden interface, participants also engaged in ad-hoc social interactions (e.g., commenting on avatar appearance) and made several connections and references to recent cultural phenomena (e.g. mention of the term 'jail breaking' in reference to the Model T). Comparing the amount of chat comments, showed that the presenter made 116 comments (of which 87 were used to tell the story) during the 50-min presentation, and other participants contributed 144 comments.

These results suggest that avatar communication in a 3D virtual environment can provide a highly interactive experience that holds participant interest for even relatively short presentations. Among the positive aspects reported, by study participants was that the session felt more like a conversation than a presentation. They suggested that the use of avatars-made presenting and interacting less intimidating than a traditional presentation, and therefore a potentially valuable tool for use in classroom and educational contexts. They also commented on the interaction tracking, recording and replay capabilities enabled in the story garden application providing an excellent way to preserve the interactions and experiences, making them accessible for participants to revisit or to share with others at a later time. Negative comments focused on initial difficulties in navigating in the 3D space and manipulating the avatar viewing angle. There was also significant IT effort required to install and properly configure museum computers to enable the museum staff to use and evaluate the application.

15.6 Summary

By using the unique programmatic capability in the 3D virtual worlds, the proposed interaction management system provides a novel mechanism to track and analyze multi-modal interactions around document contents that is not offered by any existing collaboration technologies. It provides a powerful union of document contents and social experiences in document interactions. Through the concept of document interactions, a sense of social experiences can be explicitly managed and brought back to the repository contents, which further enriches the content lifecycle. A prototype of the proposed system has been developed and demonstrated that such an interaction management system also enhances the shared 3D DocuSpace with compelling social experiences around the document contents by providing instant feedback from interactions, which in turn enables a more engaging and productive interactions overall. The proposed system offers an extensible platform approach from which many innovative and value-driven applications and services can be delivered.

References

1. Bier E, Janssen B, Wall P, Karlsson J, Sun T, Peng W, Langford Z (2011) A story-based approach to making sense of documents, IASTED human computer interface conference, Washington, DC, 15–18 May, 2011
2. Brown JS, Duguid P (2000) The social life of information. Havard Business School Press, Boston

3. Cherbakov L, Brunner R, Lu C, Smart R (2009) Enable far-reaching enterprise collaboration with virtual spaces, Part 1 and 2, IBM Report, May 29, 2009
4. Cisco telepresence. http://www.cisco.com/en/US/products/ps7060/index.html
5. Cockburn A, McKenzie B (2001) 3D or not 3D? Evaluating the effect of the third dimension in a document management system. ACM SIGCHI 27(3):39–43
6. Ding D, Zhu M (2003) A model of dynamic interest management: interaction analysis in collaborative virtual environment. In: Proceedings of the ACM symposium on virtual reality software and technology (VRST'03)
7. Held RM, Durlach NI (1992) Telepresence. Presence Teleoper Virtual Environ 1:109–112
8. Hsu P (2009) TimePlace: towards a populated 3D information place. IEEE Int Work Soc Inform, pp 23–28
9. Laga H, Amaoka T (2009) Modeling the spatial behavior of virtual agents in groups for non-verbal communication in virtual worlds, 3rd International universal communication symposium
10. Linden Research Second Life Virtual World. http://www.cisco.com/en/US/products/ps7060/index.html
11. Made in IBM Labs: secure 3D meeting service now available with lotus sametime. http://www-03.ibm.com/press/us/en/pressrelease/27831.wss
12. Meehan M, Insko B, Whitton M Jr, Brooks FP (2002) Physiological measures of presence in stressful virtual environments. ACM Trans Graph 21(3):645–652
13. Montoya M, Massey A PVP: measuring perceived virtual presence in collaborative virtual environment, submitted to MIS quarterly, special issue on new ventures in virtual worlds, September 2008
14. Open simulator, 3D virtual world open source project. http://opensimulator.org/wiki/Main_Page
15. Open Wonderland. http://openwonderland.org/
16. Sallnas E (2005) Effects of communication mode on social presence, virtual presence, and performance in collaborative virtual environment, presence Vol 4, No. 4
17. Sheridan TB (1992) Musings on telepresence and virtual presence. Presence Teleoper Virtual Environ 1:120–126
18. Sun T, Karlsson J, Peng W (2009) Collaborative 3D DocuSpace: a platform for next-generation document services in 3D virtual worlds. IEEE international conference on service operations & logistics, and informatics, Chicago, USA
19. Teleplace Enterprise Fall (2010) Adds sharepoint support, enterprise-class enhancements. http://www.teleplace.com/company/press_releases/pr-2010_09_01.php
20. Thiel U, Brocks H, Stein A (2005) Agent-based interaction management for document-centered collaboration in a digital library. In: Proceedings of HCI international 2005, Las Vegas
21. Unity3D/Jibe virtual world solutions. http://infiniteunity3d.com/visit-jibe/
22. Wadley G, Ducheneaut N (2009) "Virtual barn-raising": a study of collaborative building in second life. under review, CHI
23. Wicklund D (2010) See the Stanford libraries' island in second life. speaking of computers, issue No. 83. http://speaking.stanford.edu/highlights/See_SU_Libraries_in_Second_Life.html
24. Xerox DocuShare content management system http://docushare.xerox.com/

Chapter 16
Dempster–Shafer Theory Based Consensus Building in Collaborative Conceptual Modeling

Lili Jia, Huiping Zhou, Zhiying Wang and Chengzheng Sun

Abstract In this paper, we propose a Dempster–Shafer theory based model for belief merging in collaborative conceptual modeling (CCM). This model is a kind of iterative, mutual evaluating and role-based belief revision process. Based on concepts from belief theory, we conceive that the reliability belief participants assume to each other will affect their choice between opinions. The role people acting will also influence the belief merging process. This model is not only a vivid simulation to the real world, but also helpful to decision-pruning and speed up convergence.

Keywords Collaborative conceptual modeling · Consensus · Role · Demspter–Shafer theory

L. Jia (✉) · H. Zhou · Z. Wang
School of Computer Science, National University of Defense Technology,
Changsha, People's Republic of China
e-mail: lili.jia@nudt.edu.cn

H. Zhou
e-mail: hpzhou@nudt.edu.cn

Z. Wang
e-mail: zywang@nudt.edu.cn

C. Sun
School of Computer Engineering, Nanyang Technological University,
Singapore, Singapore
e-mail: CZSun@ntu.edu.sg

J. J. Park et al. (eds.), *Proceedings of the International Conference on Human-centric Computing 2011 and Embedded and Multimedia Computing 2011*, Lecture Notes in Electrical Engineering 102,
DOI: 10.1007/978-94-007-2105-0_16,

16.1 Introduction

In the process of the collaborative software engineering (CSE), typically multiple groups develop and maintain serial versions of a range of products, in parallel [1]. Requirement engineering is an important branch of software engineering [2]. Many researchers consider that requirement engineering is a process composed of two elements namely requirement elicitation, and requirement modeling [3]. In the modeling phase, the identified facts elicited in the former phase will be represented and organized as conceptual models (e.g. UML model, entity-relationship (ER) model, etc.). Collaborative conceptual modeling (CCM) is a challenging aspect of the requirement engineering process [4].

CCM can be regarded as an iterative decision-making process. People who have different knowledge background may help in better understanding and analyzing the problem. According to their different professional fields, experiences and authorities, they will play different roles during the decision-making process. So it is a very natural way to introduce roles into the area of Computer-supported cooperative work (CSCW) [5–9].

Although it seems so intuitive to incorporating of information from multiple sources for creating an overall conceptual model, still there are various issues needed to be addressed. First, humans inevitably make conception errors due to cognition problems or short-term memory. Second, epistemic uncertainty is an essential element of human judgments that makes them even more susceptible to inaccuracy and imprecision [10]. All above imply that not all asserted information from various sources are equally exact or trustworthy. Hence uncertainty and imprecision should be considered during the iterative process of model merging participated by many experts with different roles.

There are three major frameworks through which the problem of interval-based representation of uncertainty has been handled, namely imprecise probabilities (initial work by Walley and Fine; Kuznetsov); possibility theory (Zadeh; Dubois and Prade;Yager); and the Dempster–Shafer theory of evidence (Dempster; Shafer; Yager; Smets) [11].

In this paper, a Dempster–Shafer theory based consensus building model is proposed. It integrates the role-based iterated revision belief into CCM to simulate the dynamic belief evaluation process between experts. During the process, experts from different areas not only provide their own suggestion, but also take in others. According to individual trustworthiness evaluation they make to everyone, experts will choose to reserve their own propositions or not. Actually, the evaluation course is also in a perpetual changing way. Based on the discussion result from last round experts will modify their reputation assessment results, and inevitably according to the roles played by other experts. Such belief modification is the base of belief merging which is the base of model merging. We believe that this approach can help to accelerate the convergence speed to create a final model. We also believe it is a true portrayal of the mental activity of people in the negotiation process of CCM.

The rest of this paper is organized as follows: First, some related works and motivations are introduced. Then, the details of the proposed model are given. Finally, we conclude that the model we present can facilitate consensus building in CCM.

16.2 Background and Motivations

Our proposed model originates from the stream of research that is being pursued in the collaborative conceptual model design field. The model also borrows various operators, schemes, and ideas from Dempster–Shafer theory and consensus decision-making; therefore, we briefly introduce some relevant concepts and works in these areas in the following subsections.

16.2.1 Dempster–Shafer Theory

In the early 1980s, the Dempster–Shafer theory, which is a generalization of the Bayesian theory of subjective probability, captured the attention of AI researchers. They adapt probability theory to expert systems, e.g. in MYCIN, an early expert system developed in the early 1970s at Stanford University. Up to now the Dempster–Shafer theory has been widely studied in computer science and artificial intelligence [12].

In the Dempster–Shafer theory framework every pieces of proposition (information) from different data sources will be assigned degrees of belief. Data sources are persons, organizations, or any other entities that provide information for a question. In CCM, data sources are usually experts, who give subjective quantifiable statements. In order to cluster these propositions, Dempster's combination rule will be calculated.

The following concepts and theories are abstracted from [13].

1. The frame of discernment

In the Dempster–Shafer theory, the frame of discernment Θ consists of N mutually exclusive and exhaustive elements describing all of the sets in the hypothesis space. The frame of discernment can be defined as:

$$\Theta = \{\theta_1, \theta_2, \ldots, \theta_j, \ldots, \theta_N\}. \tag{16.1}$$

It is assumed that Θ is finite[2]. Each subset of represents a proposition of interest:

$$2^{\Theta} = \{\Phi, \{\theta_1\}, \{\theta_2\}, \ldots, \{\theta_N\}, \{\theta_1 \cup \theta_2\}, \{\theta_1 \cup \theta_3\}, \ldots, \Theta\}. \tag{16.2}$$

2. Mass function (Basic probability assignment)

The Dempster–Shafer theory assigns a belief mass to each proposition of Θ to represent the influence of a piece of evidence on this subset. It exists in the form of a probability value which is a positive number between the interval 0 and 1. A mass function m must satisfy the flowing three constrains:

$$m: \quad 2^{\Theta} \rightarrow [0, \quad 1]. \tag{16.3}$$

$$m(\Phi) = 0. \tag{16.4}$$

$$\sum m(A) = 1, \; A \subseteq 2^{\Theta}. \tag{16.5}$$

3. Belief function

The belief function in the Dempster–Shafer theory shows the amount of belief that one data source has assigned to any proposition. The relation between the mass function m and belief function is:

$$\mathrm{Bel}(A) = \sum_{B \subseteq A} m(B), \quad A \subseteq \Theta. \tag{16.6}$$

4. Dempster 's combination rule

When there are two evidences supporting a proposition with different belief assignments, Dempster's combination rule will be used to merges these belief value expressed by various independent sources of information. The combination operator acts on two basic belief assignments m_1 and m_2. The result is a integration of the collective belief of the two information sources ($m_{1,2}$).

$$m(A) = \begin{cases} 0 & (A = \phi) \\ \frac{\sum_{A_i \cap B_j = A} m_1(A_i) m_2(B_j)}{1-K} & (A \neq \phi) \end{cases} \tag{16.7}$$

$$K = \sum\nolimits_{A_i \cap B_j = \Phi} m_1(A_i) m_2(B_j) < 1. \tag{16.8}$$

K is a normalization factor that represents the degree of conflict between m_1 and m_2. Dempster's combination rule has been criticized due to its counter-intuitive results under highly conflicting belief expressions.

In addition, in Dempster–Shafer theory, there are many ways which conflict can be incorporated when combining multiple data sources of information [11]. One representative work has been done by Yager [14]. The other interesting approach is proposed by Tin Kam Ho et al. [15]. We will employ Tin Kam Ho's interpretation in the process of evaluating data sources' reliability.

16.2.2 Consensus Decision-Making in CCM

Consensus decision-making is a group decision-making process that not only seeks the agreement of most experts, but also the resolution or mitigation of minority objections. Consensus is usually defined as meaning both general agreement, and the process of getting to such agreement. Consensus decision-making is thus concerned primarily with that process.

Many researchers conceive that requirement engineering is a process composed of two elements of requirement elicitation, and requirement modeling [3]. Chen believes that "conceptual modeling and its derived software implementation of sharable information services in a global information architecture will be a major challenge" [4]. Various attempts have been made to develop all kinds of concepts, model, and tools to support CCM [16–20]. There are many new problems occurred during collaboratively merging and integrating different conceptual models in CSE. One of the most important and interesting problem is how to build consensus.

Those sub-models (propositions) are collaboratively developed by different experts with different roles. Some of them are domain experts, some are end-users, some are senior conceptual modelers, and some are novices. Different requirement analysts will have different propositions with the same software entity, even if the information they have is abstracted from the same documents or specifications. Discrepancies must be solved in the process of merging and integrating. Sometimes, they also inclined to use information from viewpoints to create their own software requirements. Viewpoints given by Nuseibeh is "loosely coupled, locally managed distributed objects which encapsulate partial knowledge about a system and its domain, specified in a particular, suitable representation scheme, and partial knowledge of the process of development" [21].

It is very natural and convincing to incorporate multiple viewpoints [22] into the conceptual model. We conceive that the conceptual model achieved in this way is much closer to the real world and more completely to fulfill end-users requirements than the one created from single viewpoint.

However, absolute confidence shouldn't always be put on to humans' ability to cognize and perceive the physical world. Some researchers think that due to problems of framing, resistance of mental models to change, risk aversion, limitations of short-term memory, and other cognitive and perceptual biases, errors will inevitably be made when humans reason. This implies that conception error is inevitable. So we shouldn't take all asserted viewpoints as correct or reliable. There is another question: is the information we capture from various data sources consistent with each other at all times? The answer is definitely no. Considering these complicate situations, one possible and sensible way to solve the uncertainty and imprecision is to incorporate them into the process of merging and integrating model. Usually in the decision-making process, the consensus building is really an iteration procedure incorporated with oppugning with each other about the accuracy of partial models created by experts, even the one created by an expert himself or herself.

16.2.3 Introduing Uncertainty into CCM

Some work related with belief merging in CCM will be discussed in this section.
1. Belief merging

Experts' beliefs are not always compatible with each other. How to form the group's beliefs is the main aim of belief merging. In most cases, the final decisions are made through the reiteration procedure of voting, belief revision, and belief merging. Sometimes, there is still a need of decisiveness by an arbitrary decision-maker. A lot of approaches have been presented to solve this problem, including the belief game model [23], belief negotiation model [24], and possibilistic belief base merging [10] etc. They are all devoted to handle inconsistent problems occurred when merging and integrating information from different viewpoints.

2. Collaborative conceptual model

According to the survey conducted by M2Research at San Diego, US in 2001, 70% questionnaires indicated that the need for companies to use collaborative design software can be foreseen. Until recent years, people begin to pay more attention to this field no matter a consensus we have made early that one of the most significant feature of software engineering is its inherent need of integrating many software engineers' efforts to collaboratively, and cooperatively produce a large software system [25]. The collaboration between different experts, including engineers, end-users, and funding sources (stakeholders) etc. is helpful to produce a series of common understanding artifacts.

The researches in this domain are concentrated on two schemes. One scheme emphasizes the importance to solve the low-level problems appearing when merging and integrating multiple submodels, such as consistency and conflict. The other one hammers at developing an ideal environment to facilitate consensus building by providing awareness, radar view, and text, audio, or video-based communication tools etc.

3. Role-based CSCW

It is Leland et al. who introduced roles into Computer-supported collaborative work (CSCW) in 1988 [26]. Recent years, little attention has been paid to role-based CSCW [6, 7]. In most cases, people take roles as interaction media in collaborative systems. What inspired us more is some researchers think that in the CSCW community there is no doubt about the importance of roles.

16.2.4 Motivations

During the process of merging and integrating collaborative conceptual models, experts' propositon possess some degree of uncertainty. To overcome this issue, different models have been presented for formally negotiating over and merging of conceptual models toward consensus. That's the fountain of our work. We concern more about the approach presented in [20] which focuses on the formalization of uncertainty and expert reliability through the employment of belief theory.

In [20], before merging submodels, propositions are annotated with varying degrees of belief, disbelief, and uncertainty. This three dimensional belief structure is based on subjective logic. Then based on this belief structure, a semi-automatic negotiation and consensus building model will further help to refine the integrated model. This is an iterative process. After initial merging and integrating, experts will revise their own submodel also the belief assignment to their propositions according to the intermediate merging result. Then the next round of negotiation will proceed based on the revision until the consensus is made. Each expert can stick to his own opinion and ignore or oppose the recommendations (the intermediate merged and integrated model) or accept them completely or partially. In the case of no final decision made within the limited timetable, the potency metric as a third party is used to prune inconsistent and conflicting propositions.

In the above works, the belief mass are only assigned to the artifacts created by the expert himself. This schema is a little different from the evaluation mechanism when people make decision in the real world. In the latter situation, people will not only assign a degree of belief, disbelief, and uncertainty to the information they catch, but also will judge the reputation of the provider. Referencing the result of last round negotiation, people will modify all the data sources' reliability. This kind of reliability assignment will facilitate people to make a more sensible, comprehensive, and mature choice during the course of consensus building.

In this paper, we want to integrate this evaluation process into the CCM. We believe it is an intuitive comprehension of the dynamic and mutual belief estimation between people in the real decision-making process.

16.3 Demspter–Shafer Theory Based Consensus Building Model in CCM

Nature is the most excellent scientist. Even if to some very knotty problem, people always can find the answer from nature. Radar, helicopter, and submarine etc. are no doubt the achievement inspired by nature. As to software systems, we still keep simulating nearly all kinds of social activities in the computer. That's also the inspiration we obtain to propose the following model.

16.3.1 The Formal Definition of Model: CBM

During the process of CCM, specifications experts made usually show some degree of uncertainty. So it is necessary for experts to negotiate with each other to build consensus. Every expert will also build his own reputation matrix in which other experts' trustworthiness values will be set and modified according to the result of last round discussion.In order to capture all the information above, Dempster–Shafer theory based consensus building model CBM is proposed in this paper.

The consensus building model CBM that we propose is based on the degree of experts' certainty towards their expressed specifications. To capture the degree of certainty, we employ two kinds of structure both based on Dempster–Shafer theory. One is a two-dimensional reputation matrix which expresses to what extent participants believe with each other. The other structure allows experts with varying roles to express their opinions with varying degrees of belief. The exploitation of these two kinds of structure can help us to reason about different specifications that are coming from multiple sources.

CBM's formal definition is defined as a 5-tuple:

CBM = (*R*, *P*, *E*, BE, BP).

1. *R*: In CBM, *R* represents the roles experts can play like specialist or novice. *R* can be defined as a 2-tuple:

$$R = (rname, bbelief),$$

wherein,

- rname is the name of a role;
- bbelief is the basic belief assignment of rname.

2. *P*: *P* represents people (experts) who involves into the CCM process. *P* is defined as a 2-tuple:

$$P = (pname, RoleSet),$$

wherein,

- pname is the name of an expert;
- RoleSet is a role set which can be defined as $\{r_1, r_2, \ldots, r_m\}$. Every r_i ($i \in \{1, 2, \ldots, m\}$) belongs to *R*.
- Here the term "role set" just has the same meaning as what sociologists define it. Consider a computer field expert for instance, involves one role as a specialist, another as an arbitrator. Although one expert can take multi-roles during the process of CCM, it is assumed that only one role is active at any given point in CBM.

3. *E*: *E* represents elements in concept model. *E* can be defined as a 2-tuple:

$$E = (ename, etype),$$

wherein,

- ename is the name of an element;
- etype is the category which the element with ename belongs to. For example, in ER model etype will include entity, attribute and relationship:

$$etype = entity, attribute, relationship$$

4. *BE*: Enlightening from the group of Sun's work in real-time cooperative editing systems [27], CBM will use the similar data structure BE to the state vector adopted in sun's work. Such state vector is used to record the belief assignment providing by *P* with rolse *R* on *E*. BE is defined as a 4-tuple:

$$BE = (E, P, R, bvalue),$$

wherein,

- bvalue represents the belief mass expert P assigned to element E with role R.

5. *BP*: BP is a structure to simulate the mutual belief evaluation activities between all experts. It is a belief matrix. If there are n experts participating into this CCM process, BP should be a n×n matrix:

$$\mathrm{BP} = \begin{pmatrix} \mathrm{BP}_1^1 & \mathrm{BP}_2^1 & \cdots & \mathrm{BP}_n^1 \\ \mathrm{BP}_1^2 & \mathrm{BP}_2^2 & \cdots & \mathrm{BP}_n^2 \\ \cdots & \cdots & \cdots & \cdots \\ \mathrm{BP}_1^n & \mathrm{BP}_2^n & \cdots & \mathrm{BP}_n^n \end{pmatrix}.$$

The cell locating at row i and column j, denoted by BP_j^i, represents to what extent expert p_i believe in expert p_j. P_i's reputation value is defined as BP_i.

- If $\mathrm{BP}_j^i = 1.0$, it means p_i fully believes p_j.
- If $\mathrm{BP}_j^i = 0$, it means p_i fully disbelieves p_j.
- If $\mathrm{BP}_j^i = x$ ($x \in [0,1]$), then means p_i partially believes p_j.
- Normalization work is used to form BP_i:

$$\mathrm{BP}_i = \frac{\sum_{j=1}^{n} \mathrm{BP}_i^j}{\sum_{k=1}^{n} \sum_{j=1}^{n} \mathrm{BP}_k^j},$$

and

$$\sum_{i=1}^{n} \mathrm{BP}_i = 1$$

Actually, BP is a reputation matrix, indicating the degree of trustworthiness between experts. Based on BP, people can decide to what extent they will identify with others.

16.3.2 The Overall Flow of CBM

The overall flow of the proposed model can be seen in Fig. 16.1. It is assumed that experts work on the same one model specifications, and annotate them with proper belief values. In order to support one model view, some kind of mechanism like locking is needed. The restriction is that a common application vocabulary is

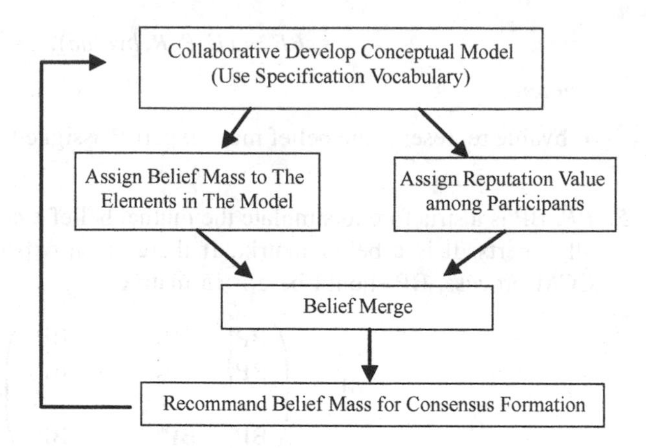

Fig. 16.1 The overall flow of CBM

needed for expressed specifications. This is required so that specification overlaps are detected. These specifications are then automatically converted into BE structure. By analyzing BE and BP structures, formal recommendations on belief adjustment are given to elements in E in order to guide them towards consensus. Aroused by merged belief values, model elements may be modified and belief mass both on elements in E and participants may be revised. Once experts believe that they have reached an acceptable level of agreement or mediator terminate such iteration process, the specifications are merged and the reliability of the expressions of each element in E is computed. Therefore, this process is repeated until a stable merged model is obtained.

In CBM, it is assumed that no task division advanced is needed and the terminology repository is incrementally developed and shared simultaneously by the group. To resolve the conflict between terminologies, a general vocabulary set should be defined so that specification overlaps can be detected, merged or pruned.

16.3.3 The Merge Procedure

In this section, we will describe how conceptual models originating from different experts can be merged on the basis of negotiation, consensus building, and model pruning.

Take database conceptual modeling as an example, analysts will collaboratively create a conceptual model (ER model) to express the relationship between teachers, students and courses.

In Fig. 16.2, conflicts can be easily detected. It is recommended that these inconsistencies should be tolerated until whenever a final decision must be made on the conclusive conceptual model. Inconsistencies and conflicts arise as a result of the difference in the belief mass assigned to different E by different P. The conflict may be resolved through the consensus building process. In model CBM, the employed consensus building process is a recommendation scheme which is defined as follows:

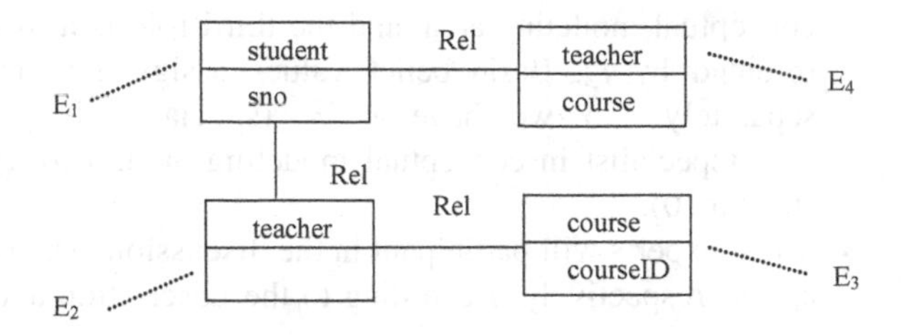

Fig. 16.2 Different opinion about a single model

Step 1 Create or modify elements in E to build a final conceptual model

Step 2 Construct BE. Expert assigns the belief mass to the elements in E defined by him

Step 3 Assign belief mass to each other. Participants' role will affect the belief mass they set to themselves. Compute everyone's reputation value by normalization method described later

Step 4 Use Dampster–Shafer theory to merge belief mass employed on elements in set E

Step 5 After compare with belief merge results, steps 1–4 will be repeated until the situation under what a ultimate conceptual model must be made. participants may modify the conceptual model, belief mass to elements in E, and reputation value to all analysts.

When every expert's reputation value is calculated, consider the role they take, we should revise $\mathrm{BP}_i^i (1 \leq i \leq n)$ before normalization:

$\mathrm{BP}_i^i = \mathrm{bbelief} \times \mathrm{BP}_i^i.$

After revision, every expert's reputation is normalized. As to the merging belief mass of element in E, we will employ Tin Kam Ho's [15] interpretation in the process of evaluating data sources' reliability. This combination rule is a kind of weighting method:

$$m(A) = \sum_{1}^{n} R_i \times m_i(A).$$

In above formula, R_i is the reliable factor of evidence and it is BP_i in the model CBM. According to the value of Bel(A), it's clear to see that to what extent every proportion is supported and participants can choose what to do in the next negotiation.

16.3.4 Example

Before describing this process, we have the flowing assumption:

- There are three roles which participants can take. Wherein, the role of specialist in computer teaching area represented by r_1, r_2 is a role of specialist in

conceptual modeling area, and the third role is a fresh graduated student represented by r_3. Basic belief values assigned to them are 1.0, 1.0, and 0 separately. So we have r_1 = (specialist in computer teaching, 1.0), r_2 = (specialist in conceptual modeling area, 1.0), and r_3 = (fresh graduated student, 0).
- Three experts will participant in the discussion. They are represented by p_1, p_2, and p_3 respectively. According to the description above, it is supposed that

$$\text{Roleset} = \{r_1,\ r_2,\ r_3\},$$

and

$$P = \{(p_1, \{r_1\}), (p_2, \{r_2\}), (p_3, \{r_3\})\}.$$

- Elements appearing in the specification will belong to E. For simplification, only three different types are assumed existing in the set etype, they are etype = {entity, relationship, attribute}.

As shown in Fig. 16.2, there are 7 elements in Set E:

$$\begin{aligned} E_1 &= (\text{student, entity}),\\ E_2 &= (\text{teacher, entity}),\\ E_3 &= (\text{course, entity}),\\ E_4 &= (\text{course, attribute}). \end{aligned}$$

And BE = $\{\text{BE}_i \mid 1 \le i \le 7, i \in N\}$. BE_i is given by three participants as follows:

$$\begin{aligned} &\text{BE}_1 = (E_1, p_1, r_1, 1), \text{BE}_2 = (E_2, p_1, r_1, 1),\\ &\text{BE}_3 = (E_3, p_1, r_1, 1), \text{BE}_4 = (\text{E}_4, p_1, r_1, 0),\\ &\text{BE}_5 = (E_1, p_2, r_2, 1), \text{BE}_6 = (E_2, p_2, r_2, 1),\\ &\text{BE}_7 = (E_3, p_2, r_2, 1), \text{BE}_8 = (E_4, p_2, r_2, 0),\\ &\text{BE}_9 = (E_1, p_3, r_3, 1), \text{BE}_{10} = (E_2, p_3, r_3, 1),\\ &\text{BE}_{11} = (E_3, p_3, r_3, 0.5), \text{BE}_{12} = (E_4, p_3, r_3, 0.5). \end{aligned}$$

In the first round negotiation, two-dimensional reputation matrix is assumed like

$$\text{BP} = \begin{pmatrix} 1.0 & 1.0 & 0.5 \\ 1.0 & 1.0 & 0.5 \\ 1.0 & 1.0 & 1.0 \end{pmatrix}$$

And after normalization, we have

$$\text{BP}_1 = 0.4,\ \text{BP}_2 = 0.4, \text{ and } \text{BP}_3 = 0.2.$$

Combination with weighting, we know

$$m(E_1) = 1.0, m(E_2) = 1.0, m(E_3) = 0.9, \text{ and } m(E_4) = 0.1.$$

so it is can be calculated that

$$\text{Bel}(E_1) = 1.0, \text{Bel}(E_2) = 1.0, \text{Bel}(E_3) = 0.9, \text{Bel}(E_4) = 0.1.$$

From these belief values, it is suggested that in the wish of the majority "course" should be taken as an entity, not an attribute. P_3 may revise his (her) belief mass assigned to E_4 or delete E_4 directly.

16.4 Conclusions

In this paper, we propose an consensus building model CBM for belief merging in CCM which is inevitably tarnished by uncertainty of multi-source information. In this merging process we integrate an iterative, mutual evaluating and role-based belief revision process into belief merging which is the base of model merging. Belief merging is based on Dampster–Shafer theory. We take participant's reputation as the weight of belief mass of elements in E. In our model, we conceive that the reliability belief we assume to others will influence our judgments to their opinions. Integrating the dynamic, mutual and role-based iterated revision belief into the CCM will be helpful to decision-pruning and speed up convergence. But problems will easily emerge if we simply pursue convergence. In that case, some information proved to be very useful later may be abandoned in the beginning. The role participant takes should be allowed to change. Another restriction is that a common application vocabulary is needed for expressed specifications. This is required so that specification overlaps are detected. These problems need us pay more attention to in our future work.

Acknowledgments The authors graciously acknowledge the support of China Scholarship Council and the funding from the National 863 Hi-tech Project (No. 2007AA12Z147), People's Republic of China.

References

1. Cook C, Churcher N (2006) Constructing real-time collaborative software engineering tools using CAISE, an architecture for supporting tool development. In: ACSC'06: Proceedings of the 29th Australasian computer science conference, Australian computer society, Inc., Hobart, Tasmania, Australia pp 267–276
2. Zave P, Jackson M (1997) Four dark corners of requirements engineering. J ACM Trans Softw Eng Method 6:1–30
3. Leite JCSP, Freeman PA (1991) Requirements validation through viewpoint resolution. J IEEE Trans Softw Eng 17:1253–1269
4. Chen PP, Thalheim B, Wong LY (1999) Future directions of conceptual modeling. J Lect Notes Comput Sci 1565:287–301
5. Zhu H, Zhou M (2008) Roles in information systems: a survey. J IEEE Trans Syst Man Cybern Part C: Appl Rev 38:377–396
6. Guzdial M, Rick J, Kerimbaev B (2000) Recognizing and supporting roles in CSCW. In: CSCW'00: Proceedings of the 2000 ACM conference on computer supported cooperative work, ACM, Philadelphia, Pennsylvania, USA, pp 261–268
7. Smith RB, Hixon R, Horan B (1998) Supporting flexible roles in a shared space. In: CSCW'98: Proceedings of the 1998 ACM conference on computer supported cooperative work, ACM, Seattle, Washington, USA, pp 197–206
8. Steimann F (2000) On the representation of roles in object-oriented and conceptual modelling. J Data Knowl Eng 35:83–106
9. Greenberg S (1991) Personalizable groupware: accommodating individual roles and group differences. In: ECSCW'91: Proceedings of the second conference on European conference

on computer-supported cooperative work, Kluwer Academic Publishers, Amsterdam, The Netherlands, pp 17–31

10. Amgoud L, Kaci S (2005) An argumentation framework for merging conflicting knowledge bases–the prioritized case. J Lect Notes Comput Sci 3571:527–538
11. Sentz K, Ferson S (2002) Combination of evidence in Dempster-Shafer theory. Technical report, Sandia National Laboratories
12. Ferson S, Kreinovich V, Ginzburg L, Myers DS, Sentz K (2003) Constructing probability boxes and Dempster-Shafer structures. Technical report, Sandia National Laboratories
13. Shafer G (1976) A mathematical theory of evidence. Princeton University Press, New Jersey
14. Yager RR (1987) On the dempster-shafer framework and new combination rules. J Inf Sci 41:93–137
15. Ho TK, Hull JJ, Srihari SN (1994) Decision combanation in multiple classifier systems. J IEEE Trans Patter Anal Mach Intell 16:66–75
16. Ram S, Ramesh V (1998) Collaborative conceptual schema design—a process model and prototype system. J ACM Trans Info Syst 16:347–371
17. Hayne S, Ram S (1995) Group data base design: addressing the view modeling problem. J Syst Softw 28:97–116
18. Pottinger R, Bernstein PA (2003) Merging models based on given correspondences. In: 29th international conference on very large data bases, VLDB endowment, vol 29, pp. 826–873
19. Altmann J, Pomberger G (1999) Cooperative software development: concepts, model and tools. In: Technology of object-oriented languages and systems (TOOLS-30), IEEE Press, Santa Barbara, CA, USA, pp 194–207
20. Bagheri E, Ghorbani AA (2008) A belief-theoretic framework for the collaborative development and integration of para-consistent conceptual models. J Syst Softw 82:707–729
21. Nuseibeh B, Kramer J, Finkelstein A (1994) A framework for expressing the relationships between multiple views in requirements specification. J IEEE Trans Softw Eng 20:760–773
22. Finkelstein A, Kramer J, Nuseibeh B, Finkelstein L, Goedicke M (1992) Viewpoints: a framework for integrating multiple perspectives in system development. Int J Softw Eng Knowl Eng 2:31–57
23. Sabetzadeh M, Easterbrook S (2006) View merging in the presence of incompleteness and inconsistency. J Requir Eng 11:174–193
24. Booth R (2006) Social contraction and belief negotiation. J Inf Fusion 7:19–34
25. Whitehead J (2007) Collaboration in software engineering: a roadmap. In: 2007 future of software engineering. IEEE Press, Minneapolis, MN, USA, pp 214–225
26. Leland MDP, Fish RS, Kraut RE (1988) Collaborative document production using quilt. In: Proceedings of the 1988 ACM conference on computer-supported cooperative work. ACM, Portland, Oregon, USA, pp 206–215
27. Sun C, Jia X, Zhang Y, Yang Y, Chen D (1998) Achieving convergence, causality preservation, and intention preservation in real-time cooperative editing systems. J ACM Trans Comput Hum Interact 5:63–108

Part IV
HumanCom 2011 Session 3: Network or Distributed Algorithms, Applications

Chapter 17
Research Application of the Internet of Things Monitor Platform in Meat Processing Industry

Tao Hu, Minghui Zheng and Li Zhu

Abstract Food safety is closely related to health. How to use information technology to ensure food safety has become a hot research topic. In this paper, we regarded the meat processing industry as the research subject. According to food safety during sausage production process, we built a monitor platform based on internet of things through video surveillance technology, sensor networks and GPS. Based on Hazard analysis critical control point, we determined the key control points of sausage production process firstly, and then we used the internet of things monitor platform monitoring all the critical control points (CCPs), to ensure the food safety during sausage production process and provide food trace function. The internet of things monitor platform is a new idea for food safety in the meat processing industry.

Keywords Internet of things · HACCP · Food safety · Video surveillance · Sensor networks · GPS · Food trace

T. Hu (✉) · M. Zheng · L. Zhu
School of Information Engineering, Hubei University for Nationalities,
Enshi, Hubei, People's Republic of China
e-mail: hutao_505@hotmail.com

M. Zheng
e-mail: mhzheng3@gmail.com

L. Zhu
e-mail: lier_zhu@163.com

J. J. Park et al. (eds.), *Proceedings of the International Conference on Human-centric Computing 2011 and Embedded and Multimedia Computing 2011*, Lecture Notes in Electrical Engineering 102,
DOI: 10.1007/978-94-007-2105-0_17,

17.1 Introduction

Many serious food safety accidents have happened in recent years. Ireland and Belgium government had informed that they detected Dioxin in fresh meat and deep processing meat products, and some China meat processing companies used the fresh meat which had been Clenbuterol pollutions to produce hams. Dioxin and Clenbuterol are serious threats to human health. Now, more and more organizations are concerned about food safety issues.

Many scientists proposed some methods to ensure the safety of meat produce and process, such as food safety traceability system. Food safety traceability system use automatic identification and IT technology to monitor and record food cultivation, processing, packaging and transportation information. Customer and management can query information those information through internet or mobile phone. Wilson and Clarke [1] used the internet to build food safety traceability system, and then Regattieri et al. [2] proposed the general framework and experimental evidence to trace food products. Jones et al. [3] used the RFID in food traceability, and some scientist's research on food safety traceability based on two-dimensional bar code.

Either RFID or two-dimensional bar code cannot monitor the food processes directly. Taking into account the widespread consumption of meat products, we propose a new monitor model to ensure the meat products safety. Firstly, we use hazard analysis critical control point (HACCP) to analysis sausage production process, and then determine the critical control points (CCPs) of sausage production. To each CCP, we use video surveillance technology, sensor networks, or GPS to monitor. Finally, we build the internet of things monitor platform in sausage production process. All key data is stored in monitor platform; the platform has the ability to trace meat from raw materials to product.

17.2 Key Technique

17.2.1 HACCP

HACCP is a management system in which food safety is addressed through the analysis and control of biological, chemical, and physical hazards from raw material production, procurement and handling, to manufacturing, distribution and consumption of the finished product [4]. HACCP is considered the most economical way to control the diseases caused by food. And it accessed recognition by FAO/WHO codex alimentarius commission (CAC). CCP is a step at which control can be applied and is essential to prevent or eliminate a food safety hazard or reduce it to an acceptable level.

The HACCP system consists of seven principles [4]:

1. Conduct a hazard analysis
2. Determine the critical control points (CCPs)

3. Establish critical limit(s)
4. Establish a system to monitor control of the CCP
5. Establish the corrective action to be taken when monitoring indicates that a particular CCP is not under control
6. Establish procedures for verification to confirm that the HACCP system is working effectively
7. Establish documentation concerning all procedures and records appropriate to these principles and their application.

17.2.2 Internet of Things Monitor Platform

Internet of things was defined as "A world where physical objects are seamlessly integrated into the information network, and where they, the physical objects, can become active participants in business processes. Services are available to interact with these 'smart objects' over the internet, query their state and any information associated with them, taking into account security and privacy issues." [5]. Lots of technologies can be used in internet of things to together internet-enable objects, such sensor network, RFID, etc.

In this paper, three technologies are used to build an internet of things monitor platform which can monitor the sausage production process, detect the temperature and humidity of sausage production facilities, and monitor the sausage production transportation. The three technologies are described as follows.

Video Surveillance Technology. Use number of camera to monitor the scenes in real time. Through the video surveillance, we can get image and voice from the scene. And we can analysis the image and voice to get some important information.

Senor Network. Senor network is composed of many front-end sensors and network protocol. Use senor to collect data from monitor scenes, such as temperature and humidity. And use network protocol to transfer data to data process system.

GPS. The GPS is a satellite-based navigation system made up of a network of 24 satellites placed into orbit. GPS satellites circle the earth twice a day in a very precise orbit and transmit signal information to earth. GPS receivers take this information and use triangulation to calculate the user's exact location.

17.2.3 Food Safety Traceability System

Food traceability refers to record-keeping systems that provide the ability to identify the path and the history of an animal, food product, or food ingredient through food supply chain.

Table 17.1 Sausage production processes

Critical control points	Is critical control points?
Inventory	Yes
Thaw	No
Cutting	No
Pickling	Yes
Ground meat	No
Chopping	Yes
Bottling	No
Cooking	Yes
Smoking	Yes
Cooling	No
Packaging	No
Transportation	Yes

Quick product recall and identification of all sourced components are vital to ensure fast product recall and help limit damage to your reputation. The ability to trace and follow a food, feed, food producing animal or substance through all stages of production and distribution is vital for consumer's safety.

Food safety traceability system can provide functions as below:

1. Traceability systems improve management on the supply side by increasing efficiency of process monitoring and assurance.
2. Traceability systems complement existing food safety and quality assurance systems by enhancing trace back and recall abilities.
3. Traceability systems enable the differentiation of food products or ingredients with subtle or undetectable content and process attributes through the need for raw material tracking and/or process identification, preventing product misrepresentation and potentially providing market advantages.

17.3 Analysis Sausage Production Processes Based on HACCP

To describe the internet of things monitor platform in meat processing industry application, we analysis sausage processes based on HACCP. Producing sausage needs a long cycle and many complex processes. If the producer cannot control the product processes well, the quality of sausage cannot be guaranteed.

Using HACCP system to analysis sausage production processes and then determining the CCPs of sausage production. Through controlling those CCPs, we can minimize the dangers of sausage production. it can avoid the product quality risks through testing final product. Control and manage the CCPs can make sausage quality to meet national food safety standards.

Sausage production processes are divided into 12 parts. But not all parts are CCPs. Using the seven principles of HACCP system, we can find out which parts are CCPs. Table 17.1 shows the result of HACCP analysis.

Table 17.2 Critical control points in sausage production

Critical control points	Critical environment requirements
Inventory	Temperature: <−10°C Relative humidity: 80–90%
Pickling	Temperature: −1–5°C Relative humidity: >85% Time: 24–72 h
Chopping	Start temperature: 5°C End temperature: 12°C
Cooking	Water temperature in beginning stage: 100°C Water temperature in intermediate stage: 85°C Time: 10–20 min
Smoking	Temperature: 50–80°C Relative humidity: 40% Time: 10 min–24 h
Transportation	Time: 24–48 h Transport routes

Each CCP has specific environmental requirements, such as temperature, humidity, and storage time. The production CCPs and corresponding environmental requirements are summarized as Table 17.2.

17.4 Internet of Things Monitor Platform for Sausage Product

To guarantee sausage product quality, we use the internet of things monitor platform to monitor and control the CCPs. Consider each CCPs have specific environment requirements; we use three technologies in our monitor platform.We use video surveillance technology to monitor CCP's scene. The management can observe whether the worker's operator meets production requirements. It can guarantee the whole production chain from outside environment too.

Using sensor network, a factory can access environment data from each CCP, such as production temperature and humidity. If CCP temperature or humidity exceeds the environment requirement of producing sausage, sensor network will release warning information in real time. When management departments get this information, they can adjust production environment timely and make the whole production chain in a secure environment.

Using GPS position system; we can ensure the whole transport process is in controlled status. Factory can record the transport routes and time through GPS. Monitoring the whole transportation process can guarantee the safety of food in circulation.

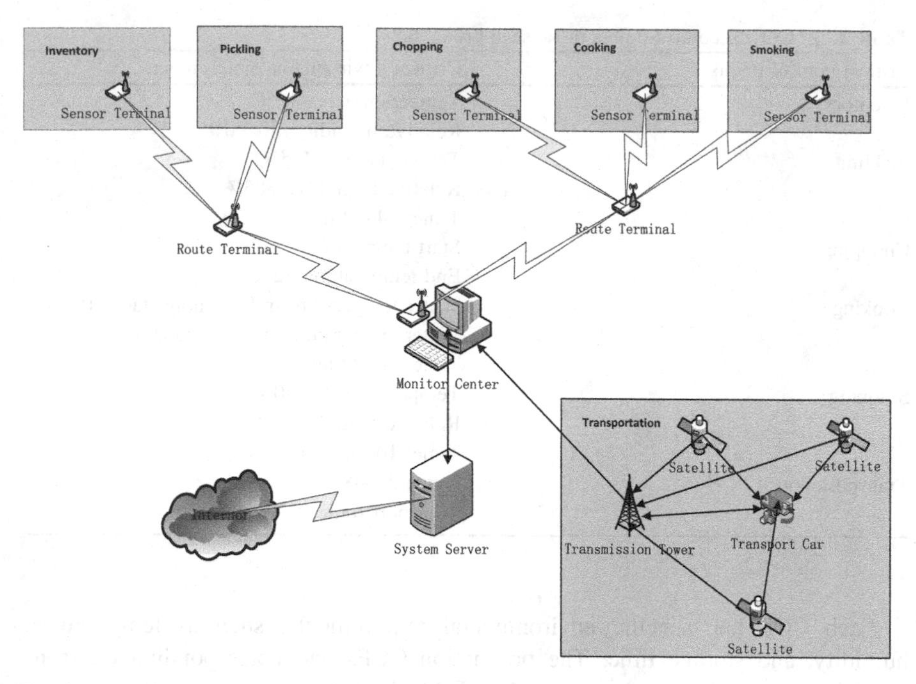

Fig. 17.1 Things of internet monitor platform block diagram

The internet of things monitor platform save the total data of sausage production, which contain raw materials, production process and transportation. If in the event of sausage product safety accident, management department can quickly position which process lead to the accident through internet of things monitor system which can do a strong support to deal with accident and improve the production process. Figure 17.1 shows internet of things monitor platform block diagram.

Monitor platform uses sensor node to access information from inventory process etc. Sensor node integrates temperature sensor, humidity sensor and camera. The sensor node also can time those processes. And platform uses GPS to monitor the transportation process. When monitor platform finds the production processes have problem, it will send warning message to management department from monitor center. All data are stored in system server. Users can trace the food information through web. Figure 17.2 shows the process of operation of the internet of things monitor platform for capturing monitor data, sending alarm message and providing food trace function.

Sensor node integrates one temperature sensor, one humidity sensor and one camera. Using sensor node can capture temperature, humidity and video information.

Sensor node use wireless module to captured data to monitor center. Figure 17.3 shows the sensor node block diagram.

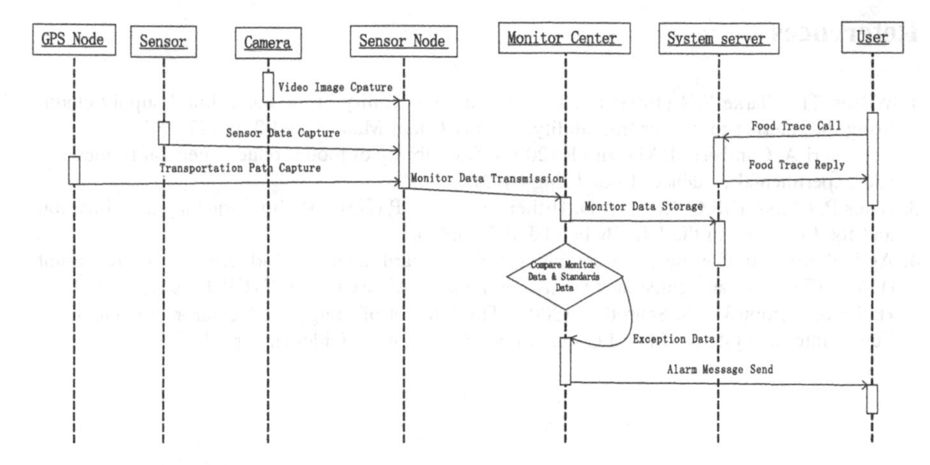

Fig. 17.2 Things of internet monitor platform monitoring operation

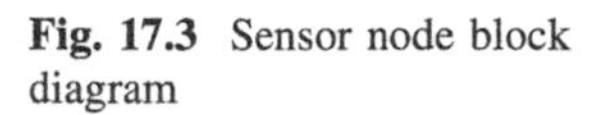

Fig. 17.3 Sensor node block diagram

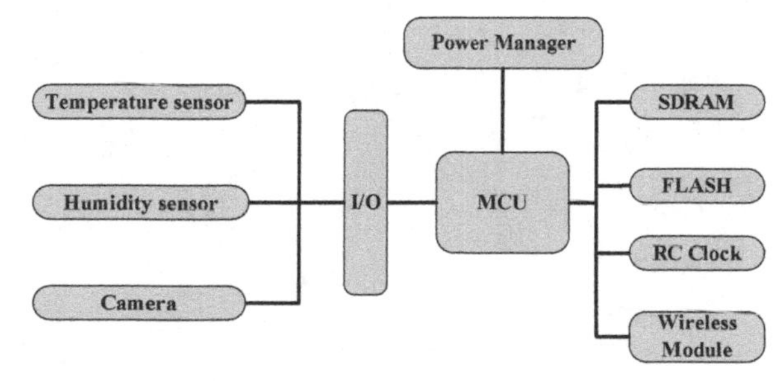

17.5 Conclusions

In internet of things monitor platform is proposed in the paper. We use HACCP system to analysis sausage production process and determine the CCPs. We also use video surveillance technology, sensor networks and GPS to build internet of things monitor platform. Monitor can capture the key data from sausage production CCPs, such as temperature, humidity and video image. Monitor platform also provides sausage product transportation route function and can trace sausage production process. Using internet of things monitor platform can guarantee sausage product quality, which provides new ideas for food safety in meat processing industry.

References

1. Wilson TP, Clarke WR (1998) Food safety and traceability in the agricultural supply chain: using the internet to deliver traceability. Supply Chain Manag Int J 3(3):127–133
2. Regattieri A, Gamberi M, Manzini R (2007) Traceability of food products: general framework and experimental evidence. J Food Eng 81(2):347–356
3. Jones P, Clarke-Hill C, Comfort D, Hillier D, Shears P (2005) Radio frequency identification and food retailing in the UK. Br Food J 107(6):356
4. Agriculture and Consumer Protection (1997) Hazard analysis and critical control point (HACCP) system and guidelines for its application, Annex to CAC/RCP 1–1969, Rev. 3
5. Haller S, Karnouskos S, Schroth C (2008) The internet of things in an enterprise context. In: Future internet systems (FIS). LCNS, vol 5468. Springer, Heidelberg, pp 14–18

Chapter 18
S-Rsync: An Efficient Differencing Algorithm with Locally Chunk Digests Generating for File Synchronization Services

Hang Zhang, Nong Xiao, Fang Liu, Huajian Mao and Xianqiang Bao

Abstract Data synchronization among different clients is common place in the information era. Ideally, we expect that the synchronization process transmits less data and achieves lower response time. To meet these requirements, synchronization systems usually employ differencing algorithms to reduce the volume of data to be synchronized. Rsync is a typical algorithm of this type, which is used widely. However, for the file synchronization services adopting Rsync, the server may always busy for generating chunk digests, which probably will be the bottleneck. We propose an improved differencing algorithm called S-Rsync. S-Rsync migrates a portion of computation from server to client by taking use of the duplication mechanism. This optimization not only eases the server, but also reduces the amount of data to transmit. As a result, synchronization system using S-Rsync algorithm achieves lower response time. Based on the rationale described above, we further improve S-Rsync in term of both spatial and time complexities. Correspondingly, we propose a dynamical selection scheme, which selects a proper algorithm between S-Rsync and Rsync depending on maintaining the versions of files. We construct a prototype to

---

H. Zhang (✉) · N. Xiao · F. Liu · H. Mao · X. Bao
Department of Computer Science and Technology, National University of Defense Technology, 410073, ChangSha, China
e-mail: zhanghang0259@gmail.com

N. Xiao
e-mail: nongxiao@nudt.edu.cn

F. Liu
e-mail: liufang@nudt.edu.cn

H. Mao
e-mail: huajianmao@nudt.edu.cn

X. Bao
e-mail: baoxianqiang@gmail.com

J. J. Park et al. (eds.), *Proceedings of the International Conference on Human-centric Computing 2011 and Embedded and Multimedia Computing 2011*, Lecture Notes in Electrical Engineering 102,
DOI: 10.1007/978-94-007-2105-0_18,

evaluate our proposed algorithm. The experimental results show that our algorithm outperforms Rsync constantly *abstract* environment.

Keywords Synchronization · Differencing algorithm · Rsync

18.1 Introduction

Driven by the developments of computers and networks, a variety of portable devices (e.g. notebooks, PDAs) enter into people's daily life. The prevalence of portable devices introduces into some substantial challenges. One important challenge is the data synchronization among these devices. Generally, portable devices interconnect with each other. User's data may be shared by them. Updating these shared data is a challenging issue that has attracted many research interests.

In order to solve the problem described above, software aiming at file synchronization is developed. These specific software tools automatically synchronize files locating at different devices. This function supplies much convenience for those people who change their work environments frequently. Supposing a scenario described as follows, Bob has not finished his work until the official time is over. He hurries to home, launches his personal host and finds that the unaccomplished work has already been synchronized to the host at home. He can continue his work immediately. File synchronization software is so useful that many corporations pay attention to it. Corresponding productions with great impacts include Dropbox [1], LiveSync [2], Everbox [3] and so on.

Taking Dropbox for an example, it designates a folder for synchronization at the client end. Files in the given folder are monitored automatically. Once the file on client is updated, corresponding program submits the new version to server. As a result, other authorized clients are able to access the updated data immediately. The implicit synchronization makes it convenient to share information.

However, one of the most important challenges of synchronization involves data transmitting. Once a file is updated, it will be synchronized to some client sooner or later. An alternative scheme is to transmit the entire file to these clients arbitrarily, even if the file only receives some trivial updates. Apparently, if files are updated frequently, this scheme requires unreasonable network bandwidth. To decrease the bandwidth consumption, most file synchronization software just transmit the difference between two versions. In other words, only the fraction of data that has been changed is transmitted from client to servers. However, this optimization must be supported by some algorithms that are used to compute the difference among different copies of data.

Rsync [4, 5] is one kind of these algorithms with high performance. Considering the copy on server, Rsync partitions it into several chunks with fixed block size. For each chunk, Rsync employs a hash function to calculate its hash value, like MD4 [6], and then send these hash values to the client, usually called chunk digests. Synchronization client makes use of the chunk digests to determine whether a chunk of the file on client had been updated. Only these updated chunks are transmitted through network. Therefore, the bandwidth requirement is reduced. Rsync is extensively adopted by data synchronization and deduplication due to its simplicity and superior performance.

However, the heavy load of the server will be a bottleneck for the synchronization system adopting Rsync algorithm. Once a file is to be synchronized, the server must calculate hash values for all chunks of the file, and send them to the client through network. Even more, a server supplies service for numerous clients. These clients may request for synchronization to the server simultaneously. So, the server will be too busy to calculate chunk digests and to distribute these data to clients in time. With the observation that, after one file is updated, the copies in both client and server are consistent of the same content. When the file receives its updates, the corresponding client is able to calculate the chunk digests locally without server's help. Based on this observation, we propose a new synchronization method, S-Rsync, which overcomes the challenges by dispensing the calculation overhead of chunk digests to these clients. S-Rsync calculates the difference between two versions by the client itself rather than the server. Only differences are transmitted through network. The server makes use of these differences to reconstruct the new copy for the file on client. This optimization will surely improve the process of data synchronization in two aspects. First, the server does not need to do so much computation of chunk digests any more. Second, transmitting of these chunk digests is evitable, so this will save a certain reasonable bandwidth.

Based on the discussion described above, we improve Rsync to achieve higher performance and further reduce the resource consumption and reduce overload of the server, we call it S-Rsync (Rsync used in file synchronization services). Our contributions are listed as following:

- With replication mechanism S-Rsync algorithm makes synchronization operation not need to receive chunk digests from server at beginning. In this way, it really reduces the response time of synchronizing and the overload of server, and then S-Rsync constructs difference of files that needed to be synchronized locally.
- To reduce space overhead which is affected by the replica, we store chunk digests of the synchronizing file replica instead of keeping the file itself.
- Additionally, the process of computing chunk digests can be integrated into last synchronization operation, reducing the local computing overhead.
- Finally, a scheme of adaptively selecting Rsync or S-Rsync algorithm has been stated base on the file version information on client.

Above all, S-Rsync we proposed has great performance in reducing the bandwidth consumption and response time by generating chunk digests locally for file synchronization services. Compared with R-sync, S-Rsync reduces synchronization operations response time and overload of server, and improves the scalability of the file synchronization services.

The rest of this paper is organized as follows. Section 18.2 reviews the related works. Section 18.3 introduces the original Rsync briefly, and states the problems brought by using it in file synchronization services. Section 18.4 proposes our S-Rsync algorithm. Optimizations for spatial and time complexities are introduced in the section as well. The last section concludes this paper and gives some future works.

18.2 Related Works

Recently, there are some differencing algorithms presented aiming at special applications or special occasion, with certain difference generated way. Rsync [4, 5] as one of them is widely used for synchronization, backup, recovery system, such as vary-sized blocking Rsync using in LBFS [7] with Rabin's fingerprinting method [8]. And there are also some other algorithms adopting some other consideration, one of which is Delta-encoding (Delta). Reference [9] uses "delta" vcdiff [10] encoding way to improve performance of HTTP traffic. The algorithm vcdiff is one of the best delta encoding algorithms, which is a general and portable way. Every differencing algorithm has its own advantages on the reasonable size of difference or computing overhead, according to its special scene [11]. Reference [12] systematically evaluates four bandwidth reduction algorithms with several types of files under some representative network connection technologies, and then shows the results that different approaches have different performance in terms of different metrics.

Because of the simplicity and efficiency, Rsync is used widely in synchronization system. There are some previous works to improve Rsync algorithm. A version milti-round [13] version of Rsync was proposed to figure out the similarity of two files in fine granularity, called Mrsync. The Mrsync performs well when files consist of short matching sequences separated by small differences [12]. How to reconstruct synchronize file in place is also studied aiming at making Rsync goes well on some mobile device where the space is limited by some work [14, 15]. The in place Rsync algorithm encodes the compressed representation of a file in a graph, which is then stored to help reconstructs the target file in place on space constrained devices.

18.3 Rsync Algorithm in File Synchronization

Rsync is used widely in synchronization system. Firstly, the process of how the Rsync algorithm runs will be introduced.

Rsync synchronizes two files, changes a file with old version on one server up to date with a file with old version on client. By detecting the common sections of the old version and the new version, Rsync sends as few data as possible, approximately only changed data, and then rebuilds the new version file using the common sections from old version.

18.3.1 A Introduction of Rsync Algorithm

Rsync algorithm [4] use a fixed-sized blocking scheme to divide files, first server divides file B to series chunks with fixed blocking size on server side, computes strong signature and weak signature for every chunk using a hash function to form a chunk digests file, and send it to A through network.

So we can use Rsync algorithm in file synchronization services to accelerate the process of updating file to server. Considering one scenario that we use Rsync algorithm, there are client C and server S grouped together as one file synchronizations system connected by a network. There is a file F in the system, which may have been updated to server before. So both client and server have copies of F. Client C gets a copy F_C, server S gets a copy F_S. F_C and F_S are likely to have some common regions. At a moment, the client found F_C changed to file F'_C as soon as user modified it. The file synchronization services needs to synchronize to server, for updating F_S to F'_S on server S.

Rsync algorithm is used here to synchronize file. It usually uses rolling checksum as weak signature, MD5 as strong signature. Only when the weak signatures of the two chunks match, then computes strong signatures and compares them. Here we use R and H to represent them respectively. Here comes the process of synchronizing:

1. S divides F_S into N equally sized blocks F_{Sj} and computes signatures R_j and H_j on each block. These signatures are sent to C
2. For each byte offset i in F'_C, C computes weak signature R'_j on each block starting at i
3. C compares R'_j to each R_j received from S
4. For each j where R'_j matches R_j, C computers H'_j and compares it to H_j
5. If H'_j matches H_j, then sends C a token to indicating a block match and which block matches. Otherwise C sends a literal byte to S
6. S receives literal bytes and tokens from C and uses these to construct $F_C(F_{Sj})$

The process of Rsync algorithm is shown (Fig. 18.1).

18.3.2 Problems Brought by Using Rsync in Synchronization Services

Rsync algorithm performs well at the copies of files which are inclined not to be changed frequently, such as updating Linux kernel from version 2.0.9 to 2.0.10. Rsync often is used to manage distributed replicas with faster and more efficient methods. However, the file modified by user is required to synchronize to server as soon as possible in file synchronization services. Sometimes, users may modify the file very often, so a lot of synchronize operation will be committed in a short time,

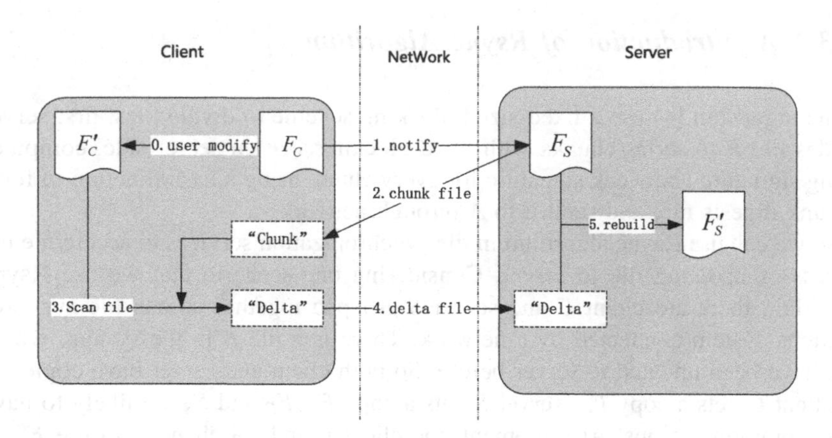

Fig. 18.1 Sychronizing file from client to server with Rsync. Rsync needs computing chunk digests first at server, and then send it to client, to help reconstucting difference

such as user is working on writing a file of Microsoft Word type, which is inclined to be saved frequently.

Let's watch the steps of Rsync carefully, we should divide file into chunks and computes both weak and strong signatures for each chunks at server side of every synchronization operation, and send them to client. However, this step is very costly for some centralized servers, especially when there are great numbers of users requiring service. Server should do this operation for every file which is required by many users. The transmission of chunk digests transferred will cost much bandwidth, and increase response time of synchronizing operation.

For the issue that files changed frequently in file synchronize services bringing great overhead, we exploit the potential to migrate some computing burden from the client to server, expecting to reduce the overload of server.

18.4 S-Rsync Algorithm

In this section, we propose S-Rsync algorithm to solve the problems brought by using Rsync in file synchronization services, taking advantage of replication mechanism.

Firstly, we will introduce the replication mechanism in S-Rsync. With this mechanism, it is possible to construct difference, "delta", locally at client, not need to get chunk digests from server.

Repliacation mechanism. In order not to get chunk digests file from sever, generating it locally, when user has modified the file F on client, from the old version F_C to new version F'_C, the old copy and new copy, should both be able to access at same time.

Since we cannot predict when user will change a file, it is necessary to replicate every file to get a replica, which is monitored for synchronization. Once user change the file for F to F', it is possible to have both F and F'. Then old version F could be used to get chunk digests locally on the client.

It is obviously that using replica will cost some extra space of client's disk, decreasing the space utility. We will further introduce some method to optimize space consuming problem later. The experiment results show the overhead of space consuming is not much with the optimization on special complexity.

18.4.1 S-Rsync Algorithm

S-Rsync has some very similar steps with Rsync, except for maintaining and taking use of replica in client. S-Rsync locally generates the difference between old version and new version, no need to get differencing with receiving chunk digests file from server first.

Assuming that after one certain synchronization operation, the file F_C on client C is the same as F_S on server . When user has changes one file from F_C to F'_C, we use S-Rsync to synchronize the file to server as following:

1. C divides F_C into equally N sized blocks F_{Cj} and computes signatures R_j and H_j on each block
2. For each byte offset i in F'_C, C computes weak signature R'_j on each block starting at i
3. C compares R'_j to each R_j received from S
4. For each j where R'_j matches R_j, C computers H'_j and compares it to H_j
5. If H'_j matches H_j, then sends C a token to indicating a block match and which block matches. Otherwise C sends a literal byte to S
6. S receives literal bytes and tokens from C and uses these to construct $F_C(F_{Sj})$
7. Rename F_C to F'_C, and make a replica of F_C

It is worthy of noticing that at a step of S-Rsync is used to generate the signatures, which is the replica of old version of file. The signatures are generated locally instead of receiving them from the server. The process of S-Rsync algorithm goes as following in Fig. 18.2:

Comparing with Rsync, S-Rsync uses the two files, one is the replica of and the other one is the file which is changed, to generate difference locally, not need to get chunk digests from server. In this way, some bandwidth can be saved and the response time of synchronization is also reduced, especially when S-Rsync runs on low-bandwidth link. S-Rsync use replication mechanism to achieve the goal, but it definitely will reduce the space utility rate, which needs 50.

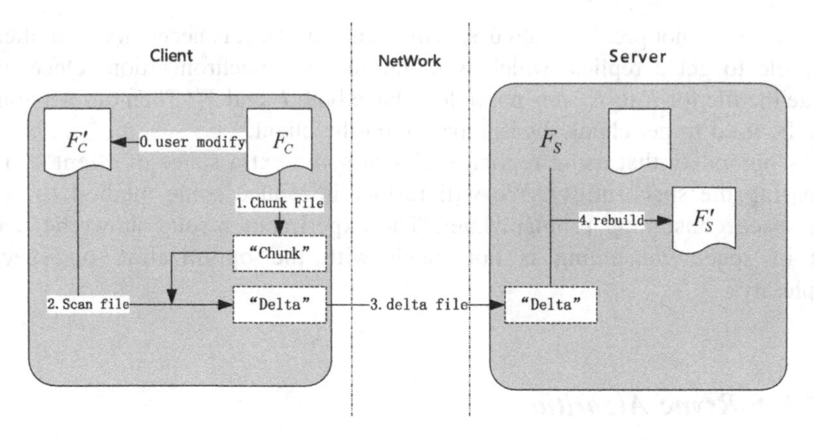

Fig. 18.2 Sychronizing file from client to server with S-Rsync. The difference, "delta" was generated locally, no need to receive the chunk digests form server first

18.4.2 Time and Spatial Complexities Optimization for S-Rysnc

The replication mechanism is the base that S-Rsync algorithm runs on. However, the extra space consuming does come along with replication. If we could dig some potential on saving enough information with less capacity from S-Rsync to reduce the capacity of replica, it is possible to increase the space utility.

Spatial complexity optimization. There should be two copies of to insure S-Rsync goes well after very synchronization operation. That means we must keep both old version F_C and new version F_C' at the same time, whenever user has modified file from F_C to F_C'. At that time the chunk digests could be constructed from F_C, and used to generating the difference together with F_C'.

Here we can see that the chunk digests file of F_C is used in the steps of S-Rsync, not the file F_C itself. So just storing the chunk digests file F_C of is already enough. As usual, the chunk digests file takes approximately 100 times less space and bandwidth than the file. Then the extra space used to store replica can be apparently decreased.

S-Rsync stores the chunk digests file F_C to get enough information when generating difference, instead of storing the file F_C itself. Usually the digests of one file is much smaller than the file itself, since the chunk file only contains the offset, weak and strong signatures information. With this improvement, storing digests instead of replica itself, the extra space consumed by replica can be dramatically reduced. That will increase the space utility rate of S-Rsync.

Time optimization. After re-examining steps of S-Rsync carefully, we found that there is one process of S-Rsync contains after one synchronization operation: deleting the copy F_C, renaming the newest version F_C' to F_C, constructing chunk digests file F_{chunk} from F_C, and waiting for next user changing operation.

If the construction of after F_{chunk} synchronized could be integrated into the process of generating difference, the time consuming by construction of F_{chunk} can be saved. S-Rsync has one step: for each byte offset i in F'_C, C computes both weak and strong signatures on each block starting at i. If the byte offset i is one of the block boundaries offsets set $\{k \mid k = File\_Size \bmod Block\_Size\}$, that indicates i will be one block boundary in the F'_C(future F_C), when generating the chunk digests of the file F'_C. Here the weak signature, strong signature of block starting at i has been omputed already, so these signatures can be stored for the chunk digests file F'_{chunk} of F'_C. That is to say, the chunk digests file F'_{chunk} of F'_C could be constructed as soon as the process of checking every offset of F_C finished.

S-Rsync with time and space optimization. The S-Rsync goes with time and space optimization:

1. Read from F_{chunk}, the chunk digests file of F_{chunk}, and get R_j and H_j for each block starting at i, then form the search table
2. create F'_{chunk}
3. For each byte offset in i, C computes weak signature R'_j on each block starting at i.
4. If H'_j matches H_j, then sends C a token to indicating a block match and which block matches. Otherwise C sends a literal byte to S
5. S receives literal bytes and tokens from C and uses these to construct $F_C(F_{Sj})$
6. If $i \in \{k \mid k = File\_Size \bmod Block\_Size\}$, then write (i, R'_j, H'_j) into F'_{chunk}
7. S receives literal bytes and tokens from C and uses these to construct $F(F'_S)$
8. Rename F'_{chunk} to F_{chunk}

One step of S-Rsync is worth paying attention on, that is when i is just in set $\{k \mid k = File\_Size \bmod Block\_Size\}$, writing the signatures R'_j and H'_j into F'_{chunk}, which have been computed already. However, one case R'_j and H'_j have been computed already is just in the situation that $block_i$ matches one chunk in the search table. In this case, S-Rsync with time optimization outperforms original S-Rsync, because of saving the computing time by avoiding calculating these already signatures. Fortunately, since the files are often changed frequently in file synchronization services, there always are many similarities between the old version and the new version. There will be plenty of common sections between the copies. Having so many chunks signatures match will make these signatures computed previously. With the computed R'_j and H'_j, S-Rsync will perform an well in these cases.

18.5 Experiment and Evaluation

Our experiment compares traditional Rsync with S-Rsync and S-Rsync improved with time and spatial optimization, by evaluating the amount of data transmission affected by different fixed block size. Results indicate that S-Rsync upgrades performance in the files changed frequently synchronization environments as we expected since it always transmits less data over the network. Although the replication mechanism decreases the capacity utility rate used by S-Rsync, our results show that it would not cost too much extra space wasting with the spatial optimization. S-Rsync outperforms traditional Rsync in easing the load of server and reducing response time of synchronization operation.

18.5.1 The Test Enviroment

We test S-Rsync in real network environment. The client and server are connected via LAN with ReiJie switch. The entire algorithms run on a desktop, with Intel 2.6 GHz CPU, 2 GB RAM, 320 GB disk, and Microsoft Windows 7 OS. The back-end server service runs on a HP DL180G6 sever, with Intel Xeon E5506 Quad processor, with 8 GB RAM, 500 GB disk, and ubantu 9.04 operating system.

We use C++ to implement algorithms of Rsync, S-Rsync and S-Rsync with optimization. Some different metrics are measured, such as the extra space consuming, the overhead of computing time, delay of synchronization, and bandwidth saving. Several types of files are synchronized with the three algorithms above.

18.5.2 Experiment Results

We use 2 KB as the fixed-blocking chunk size to test three algorithms, which is unusually chosen in traditional Rsync, as an experience value. In the Table 18.1, it lists the different file types, and the size of the files, and the size of chunk digests file. It is obviously that all the chunk digest files are much smaller than the original files.

The space requirement has been tested firstly, about the three algorithms, Rsync, S-Rsync, and S-Rsync with spatial optimization. The result is shown in Fig. 18.3. Rsync just requires the space as the size of file to be synchronized, no more extra capacity needed. To keep the replica, S-Rsync should use twice times than the size of synchronizing file. However, S-Rsync with space optimization needs much less extra space, which can be inferred from the Table 18.1. Since the chunk digests file information of one file is much smaller than itself, S-Rsync does not bring so much space overhead. The synchronization operation delay of three algorithms also has been tested, which is composed of local computing time overhead and the data transmission time overhead. The test result is shown in Fig. 18.4. It is easily to see S-Rsync outperforming Rsync, since the chunk digest file is no need to be transferred in S-Rsync. This part of delay can be avoided. With time optimization, S-Rsync additionally saves more time, for the local computing time has been shorten. So the overall delay is the smallest of the three cases. There is an interesting observation here, the fixed-blocking chunk size using to divide file

Table 18.1 The size origin file and chunk digests file of different file types

File Type	doc	ppt	pdf	source	latex
Size (KB)	2,209	14,320	855	1,294	4,724
Chunk file size (KB)	72	252	16	23	84

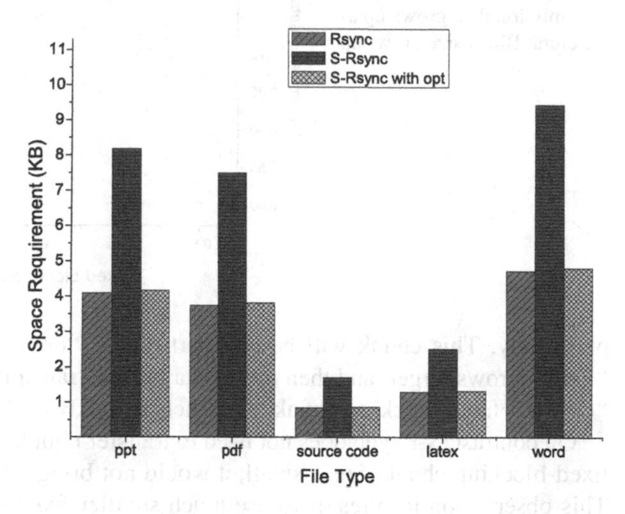

Fig. 18.3 Space requirement by three algorithms, Rsync, S-Rsync, and S-Rsync with space optimization (S-Rsync with opt)

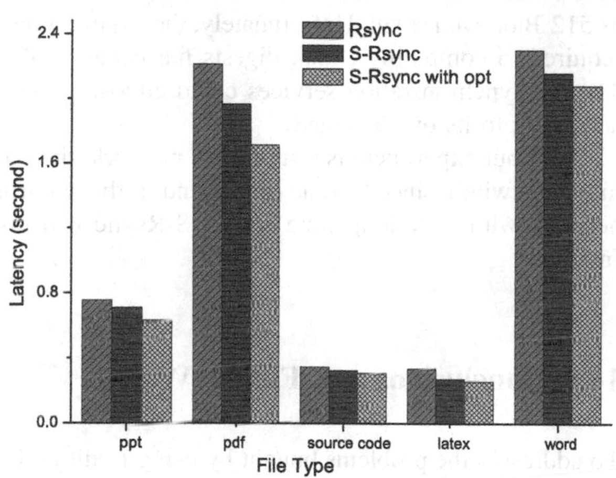

Fig. 18.4 Delay of synchronization about the three algorithms, Rsync, S-Rsync, and S-Rsync

into series blocks affects the size of both chunk digests file and "delta" difference file, as we know. So in Rsync algorithm the fixed-blocking chunk also will affect the amount of data transmission, shown in Fig. 18.5. As the fixed-blocking chunk grows larger, the amount of data transmission will decrease at first. Since the number of chunks decreases as the chunk file size grows, and the size of chunk digests file is getting smaller too.

However, when the size of fixed-blocking chunk size grows, for one chunk, it will probably contains difference of neighbor chunk, which it does not have

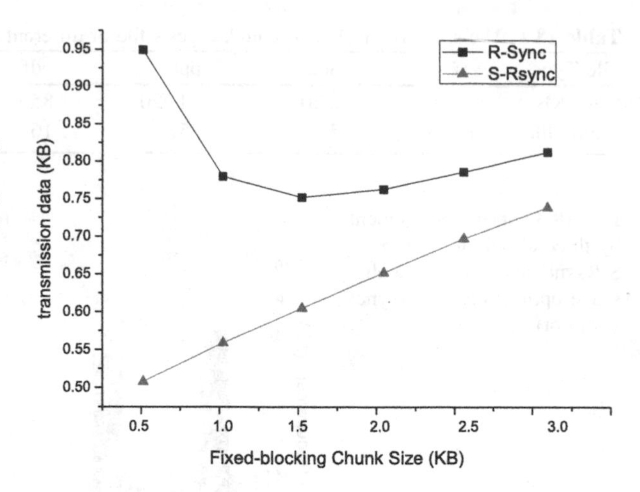

Fig. 18.5 How the size of fixed-blocking chunk affects communication amount. Fixed-blocking chunk size will affect both size of chunk file and "delta". The transmission data grows up as the clunk Block size grows up

previously. This chunk will be one part of the "delta" file, causing the size of "delta" grows larger, and then more data will be transmitted. So Rsync often use a "classic" fixed-blocking chunk size, such as 2 KB or 3 KB.

On contrast, S-Rsync does not need to transfer chunk file via network. When the fixed-blocking chunk size is small, it would not bring much more data to transmit. This observation inspires us to use much smaller fixed-blocking chunk size, such as 512 B or even small. Unfortunately, the smaller one chunk is, the more time it is required to computing chunk digests file because of the larger numbers chunks. The file synchronization services could choose proper fixed-blocking chunk size according to its own demand.

Since our experiment is test in LAN network, the improvement is limited by the high bandwidth, since the one time round in this environment is small. For the real network with very long time round, S-Rsync can do better in real world-wide Internet.

18.6 Conclusions and Future Work

To addresses the problems bought by using tradition Rsync in file synchronization services, we have described the design, implementation, and performance of S-Rsync for synchronizing files. S-Rsync is a modification to the open-source Rsync utility so that the difference of files between client and server may be constructed locally not needing to receive chunk digests file from sever previously. This could surely reduce the response time of synchronization, save bandwidth consumption for data transmission, and reduce overload of sever.

In the future, we will further improve the differencing algorithm for the file synchronization services with copy-on-write mechanism.

Acknowledgments We appreciate Professor Weisong Shi for the discussions with him, who is from Wayne State University, USA. We are grateful to the anonymous reviewers for their valuable suggestions to improve this paper. This work is partially supported by National Natural Science Foundation of China grants NSFC 60736013, NSFC 60903040, NSFC 61025009 and Program for New Century Excellent Talents in University NCET-08-0145.

References

1. Dropbox. www.dropbox.com/
2. LiveSync. https://www.foldershare.com/
3. Everbox. http://www.everbox.com/
4. Tridgell A, Mackerras P (2004) The rsync algorithm. Technical report
5. Tridgell A (1999) Efficient algorithms for sorting and synchronization. Doktorarbeit, Australian National University
6. Rivest RL (1990) The MD4 message digest algorithm. In CRYPTO 303–311
7. Muthitacharoen A, Chen B, Mazires D (2001) A low-bandwidth network file system. In: SOSP, pp 174–187
8. Rabin M (1981) Fingerprinting by random polynomials. Center for Research in Computing Techn., Aiken Computation Laboratory, Univ
9. Mogul JC, Douglis F, Feldmann A, Krishnamurthy B (1997) Potential benefits of delta encoding and data compression for HTTP. In: SIGCOMM, pp 181–194
10. Korn DG, Vo K (2002) Engineering a differencing and compression data format. In: USENIX annual technical conference, general track 219–228
11. Lufei H, Shi W, Zamorano L (2004) On the effects of bandwidth reduction techniques in distributed applications. In: EUC, pp 796–806
12. Rasch D, Burns R (2003) In-place rsync: file synchronization for mobile and wireless devices. In: Proceedings of the annual conference on USENIX annual technical conference. pp 15–15. USENIX Association, Berkeley, CA, USA
13. Langford J (2001) Multiround rsync
14. Burns RC, Long DDE (1998) In-place reconstruction of delta compressed files. In: PODC, pp 267–275
15. Burns RC, Stockmeyer LJ, Long DDE (2003) In-place reconstruction of version differences. IEEE Trans Knowl Data Eng 15:973–984
16. Shi W, Santhosh S, Lufei H (2004) Cegor: An adaptive distributed file system for heterogeneous network environments. In: ICPADS, pp 145–152

Chapter 19
VESS: An Unstructured Data-Oriented Storage System for Multi-Disciplined Virtual Experiment Platform

Wenbin Jiang, Hao Li, Hai Jin, Lei Zhang and Yaqiong Peng

Abstract The maintenance of unstructured data for a large data-related system is still a hard nut to crack, especially in the *multi-disciplined virtual experiment platform* (MDVEP), where lots of unstructured data are produced, small in size and coming from various applications. This paper presents a new storage system named VESS for unstructured data from MDVEP, which can improve the performance of the system and simplify the development of upper-layer applications. It provides HTTP-based data request method as system user-friendly interface. Index table relying on a universal metadata file is used to ensure the consistency and integrity of replicas. Message queue is adopted to sequence modification operations and guarantee consistency, while signature-based authentication guarantees the security. Experimental results show that VESS can serve as a data storage service for virtual experiment with high performance.

Keywords Unstructured data · Virtual experiment · Data storage · Service

19.1 Introduction

With the explosive development of the information technology all over the world, the data produced by this trend become more and more various in forms, complicated in content, and enormous in quantity. They include a variety of operational logs, statistic results, documents, emails, hypertexts, pictures, audio, video files and other un-interpretable data for machines. These kinds of data

W. Jiang (✉) · H. Li · H. Jin · L. Zhang · Y. Peng
Services Computing Technology and System Lab,
Cluster and Grid Computing Lab, School of Computer Science and Technology,
Huazhong University of Science and Technology, 430074 Wuhan, China
e-mail: wenbinjiang@hust.edu.cn

J. J. Park et al. (eds.), *Proceedings of the International Conference on Human-centric Computing 2011 and Embedded and Multimedia Computing 2011*, Lecture Notes in Electrical Engineering 102,
DOI: 10.1007/978-94-007-2105-0_19,

themselves are meaningless to computers, which are treated as binary streams to transfer, process and store. They can all be called as unstructured data [1], which neither have specified data models nor obvious data semantics that can be easily used by computers. According to the survey of the Gartner Group [2], 80 percent of business is conducted on unstructured information and unstructured data doubles every three months. How to manage such massive unstructured data and construct an efficient, user-friendly data storage platform is still a hard nut to crack. Many companies and communities have been suffering from the unexpectedly growth [3, 4]. It is urgent to improve the capability of data accessing and transaction processing to provide a uniform, stable unstructured data storage platform with high throughput. It is also required by the *multi-disciplined virtual experiment platform* (MDVEP). The MDVEP contains large amounts of unstructured data such as virtual experiment components (in XML) for different disciplines, experiment scenes (in XML), documents (in TXT, DOC, etc.), videos (in mp4, avi, etc.). Since there are millions of users and more than 100 experiments in this platform, an efficient storage system with high performance is desirable.

Typically, there are two ways to store unstructured data using existing technologies or the combination of them.

The first one is storing unstructured data in a local file system, with maintaining an index table in memory or a relational database. This method can speed up read/write operations and provide high throughput. However, maintaining the index table is a tough task. It is hard to guarantee the integrity and consistency between the original data and their index information. Moreover, some additional authentication is needed to guarantee the security of a local file system.

The second one is storing unstructured data in a relational database system, always represented as Binary Large Object (BLOB) field. This method fully takes the advantages of modern database technologies, such as transaction processing, data arrangement and security issues. While under the circumstance of large amount of small files roll-in and roll-out, the relational database has limited performance of query and read/write. The optimization space is also relatively small. On the other hand, the relational database is hard to scale-out, for complex table designs and many join operations.

The objective of this research is to design and implement an efficient, user-friendly and always-available storage system for virtual experiment (VESS). The system is motivated by the following needs: developers use different operating systems and prefer using mixed languages and tools to develop end-user applications. All of these applications need to read/write unstructured data from/to backend storage system. In this circumstance, there are large amounts of unstructured data. Most of them are XML or image files, the sizes of which are relatively small.

The storage system serves as a data storage service to provide RESTful [5] uniform interfaces for reading/writing and advanced query mechanism. It is written in python which is portable and easy to maintain and deploy, providing HTTP based data request methods which are platform and developing language independent.

The rest of this paper is orgnized as followings. Section 19.2 gives the background knowledge related to this work including an overview of the data model, basic theories and the *Representational State Transfer* (REST) architecture. Section 19.3 discusses some related works. Section 19.4 describes the fundamental issues of VESS design and implementation. Features and key technologies of VESS are presented. In Sect. 19.5, measurements of VESS both in functionality and performance are given, and comparison results with file system and relational database are shown. Finally, Sect. 19.6 draws some conclusions and gives some directions for future work.

19.2 Background

19.2.1 Data Model

Inspired by Bigtable's [6] data model, we fit it into VESS using a flat map table. The map is also indexed by row keys; each value in the map is always un-interpreted bytes.

(id:string, active:int, fnv:string, class:string, timestamp:string, sysid:string, filename:string, fileext:string) → (content:string)

The key *id* is used to identify a data file in VESS. It is assigned by the system automatically. *active* indicates the data file is available or not. *fnv* is the digest of data file to verify file's integrity. *class* is the classification of data file's type. *timestamp* is used to indicate when the data file are created. *sysid* is the universal identification of a data file, which would be helpful when you migrate data. *filename* and *fileext* are not mandatory.

19.2.2 Basic Theories

Traditional database provides ACID (*Atomicity, Consistency, Isolation, and Durability*) guarantees which have poor availability [7]. According to Eric Brewer's CAP [8] (*Consistency, Availability and Partition Tolerance*) theorem, it is impossible for a distributed system to provide all the three guarantees (CAP) simultaneously. In a distributed system, partition tolerance is a basic requirement. Therefore, the design process of a distributed system is a balance between consistency and availability. VESS is designed to be an always-available service. We choose Eventually Consistent [9], since there is a time window between users' each operation and it is hard for end-users to be aware of the inconsistency. We simplify the merge of concurrent write/update of identical data using *last write wins* policy.

19.2.3 REST Architecture

REST is a style of software architecture for distributed hypermedia system, the largest REST architectural system is World Wide Web. VESS is also designed as a REST architectural system. RESTful interfaces provide uniform and user-friendly access methods for system users and end-users. Meanwhile, RESTful interfaces are stateless, and there is no context stored on the server between requests. The HTTP methods corresponding to relational database's Create, Read, Update, and Delete (CRUD) operations are:

- GET method: retrieve a representation of addressed data, expressed in an appropriate Internet media type.
- PUT method: update addressed data, if it does not exist, create it.
- POST method: create a new entry.
- DELETE method: delete the addressed data.

19.3 Related Work

Dynamo [3] is a highly available key-value storage system introduced by Amazon to provide an *always-on* experience. The design of Dynamo is to avoid centralized control as much as possible, which brought lots of consistency-related problems.

Bigtable [6] is a well-known infrastructure introduced by Google, which is a distributed storage system for managing structured data. It maintains a sparse, distributed, persistent multi-dimensional sorted map. However, Bigtable is owned by Google. It is hard for the third part to use it. Fortunately, There are some Bigtable-base open source projects available.

Apache Cassandra [4] is an open source distributed storage system for unstructured data, which was originally developed by Facebook. It combines the data model of Bigtable and the distributed architecture of Amazon Dynamo. This combination provides more flexibility to users. However, its stability is not perfect yet.

There also exist diversified document oriented storage systems, such as CouchDB [10] and MongoDB [11]. CouchDB is a document-oriented database that can be queried and indexed in a MapReduce fashion using JavaScript. It is written in Erlang. While MongoDB is a scalable, high-performance, open source, document-oriented database, which is written in C++ and base on Bigtable. Both of above are schema-free. This kind of architecture is suitable for sparse and document-like data storage. But CouchDB is not mature enough. So we choose MongoDB as our reseach base.

Fig. 19.1 VESS architecture

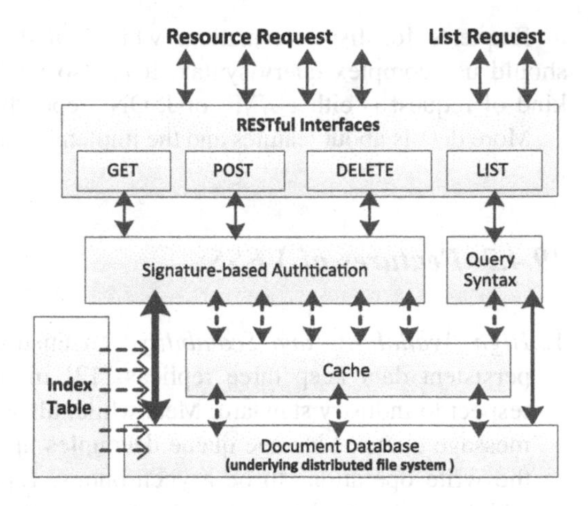

19.4 Design and Implementation

19.4.1 VESS Architecture

Based on analyzing users' requirements deeply and comprehensively, this paper aims to present an efficient, user-friendly and always-available storage system (VESS) for MDVEP. This system improves and combines some different technologies to satisfy users' personalized requirements.

The architecture of VESS is illustrated in Fig. 19.1. It provides RESTful interfaces with GET, POST, DELETE, and LIST methods. The component of Signature-based Authentication provides security protection for data access. All resource requests need to be authorized. The request URI contains signature information generated by digesting request content.The Cache component provide multi-level cache strategy to make users to be closer to backend data. The granularity of each level is well-made to guarantee the efficiency of data access. The Index Table maintains consistency among cache, document database, and distributed file system. The Document Database contains the concrete content of the unstructured data, which is based on MongoDB.

In VESS, each unstructured data resource is assigned a *Universal Resource Identify* (URI). Through resource's URI, users can query resources with different filter conditions and manipulate resources.

Some schemas are used to extract metadata which will be inserted into in-memory index table. When the size of unstructured data reaches a certain scale, the database and index table will be split horizontally. All the data in distributed file system will be duplicated among different disks to guarantee the availability.

Requests for a single unstructured data are provided by three methods: POST (Create/Update item), GET (Read item), and DELETE (Delete item).

Requests for list information, which satisfy some certain filter conditions, should use complex query syntax. It is also URI-based. The return value of this kind of request is either XML or JSON, depending on users' requirements.

More details about features and the implementation of VESS are discussed below.

19.4.2 Features of VESS

1. *High Availability and Scalability*: To ensure data service's availability, all persistent data keep three replicas [12] in different physical machines with respect to industry standard. Meanwhile, all write/update operations are sent to message queue. Message queue decouples applications from database, making the write operations to be asynchronous. Dynamo introduces a NWR model which can be used to make adjustment between availability and consistency. Finally it gets to be eventually consistent. As a read-intensive system, we minimize R (default value is 1) and maximize W (default value is N). Database is set to be master–slave model. The performance of read operations can be improved by adding slave nodes.
2. *Transparency*: REST architecture abstracts all unstructured data as resources which are mapped to each unique URI with CRUD (*Create, Retrieve, Update and Delete*) operations. What users only need to know is how to manipulate RESTful interfaces, without being aware of how VESS organizes and processes resources.
3. *Security*: In the REST architecture, all operations are stateless. The processing of requests does not bring any session or cookie. Based on this feature, we can follow the way of HTTP request and define format of URI to eliminate unauthorized access.
4. *Complex query*: Combining with existing SQL syntax and query requirements of unstructured data, we convert URI in the form of {*key: value*} into SQL. Keywords supported in VESS are WHERE, SKIP, LIMIT, and SORT. Query URI can be combined by any of the four keywords:

- WHERE: the syntax is *WHERE = field: filter: condition*. Supported filters are listed in Table 19.1. Return list matches the filter condition.
- SKIP: the syntax is *SKIP* = *N*. Ignore first *N* items of the list. In other word, these items will not be returned.
- LIMIT: the syntax is *LIMIT* = *M*. Limit the returned list to be the first *M* items.
- SORT: the syntax is *SORT = field: [ASC|DESC]*. Sort the returned list in ascending or descending order of the specified field.

The above easy-to-use query interfaces provide users a set of user-friendly and elastic accessing methods to the data.

Table 19.1 Complex query filters

Filter	Meaning	Syntax
Lt	Less	Field:lt:condition
Lte	Less equal	Field:lte:condition
Gt	Great	Field:gt:condition
Gte	Great equal	Field:gte:condition
Eq	Equal	Field:eq:condition
Ne	Not equal	Field:ne:condition
Re	Regex	Field:re:condition

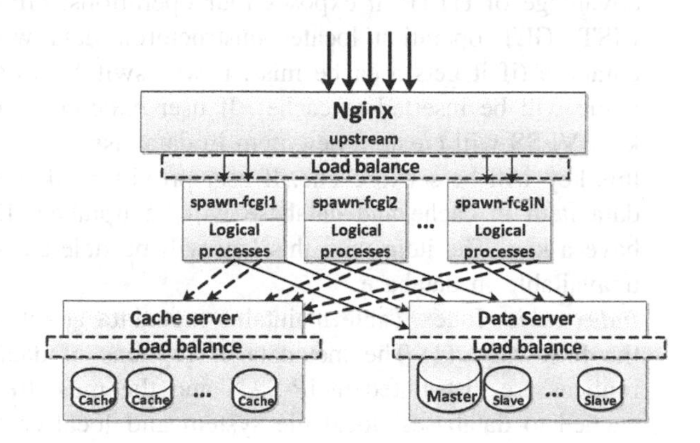

Fig. 19.2 VESS data flow and implementation

19.4.3 VESS Implementation

According to the requirements of Virtual Experiment (VE) platform, it is reasonable to partially adopt some open source technologies in VESS. Its data flow and implementation details are shown in Fig. 19.2.

Nginx is chosen to distribute requests from users to spawn-fcgi that is responsible for maintaining the lifecycle of logical processes. The distribution is based on round-robin algorithm. Logical processes are used to process concrete data resource requests including POST, GET, and DELETE, which are implemented in Python language.

Cache servers are independent memory cache systems, which are responsible for different partitions of data resources. Its load balance is based on the hash of resources' keys. Unstructured data items in cache are stored in {*key:value*} format using *Least Recently Used* (LRU) algorithm for age-out. The *key* is assigned as a universal ID, while the *value* is the unstructured data item. Items which are read/inserted/updated recently will be inserted into cache.

Meanwhile, master–slave structure is introduced to gain high availability and read performance of data servers. Read and write operations are delivered to slaves and master respectively. The data of the master is mirrored to slaves. VESS uses message queue to sequence insert, update, and delete operations, which is also used to keep data consistency between cache and data servers.

Key technologies of VESS include: system interface that is used to provide HTTP-based data request method, index table relying on a universal metadata file to ensure the consistency and integrity of replicas, message queue that is adopted to sequence modification operations and guarantee consistency, while signature-based authentication is responsible for authorizing each request and guarantee the security.

1. *System Interface*: The access of unstructured data in VESS fully takes advantage of HTTP, it exposes four operations: GET, POST, DELETE, and LIST. GET operation locates unstructured data with the key in cache or database (if it gets a cache miss, it will switch to database and the returned value will be inserted to cache). If user executes a POST operation without key, VESS will create a new item in database and return a key value to user, this key will be set to cache. If user provides a key for POST operation, the data item in cache and database will be updated. DELETE operation must have a key. The item with this key will be deleted from cache and set to be unavailable in database.
2. *Index table*: Index Table maintains metadata about data replicas. It includes the following: (1) The metadata of replicas of databases' data files. These replicas are distributed in IP-SAN and the most frequently visited ones are cached to databases' local file system and local cache. (2) The metadata of cached data items. Based on the correlation of subjects in virtual experiment, data items of the same or similar subjects are pre-fetched into cache to accelerate requests.

   The current implementation of the index table stores its metadata in the memory, which is also persisted into file system as an independent file.
3. *Message queue*: Message queue can decouple the logical process from the persistent storage, and make modification operations sequential and asynchronous. The processes are illustrated in Fig. 19.3. It fully takes advantage of master–slave structure, that all of data modification operations are only executed through master server. It dramtically simplifies the implementation of data merge and consistency maintenance.
4. *Signature-based authentication*: As RESTful interface is stateless, it is impossible to authorize data access through session or cookie. The only way left is to consider URI-based digital signature. Users and VESS have an agreement on secret key creation and mechanism on how to generate the signature and authenticate it. The mechanism is shown in Fig. 19.4. The secret key and the token are known to both user and VESS. The difference between them is that the secret key is a string to identify unique user and the token is a string to identify a single request. MD5 hash is imported to generate signature.

   Once users need to request data, the first thing is to get TOKEN from TOKEN DB, which will be combined with resource's URI and the secret key that can be

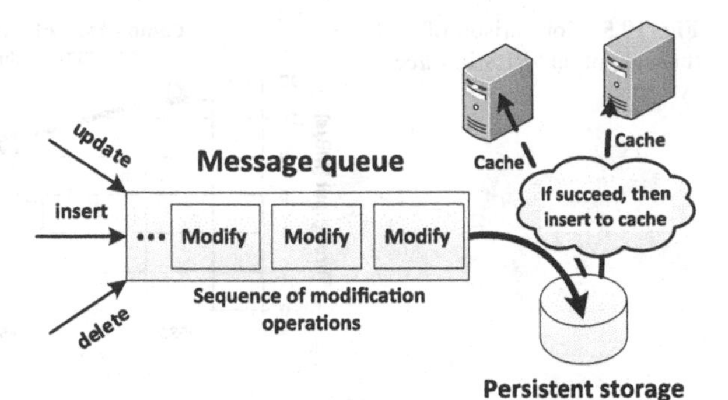

Fig. 19.3 Message queue

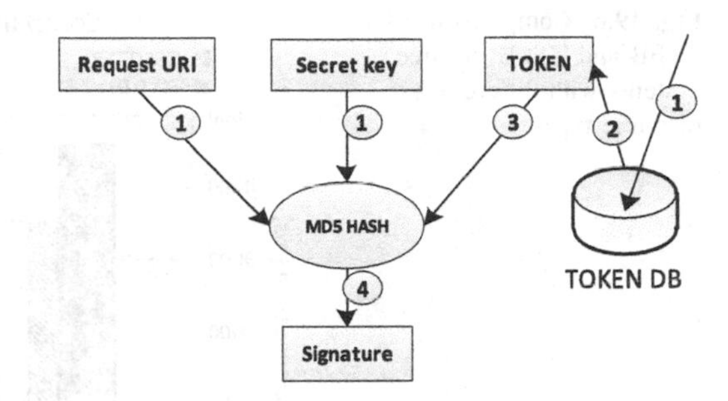

Fig. 19.4 The process of generating signature

obtained from web interface. The next step is to use the TOKEN, the request URI and the secret key to generate digest signature string with MD5 hash method. With the signature, TOKEN and request URI, a new authorized request URI can be formed. On the other hand, server side of VESS will use the same method to authorize the digest signature and data request.

19.5 Performance Evaluation

In this section, we compare VESS with the solutions of file system and relational database. This evaluation consists of four tasks. The first one is to evaluate VESS's average throughput of read operations. The next is aimed to get average throughput of write operations. The third one is a performance comparison with different storage patterns. Finally, we evaluate the scalability of VESS.

We deploy VESS on an application node following three database nodes using Global File System from Redhat based on IP-SAN. Another node is set to serve as load generator used to generate concurrent requests. All the servers are located in a LAN. Three cache servers are set on each database node and one cache server on application node. The memory usage for all of them is 1 GB. Each node is equipped with two quad-core 2.26 GHz Xeon processors, 16 GB of memory, two gigabyte Ethernet links and two 146 GB SAS disk, runs RHEL5.3 with kernel 2.6.18. Microsoft Web Application Stress Tool is used to simulate massive data requests.

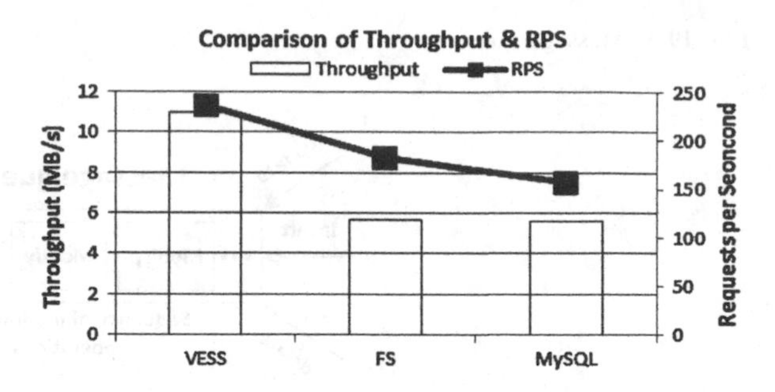

Fig. 19.5 Comparison of throughput and RPS in three systems

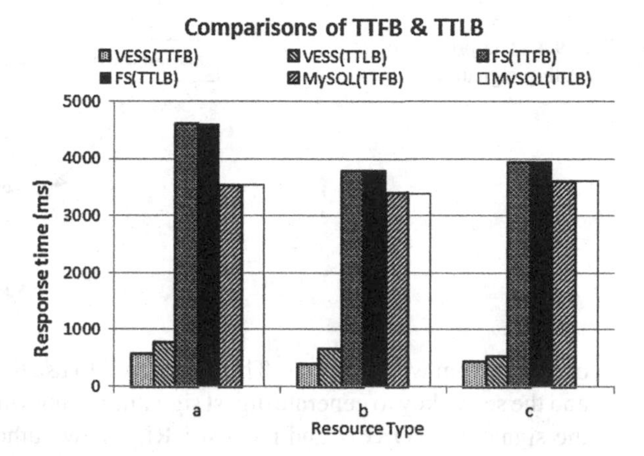

Fig. 19.6 Comparisons of TTFB and TTLB in three systems with different resource types

Dataset for read and comparison evaluation consists of variety xml files with sizes between 3 KB and 600 KB. The amount of data items is 700,000, and total size is about 36 GB. The throughput of loading this dataset to VESS is nearly 6 MB/s. The number of requests is 125 per second. We keep running VESS for 7×24 h under a heavy load by making simulation that 60,000 users generate requests within randomly delay between 0 ms to 500 ms. VESS performs stable enough both in functionality and performance. As shown in Fig. 19.5, the average throughput of read operations is nearly 11 MB/s, and the number of requests per second (RPS) are 236. Here, we compare VESS with the ext3 file system and the master–slave structure MySQL relational database.

In order to compare three systems' response time, we evaluate two factors: (1) total time the first byte is received (TTFB), (2) total time the last byte is received (TTLB). As shown in Fig. 19.6, where *a*, *b*, and *c* are three different types of resources.

We can see that VESS has a dramatic improvement on response time and much higher throughput from Figs. 19.5 and 19.6. Additionally, we notice that the waiting for response from server spends most time of a request. Receiving data from server is rather quick. The three storage systems are all bounded to RESTful interfaces.

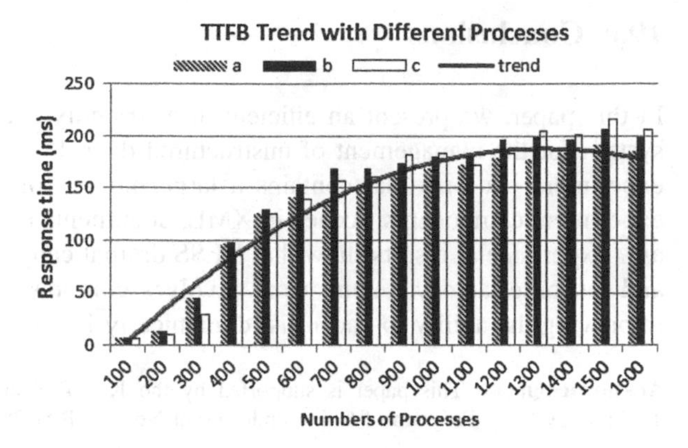

Fig. 19.7 TTFB trend with different processes

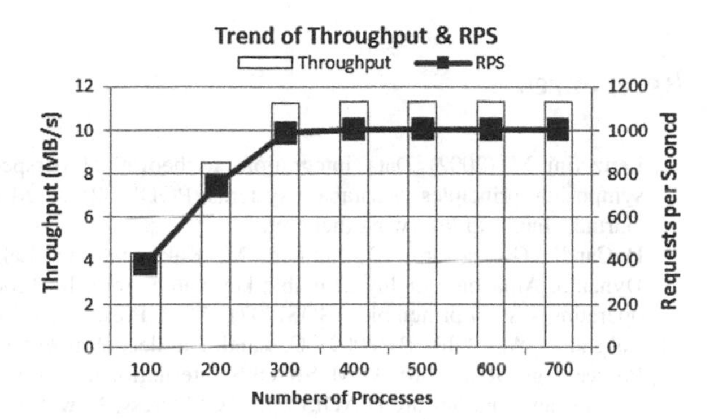

Fig. 19.8 Throughput and RPS with different processes

For the scalability of VESS, we simulate users' concurrent requests with different numbers of processes increasingly. Figures 19.7 and 19.8 show the trends of TTFB, throughput, and RPS, respectively.

In Fig. 19.7, the response time increases almost linearly with the growth of the amount of processes used to generate concurrent requests, when it is less than 1,000. However, when the amount of processes is more than 1,000, the response time almost does not change and be stable around 200 ms. Moreover, we can also get similar results from Fig. 19.8 that throughput and RPS will not change a lot after the amount of processes reaches a certain threshold, regardless of the increment of request processes.

In a summary, VESS has a dramatic improvement on response time and much higher throughput, which can approve its efficiency clearly. When it reaches its peak capability, the throughput and RPS performs stable and the average response time is around some stable values. As the increment of backend application processes, cache servers and data servers, VESS can get almost linear performance improvement. Moreover, it is easy for VESS to scale-out.

Additionally, in the process of designing, implementing and maintaining VESS, we gained some useful experiences. One of them is to use appropriate granularity of cache within different layers of the system and make users close to data where it is.

19.6 Conclusions

In this paper, we present an efficient, user-friendly and always-available storage system for the management of unstructured data. It has been applied in a virtual experiment platform that contains a large number of unstructured data such as experiment components, scenes in XML, documents in TXT, doc, videos in mp4, avi. Experimental results show that VESS dramatically improves throughput, RPS and response latency. Future work involves compressing data files in cache and improving the ability to cache list dynamically in consideration of consistency.

Acknowledgment This paper is supported by the Key Project in the National Science and Technology Pillar Program of China under grant No.2008BAH29B00.

References

1. Lenzerini M (2002) Data integration: a theoretical perspective. In: Proceedings ACM symposium principles of database systems (PODS'02). ACM Press, New York, pp 233–246
2. Gartner, Inc. http://www.gartner.com/
3. DeCandia G, Hastorun D, Jampani M, Kakulapati G, Lakshman A, Pilchin A (2007) Dynamo: Amazon's highly available key-value store. In: Proceedings of ACM symposium operating system princinples (SOSP'07). ACM Press, New York, pp 205–220
4. Lakshman A, Malik P (2009) Cassandra–a decentralized structured storage system. In: Proceedings of the 3rd ACM SIGOPS international workshop on large scale distributed systems and middleware (LADIS'09). ACM Press, New York, pp 35–40
5. Fielding R, Taylor R (2002) Principled design of the modern web architecture. ACM Trans Internet Technol 2(2):115–150
6. Chang F, Deam J, Ghemawat S, Hsieh W, Wallach D, Burrows M, Chandra T, Fikes A, Gruber RE (2006) Bigtable: a distributed storage system for structured data. ACM Trans Comput Syst 26(2):1–26
7. Fox A, Gribble SD, Chawathe Y, Brewer EA, Gauthier P (1997) Cluster-based scalable network services. ACM SIGOPS Oper Syst Rev 31(5):78–91
8. Brewer EA (2000) Towards robust distributed systems. In: Proceedings of ACM symposium princinples of distributed computing (PODC'00). ACM Press, New York, p 7
9. Carstoiu B, Carstoiu D (2010) High performance eventually consistent distributed database Zatara. In: Proceedings of 2010 6th international conference on networked computing, (INC'10). IEEE Press, Piscataway, pp 54–59
10. Yu Q (2010) Metadata integration architecture in enterprise data warehouse system. In: Proceedings of 2010 2nd international conference on information science and engineering (ICISE 2010). IEEE Press, Piscataway, pp 340–343
11. Verma A, Llora X, Venkataraman S, Goldberg DE, Campbell RH (2010) Scaling eCGA model building via data-intensive computing. In: Proceedings of 2010 IEEE congress on evolutionary computation (CEC'10). IEEE Press, Piscataway, pp 1–8
12. Ghemawat S, Gobioff H, Leung ST (2003) The Google file system. In: Proceedings of ACM symposium operating system principles (SOSP'03). ACM Press, New York, pp 29–43

Chapter 20
PKLIVE: A Peer-to-Peer Game Observing System

Hong Yao and Lenchen Yu

Abstract Gaming services are popular for Internet users. How to broadcast live gaming scenes to large-scale users over internet is a big problem. Traditional schemes have poor user experiences, cost large bandwidth consumption. In order to lower down the broadcast dataset, some solutions propose an alternative approach based on game data content delivery instead of video content. At present, these works are based on client/server model, which is hard to scale. In this paper, we introduce PKLIVE, a game observing system based on peer-to-peer (P2P) model. PKLIVE system broadcasts game data content to reduce bandwidth consumption, maintains a P2P overlay for each gaming channel and adopt a reliable UDP protocol to break the limitation of the TCP connections, decentralizes servers to enhance system's availability and scalability. Preliminary tests show that PKLIVE works well, low-latency, low bandwidth consumption and very considerable bandwidth savings for game service providers.

Keywords Multiplayer online games · Peer-to-peer · Live

20.1 Introduction

Peer-to-peer (P2P) can support service for lots of end users with little hardware investment, which suits for the efforts to lower down the service cost for Multiplayer Online Games (MOG) and Massively Multiplayer Online Games (MMOG) [1]. When international e-sports game is held, such as World Cyber Games

H. Yao (✉) · L. Yu
School of Computer Science and Technology,
China University of Geosciences, 430074 Wuhan, China
e-mail: yaohong@cug.edu.cn

J. J. Park et al. (eds.), *Proceedings of the International Conference on Human-centric Computing 2011 and Embedded and Multimedia Computing 2011*, Lecture Notes in Electrical Engineering 102,
DOI: 10.1007/978-94-007-2105-0_20,

(WCG), there are a large number of people who will concern about. How to broadcast live gaming scenes to large-scale Internet users is a big problem.

Popular P2P live streaming, VOD system, such as PPLive [2], Anysee [3], GridCast [4], can help significantly reduce the deployment cost, but game video data (by camera) is much more bigger than game message data (by game software). There is another way to lower the cost.

Two battle observing system WTV [5] and GGTV [6] adopt the approach based on game data content. With this technology, the bandwidth consumption is generally less than 20 KB/s. But the centralized architecture of WTV or GGTV makes the system scales hard.

This paper introduces PKLIVE, a large-scaled game observing system. Any peer in PKLIVE transmits game data to reduce bandwidth consumption, maintains a P2P overlay for each channel and uses a reliable UDP protocol rather than TCP to break the limitation of the TCP connection number and bandwidth at server side. PKLIVE decentralizes servers to enhance system's availability and scalability.

PKLIVE is used for game live/observing, we chose War Craft III as a representation MOG, without any changes on the binary code of this game to validate the system. PKLIVE captures all network packets of the source game host and transmits them to a game server which is bound with a tracker server. The tracker server maintains a P2P overlay network. Peers in the same overlay pull the game data from other peers or from game server when necessary. Preliminary tests show that PKLIVE works well.

The rest of this paper is organized as follows. Section 20.2 discusses the related work. Section 20.3 describes some key design issues in detail. Section 20.4 shows the tests results and presents the performance evaluation. Section 20.5 concludes this paper.

20.2 Related Works

There are mainly three areas of research related to PKLIVE. They are data-based playback of entire game scenes, Network Address Translation (NAT) and the reliable transfer protocol over UDP, P2P multicast/live-streaming algorithm.

20.2.1 Data-Based Playback of Game

The essential issue of playback from game data is that the whole game process can be replayed or recovered from the previous data, especially the data transmitted in the network or stored in a file. There are so many games supporting observers and replays, such as Star Craft, War Craft III, Rome: Total War, Age of Empires III, and there will be much more in the future. In this way, a third party player who is

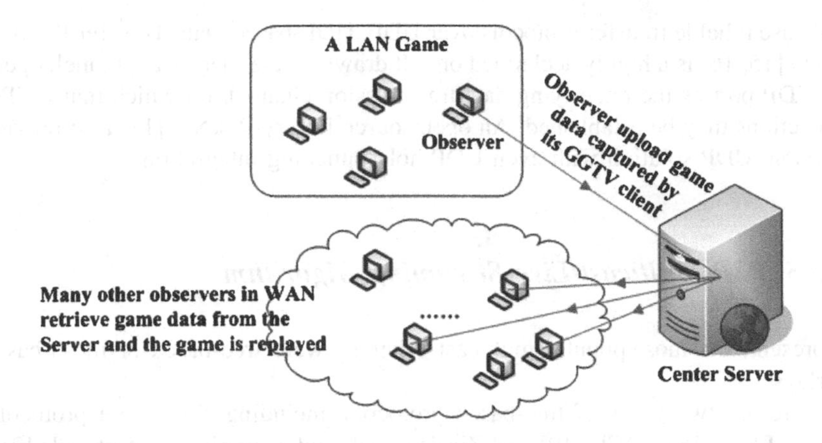

Fig. 20.1 Architecture of GGTV

not involved in a game scene will be able to recall the whole game process recovered by the game data. However, available game limits the number of observers now. How to break the limitation and extend the services to the wide area network is the motivation of PKLIVE.

The mainly game observing systems in the world are GGTV, WTV, and VSLive, the implementations are more or less the same. The architecture of GGTV is illustrated in Fig. 20.1. The observer in a LAN game starts a GGTV client. When the game is running, all packets went through the observer are captured by the GGTV client and uploaded to a center server.

20.2.2 NAT and Reliable UDP

Due to some technical reasons, users behind NAT [7, 8, 9], do not know their IP-Port outside of NAT, so they can not be connected actively.

STUN [10] is a simple and effective way to make UDP communication behind NAT possible. NatTrav [11] implements asymmetric TCP hole punching procedure. Nutss [12] and NatBlaster [13] introduce a more complex TCP hole punching trick that can work around some of the bad NAT behaviors, but they require the rendezvous server to spoof source IP addresses, and they also require user applications to access RAW sockets, usually available only at root or administrator privilege levels. As a result, these solutions are not widely adopted by current existing P2P applications.

While TCPBridge [14] introduced a more sophisticated mechanism that help hosts behind NAT communicate with TCP. In its mediate layer, a reliable UDP is fulfilled replacing TCP. This substitution is completely transparent to the upper application. It is developed particularly for a P2P game platform PKTown [1].

To use reliable transfer protocols over UDP, UDP-based Data Transfer Protocol (UDT) [15, 16] is a highly acclaimed one. It draws on the idea of the tunnel, opens one UDP port as the underlying data transmission channel, on which many UDT connections may be established. An open source library RakNet [17] also provide a reliable UDP solution with even UDP hole punching integration.

20.2.3 P2P Multicast/Live-Streaming Algorithm

At present, the most popular multicast schemes were tree-based or mesh-based overlays.

There are two types of tree-based protocols, including single tree protocols, such as ESM [18], NICE [19] and ZigZag [20], and multiple tree protocols [21]. The key point is how to build a stable and scalable multicast tree with high efficiency. Each joining node tries to find a near node with sufficient capability to contribute data, and become its child. However, the leaving or crash behavior of nodes in the upper layers often causes instability or thrash.

To improve the stability of overlay, mesh-based protocols have been proposed. Mesh-based overlay is a kind of unstructured overlay, where there is no affiliation between peers. Each peer can retrieve data from multiple peers as well as contributing services for other peers. The resource utilization of a mesh is higher than that of a tree. Meshes based on Gossip protocol can find fresh peers in the single mesh with low management overhead, but not in global P2P networks. Due to the random selection algorithm, the quality of service cannot be guaranteed, such as the startup delay. Also, to decrease the impact of autonomy of peers on streaming services, a very large buffer is necessary. When there are large quantities of nodes, some of them may be promoted to super nodes, among which another overlay can be established and maintained, like Skype [22].

20.3 Design of PKLive

Network games in general are highly latency-sensitive. The game experience will be poor when there is a high latency. Network games also generates small amount of data, for example, a 40 min game of War Craft III is merely about 1 MB. Moreover, flow control and congestion control mechanism are not needed by the same reason.

To achieve the reliability of game data, each packet of one game is time stamped and numbered from zero followed by an increment of one. By this, nodes internally need to maintain a sliding window as in TCP so that missing packets will be re-requested at the appropriate time. At the same time, the PKLIVE client fetches packets continuously from the sliding window to recover the real game process.

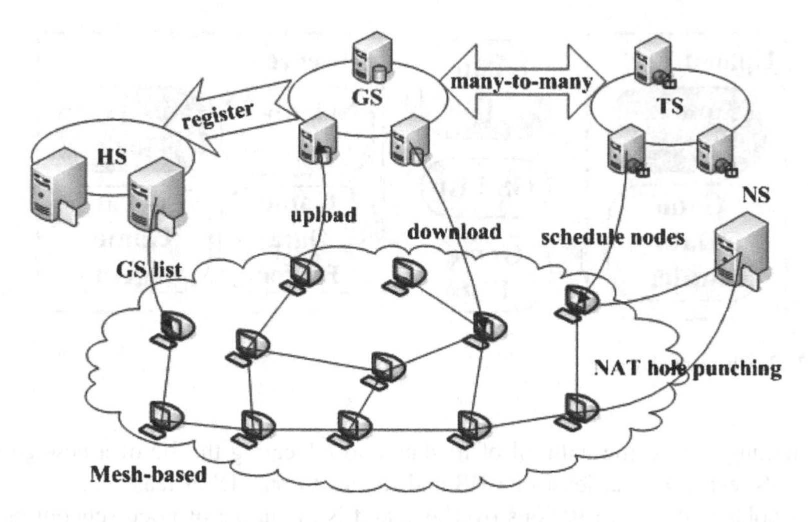

Fig. 20.2 Architecture of PKLIVE

Considering that an observer must watch from the beginning of the game, which is determined by the observing mode. Nodes in overlay also have to obtain packets from No.0 to the following. Thus, the situation that how many packets a node holds is laconically represented by a pair: [Game ID, Available Packets]. The former is the unique ID of a game, while the latter is the number of available packets that is an important criterion for peer selection. For example, a node watching Game 666 with packet 0123467 is abstracted as [666, 5].

The design of PKLIVE is about to consider the following issues: a reliable transfer protocol over UDP for convenient NAT hole punching, a mesh-based P2P overlay and algorithm to ensure low-latency, characteristics that the amount of game data is small and the game process must be watched from the initiate rather than any time in the middle.

20.3.1 System Architecture

As shown in Fig. 20.2, the system is composed by Hall Server (HS), Game Server (GS), Tracker Server (TS), an optional NAT Server (NS) and many clients.

HS is the login portal of the system. It accepts the registration and un-registration of GSs, and holds a list of GS to reply the request from clients. There may be several HSs among different ISP to adapt the regional network differences. Overall, HS is a mediate layer between GSs and clients. The prohibition of client's direct access to GS is in consideration of security, load balance and scalability.

GS is the data source server. Its responsibilities are receiving game data from uploading nodes, distributing the data to the observing ones who actively pull,

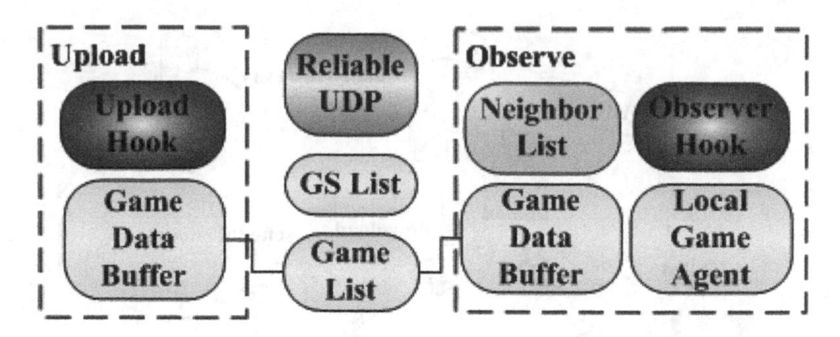

Fig. 20.3 Client modules

maintaining each game with all of its data and allocating the ID of a new game. Each GS registers to at least one HS and relates to one TS at least.

TS holds nodes from various overlays, and is in charge of node scheduling to the client that is just joining overlay.

NS helps the process of UDP hole punching between two nodes. It just plays the role of server in STUN [10]. All network communications in PKLIVE implemented by UDP.

Client is a relatively sophisticated module in PKLIVE, of which the inner modules is shown in Fig. 20.3.

The upload module is consist of a game upload hook, which will be injected to the game process to capture all the packets game sends and receives, and a game data buffer, which stores all packets and upload them to GS. The observe module has a hook, a local game agent, a neighbor list and a game data buffer. The replay of a game is performed by the local game agent that acts like a remote game host and the hook that redirects all packets that were sent to a remote address to the local game agent. The game process is running with the illusion that it joins a real host but actually the local game agent. The neighbor list and game data buffer work together to carry out the mesh-based P2P algorithm. All game data sending between nodes are wrapped with reliable UDP.

20.3.2 Key Terms

PKLIVE is not only a game live software but also can provide VOD services. The essential distinction is how much is the time difference from the beginning of the living game to the moment that a node starts to observe this game (starting time difference). The low-latency requirement is obviously most high when the game is live. On the contrary, it is not so latency sensitive while a node is enjoying a game VOD, because a plenty of data and copies required in a few minutes are residing in the overlay. These data will be retrieved by the node as soon as possible and last for a long time.

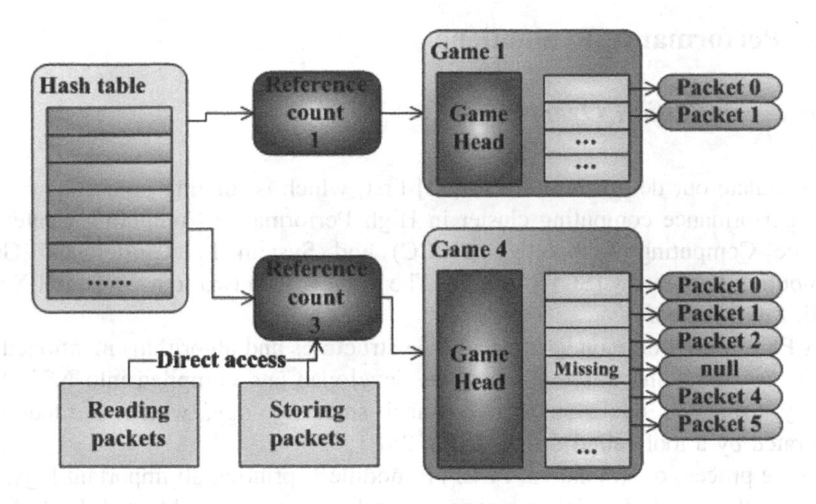

Fig. 20.4 Illustration of game data buffer

A 'zero latency sensitive threshold' is proposed. When the starting time difference is greater than this threshold, all P2P algorithms ignore the latency factor. Another important term called 'latency sensitive factor' is defined as below:

latency sensitive factor = 100 × (zero latency sensitive threshold − starting time difference)/zero latency sensitive threshold

It reflects the degree of latency sensitivity.

To quantify the priority of node selection to communicate, 'node point' is proposed. It is one-way relationship, which is defined as below:

Node point = (100 − latency sensitive factor) × (peer's available packets − available packets) − latency sensitive factor × RTT + static service quality + dynamic service quality

Note that RTT in the formula above is measured by milliseconds and two types of service quality are calculated by node's state of hardware.

20.3.3 Game Data Buffer Management

Figure 20.4 shows the schematic diagram of the game data buffer. The game data buffer caches all data of different games in a hash table. Hash table does not directly hold a game block, but a reference count pointing to a game. Thus, data copy is reduced to least and memory management is automatically performed. Another feature of PKLIVE client is a single-thread modal, which makes the program not lock any data to improve the efficiency.

There are two typical procedures in sliding window: packet process algorithm that will be invoked at the arrival of a data packet, data management algorithm that is running in the background all the time.

20.4 Performance Evaluations

20.4.1 Simulation Overview

We simulate our design in the NS2 [23] first, which is running on a node of the high performance computing cluster in High Performance Computing Center in Service Computing Technology (HPCC) and System Lab/Cluster and Grid Computing Laboratory (SCTS/CGCL). The node is with two four-core Intel Xeon CPU, 8 GB memory.

A PKLIVE module integrating all data structures and algorithms mentioned in this paper and some miscellaneous was developed and compiled into NS2. We mainly evaluated our system in a transit-stub topology, which is randomly generated by a tool called GT-ITM [24, 25].

In the process of simulation, PKLIVE module is printing all important logs, by checking the result of which, the experimental results can be obtained. Last, three performance indicators, which are discontinuity, the overhead of node's bandwidth and the bandwidth saving rate of servers, are analyzed from the previous text data by Gawk [26].

20.4.2 Preliminary Test Results and Analysis

Discontinuity, which is the direct reflection of user experience, shows how much seconds delay there are in the process of watching the entire game. The formal definition of discontinuity is given:

$$\mathrm{Dis} = \sum(\mathrm{TAi} - \mathrm{TPi}), \quad i = 0, 1, 2, \ldots \qquad (20.1)$$

Dis is the abbreviation of Discontinuity. TAi is the actual time when game packet i is submitted to the game. TPi is the punctual time when game packet i shall be submitted.

As shown in Fig. 20.5, there are four groups of data, which is gathered in four separate simulations at the scale of 300, 500, 1,000 and 2,000 nodes. The horizontal line called alerting line in the middle of each sub-graph indicates a 10 s discontinuity. The abscissa is the node sequence to identify of different nodes, while the ordinate is the discontinuity value.

The vast majority of discontinuity is under the alerting line, but few dots beyond it. This mean, nodes watching games, only have to wait for up to 10 s, which is a quite satisfactory result compared with the probable intermittent in a P2P live-streaming system.

As is depicted in Fig. 20.6, average bandwidth costs are between 5 to 15 KB per second. There is a little fluctuation at 50 nodes. After that, the curve is smooth like a straight line. When there are more than 100 nodes, bandwidth costs become less than 10 KB/s, which is only 58% of the best P2P live-streaming system.

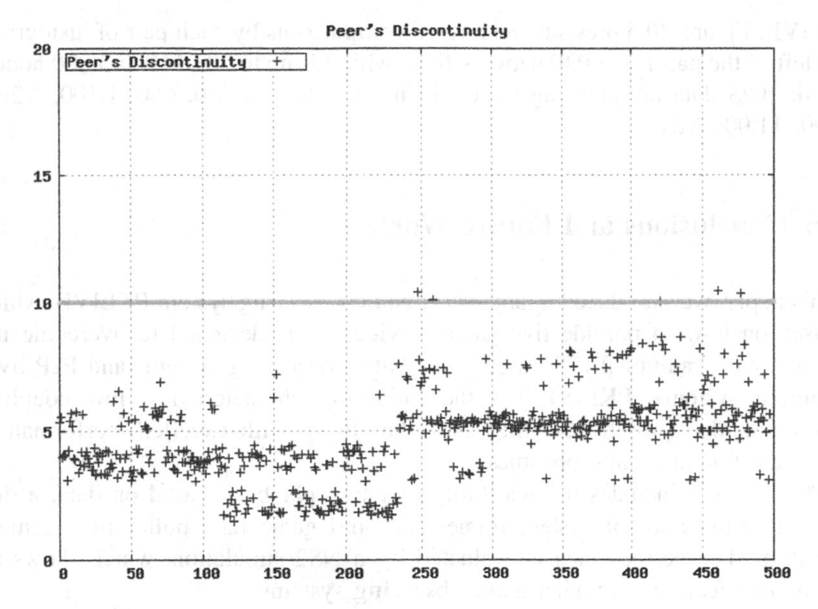

Fig. 20.5 Discontinuities in four node scales

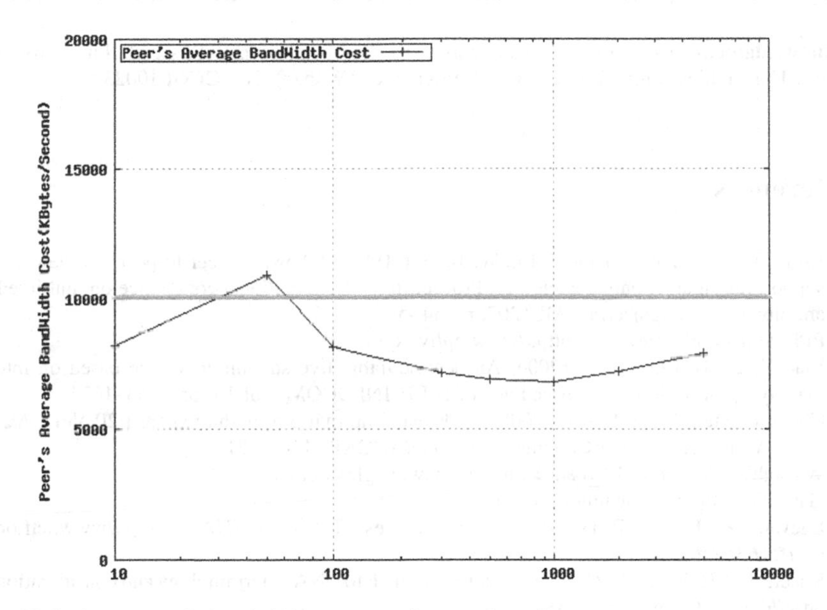

Fig. 20.6 Average node bandwidth cost at various node scales

The bottleneck of WTV or GGTV is generally the insufficient bandwidth of servers. To address the amount of bandwidth savings, the simulation simply calculates a C/S statistics of bandwidth according to the corresponding data flow in a

PKLIVE. Figure 20.6 presents the intense comparisons by each pair of histogram. The left of the pair is the P2P network flow, which is up to 204 MB at 5,000 nodes. But the C/S data are amazing large which are 22, 110, 220, 660, 1,100, 2,200, 4,400, 11,000 MB.

20.5 Conclusions and Future Work

In this paper, we introduced a large-scale game observing system PKLIVE, which focuses on how to provide live game services. It is designed to overcome the deficiency of traditional C/S based game battle observing systems and P2P live-streaming systems. PKLIVE has the following characteristics: low coupling between modules, high flexibility and availability with excellent performance, good fault-tolerance and robustness.

The research includes the feasibility of game playback based on data, a distributed architecture of system framework, and game data buffer management algorithm. The performance is evaluated by a NS2 simulation, which shows an outstanding feature over other game observing systems.

In the future, we will try to do an inter-overlay optimization, and continue to study deeply the playback mechanism to accommodate more games in PKLIVE.

Acknowledgment The Project was Supported by the Fundamental Research Funds for the Central Universities, China University of Geosciences (Wuhan). No. CUGL100232.

References

1. Jin H, Yao H, Liao X, Yang S, Liu W, Jia Y (2007) PKTown: a peer-to-peer middleware to support multiplayer online games. In: Proceedings of international conference on multimedia and ubiquitous engineering, MUE'07, pp 54–59
2. PPLive network television. http://www.pplive.com
3. Liao X, Jin H, Liu Y et al (2006) Anysee: scalable live streaming service based on inter-overlay optimization. In: Proceedings of IEEE INFOCOM, vol 18, pp 1663–1674
4. Cheng B, Stein L, Jin H et al (2008) GridCast: improving peer sharing for P2P VoD. ACM Trans Multiméd Comput Commun Appl (TOMCCAP) 4(4):1–31
5. WaaaghTV Warcraft 3 Broadcasting. www.waaaghtv.com
6. GarenaTV. http://www.jingwutai.com
7. Egevang K, Francis P The IP Network Address Translator (NAT). http://www.ietf.org/rfc/rfc1631.txt
8. Srisuresh P, Holdreg M IP Network Address Translator (NAT) terminology and considerations. http://www.ietf.org/rfc/rfc2663.txt
9. Senie D Network Address Translator (NAT)-friendly application design guidelines. http://www.ietf.org/rfc/rfc3235.txt
10. Rosenberg J, Weinberger J, Huitema C et al STUN—simple traversal of user datagram protocol (UDP) through Network Address Translators (NAT). http://www.ietf.org/rfc/rfc3489.txt
11. Eppinger J L (2005) TCP connections for P2P apps: a software approach to solving the NAT problem. Technical Report CMU-ISRI-05-104, January 2005

12. Guha S, Takeday Y, Francis P (2004) NUTSS: a SIP-based approach to UDP and TCP network connectivity. In: Proceedings of SIGCOMM 2004 workshops, August 2004
13. Biggadike A, Ferullo D, Wilson G, Perrig A (2005) NATBLASTER: establishing TCP connections between hosts behind NATs. In: Proceedings of ACM SIGCOMM ASIA workshop, April 2005
14. Liu S, Jin H, Liao X et al (2008) TCPBridge: a software approach to establish direct communications for NAT hosts. In: Proceedings of the 6th ACS/IEEE international conference on computer systems and applications (AICCSA'08), pp 247–252
15. Gu Y, Grossman RL (2007) UDT: UDP-based data transfer for high-speed wide area networks. Computer networks, special issue on hot topics in transport protocols for very fast and very long distance networks, vol 51, pp 1777–1799
16. Gu Y, Grossman RL UDT: breaking the data transfer bottleneck. http://udt.sourceforge.net
17. Jenkins software. Raknet manual. http://www.rakkarsoft.com/raknet/manual
18. Chu Y-H, Rao SG, Seshan S et al (2002) A case for end system multicast. IEEE J sel areas commun 20(8):1456–1471
19. Banerjee S, Bhattacharjee B, Kommareddy C (2002) Scalable application layer multicast. In: Proceedings of ACM SIGCOMM, pp 205–217
20. Tran D, Hua K, Sheu S (2003) Zigzag: an efficient peer-to-peer scheme for media streaming. In: Proceedings of IEEE INFOCOM, pp 1283–1292
21. Castro M, Druschel P, Kermarrec A et al (2003) SplitStream : high-bandwidth multicast in cooperative environments. In: Proceedings of ACM SOSP, pp 298–313
22. Guha S, Daswani N, Jain R (2006) An experimental study of the skype peer-to-peer VoIP system. In: Proceedings of IPTPS
23. The NS manual. http://www.isi.edu/nsnam/ns/ns-documentation.html
24. Zegura EW, Calvert K, Bhattacharjee S (1996) How to model an internetwork. In: Proceedings of IEEE Infocom'96, San Francisco, CA, pp 594–602
25. Calvert K, Doar M, Zegura EW (1997) Modeling internet topology. IEEE Communications Magazine, June 1997, pp 160–163
26. Gawk—GNU project—Free Software Foundation (FSF). http://www.gnu.org/software/gaw

Part V
HumanCom 2011 Session 4: Privacy, Security and Trust Management

Chapter 21
Provably Secure Two-Party Password-Based Key Agreement Protocol

Huihua Zhou, Tianjiang Wang and Minghui Zheng

Abstract This paper considers the issue on authenticated two-party key agreement protocol over an insecure public network. Many authenticated key agreement protocols have been proposed to meet the challenges. However, existing protocols are either limited by the use of public key infrastructure or by their security, suffering dictionary attack. To overcome these disadvantages, we propose an efficient two-party password-based key agreement protocol resistant to the dictionary attacks by adding password-authentication services. Under the Computation Gap Diffie-Hellman assumption, we will show the proposed protocol is provably secure in both the ideal-cipher model and the random-oracle model.

Keywords Key agreement · Password-based authentication · Provable security · Computation Gap Diffie-Hellman assumption

21.1 Introduction

Two-party authenticated key agreement (AKA) protocols enable two users communicating over an insecure, open network to establish a shared secret called session key and furthermore to be guaranteed that they are indeed sharing this session key with each other. The session key may be subsequently used to achieve some cryptographic goals such as confidentiality or data integrity.

H. Zhou · T. Wang
College of Computer Science, Huazhong University of Science and Technology,
430074 Wuhan, Hubei, China
e-mail: hhzhou2010@gmail.com

H. Zhou · M. Zheng (✉)
Department of Computer Science, Hubei University for Nationalities,
445000 Enshi, Hubei, China
e-mail: mhzheng3@gmail.com

J. J. Park et al. (eds.), *Proceedings of the International Conference on Human-centric Computing 2011 and Embedded and Multimedia Computing 2011*, Lecture Notes in Electrical Engineering 102,
DOI: 10.1007/978-94-007-2105-0_21,

Password is one of the ideal authentication approaches to agree a session key in the absence of Public Key Infrastructure (PKI) or pre-distributed symmetric keys. Low-entropy passwords are easy for humans to remember but cannot guarantee the same level of security as high-entropy secrets such as symmetric or asymmetric keys, so a password-based key agreement protocol could suffer from the so-called dictionary attacks [1]. Usually dictionary attacks are classified into two classes: on-line and off-line dictionary attacks. The on-line dictionary attacks are always possible, but these attacks can not become a serious threat because the on-line attacks can be easily detected and thwarted by counting access failures. However, off-line dictionary attacks are more difficult to prevent. The minimum required from a key agreement protocol is secure against this attack. An encrypted key agreement protocol that enables two communicating entities authenticating each other and establishing a session key via a shared weak password was proposed in [2]. Since then, some Two-party Password-based Authenticated Key Agreement (TPAKA) protocols were proposed in improving the security and performance. However, existing protocols are either limited by the use of Public Key Infrastructure (PKI) or by their security, suffering dictionary attack [3].

Recently, Joux [4] discovered the Gap Diffie-Hellman problem over elliptic curve. This hard problem can be implemented using bilinear pairing. In this paper, we propose a TPAKA protocol based on bilinear pairing. In our protocol, the legitimate user A and user B can share only a low-entropy human-memorable password and communicating over an insecure channel controlled by the active adversary, to agree upon a high-entropy session key among themselves. We emphasize that our password-authenticated protocol do not need any fixed PKI. Our protocol is provably secure in the random-oracle and ideal-cipher models under the Computation Gap Diffie-Hellman (CGDH) assumption [4].

The remainder of this paper is organized as follows: In Sect.21.2, we propose a two-party password-based key agreement protocol in details. In Sect.21.3, we discuss provably security of proposed protocol in the random-oracle and ideal-cipher models under the CGDH assumption. Finally, in Sect.21.4 we conclude.

21.2 Proposed Two-Party Password-Based Authenticated Key Agreement Protocol

In this section, we propose a two-party authenticated key agreement protocols enable two users communicating over an insecure using bilinear paring.

21.2.1 Notations

At first, we present the following notations are used throughout this paper:

- q: a secure large prime.
- p: a large prime such that $p = 2 \bmod 3$ and $p = 6q-1$.

- E: a supersingular curve defined by $y^2 = x^3 + 1$ over $\mathbb{F}_p$.
- $\mathbb{G}_1$: the subgroup of $E/\mathbb{F}_p$ of order q.
- P: a primitive generator for the group $E/\mathbb{F}_p$.
- $\mathbb{G}_2$: the subgroup of $\mathbb{F}^*_{q^2}$ of order q.
- $\hat{e} : \mathbb{G}_1 \times \mathbb{G}_1 \rightarrow \mathbb{G}_2$ be a bilinear pairing as defined in [4].
- ID_A, ID_B: the identities of user A and user B.
- pw_{AB}: the share password of user A and user B.
- $(\mathrm{E}_k, \mathcal{D}_k)$: an ideal-cipher system. E_k is a keyed permutation over G_q and $\mathcal{D}_k$ is inverse of E_k. k is the symmetric key.
- $\mathcal{H} : \{0,1\}^* \rightarrow \{0,1\}^{l_{\mathcal{H}}}$, a one-way hash function which maps a string to an element of $\mathbb{F}_p$.
- $\mathcal{F} : \{0,1\}^* \rightarrow \{0,1\}^{l_{\mathcal{F}}}$, a one-way hash function which maps a string to an element of $\mathbb{G}_1$.

21.2.2 Our Two-Party Password-Based Authenticated Key Agreement Protocol

In our protocol, the legitimate user A and user B can share only a low-entropy human-memorable password pw_{AB} and communicating over an insecure channel controlled by the active adversary, to agree upon a high-entropy session key k among themselves. The protocol was described as follows (Fig. 21.1):

Step 1. User A selects a number $a \in Z_p^*$ randomly, then computes $\eta_a = aP + pw_{AB}P$ and sends $\{ID_A, \eta_a\}$ to user B.

Step 2. Upon receiving $\{ID_A, \eta_a\}$, User B computes $aP = \eta_a - pw_{AB}P$ and $R = \mathcal{F}(ID_A, ID_B)$. Then B selects a number $b \in Z_p^*$ randomly, computes

$$s = \hat{e}(6aP, 6bR) = \hat{e}(6P, 6R)^{ab}, \text{ and}$$

$$\eta_b = bP + pw_{AB}P.$$

After that, B sends $\{ID_B, \eta_b\}$ to A.

Step 3. Upon receiving $\{ID_B, \eta_b\}$, user A computes

$$bP = \eta_b - pw_{AB}P,$$

$$R = \mathcal{F}(ID_A, ID_B) \text{ and}$$

$$s = \hat{e}(6aR, 6bP) = \hat{e}(6P, 6R)^{ab}.$$

Step 4. Finally, user A and user B can compute the session key $k = \mathcal{H}(aP, bP, R, s)$ simultaneously.

User A (ID_A, pw_{AB}) **User B** (ID_B, pw_{AB})

1. $a \in Z_p^*$

   $\eta_a = aP + pw_{AB}P$

   $\xrightarrow{ID_A, \eta_a}$

2. $aP = \eta_a - pw_{AB}P$

   $R = \mathcal{F}(ID_A, ID_B)$

   $b \in Z_p^*$

   $s = \hat{e}(6P, 6R)^{ab}$

   $\eta_b = bP + pw_{AB}P$

   $\xleftarrow{ID_B, \eta_b}$

3. $bP = \eta_b - pw_{AB}P$

   $R = \mathcal{F}(ID_A, ID_B)$

   $s = \hat{e}(6P, 6R)^{ab}$

4. $k = \mathcal{H}(aP, bP, R, s)$ $k = \mathcal{H}(aP, bP, R, s)$

Fig. 21.1 The proposed TPAKA protocol

Note that $6P$ in step 2 and step 3 is a generator of $\mathbb{G}_1$ because P is a primitive generator of $E/\mathbb{F}_p$ and $p = 6q - 1$. Accordingly, $6aP$ and $6bR$ belong to the group $\mathbb{G}_1$and can be the inputs of the bilinear pairing $\widehat{e}$.

21.3 Security Analysis of the Proposed Protocol

This section demonstrates the security of the proposed protocol. At first, a security model that is used to formalize the protocol and the adversary's capabilities was described. The theorem 1 in Sect.21.3.2 presents the main security result of the proposed key agreement protocol. This theorem describes that the security of the session key is protected against dictionary attacks.

21.3.1 Security Model

This model is used to formalize the protocol and the adversary's capabilities. We describe below our security model following closely the *Real-or-Random* (ROR) model of Abdalla et al. [5], instead of the *Find-then-Guess* (FTG) model of Bellare et al. [1] as standardized by Bresson et al. [6]. The ROR model seems more suitable for the password-based setting.

A protocol $\mathcal{P}$ for two-party password-based key agreement protocol assumes that there are two users A and B, who share a low-entropy secret password pw_{AB} which is uniformly drawn from a small dictionary of size N. This security model allows concurrent execution of the protocol among two users, so each of users may have several instances (or called oracles) involved in distinct ones. We denote the i-th instance of U_A by U_A^i. During the execution of protocol, the adversary is given control over all communication in the external network. The interaction between the adversary and users occur only via oracle queries, which model the adversary's capabilities in a real attack. These queries are as follows:

Send(U_A^i, m): The adversary $\mathcal{A}$ can carry out an active attack by this query. The output of the query is the response generated by the instance U_A^i upon receipt of message m according to the execution of protocol $\mathcal{P}$. The $\mathcal{A}$ is allowed to prompt the unused instance U_A^i to initiate the protocol by invoking Send (U_A^i, "*start*").

Reveal(U_A^i): This query models known key attacks. The adversary $\mathcal{A}$ is allowed to send this query to a user oracle U_A^i. If U_A^i has accepted a session key k, then it returns k to $\mathcal{A}$. Otherwise it returns NULL to $\mathcal{A}$.

Execute(U_A^i, U_B^j): This query models off-line dictionary attacks. This adversary $\mathcal{A}$ sends this query to the oracles U_A^i and U_B^j, and then gets back the transcript (all message flows) of a session of the protocol $\mathcal{P}$.

H(m): This query allows adversary $\mathcal{A}$ accessing to the oracle $\mathcal{H}$. If the input string m has not been asked, $\mathcal{H}$generates a random number r and returns r to $\mathcal{A}$. At same time, $\mathcal{H}$ records (m, r) into a public table $\mathcal{H} - table$. Otherwise, $\mathcal{H}$ researches the $\mathcal{H} - table$ to find the records $(m, r')$and returns r' to $\mathcal{A}$.

Test (U_A^i): This query models the misuse of the session key k by instance U_A^i. Once the instance U_A^i has accepted a session key, the adversary $\mathcal{A}$ can attempt to distinguish it from a random key as the basis of determining security of the protocol. A random bit b is chosen; if $b = 0$ then session key is returned while if $b = 1$ then a random key is returned. Note that this query is only available when U_A^i is fresh [3].

Finally adversary output a guess bit b'. Such an adversary is said to win the game if $b = b'$ where b is the hidden bit used by the Test oracle. Let Suc denote the event that the adversary $\mathcal{A}$ wins the game. We define $Adv_{\mathcal{P}}^{AKA}(\mathcal{A}, t) = |2\Pr[\text{Suc}] - 1|$ to be the advantage of the active adversary $\mathcal{A}$ in attacking the protocol $\mathcal{P}$. Protocol $\mathcal{P}$ is a secure TPAKA protocol resistant to the dictionary attacks if $\mathcal{A}, s$ advantage is a negligible function for any adversary $\mathcal{A}$ running in time at most t.

21.3.2 Security Proof of the Proposed Protocol

The following Theorem 1 presents the main security result of the proposed TPAKA protocol. This theorem describes that the protocol achieves the AKA security [7] of the session key. In this section, we will prove the security theorem in the random oracle model.

Theorem 1 *Let* Adv^{AKA} *be the advantage that an adversary* $\mathcal{A}$ *breaks the AKA security of the TPAKA protocol within time t. Let N be the size of all possible passwords. Assume* $\mathcal{A}$ *breaks the AKA security of the TPAKA protocol by running* q_s Send *queries,* q_s Execut e *queries and* q_h H *queries. Then we have*

$$\text{Adv}_{\mathcal{P}}^{\text{AKA}}(t, q_s, q_h) \leq q_h q_s \text{Adv}_{\hat{e}}^{\text{CGDH}}(t') + q_s/N$$

where $t' \leq (2q_e + q_s)T_p + (q_e + q_s)T_e + t$, T_p *is the time to generate a random point in* E_q, T_e, *is the time to perform a bilinear pairing operation.*

Proof Let Eve_0 is ht event that the adversary $\mathcal{A}$ breaks the AKA security of the TPAKA protocol. Let Eve_1 is the event that $\mathcal{A}$ breaks the AKA security of the TPAKA protocol without breaking the password security and Eve_2 is the event that $\mathcal{A}$ breaks the AKA security of the TPAKA protocol by breaking the password security. Then, we have

$$\Pr[\text{Eve}_0] = \Pr[\text{Eve}_1] + \Pr[\text{Eve}_2].$$

□

In the following, we demonstrate the probability $\Pr[\text{Eve}_0]$ from $\Pr[\text{Eve}_1]$ and $\Pr[\text{Eve}_2]$.

1) The probability $\Pr[\text{Eve}_1]$

Let $\Pr[\text{Eve}_1]$ is the probability that the adversity $\mathcal{A}$ breaks the AKA security without breaking the password security within time t. We construct form it a (t', ε') – CGDH adversity $\mathcal{A}'$ which can break CGDH assumption with probability at least ε' within time t'.

$\mathcal{A}'$ receives $(\hat{e}, P, xP, yP, zP)$ as an input, where $zP = \mathcal{F}(ID_A, ID_B)$. First, $\mathcal{A}'$ selects the shared password pw_{AB} for user A and user B. $\mathcal{A}'$ choices a number $r \in [1, q_s]$ randomly, subsequently. After that, $\mathcal{A}'$ starts running $\mathcal{A}$ as a subroutine and answers the oracle queries made by $\mathcal{A}$.

$\mathcal{A}'$ can answer the oracle queries as the real protocol because it knows all secrets of the protocol participants. However, if $\mathcal{A}$ asks Send (U_A^i, "*start*") at the *ith* Send query, $\mathcal{A}'$ computes $e_a = \text{E}_{pw_{AB}}(xP)$ and returns the first message flow of the TPAKA protocol to$\mathcal{A}$. When $\mathcal{A}$ asks $\text{Send}(U_B^*, ID_A, e_a)$, $\mathcal{A}'$ checks whether

$$e_a \stackrel{?}{=} \text{E}_{pw_{AB}}(xP).$$

If the equality holds, $\mathcal{A}'$ computes $e_b = \text{E}_{pw_{AB}}(yP)$ and returns the second message flow of the TPAKA protocol to $\mathcal{A}$. Note that if $\mathcal{A}$ asks a Reveal query of the session key constructed by xP and yP, $\mathcal{A}'$ terminates and outputs "$\perp$".

When $\mathcal{A}$ terminates, $\mathcal{A}'$ looks up the $\mathcal{H}$ – *table* to see if some queries of the form $\text{H}(xP, yP, zP, *)$ have been asked. If so, $\mathcal{A}'$ chooses at random one of them and outputs "$*$"; otherwise $\mathcal{A}'$ terminates and outputs "$\perp$".

Since the one-way function $\mathcal{H}$ is regarded as a random oracle, if $\mathcal{A}$ knows the session key k corresponded to the *i*th Send query, $\mathcal{A}$ must ask a $\text{H}(xP, yP, zP, \hat{e}(P, P)^{xyz})$ query which is recorded in the $\mathcal{H}$ – *table*. Let λ is the probability that $\mathcal{A}'$ correctly chooses among the possible $\mathrm{H}(xP, yP, zP, *)$ queries from the $\mathcal{H}-table$. It is straightforward that

$$\lambda \geq 1/q_h.$$

The probability that $\mathcal{A}'$ correctly guesses the moment at which $\mathcal{A}$ breaks the AKA security is equivalent to the probability that $\mathcal{A}'$ correctly guesses the value i, denoted by φ. Thus, we have

$$\varphi \geq 1/q_s.$$

From above, we know that the probability that $\mathcal{A}'$ breaks CGDH assumption is equivalent to the probability that $\mathcal{A}$ breaks the AKA security without breaking the password security multiplied by the probability that $\mathcal{A}'$ correctly guesses the moment at which $\mathcal{A}$ breaks the AKA security multiplied by the probability that $\mathcal{A}'$ correctly chooses among the possible $\mathrm{H}(xP, yP, zP, *)$ queries:

$$\begin{aligned} &\mathrm{Adv}_{\hat{e}}^{\mathrm{CGDH}}(\mathcal{A}) = \varepsilon' \\ &= \Pr[\mathrm{Eve}_1] \times \delta \times \varphi \\ &\geq \Pr[\mathrm{Eve}_1]/(q_s \times q_h) \end{aligned}$$

Therefore,

$$\Pr[\mathrm{Eve}_1] \leq q_s \times q_h \times \mathrm{Adv}_{\hat{e}}^{\mathrm{CGDH}}(\mathcal{A}).$$

Let T_P is the time to generate a random point in $\mathbb{G}_1$. Let $T_e$is the time to perform a bilinear pairing. In each execution of the TPAKA protocol, $\mathcal{A}'$ will generate two random points (aP, bP) and compute one bilinear pairings k. Thus, the running time of $\mathcal{A}'$ is:

$$t' \leq (2q_e + q_s)T_p + (q_e + q_s)T_e + t.$$

2) The Probability $\Pr[\mathrm{Eve}_2]$

The adversary $\mathcal{A}$ breaks the password security of the TPAKA protocol by the following two attacks: on-line dictionary attack and off-line dictionary attack.

Case 1 On-line Dictionary Attack

The adversary $\mathcal{A}$ can determine the correctness of a guessed password by sending it to a user in the TPAKA protocol. If the guessed password is correct, $\mathcal{A}$ get session key$k$. The probability that $\mathcal{A}$learns the user's password by on-line dictionary attack, denoted by $\Pr[\mathrm{Eve}_2^{on}]$, is bounded by q_s and N. That is

$$\Pr[\mathrm{Eve}_2^{on}] \leq q_s/N.$$

Case 2 Off-line Dictionary Attack

Note that the only messages contain the information of password pw_{AB} are $\eta_A = aP + pw_{AB}P$ and $\eta_B = bP + pw_{AB}P$. Since aP and bP are uniformly distributed in $E/\mathbb{F}_p$, the adversary $\mathcal{A}$ cannot get any information to verify the validity of a guessed password pw'_{AB} from η_A and η_B. Thus, the probability that

$\mathcal{A}$ learns a user's password by off-line dictionary attack, denoted by $\Pr[\mathrm{Eve}_2^{\mathit{off}}]$, is negligible.

By definition in Sect.21.3.1, the probability that $\mathcal{A}$ breaks the password security is the probability that $\mathcal{A}$ learns a user's password by on-line dictionary attacks added by the probability that $\mathcal{A}$ learns a user's password by off-line dictionary attacks. That is

$$\Pr[\mathrm{Eve}_2] = \Pr[\mathrm{Eve}_2^{on}] + \Pr[\mathrm{Eve}_2^{\mathit{off}}] \le q_s/N.$$

By above equations, we have

$$\begin{aligned}\mathrm{Adv}_{\mathrm{P}}^{\mathrm{AKA}}(\mathcal{A}) &= \Pr[\mathrm{Eve}_0] = \Pr[\mathrm{Eve}_1] + \Pr[\mathrm{Eve}_2] \\ &\le q_s \times q_h \times \mathrm{Adv}_{\hat{e}}^{\mathrm{CGDH}}(\mathcal{A}') + q_s/N.\end{aligned}$$

21.4 Conclusions and Further Work

This paper has proposed an efficient and secure two-party password-based key agreement protocol, derived form the authenticated key agreement protocol. Using the proof technique of Abdalla et al., the proposed protocol is proven to be secure against dictionary attacks under CGDH assumption in both the random oracle model and the ideal cipher model. To obtain secure and efficient two-party password-based key agreement protocol under the standard model instead of random oracle model is interesting topic and this area requires to be researched for further improvement.

Acknowledgments This work was partially supported by the Natural Science Foundation of Huhei Province under Grant No. 2009CDA143 and D20111901.

References

1. Bellare M, Pointcheval D, Rogaway P (2000) Authenticated key agreement secure against dictionary attacks. In: Proceedings of EUROCRYT'00, LNCS, vol 1807. Springer, Berlin, pp 139–155
2. Bellovin S, Merritt M (1999) Encrypted key exchange: password-based protocols secure against dictionary attacks. In: Proceedings of IEEE computer society symposium on research in security and privacy. IEEE Press, New York, pp 72–84
3. Zheng MH, Zhou HH, Li J, Cui GH (2009) Efficient and provably secure password-based group key agreement protocol. Comput Stand Interfaces 31(5):948–953
4. Joux A (2000) One round protocol for tripartite Diffie-Hellman. In: Proceedings of ANTS, LNCS, vol 1838. Springer, Berlin, pp 385–394
5. Abdalla M, Fouque PA, Pointcheval D (2005) Password-based authenticated key exchange in the three-party setting. Proceedings of PKC'05, LNCS, vol 3386. Springer, Berlin, pp 65–84

6. Bresson E, Chevassut O, Pointcheval D (2002) Group Diffie-Hellman key exchange secure against dictionary attack. Proceedings of ASIACRYPT'02, LNCS, vol 2501. Springer, Berlin, pp 497–514
7. Byun JW, Lee DH, Lim JI (2007) EC2C-PAKA: an efficient client-to-client password-authenticated key agreement. Info Sci 177(19):3995–4013

6. Bresson E, Chevassut O, Pointcheval D (2002) Group Diffie-Hellman [illegible] secure against [illegible] attacks. Proceedings of ASIACRYPT [illegible] LNCS vol [illegible] Springer, Berlin, [illegible]
7. Byun JW, Lee DH, Lim JI (2006) [illegible] client-to-client password-[illegible] key agreement [illegible]

Chapter 22
Subjective Logic-Based Trust Update for Wireless Sensor Networks

Yong Zhang, Keqiu Li, Wenyu Qu, Hong Gu and Dongwei Yang

Abstract Wireless sensor networks are drawing lots of attention as a method for realizing a ubiquitous society, which needs more than one sensor to accomplish the task together. Security is one of the critical problems. Trust and reputation management is considered as the effective supplement of traditional cryptography security for wireless sensor networks. This paper proposes the trust update algorithm based on subjective logic. This algorithm solves the invalidity of the trust update based on the posterior distribution of reputation random variable. The extensive simulations by J-Sim show the proposed algorithm can evaluate the validity of the trust update using belief of subjective logic opinion.

Keywords Wireless sensor network · Subject logic · Conditional reasoning · Trust updating

Y. Zhang · D. Yang
School of Computer and Information Technology, Liaoning Normal University, 116081 Dalian, China
e-mail: cony678@gmail.com

Y. Zhang · K. Li (✉)
School of Computer Science and Technology, Dalian University of Technology, 116024 Dalian, China
e-mail: keqiu@dlut.edu.cn

W. Qu
School of Information Science and Technology, Dalian Maritime University, 116026 Dalian, China

H. Gu
School of Control Science and Engineering, Dalian University of Technology, 116024 Dalian, China

J. J. Park et al. (eds.), *Proceedings of the International Conference on Human-centric Computing 2011 and Embedded and Multimedia Computing 2011*, Lecture Notes in Electrical Engineering 102,
DOI: 10.1007/978-94-007-2105-0_22,

22.1 Introduction

A wireless sensor network (WSN) typically comprises of a large number of sensor nodes deployed in high density in an area. These nodes have limited processing capability, limited memory and are powered by very limited energy batteries. The basis task of sensor nodes is to sense and process the information of the environment. Besides, they can use a wireless channel to collaborate among themselves and send the information to base stations. Currently, there exists a wide range of applications that make use of wireless sensor wireless [1, 2]. However, sensor nodes are implemented using inexpensive hardware components which are highly unreliable. Faults can be caused by physical damages, software mistakes, and resource depletion. Faulty nodes generate untrustworthy data readings that decrease the service quality of the network [3].

Typical methods to deal with the trust problem in WSNs concentrate on cryptographic technology. However, a cryptogram can only ensure trust data relaying but is helpless if the delivered data are untrustworthy. The trust and reputation system, on the other hand, can settle this problem. Trust and reputation management is considered as the effective supplement of traditional cryptography security [4]. It has been widely studied in peer-to-peer network, wireless communication, and the pervasive computing environment.

The field of trust management or reputation systems for WSN is becoming of interest in the recent years. A lot of effort has been done in the area of trust management systems for P2P, ad hoc networks and WSNs. Reputation system is used to help a node evaluate the trustworthiness of other sensor nodes and make decisions within the network. Ganeriwal et al. [5] propose a reputation-based framework for sensor networks (RFSN) where each sensor node maintains reputation metrics which both represent past behavior of other nodes and are used as an inherent aspect in predicting their future behavior. RFSN employs a beta reputation system for reputation representation, update, integration and trust evolution.

However, RFSN [5] ignores node's trend of current behavior, due to using reputation random expectation to represent trust. To overcome the above deficiencies, this paper presents a novel trust updating model based on subjective logic. In our model, we use subjective opinions to evaluate if expectation-based trust can reflect node's trend of current behavior. The simulation experiments show that the proposed model can reflect node's trend of long behavior and current behavior.

The rest of this paper is organized as follows. We first discuss some related works in Sect. 22.2. We present our subjective logic-based trust updating model in Sect. 22.3. In Sect. 22.4, we demonstrate the simulation scenarios, and give the analysis and provide the results. Section 22.5 presents the concluding remarks.

22.2 Related Work

22.2.1 Trust Management in WSNs

In previous years, a number of research works on trust management for wireless sensor networks have been done.

Shaikh et al. [6, 7] proposed a new lightweight group-based trust management scheme (GTMS) for wireless sensor networks, which employs clustering. Theoretical as well as simulation results show that GTMS is more suitable for large-scale sensor networks, and can detect and prevent malicious, selfish, and faulty nodes.

Sun et al. [8] describes trust evaluation mechanisms in distributed networks such as MANETs and sensor networks, with a focus on protecting such systems against malicious attacks.

Boukerche et al. [9] have proposed an ATRM scheme for WSNs. ATRM is based on a clustered WSN and calculates trust in a fully distributed manner. ATRM works on specific agent-based platform.

Sun et al. [10] have presented an information theoretic framework to quantitatively measure trust and model trust propagation in ad hoc networks. In the proposed framework, trust is a measure of uncertainty with its value represented by entropy. Authors develop four axioms that address the basic understanding of trust and the rules for trust propagation. Based on these axioms, Authors present two trust models: entropy-based model and probability-based model.

Ozdemir [11] has presented a reliable data aggregation and transmission protocol, called RDAT, which is based on the concept of functional reputation. Protocol RDAT improves the reliability of data aggregation and transmission by evaluating each type of sensor node action using a respective functional reputation. In addition, protocol RDAT employs a fault tolerant Reed–Solomon coding scheme based multi path data transmission algorithm to ensure the reliable data transmission to the base station.

Crosby and Pissinou [12] proposed a reputation and trust-based cluster head election algorithm for secure cluster formation. They considered reputation as a probabilistic distribution and calculates it using the beta distribution of node's past behavior, considering the number of successful and unsuccessful interactions. Boukerche and Li proposed an agent based trust and reputation management scheme for WSN [13]. It assumes clustered WSN with backbone and is based on a mobile agent system. It uses the mobile agent which maintains the trust and reputation locally with each node.

Wang et al. [3] investigated the benefits of a distributed reputation system in target localization, defined a node reputation as its measurement performance and computed the reputation by the Dirichlet distribution. By assuming the sensing model of each node to be mixed Gaussian, they used reputation to estimate parameters *f* the sensing model and modified a node's original measurement. In addition, they also developed a reputation-based local voting algorithm to filter the untrustworthy data and then estimated the target location by a particle swarm optimization algorithm.

Lopez et al. [14] made an analysis of different approaches for trust and reputation management systems for wireless sensor networks, and identified a set of best practices in the design of a trust management system for WSN.

Liu et al. [15] first give a description of familiarity value, which represents a node's familiar degree with another individual node, and then present a novel reputation computation model to discover and prevent selfish behaviors by combining familiarity values with subjective opinions.

22.2.2 Subjective Logic

Subjective logic [16] represents a specific belief calculus that uses a belief metric called *opinion* to express beliefs. An opinion is denoted as

$$\omega_x^A = (b_x, d_x, u_x, a_x) \tag{22.1}$$

where b_x, d_x, and u_x represent belief, disbelief and uncertainty, respectively. $b_x, d_x, u_x \in [0,1]$ and $b_x + d_x + u_x = 1$. ω_x^A expresses the relying party A's belief in the truth of statement x. The parameter $a_x \in [0,1]$ is called the base rate, and is used for computing an opinion's probability expectation value that can be determined as

$$E(\omega_x^A) = b_x + a_x u_x \tag{22.2}$$

More precisely, parameter a_x determines how uncertainty shall contribute to the probability expectation value $E(\omega_x^A)$. In the absence of any specific evidence about a given party, the base rate determines priori trust that would be put in any member of the community.

22.3 Subjective Logic-Based Trust Updating Model

22.3.1 Trust Update Based on the Posterior Distribution

RFSN [5] updates reputation using the posterior distribution. Let θ denote the reputation of node j held by node i. RFSN assigns to θ a prior distribution $p(\theta)$ that reflects uncertainty about the behavior of node j before any transactions with i take place. Prior distribution $p(\theta)$ with two parameters can be expressed

$$\begin{gathered} p(\theta) = \frac{\Gamma(\alpha+\beta)}{\Gamma(\alpha)\Gamma(\beta)}\theta^{\alpha-1}(1-\theta)^{\beta-1} \\ \forall\, 0 \le \theta \le 1, \quad \alpha \ge 0, \quad \beta \ge 0 \end{gathered} \tag{22.3}$$

Table 22.1 Conditional probability

	x_1	x_2
$p(X\|\theta)$	$p(x_1\|\theta_1) = 0.25$	$p(x_2\|\theta_1) = 0.75$
	$p(x_1\|\theta_2) = 0.75$	$p(x_2\|\theta_2) = 0.25$

where $\Gamma(\cdot)$ is the gamma function. The mean of a beta distribution with parameters (α, β) is $\alpha/(\alpha+\beta)$ and its variance is $\alpha\beta/(\alpha+\beta)^2(\alpha+\beta+1)$.

Let $X \in \{0, 1\}$ denote node i's rating of node j for a single transaction. Then, given j's reputation θ, the probability that node j will be cooperative is

$$p(X|\theta) = \theta^X(1-\theta)^{1-X}. \quad (22.4)$$

Once the transaction is complete, RFSN updates reputation using the posterior distribution for θ

$$p(\theta|X) = \frac{p(X|\theta)p(\theta)}{\int_{[0,1]} p(X|\theta)p(\theta)d\theta} = \frac{\Gamma(\alpha+\beta)}{\Gamma(\alpha)\Gamma(\beta)} \cdot \theta^{\alpha+X-1}(1-\theta)^{\beta+1-X-1}. \quad (22.5)$$

Then, the beta parameter updates would be,

$$\alpha_j^{\text{new}} = \alpha_j + p \quad \beta_j^{\text{new}} = \beta_j + 1 - p. \quad (22.6)$$

Let *trust metric* T_{ij} denote node i's prediction of the expected future behavior of node j. T_{ij} is obtained by taking a statistical expectation of this prediction:

$$T_{ij}^{\text{new}} = E[\theta_{ij}^{\text{new}}] = E[Beta(\alpha_j^{\text{new}}, \beta_j^{\text{new}})] = \frac{\alpha_j^{\text{new}}}{\alpha_j^{\text{new}} + \beta_j^{\text{new}}}. \quad (22.7)$$

In order to explain reputation update of RFSN, we give an example as follow. When nodes i and j transact, node i evaluates the reputation of node j according to result of transaction with node j. The result of transaction X includes two states x_1 and x_2, which represent transaction success and transaction unsuccess, respectively. Reputation value of node j is divided into two intervals, $\theta_1 = [0.0, 0.5]$ and $\theta_2 = [0.6, 1.0]$. Table 22.1 gives conditional probability $p(X|\theta)$ in the case of given reputation intervals.

Given a statistical base rate $a(\theta_1) = 0.2$, $a(\theta_2) = 0.8$. We can calculate the posterior probability according to (22.5) as shown in Table 22.2.

As shown in Table 22.2, $p(\theta_2|x_1)$ is 0.92 according to (22.5). It shows the probability which node's reputation value is in interval [0.6, 1.0], in the case of the latest successful transaction. Accordingly, $p(\theta_1|x_2)$ is 0.43 shows the probability that node's reputation value θ is in interval [0.0, 0.5], in the case of the latest unsuccessful transaction.

RFSN updates reputation expectation using the posterior distribution of θ. However, this update method based on posterior distribution would result in misdirection according to Table 22.2, due to hiding some uncertainty. For example, in the case of the latest unsuccessful transaction, $p(\theta_2|x_2)$ is 0.57 from

Table 22.2 Transaction results given reputation scope

Reputation intervals	$p(\theta\|x_1)$	$p(\theta\|x_2)$
θ_1: 0.0–0.5	$p(\theta_1\|x_1) = 0.08$	$p(\theta_1\|x_2) = 0.43$
θ_2: 0.6–1.0	$p(\theta_2\|x_1) = 0.92$	$p(\theta_2\|x_2) = 0.57$

Table 22.2 that indicates the posterior distribution of transaction is higher, which probably leads to a successful transaction while the latest transaction is unsuccessful. It is obvious that the reputation update method in RFSN does not reflect node's current behavior trend. In the other words, reputation update of RFSN is ineffective in this case.

Aiming at this shortage, this paper proposes a trust update method based on subjective logic, which uses support degree to evaluate the posterior distribution of reputation variable θ so as to judge if reputation update reflects node's current behavior trend.

22.3.2 *Trust Updating Evaluation Based on Subjective Logic*

Subjective logic uses belief to evaluate the posterior distribution of reputation variable. Subjective logic can express the uncertainty of the posterior distribution, and give the belief to the posterior distribution. According to subjective opinion, we evaluate the effectiveness of reputation update.

Given a subjective opinion $\omega_x = (b_x, d_x, u_x, a_x)$. A probability expectation value is calculated according to (22.8),

$$E(x_i|\theta_j) = p(X|\Theta) = b_{x_i|\theta_j}(x_i) + a_{x_i|\theta_j}(x_i)u_{x_i|\theta_j} \tag{22.8}$$

Compute the probability expectation values of each inverted conditional reasoning, according to (22.5) as:

$$E(\theta_j|x_i) = \frac{a(\theta_j)E(\omega_{X|\theta_j}(x_i))}{\sum_{j=1}^{l} a(\theta_j)E(\omega_{X|\theta_j}(x_i))} \tag{22.9}$$

where $a(\theta_j)$ denotes the base rate of θ_j.

In order to find the dimension(s) that can have zero belief, the belief will be set to zero in (22.8) successively for each $\theta_j \in \Theta$, resulting in different uncertainty values defined as:

$$u_{\Theta|x_i} = \frac{E_{\theta_j|x_i}(\theta_j|x_i)}{a_{\theta_j|x_i}(\theta_j)}, \quad j = 1, 2. \tag{22.10}$$

Table 22.3 Transaction result given the reputation interval

	x_1	x_2	X any
$p(X\|\theta_1)$:	$b(x_1) = 0.25$	$b(x_2) = 0.75$	$u = 0$
$p(X\|\theta_2)$:	$b(x_1) = 0.75$	$b(x_2) = 0.25$	$u = 0$

Table 22.4 Belief and uncertainty given the transaction result

Reputation interval	$p(\theta\|x_1)$	$p(\theta\|x_2)$
θ_1(0.0–0.5):	$b(\theta_1) = 0.00$	$b(\theta_1) = 0.29$
θ_2(0.5–1.0):	$b(\theta_2) = 0.60$	$b(\theta_2) = 0.00$
Θ:Any	$u = 0.40$	$u = 0.71$

Assume that θ_t is the dimension for which the uncertainty is minimum. The uncertainty-maximised opinion can then be determined as:

$$\omega_{\Theta|x_i}: \begin{cases} b_{\Theta|x_i}(\theta_j) = E_{\Theta|X}(\theta_j|x_i) - a_{\Theta|x_i}(\theta_j)u_{\Theta|x_i} \\ u_{\Theta|x_i} = u^t_{\Theta|x_i} \\ a_{\Theta|x_i} = a_{x_i} \end{cases} \tag{22.11}$$

Subjective logic trust update clearly shows the uncertainty hided in posteriori probability reasoning. The following elaborates the above example using subjective logic reasoning.

According to (22.8), the transaction result given the reputation interval is shown as Table 22.3.

According to Table 22.3 and (22.9–22.11), the belief and uncertainty given the transaction result are shown as Table 22.4.

According to Table 22.2, the $p(\theta_2|x_1) = 0.92$ is the probability of the expectation of random variable θ is in the interval θ_2 given the successful transaction. According to Table 22.4, $b(\theta_2) = 0.60$ for $p(\theta_2|x_1) = 0.92$. Compared with the $p(\theta_1|x_1) = 0.08$ with $b(\theta_1) = 0.00$, $b(\theta_2) = 0.60$ is higher, which shows the trust update is effective.

Similarly, according to Table 22.2, the $p(\theta_2|x_2) = 0.57$ is the probability of the expectation of random variable θ is in the interval θ_1 given the unsuccessful transaction. According to Table 22.4, $b(\theta_2) = 0.00$ for $p(\theta_2|x_2) = 0.57$. Compared with the $p(\theta_1|x_2) = 0.43$ with $b(\theta_1) = 0.29$, $b(\theta_2) = 0.00$ is lower, which shows the trust update is ineffective.

When the trust update is ineffective, we introduce a new method to update trust value. When the trust update is ineffective and the times of ineffective updates are more than k, the trust update could be modified as

$$T^{\text{near}}_{ij} = \frac{1}{2}|(t_1 - t_2)| \tag{22.12}$$

where t_1, t_2 is the boundary of the reputation interval with higher belief.

If the expectation of the reputation is in the reputation interval with lower belief due to the base rate, choosing the middle value of the reputation interval with the higher belief as the trust update will better reflect the uncertainty resulting from the posterior probability reasoning.

In addition, we can use the skewness and the kurtosis to update trust values when the trust update is ineffective, because Beta distribution is the single peak distribution. Single peak distribution bears two eigenfunction, such as the skewness and the kurtosis. Coefficient skewness is calculated from the ratio of the third central moment to the third power of the standard deviation. Coefficient of kurtosis is calculated from the ratio of the fourth central moment to the fourth power of the standard deviation. When the skewness is positive value, the distribution will be inclined to the value higher than expectation. When the skewness is negative value, the distribution will be inclined to the value lower than expectation. If the skewness >2, the inclination will be serious. When the kurtosis >3, the distribution is steep.

Having updated the reputation, the kurtosis is calculated from the posterior distribution. If the belief is higher and the kurtosis ≥3, the posterior distribution concentrates around the expectation. In other words, the trust value described by posterior distribution will effectively represent the trust of the bargainer.

22.3.3 Proposed Algorithm Description

The proposed reputation updating algorithm based on subjective logic is presented as follows.

Algorithm *Subjective Logic Trust Update*
Input: Transaction records
Output: Trust Value T_{ij}^{new} and its evaluation
For $i = 1$ to n do
Step 1: Calculate subjective logic opinions according to transactions.
  1.1 Calculate the posterior distribution according to Eqs. 4 and 5;
  1.2 Calculate the expectation of subjective logic using Eq. 8;
  1.3 Calculate new subjective logic opinions using Eqs. 9-11.
Step 2: Evaluate the effectiveness of reputation update using subjective logic.
  2.1 Update two parameters of Beta distribution according to Eq. 6, then update trust value T_{ij}^{new} according to Eq. 7;
  2.2 The value of T_{ij}^{new} is in interval θ_1 or θ_2. If its corresponding posterior probability is bigger and its belief is higher, current trust update T_{ij}^{new} is effective. If its corresponding posterior probability is bigger and its belief is lower, current trust update T_{ij}^{new} is ineffective.
  2.3 If T_{ij}^{new} is ineffective and the times of ineffective updates are more than k, choosing the median of the reputation interval with the higher belief as the trust update according to Eq. 12.
End for

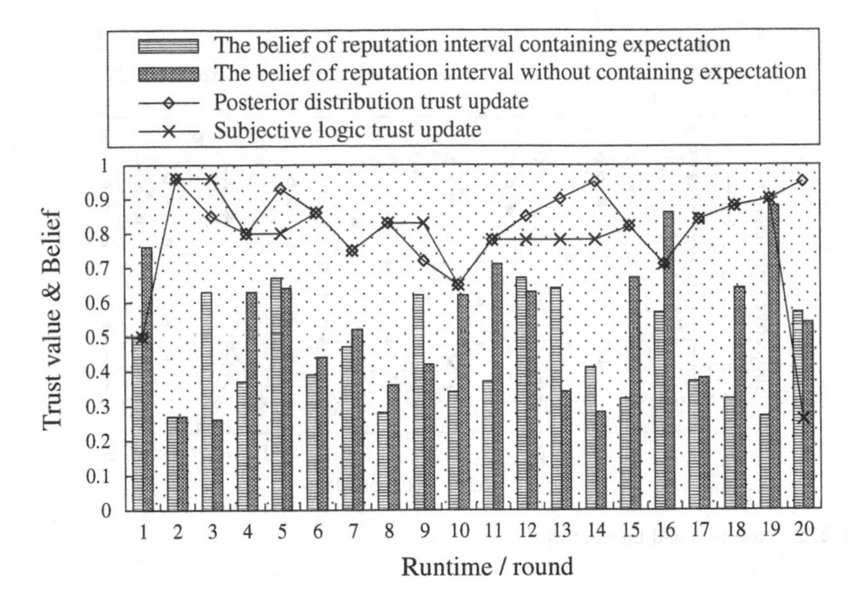

Fig. 22.1 Trust update and belief

22.4 Simulation and Analysis

The two trust update algorithms have been implemented in J-Sim 1.3 patch 4 [16]. In the simulation, 150 nodes are deployed randomly in the 100 × 200 area. The initial energy for each node is 1.0. MAC layer adopts the 802.15.4 protocol and network layer uses LEACH. Trust manage system adopts RFSN. The experiments compare the posterior distribution trust update with the subjective logic trust update in both trust update track and the energy consumption. Furthermore, we analyze the change of various elements in subjective logic opinion. The simulation runs 240 rounds, The result is given as follows.

22.4.1 Trust Value and Belief

Trust value and belief of both algorithms are shown as Fig. 22.1. The right data cylinder of each pair of cylinders represents the belief of the reputation interval which the expectation of reputation random falls down. The left one represents the belief of other reputation interval. If the right data cylinder is higher than the left one, the current trust update bears higher belief, and is effective. If the right data cylinder is lower than the left one, the current trust update bears lower belief, thus the trust update will be the median of the reputation interval with higher belief. The comparison between the two kinds of trust update shows that the subjective logic trust update better reflects the uncertainty changes resulting from the changes

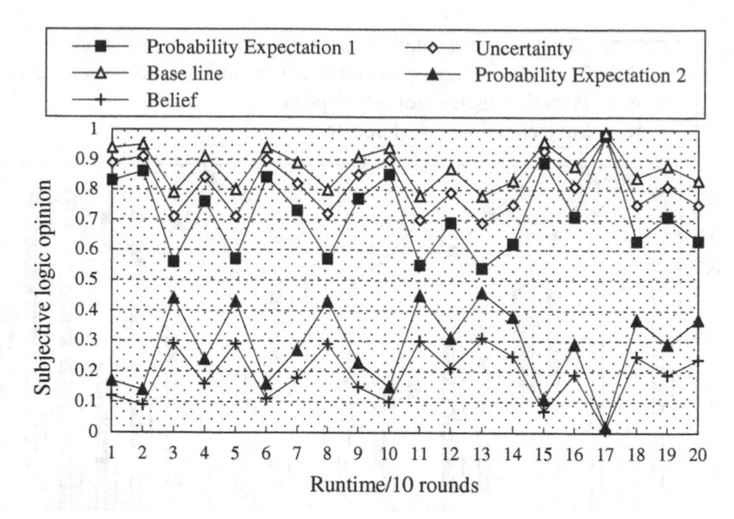

Fig. 22.2 Base rate and uncertainty

of the base rate, so as to effectively evaluate trust update based on the posterior probability reasoning.

22.4.2 Changes of Uncertainty Resulting from Changes of the Base Rate

As is shown in Fig. 22.2, probability expectation one is used to calculate the uncertainty in (16). Probability expectation two is used to calculate the probability expectation with the non-zero belief. The change of subjective logic opinion is shown in Fig. 22.2. The probability expectation 1 increases along with both the base rate and the uncertainty. In other words, the increase of the base rate will result in the increase of uncertainty. The belief is inversely proportional to uncertainty, which shows the belief can effectively measure the validity of trust update base on posterior distribution as shown in Fig. 22.2.

22.4.3 Energy Consumption Contrast

As is shown in Fig. 22.3, the both trust update algorithms consume almost the same amount of energy. The energy consumption in subjective logic is higher because evaluating the trust update will consume extra energy. However, the amount of calculation is comparatively small, so the energy consumption is almost the same with trust update based on the posterior distribution.

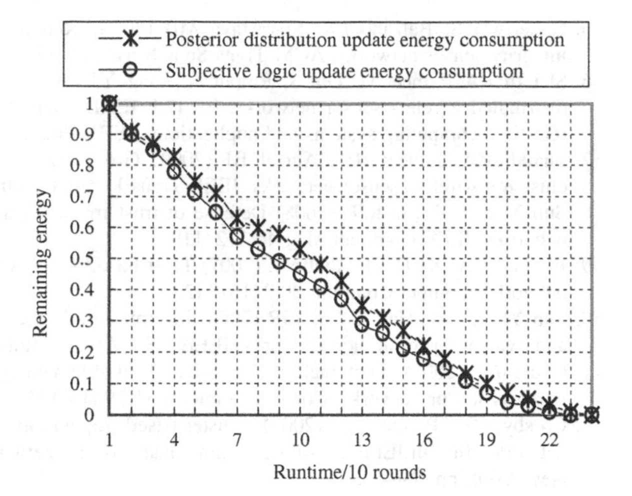

Fig. 22.3 Energy consumption

22.5 Conclusion

Aiming at the invalidity of trust update based on the posterior distribution, this paper proposed the trust update algorithm based on subjective logic. This algorithm solves the invalidity of the trust update based on the posterior distribution of reputation random variable. The simulation shows that the proposed algorithm can evaluate the validity of the trust update using belief of subjective logic opinion. However, the algorithm may not evaluate the validity of trust update under some conditions, such as the higher belief and lower skewness, or lower belief and higher skewness. Therefore, how to evaluate the validity of trust update using the subjective logic reasoning in above cases is the key area of our future work.

Acknowledgments This work is supported by NSFC under grant nos of 60973117, 60973116, 90818002, and 60973115, and New Century Excellent Talents in University (NCET) of Ministry of Education of China. This work is also supported by Liaoning Doctoral Research Foundation of China (Grant No. 20081079), and Dalian Science and Technology Planning Project of China (Grant No. 2010J21DW019).

References

1. Akyildiz IF, Su W, Sankarasubramaniam Y, Cayirci E (2002) A survey on sensor networks. IEEE Commun Mag 40:102–114
2. Li M, Liu YH (2009) Underground coal mine monitoring with wireless sensor networks. ACM Trans Sens Netw 5:1–29
3. Wang X, Ding L, Bi DW (2010) Reputation-enabled self-modification for target sensing in wireless sensor networks. IEEE Trans Instrum Meas 59:171–179
4. Yu H, Shen ZQ, Miao CY, Leung C, Niyato D (2010) A survey of trust and reputation management systems in wireless communications. Proc IEEE 98:1755–1772

5. Ganeriwal S, Balzano LK, Srivastava MB (2008) Reputation-based framework for high integrity sensor networks. ACM Trans Sens Netw 4:1–37
6. Shaikh RA, Jameel H, Lee S, Rajput S, Song YJ (2006) Trust management problem in distributed wireless sensor networks. In: 12th IEEE international conference on embedded real-time computing systems and applications, IEEE Press, New York, pp 411–414
7. Shaikh RA, Jameel H, d'Auriol BJ (2009) Group-based trust management scheme for clustered wireless sensor networks. IEEE Trans Parallel Distrib Syst 20:1698–1712
8. Sun Y, Han Z, Liu KJR (2008) Defense of trust management vulnerabilities in distributed networks. IEEE Commun Mag 46:112–119
9. Boukerch A, Xu L, EL-Khatib K (2007) Trust-based security for wireless ad hoc and sensor networks. Comput Commun 30:2413–2427
10. Sun YL, Yu W, Han Z, Liu KJR (2006) Information theoretic framework of trust modeling and evaluation for ad hoc networks. IEEE J Sel Areas Commun 24:305–317
11. Ozdemir S (2008) Functional reputation based reliable data aggregation and transmission for wireless sensor networks. Comput Commun 31:3941–3953
12. Crosby GV, Pissinou N (2006) Cluster-based reputation and trust for wireless sensor networks. In: 4th IEEE consumer communications and networking conference, IEEE Press, New York, pp 604–608
13. Boukerche A, Li X (2005) An agent-based trust and reputation management scheme for wireless sensor networks. In: 2005 IEEE global telecommunications conference, IEEE Press, New York, pp 1857–1861
14. Lopez J, Roman R, Agudo I, Fernandez-Gago C (2010) Trust management systems for wireless sensor networks: best practices. Comput Commun 33:1086–1093
15. Liu YL, Li KQ, Jin YW, Zhang Y, Qu WY (2011) A novel reputation computation model based on subjective logic for mobile ad hoc networks. Future Gener Comput Syst 27:547–554
16. Josang A (2008) Conditional reasoning with subjective logic. J Multiple-Valued Log Soft Comput 15:5–38
17. J-Sim simulator. http://j-sim.cs.uiuc.edu/

Chapter 23
A Detection Method for Replication Attacks in Wireless Sensor Networks

Cheng-Yuan Ku, Yi-Chang Lee, Yi-Hsien Lin and Yi-Wen Chang

Abstract Nowadays the wireless sensor networks are widely used in many applications, such as home security monitoring, healthcare applications, and traffic control. However, a new-type attack named node replication attack has been proved to be a harmful attack for wireless sensor networks (WSNs). In this study, we try to propose an improved method which needs less memory overhead and communication cost to detect the so-called node replication attack. Simple analysis shows that this method is efficient. More analyses and detailed simulations will be done in the near future.

Keywords Wireless sensor networks (WSNs) · Node replication attacks · Intrusion detection

23.1 Introduction

The characteristics of low computational capacities, less memory, wireless communications, and extremely limited energy supply (usually a battery) of wireless sensor network (WSN) are much different from wired networks. Thus, the WSN is

---

C.-Y. Ku (✉) · Y.-C. Lee · Y.-H. Lin · Y.-W. Chang
Department of Information Management, National Chung Cheng University, 168, University Rd, Min-Hsiung, Chia-Yi County, Taiwan, China
e-mail: cooper.c.y.ku@gmail.com

Y.-C. Lee
e-mail: huyahuku2@yahoo.com.tw

Y.-H. Lin
e-mail: lego2961@gmail.com

Y.-W. Chang
e-mail: azure@mis.ccu.edu.tw

J. J. Park et al. (eds.), *Proceedings of the International Conference on Human-centric Computing 2011 and Embedded and Multimedia Computing 2011*, Lecture Notes in Electrical Engineering 102,
DOI: 10.1007/978-94-007-2105-0_23,

unsuitable to adopt too many rounds of the public key computations for security concerns, such as RSA or elliptic curve cryptography (ECC) due to the high energy consumption and computational complexity.

Unfortunately, a novel attack named node replication attack has been proved to be precarious. Thousands of low-cost sensor nodes are often deployed in unshielded places and don't have enough capability to avoid physical attacks. An adversary can easily capture the nodes and replicate them by using their legitimate secret key instead of compromising the cryptosystem. With several clones of the sensor node, the adversary can strategically add them into the network in order to steal some important information or control the network by distributing the false data.

As far as detecting node replication is concerned, there are two different strategies: Centralized Detection and Local Detection. Nevertheless, these approaches are not satisfactory. Firstly, they require large memory capacity and consume volumes of energy. Secondly, most previous protocols are restricted in a specific application of WSNs. They do not have contingent capability to be adopted in a real and variable network. Namely, they are short of adaptability. Last but not least, some assumptions of previous protocols make it impractical in a real world. Firstly, they assume that nodes are primarily stationary, which is a serious drawback. Secondly, each node has geographic location information about itself by using a GPS device. Thirdly, the security is based on many rounds of computational complexity such as ECC.

Due to the above-mentioned problems, a simpler detection protocol for node replication attack is necessary. To satisfy these requirements of users, this study proposes a detection protocol which can work against node replication attacks. We assume that the sensor nodes are not stationary, but mobile. Our protocol can detect replication with high probability and only needs lower communication overhead and memory. These characteristics are very useful for different applications of WSNs.

23.2 Literature Review

Node replication attack is an attack that adversaries capture nodes and extract some secret information (legitimate secret key) instead of compromising the cryptosystem. With several clones of the sensor node, adversaries could control the network gradually. Thus, the detection of node replication attacks in a wireless sensor networks is an important problem [1]. Fu et al. also mentioned that the node replication attack was a significant threat to the proposed schemes in their paper [2]. This is because the low-cost sensor nodes are often limited by computational capacities, less memory and limited energy.

Recently, two simple approaches against node replication attacks named Centralized Detection and Local Detection were proposed. In 2005, Parno et al. proposed an innovative scheme and brought about many researchers in discussion

[3]. The Centralized Detection relies on a central base station for detecting network activity, which is the easiest and most straightforward detection scheme. The base station examines every neighbor list of sensor nodes. If the replicated nodes were discovered, it floods the message to the whole network. However, there are usually high volumes of traffic near the base station. Once the base station fails, the whole network will also fail.

Therefore, to avoid relying on central base stations, many applications with voting mechanism named Local Detection were proposed [4]. The idea of local detection is that every sensor is responsible for its neighbors. If any node finds suspicious nodes, the voting mechanism will get rid of it. Unfortunately, this local detection fails to detect two replicated nodes which locate faraway more than two hops. Moreover, several compromised nodes might eliminate legitimate nodes by voting mechanism and legitimate nodes sometimes are eliminated due to the probabilistic error.

In view of the drawbacks for Centralized Detection and Local Detection approaches, Pamo et al. suggested two protocols: Randomized Multicast protocol and Line Selected Multicast protocol which are improved from the previous solutions [3]. In Randomized Multicast protocol, every node announces its location claim, and their neighbors randomly select the witness nodes to forward this claim for verification. The randomization function makes the adversary hard to predict the witness. However, this protocol needs a relatively high communication cost since the claim messages are flooded throughout the network. As for Selected Multicast Protocol, the neighbors send the location claim to the witnesses who are randomly selected and every node stores the location claim along the way. Once the intermediate node receives conflicting claim message, it broadcasts the alert message to the whole network. The simulation showed that the Line Selected Multicast protocol can detect more than 80% of a replicated node. However, some assumptions make it impractical for a real network. They assume that nodes are primarily stationary, which is a serious drawback for the applications of WSNs.

23.3 Design of Detection Method

23.3.1 Assumptions

As discussed in Sect. 23.2, this study tries to improve the Line Selected Multicast protocols. Therefore, some assumptions are referred to the detection protocol of Parno et al. [3].

Assumption 1 Nodes are mobile instead of primarily stationary

We improve the protocol by relaxing the limits of stationary.

Assumption 2 The protocol does not rely on location claims

GPS devices on each sensor node are not required. We do not use additional communication for location information but use a series of random numbers and time information instead to keep the transmission overheads as low as possible.

Assumption 3 Other attacks are not in the scope of this paper

What we concern is that the adversary captures a node and copies its ID and legitimate key instead of creating a new legitimate ID. If the adversary could create a new ID then it is of Sybil Attack. Other Sybil defenses discussed in Newsome et al. [5] are not in the scope of this paper.

Assumption 4 Each node performs signature verification by public key systems

We decide to only adopt the public-key certificate instead of introducing the whole public key cryptosystem to keep the corresponding computations as few as possible. As indicated by previous research result [6], occasional use of public-key signature verification is reasonable and the public key infrastructure can be easily replaced.

Assumption 5 The routing protocol will not fail and each neighbor of nodes performs detection protocol properly without time synchronization problem

Time synchronization is a critical issue for any distributed system.

23.3.2 Protocol Description

In this section, we present the proposed method for detection. Every node would execute detection periodly. The pseudo-code for every node is described below:

```
1. R(t)<-ReceiveBroadcastedRand();
     // R(t) is a random number generated at time t
2. Set timer T;
3. α->NeighborsOf(α);
   Message(α): Signature[ID(α), T(α), #seq(βm)]PK(α), ID(α), R(t);
4. While timer T not time-out and Receive Message(α)
5.   If IsNotForwardedMessage(α) then
6.     WitnessSelect<-Random_Fun[ID(α), R(t)] with probability Pλ
7.   For all WitnessSetOf(α) do
         Forward Message(α):
         <Sig[ID(α), T(α), #seq(βm)]PK(α), ID(α), R(t)>;
8.   end;
```

```
9. If IsForwardedMessage(α) then
        <Sig[ID(α), T(α), #seq(βm)]PK(α), ID(α), R(t)>;
10.     If IsNotDestination then
            IDdestination<-Forward Message(α)
11.     If IsNotRecent or IsDuplicate then
            Process the exception;
12.     If IsNotDuplicate then
            Store Message(α) and continue next loop;
13.     Else
14.         Verifies the signature<-PublicKey(α)<-f[ID(α)];
15.         If NotCoherent(<T(α)of ID(α)>,and<T(α') of ID(α)>)
               then Flood Sig[ID(α),Message(α),
16.                        ,Message(α'),WitnessID]PK(witnessID);
17.End;
18.Fresh Memory;
```

Furthermore, our protocol could be divided into four steps:

Step 1: *random number broadcasting*.

A random number is generated at time t. We use R(t) to represent it. R(t) is shared among all nodes (pseudo-code, line 1). This process is performed by centralized broadcast. The central-based station is responsible for this mission.

In our research, we use the term "coordinator" to represent the node which generates random number. Central-based station is not necessary because the coordinator could be a general sensor node, which broadcasts the generated random number in one round.

Step 2: *signature authentication*.

As a node (α) receives R(t) at time T(α), this node enters state of signature authentication. T(α) represents the time when node (α) receives R(t). T(α) would be different if coming from different nodes because this value is related to not only the distance from the coordinator but also the routing path. This makes the adversary difficult to predict T (α)'s value because the routing path is much different. It gives T(α) an important role in verifying authentication. If there are two different T(α)'s in different signature authentication message, and the two IDs of nodes are identical, then this node is suspicious, probably replicated.

As for the node (α) which receives R(t), it must sign authentication with private key. Here PK(α) represents a private key of node (α). The node (α) must pass this authentication Message (α) to every neighbor (pseudo-code, lines 2–3). The #seq(β_m) is a sequence number correlated with the neighbor list and node (α).

Step 3: *select the witness.*

Each neighbor of node (α) will receive Message (α) and send it to a set of witnesses with probability P_λ (pseudo-code, lines 5–6). The argument P_λ is important. If P_λ is high, the detection rate of node replication attack will be high but it also results in high communicational cost. Hence, we design the detection protocol as flexible to respond to different requirements in variable network. Therefore the argument P_λ is variable, not constant.

The set of witnesses W(α) is also generated from a random function, Random_Fun [ID (α), R(t)] (pseudo-code, lines 6–7). Therefore, the witnesses would be anywhere in network. The random number in every round is generated in an unpredictable way. It makes the adversary hard to comprise all ubiquitous witnesses in such a short period.

Step 4: *detection.*

For each Message (α), the previous processes ensure that authentication message is successfully received by witness (pseudo-code, lines 9–11). When the witness receives the Message (α) at first time, it simply stores the Message (α) in memory and extracts the information ID(α) into cache for duplicate verification (pseudo-code, line 12). If other Message (α) is received with the same ID (such as ID(α) == ID(α')), the witness will check whether T(α) == T(α'). If T(α) and T(α') are not identical, then the witness would alarm all nodes by flooding the network (pseudo-code, line 15). In fact, there is usually a set of witness responsible for detecting a node. It would take a very short time until the entire network receives the alarm. To compare with previous protocols which rely on a single witness or central-based station to flood the alarm, our protocol is more reliable.

The witness is area-based. All nodes in a "cell" are a group of witness. The cell could be imagined as a hexagon and the size is correlated with communication radius. We can choose a suitable hexagon size to be with 95% more than two witnesses in a cell.

23.3.3 An Example

We use the following example to demonstrate our protocol. Assume that the adversary replicates a node with ID(α). We use node (α) to represent it (Figs. 23.1, 23.2, 23.3, 23.4). The clone is node (α') and β_1, β_2,…,βn are the neighbors. The coordinator is the generator of random number R(t). It triggers the detection of node replication. The witnesses W (α) are responsible for verifying the node (α).

1. Coordinator generates and broadcasts a Random Number at time t : R(t)
   - Coordinator triggers the detection of node replication.
   - Our detection can be performed with centralized broadcast. If the other WSN does not have a base-station, the random number election with distributed mechanisms is useful.

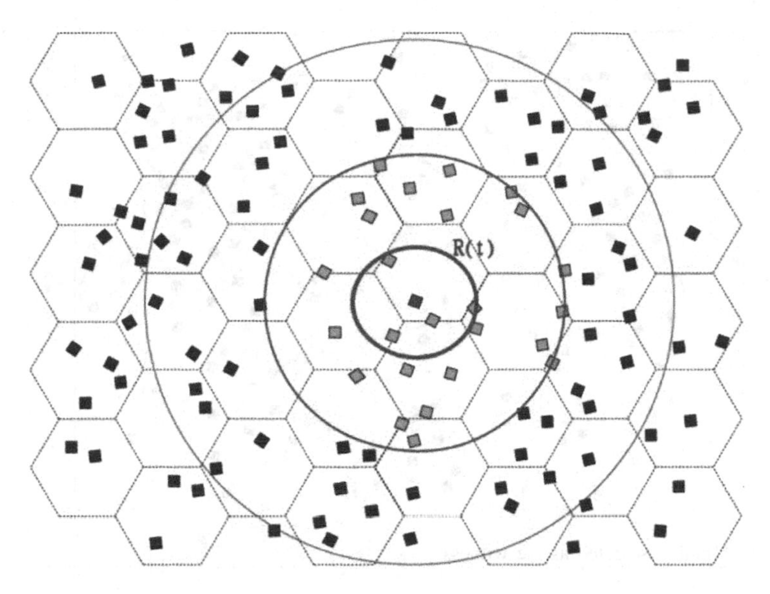

Fig. 23.1 Random Number R(t) is broadcasted

Fig. 23.2 β_1, β_2 ...βn would select the witness and send the Message (α)

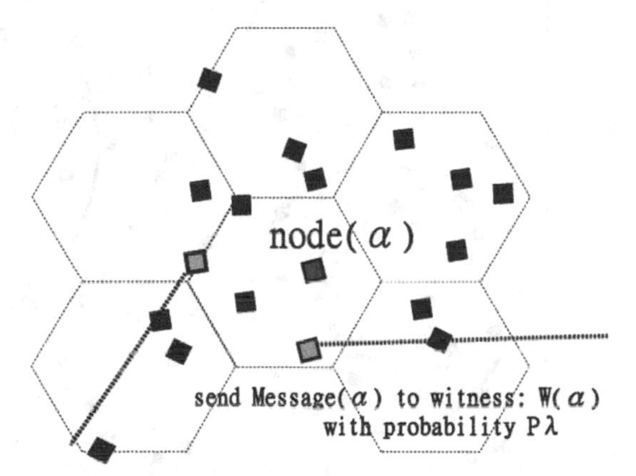

2. As node (α) receives R(t) at time T(α), the node (α) signs Message (α) with private key and passes it to every neighbor.
   - Message (α): Sig[ID(α), T(α), #seq(β_m)]$_{PK(\alpha)}$, ID(α), R(t). The neighbors of node (α), β_1, β_2 ...βn wait for Message (α) and add a parameter #frame(β_m) to a set of witnesses W(α) with probability P_{λ}.
   - β_1, β_2 ...βn do not communicate with node (α) until receiving Message (α).
   - #seq(β_m) is a random number corresponding with every neighbor of node(α).
3. The set of witnesses W(α) is generated from Random_Fun[ID(α), R(t)] and every node receiving Message (α) will check if it is one of witness or not. If yes

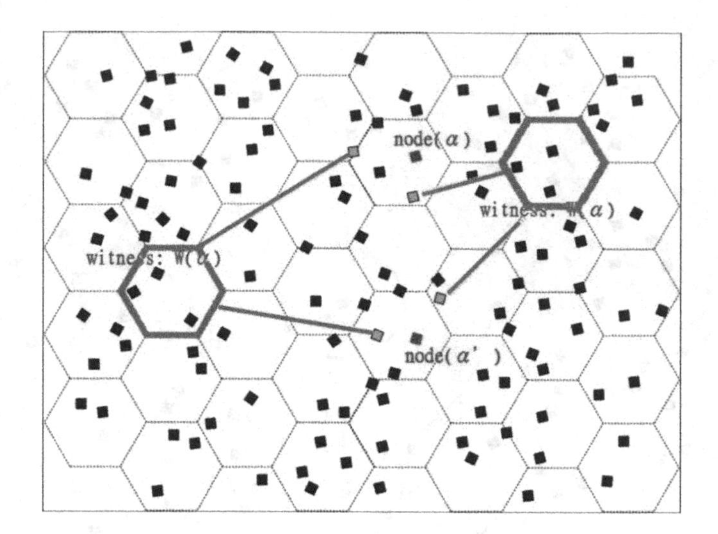

Fig. 23.3 A cell is selected as the witness

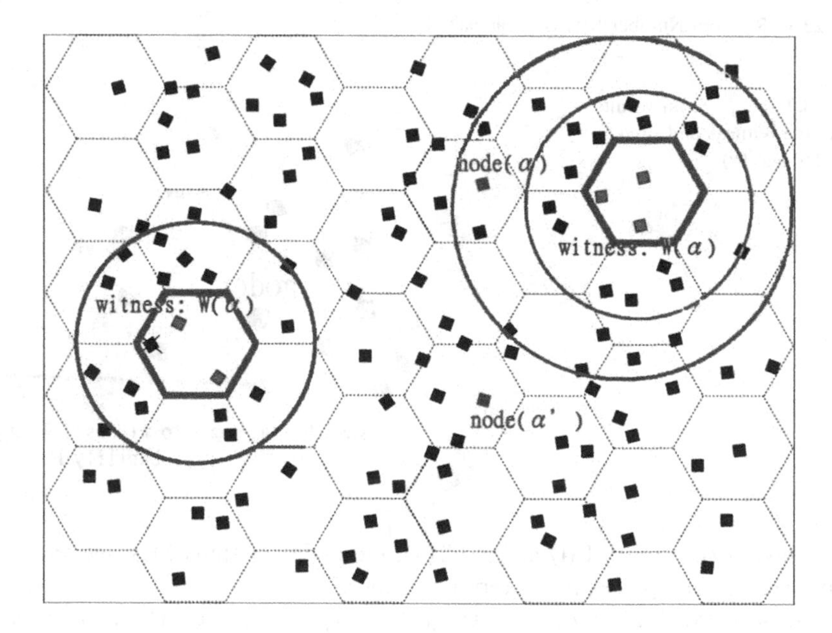

Fig. 23.4 The witness alarms all nodes by flooding network

then storing Message (α) in memory for detection. If no, passing Message (α) to next node and not storing Message (α) in memory to save space.

4. Witness check:
   If any ID(α) == ID(α') and T(α) is not equal to T(α') then node (α) and node (α') are suspicious.

Table 23.1 Success rate for detecting replication under different P_{λ}

Prl	P_{λ}	d	Prl	P_{λ}	d
0.996998	0.15	40	0.993240	0.15	35
0.999734	0.2	40	0.999915	0.25	35
0.999980	0.25	40	0.999189	0.2	35
0.999999	0.3	40	0.999992	0.3	35
0.984797	0.15	30	0.965900	0.15	25
0.997526	0.2	30	0.992458	0.2	25
0.999643	0.25	30	0.998495	0.25	25
0.999955	0.3	30	0.999732	0.3	25
0.923983	0.15	20	0.832922	0.15	15
0.977074	0.2	20	0.930869	0.2	15
0.993668	0.25	20	0.973452	0.25	15
0.998405	0.3	20	0.990527	0.3	15

If any ID(α) == ID(α') and #frame(β_m) is not equal to #frame($\beta_{m'}$) but #seq(β_m) == #seq($\beta_{m'}$) then node (α) and node (α') are suspicious.
It means Message (α) may be eavesdropped.

5. Coordinator broadcasts next Random Number at time t+1 : R(t+1)
   Coordinator broadcasts R(t+2), R(t+3), …, R(t+n) until the system is considered secure.

23.4 Protocol Analysis

The worst case is that there is only one replicate node in network. In fact, there are many replicate nodes in the network and they can be easily detected by our protocol. The probability of detection rate is related with the number of neighbors d and P_{λ}:

$$E[\text{Detection Rate}] = G\left(1-(1-P_{\lambda})^{d}\right)^{2} \tag{23.1}$$

We assume only one witness for detecting (G = 1). The (23.1) is simplified as follows:

$$P_{r1} = \left(1-(1-P_{\lambda})^{d}\right)^{2} \tag{23.2}$$

In Table 23.1, we demonstrate the estimated success rate.

Actually, more analyses and simulations are necessary to validate this protocol and we are still processing these data.

23.5 Conclusions

The detection of node replication attacks in a wireless sensor networks is a fundamental problem. However, only few feasible solutions have been proposed. The main contribution of this paper is try to thwart node replication attacks. We have discussed pervious research results about the node replication attacks and suggested an efficient detection protocol to secure wireless sensor networks.

In our protocol, the mobile sensor nodes are considered since it is very possible for the sensor nodes to be mobile, not stationary, in the real world. Besides, our protocol provides a good communication overhead, has a lower memory needs compared with earlier protocols and the probability of successful detection is still very high.

Our approach which originates from the idea of time stamp is innovative on mobile networks and makes the GPS (Location Claim Device) not a necessity, but an alternative. In a near future, more corresponding simulations will be summarized for our protocol to show the efficiency and improvement in detection capability.

References

1. Chan H, Perrig A, Song D (2003) Random key predistribution schemes for sensor networks. In: Proceedings of 2003 IEEE symposium on security and privacy, pp 197–213
2. Fu H, Kawamura S, Zhang M, Zhang L (2008) Replication attack on random key predistribution schemes for wireless sensor networks. Comput Commun 31:842–857
3. Parno B, Perrig A, Gligor V (2005) Distributed detection of node replication attacks in sensor networks. In: Proceedings of 2005 IEEE symposium on security and privacy, pp 49–63
4. Ssu KF, Chou CH, Cheng LW (2007) Using overhearing technique to detect malicious packet-modifying attacks in wireless sensor networks. Comput Commun 30:2342–2352
5. Newsome J, Shi E, Song D, Perrig A (2004) The sybil attack in sensor networks: analysis and defenses. In: Proceedings of the third international symposium on information processing in sensor networks, ACM Press, pp 259–268
6. Malan DJ, Welsh M, Smith MD (2004) A public-key infrastructure for key distribution in TinyOS based on elliptic curve cryptography. In: Proceedings of first annual IEEE communications society conference on sensor and ad hoc communications and networks, pp 71–80

Chapter 24
VIsolator: An Intel VMX-Based Isolation Mechanism

Liuxing He, Xiao Li, Wenchang Shi, Zhaohui Liang and Bin Liang

Abstract Integrity Measurement Mechanisms (IMMs) can be used to detect tampering attacks to integrity of system components, so as to ensure trustworthiness of a system. If an IMM has been compromised, measurement results are untrustworthy. Therefore, IMMs must be protected to provide credible measurement results. In this paper, we propose an isolation mechanism based on the Intel VMX technology to protect an IMM from bcing tampered with, even if the whole operating system (OS) is untrusted. The isolation mechanism we proposed can be divided into two parts, one of which is a module running inside an OS while the other one is a hypervisor running as the basis of this OS. As an IMM may be attacked by untrusted software in the way of writing its memory, the module of our isolation mechanism is used to modify the access permission of the IMM. Nevertheless, the threat is not disappeared as untrusted software may run in kernel

The work of this paper was supported in part by National Natural Science Foundation of China (61070192, 91018008, 60873213), National 863 High-Tech Research Development Program of China (2007AA01Z414), Natural Science Foundation of Bcijing (4082018) and Open Project of Shanghai Key Laboratory of Intelligent Information Processing (IIPL-09-006).

L. He · X. Li · W. Shi · Z. Liang · B. Liang
Key Laboratory of Data Engineering and Knowledge Engineering (Renmin University), Ministry of Education, 100872 Beijing, China
e-mail: heliuxing24@ruc.edu.cn

X. Li
e-mail: lixiao2008102692@ruc.edu.cn

Z. Liang
e-mail: lzh@ruc.edu.cn

B. Liang
e-mail: liangb@ruc.edu.cn

L. He · X. Li · W. Shi (✉) · Z. Liang · B. Liang
School of Information, Renmin University of China, 100872 Beijing, China
e-mail: wenchang@ruc.edu.cn

J. J. Park et al. (eds.), *Proceedings of the International Conference on Human-centric Computing 2011 and Embedded and Multimedia Computing 2011*, Lecture Notes in Electrical Engineering 102,
DOI: 10.1007/978-94-007-2105-0_24,

mode and thus they can also modify the access permission of an IMM. Benefiting from the Intel VMX technology, the hypervisor of our isolation mechanism can monitor and stop these abnormal behaviors of untrusted software. To evaluate our approach, we implement a prototype system named VIsolator. Experimental results indicate that it can effectively and efficiently protect an IMM from being tampered with.

Keywords Isolation · Protection · Integrity measurement · Operating system · Intel VMX

24.1 Introduction

Protecting integrity measurement mechanisms is one of the hot topics of information security [1]. An integrity measurement mechanism faces great amount of threats and needs to be protected [2]. A common approach to protecting an integrity measurement mechanism is to create an isolation mechanism.

Various isolation mechanisms have been proposed. In general, most of them still have some disadvantages. Some isolation mechanisms are implanted into operating system (OS) kernels or applications in order to rely on OS kernels more conveniently [3, 4], but they lack self-protection, if OS kernels crash, they will be compromised. Some isolation mechanisms utilizing special hardware [5–7] can ensure its own security, but they introduce other costs such as losing system compatibility and flexibility. More and more approaches use some existing Virtual Machine Monitors (VMM) [8] to implement isolation mechanisms [1, 9]. Although these VMM-based isolation mechanisms can ensure their own security and eliminate those costs caused by special hardware, complexity of these existing VMMs will cause some security problems [10, 11].

CPU virtualization hardware technologies [12], such as Intel Virtual Machine Extensions, or Intel VMX [13, 14], can provide a chance to avoid effectively those above disadvantages. Recently, many isolation mechanisms are implemented based on CPU virtualization hardware technologies. Most of these isolation mechanisms focuse on protecting applications [15], while an integrity measurement mechanism that needs to be protected is a part of an OS kernel in this paper. Meanwhile, some of these isolation mechanisms focuse on protecting OS kernels, but they either still have great complexity [9, 16] or are unsuitable for protecting a part of an OS kernel [17].

In this paper, we propose an isolation mechanism, called VIsolator, based on Intel VMX to protect an integrity measurement mechanism. Comparing with existing isolation mechanisms, we summarize contributions of our work below:

- Comparing with those isolation mechanisms not relying on CPU virtualization hardware technologies, VIsolator can avoid their disadvantages: (1) it can

guarantee its own security by using a hypervisor based on Intel VMX as a trusted base, (2) utilizing Intel VMX won't introduce those costs caused by special hardware and (3) its total code size is approximately 1,000 lines, so its complexity is far less than existing VMMs.
- Comparing with those isolation mechanisms based on CPU virtualization hardware technologies, VIsolator is designed according to the structure feature of a given integrity measurement mechanism which is a part of an OS kernel. VIsolator is divided into a OS kernel component and a hypervisor, and this design is significant for two reasons: (1) that allows us to protect any given part of an OS kernel, while protection for the whole OS kernel still needs some improvements [17], and (2) protection for a part of an OS kernel gets attention recently [9], but these existing approaches still have some disadvantages (e.g., having great complexity) we try to avoid.

The rest of this paper is organized as follows. Section 24.2 describes the design of VIsolator by stating two design goals. Section 24.3 shows how we can implement VIsolator based on Intel VMX. Section 24.4 presents our experimental evaluation. Section 24.5 discusses some related work, and Sect. 24.6 concludes and points out future research direction.

24.2 VIsolator Design

In this section, we first state two design goals of VIsolator and then describe the design of the VIsolator framework.

24.2.1 Design Goals

To ensure feasibility of designing VIsolator, we need a real integrity measurement mechanism as a protected object. Patos-RIP, a lightweight integrity measurement mechanism in a Linux operating system kernel, can ensure that integrity of a process is not compromised during its whole lifetime by conducting integrity measurement at appropriate points [18]. In this paper, we will design VIsolator to protect Patos-RIP. Except Patos-RIP and VIsolator, other software are untrusted.

We state two design goals before designing VIsolator:

- (G1) When untrusted software is running, Patos-RIP has been set write-protection by VIsolator through modifying the page table entry (PTE) of Patos-RIP's pages. When Patos-RIP is running, Patos-RIP has been revoked write-protection by VIsolator through modifying the PTE of Patos-RIP's pages.
- (G2) Though Patos-RIP can be set read-only by modifying PTE in G1, untrusted software in kernel mode also can modify PTE to set Patos-RIP writable. Therefore, VIsolator must prevent these untrusted software from modifying the PTE of Patos-RIP's pages.

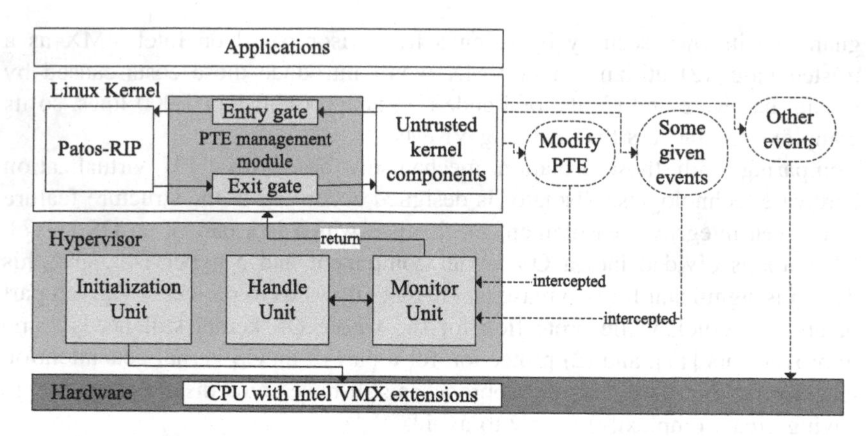

Fig. 24.1 An architectural overview of VIsolator

24.2.2 Architecture Overview of VIsolator

Figure 24.1 shows the architecture of VIsolator. We divide VIsolator into two parts: a PTE management module and a hypervisor. The PTE management module achieves G1, and G2 is achieved by the hypervisor. Like Patos-RIP, the PTE management module who contains *entry gates* and *exit gates*, is a part of a Linux kernel. The hypervisor who contains an *initialization unit*, a *handle unit* and a *monitor unit*, is independent of Linux kernels or applications.

Next we will demonstrate the working principles of all components of VIsolator to describe how VIsolator is initialized and achieves G1 $\sim$ G2.

24.2.2.1 Procedure of Initializing VIsolator

As a part of a Linux kernel, the PTE management module is initialized as a Linux kernel is initialized. On the other hand, the hypervisor is initialized at Linux runtime and the procedure is: we execute a Linux kernel module to load the image file of the hypervisor into memory, and then the *initialization unit* of the hypervisor will complete the whole initialization of the hypervisor and return to Linux.

24.2.2.2 Procedure of Achieving G1

When untrusted kernel components are to switch to Patos-RIP, it executes the *entry gate* of the PTE management module, and the *entry gate* revokes write-protection for Patos-RIP by modifying the PTE of Patos-RIP's pages. When Patos-RIP is to switch to untrusted kernel components after it completes measurement, it executes the *exit gate* of the PTE management module, and the *exit gate* sets write-protection for Patos-RIP by modifying the PTE of Patos-RIP's pages.

24.2.2.3 Procedure of Achieving G2

Once a modification of PTE occurs in Linux, this modification will be intercepted immediately by the *monitor unit* of the hypervisor, and then this unit will require the *handle unit* of the hypervisor to handle this modification. If it is an untrusted software to modify the PTE of Patos-RIP's pages, the *handle unit* will refuse this modification, or else it means the PTE management module attempts to modify the PTE of Patos-RIP's pages or an untrusted software attempts to modify other PTEs, the *handle unit* will permit this modification. It will not cause great overhead even if all modifications of PTE are intercepted, our performance experiment results show that the maximum overhead is 0.78% and the minimum overhead is 0.032%.

24.3 Implementation

To validate our design, we implement a prototype system of VIsolator. The implementations of the PTE management module, the hypervisor and their self-protection will be described in what follows.

24.3.1 PTE Management Module

The PTE management module is responsible for achieving G1 by modifying the PTE of Patos-RIP's pages to control the access permission of Patos-RIP. In order to obtain the PTE of Patos-RIP's pages, the PTE management module must obtain Patos-RIP's virtual address. In this paper, all components of Patos-RIP we need to protect are the *Integrity Measurement component*, the *OnBank* and the *measurement result* [18]. How we obtain the virtual addresses of these three components is showed in Table 24.1. Once the PTE management module obtains Patos-RIP's virtual address, it can obtain the relative PTE by invoking several kernel functions including the *pgd_offset_k*(), the *pud_offset*(), the *pmd_offset*() and the *pte_offset_kernel*(). And these kernel functions must also be protected by VIsolator.

The PTE management module consists of *entry gates* and *exit gates*. Whenever switching between Patos-RIP and untrusted kernel components happens, the *entry gates* or the *exit gates* must be executed. Therefore, as shown in Fig. 24.2, we insert an *entry gate* or an *exit gate* between each Patos-RIP function and untrusted kernel components. The PTE management module modifies a Linux kernel, so we make an effort to build the PTE management module as tiny as possible. The total C code size of the PTE management module is 144 lines.

Next we will introduce what work *entry gates* and *exit gates* do.

Table 24.1 Approaches to obtaining the virtual addresses of Patos-RIP

Patos-RIP's components	Type	Formation	Approach to obtaining the virtual address
Integrity measurement component	Code	Several kernel functions modified slightly	The virtual address can be obtained by gdb, and saved into a long array *vaddr_PatosRIP*.
OnBank	Data	A struct who occupies an entire page	The virtual address can be obtained by the pointer of struct.
Measurement result	Data	A member variable of OnBank struct	The virtual address can be obtained by the pointer of variable.

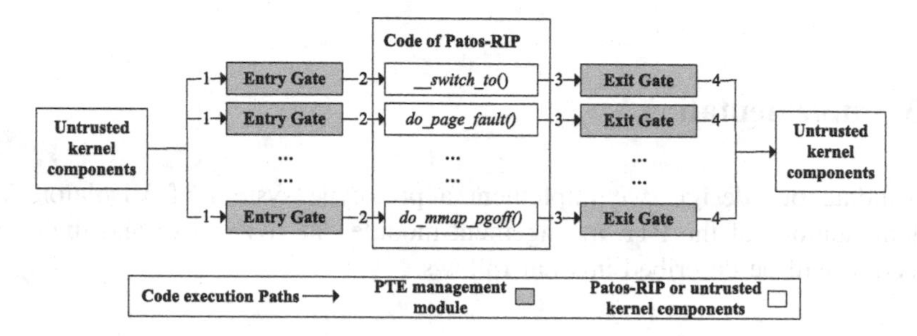

Fig. 24.2 Layout of the PTE management module

24.3.1.1 Entry Gate

The *entry gate* is responsible for revoking write-protection for Patos-RIP. The *entry gate* first obtains the relative PTE through the Patos-RIP's virtual address which can be obtained by the approaches in Table 24.1, and then revokes write-protection for Patos-RIP by replacing the current PTE with a pte_t array *pte_PatosRIP* containing the PTE which can support Patos-RIP to run normally. After that, the code of Patos-RIP becomes executable, the data of Patos-RIP becomes writable, and thus Patos-RIP will be able to run normally.

24.3.1.2 Exit Gate

The *exit gate* is responsible for setting write-protection for Patos-RIP. The *exit gate* first obtains the relative PTE like the *entry gate*, and then sets write-protection for Patos-RIP by invoking the function *ptep_set_wrprotect*() with &init_mm as the first parameter. After that, the code and the data of Patos-RIP becomes non-writable, and thus untrusted software will be unable to tamper with Patos-RIP directly if untrusted software don't modify the PTE of Patos-RIP's pages.

24.3.2 Hypervisor

The hypervisor is responsible for achieving G2 by monitoring each modification of PTE in Linux to refuse untrusted software's modification of the PTE of Patos-RIP's pages.

The hypervisor based on Intel VMX, consists of the *initialization unit*, the *monitor unit* and the *handle unit*. Next we will introduce what work the *initialization unit*, the *monitor unit* and the *handle unit* do.

24.3.2.1 Initialization Unit

The *initialization unit* is responsible for initializing the hypervisor. When the hypervisor needs to be launched, we execute a Linux kernel module to load the image file of the hypervisor into memory, and the *initialization unit* will be invoked to complete the whole initialization of the hypervisor.

Note that the VT-x option of BIOS must be enable before initializing the hypervisor, and this option can be checked by some relevant software (e.g., SecurAble [19]).

The *initialization unit* first sets several registers, then executes a series of special instructions to activate Intel VMX. Intel VMX separates CPU execution into two modes called root mode and non-root mode. After that, the hypervisor and Linux will run at different modes. Figure 24.3 shows that the hypervisor is running at ring 0 of root mode, and a Linux kernel is running at ring 0 of non-root mode while applications are running at ring 3 of non-root mode.

Next, the *initialization unit* needs to configurate a data structure called Virtual Machine Control Structure (VMCS) whose size is about 4 KB. This data structure can decide which events from non-root mode can be intercepted by root mode. Modification of PTE must execute *INVLPG* instruction. Therefore, to intercept the modification of PTE in Linux, *initialization unit* sets the INVLPG exiting bit of the data structure to allow executing *INVLPG* instruction in Linux to be intercepted by the hypervisor. Figure 24.3 illustrates that *INVLPG* instruction can be intercepted through configurating VMCS, while some given events should be intercepted unconditionally and other events can be handled directly upon hardware. After that, the *initialization unit* returns to Linux.

24.3.2.2 Monitor Unit

The *monitor unit* is responsible for intercepting the events from Linux, transferring to the *handle unit* and returning to Linux. As shown in Fig.24.4, once a certain event occurs in Linux, it will be intercepted immediately by the *monitor unit*. The *monitor unit* first confirms if this event is executing *INVLPG* instruction by reading the Virtual Machine Control Structure, if so, it will invoke the *handle unit* to

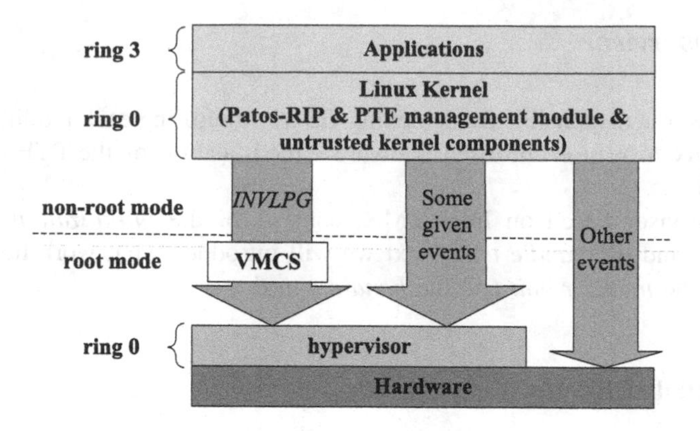

Fig. 24.3 Layout of the whole system after activating Intel VMX

handle this event, or else it means this event is a given event intercepted unconditionally so that the *monitor unit* will execute *VMRESUME* instruction to return to Linux.

24.3.2.3 Handle Unit

The *handle unit* is responsible for handling the event of modifying PTE from Linux. As shown in Fig. 24.4, if the event intercepted is executing *INVLPG* instruction, that means a modification of PTE happens in Linux, and the *monitor unit* will invoke the *handle unit* to handle this event. The *handle unit* first check if the PTE which will be modified is the PTE of Patos-RIP's pages by scanning a long array vaddr_pte_PatosRIP, which contains the virtual address of the PTE of Patos-RIP's pages. If not, the *handle unit* executes *INVLPG* instruction to permit this modification, or else the *handle unit* continues to check if the PTE management module attempts to modify the PTE of Patos-RIP's pages. If so, the *handle unit* permits this modification, or else that means untrusted software attempt to modify the PTE of Patos-RIP's pages, the *handle unit* do nothing to refuse this modification. After that, the *handle unit* returns to the *monitor unit* and then the *monitor unit* executes *VMRESUME* instruction to return to Linux.

24.3.3 Self-Protection of VIsolator

VIsolator is designed to protect Patos-RIP, but its own security must also be ensured. The PTE management module is a part of a Linux kernel like Patos-RIP, so we use the same protection approach for Patos-RIP to protect the PTE management module. On the other hand, the hypervisor and Linux run at the separated modes, therefore they own respective available memories and the memory of the hypervisor is transparent to Linux which is the only threat of the hypervisor in this paper.

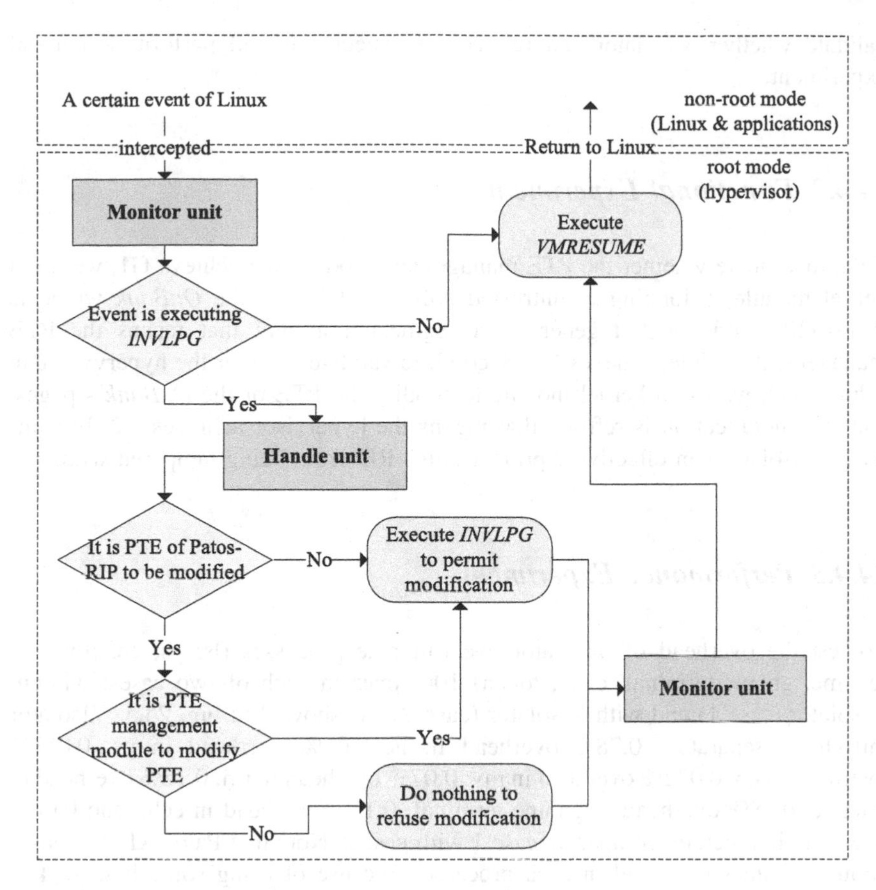

Fig. 24.4 Flow diagram of the *monitor unit* and the *handle unit*

24.4 Experimental Evaluation

We perform experiments to evaluate usefulness, function and performance of our prototype system. In this section, we present these experiments and relative results. Our experimental platform is a ThinkPad T410i PC, which has a 2.53 GHz Intel Core 2 i5 CPU with VMX extensions and 2 GB RAM. The PC runs 32-bit Fedora 13 with version of Linux kernel 2.6.37.3.

24.4.1 Usefulness Experiment

First, to validate whether Patos-RIP in our prototype system can run as before, we execute Patos-RIP's test cases for measurement, and all cases are correct that means our prototype system don't cause adverse effects for Patos-RIP. Second, to validate whether VIsolator can run as we expect, we will perform functional experiment.

24.4.2 Functional Experiment

First, to validate whether the PTE management module can achieve G1, we use a kernel module, belonging to untrusted software, to write the *OnBank* which is Patos-RIP's code, and it generates a segmentation fault that means the PTE management module achieves G1. Second, to validate whether the hypervisor can achieve G2, we use a kernel module to modify the PTE of the *OnBank*'s pages, and this modification is refused that means the hypervisor achieves G2. In summary, VIsolator can effectively protect Patos-RIP from being tampered with.

24.4.3 Performance Experiment

To test the overhead of VIsolator, we run nine processes (ls, ps, cp, mv, rm, rename, gnome-terminal, echo, touch) 100 times in each of two cases: without VIsolator (case 1) and with VIsolator (case 2). As showed in Fig. 24.5, VIsolator introduces separately 0.78% overhead in ls, 0.06% overhead in ps, 0.057% overhead in cp, 0.032% overhead in mv, 0.07% overhead in rm, 0.12% overhead in rename, 0.08% overhead in gnome-terminal, 0.16% overhead in echo and 0.08% overhead in touch by comparing case 1 with case 2. Note that Patos-RIP increases about ten times overhead in each processes because of using some hash or key calculations.

24.5 Related Work

Besides those isolation mechanisms for protecting integrity measurement mechanisms, we also discuss isolation mechanisms for protecting other objects, OS kernels specially.

Some recent approaches consider to implant isolation mechanisms into OS kernels or applications in order to rely on OS kernels more conveniently. Nooks implants an isolation mechanism into a Linux kernel to isolate untrusted extensions within lightweight protection domains inside kernel address space [3]. In-line reference monitor mentioned in [4] integrates isolation mechanism which is access policy within application codes and the isolated applications will run in a separated memory region. The main disadvantage of these isolation mechanisms is lacking self-protection. They will be compromised if OS kernels crash. On the contrary, VIsolator can ensure its own security. In fact, we have tried to run

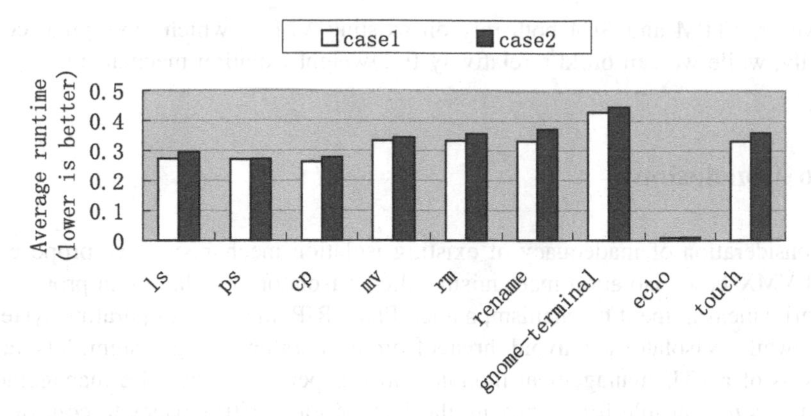

Fig. 24.5 Performance comparison using nine processes. All times are in s

distributely VIsolator and Linux in separated machines, but it is hard to protect Patos-RIP because VIsolator can't effectively intercept events from Linux.

A few isolation mechanisms focuse on special hardware that can improve security of isolation mechanisms. AEGIS implements integrity measurement mechanisms inside a secure processor [5]. Flicker [6] measures and isolates security-sensitive codes by leveraging the late launch capability of AMD's Secure Virtual Machine (SVM) extensions [20]. In [5] and [6], hardware can be regard as isolation mechanisms, but launching hardware depends on a Linux kernel or module, it cannot work correctly if Linux crashes. IBM 4758 secure coprocessor can provide a high level of security, but it will lose system compatibility and flexibility since it cannot support dynamic changes kernel codes [7]. By contrast, utilizing Intel VMX can still keep system compatibility and flexibility.

Considerable works have been conducted to provide isolation based on existing VMMs, such as Xen [21] and KVM [22]. HIMA protects an integrity measurement mechanism from measurement targets based on Xen [1]. SIM can isolate security tools from other software in a single virtual machine with Windows OS based on KVM [9]. Great complexity is always shortcoming of existing VMMs, while VIsolator has relatively less complexity because its main functions are just to modify relative PTE and monitor the modifications of PTE.

CPU virtualization hardware technologies can provide us with a chance to effectively avoid the disadvantages mentioned above. TrustVisor protects security-sensitive codes based on the CPU virtualization hardware technology of AMD SVM [15] and it is an improvement for Flicker, but the object protected by TrustVisor is application. SecVisor protects an OS kernel through ensuring that only user-approved codes can execute in kernel mode based on the CPU virtualization hardware technology of AMD SVM [17], but SecVisor works in the grain of a whole OS kernel rather than a given part of an OS kernel. vTPM protects the vTPM manager and the vTPM instances based on Intel VMX [16]. Similar to VIsolator, SIM also protects a given part of an OS kernel based on Intel VMX.

However, vTPM and SIM both rely on existing VMMs which have great complexity, while we can build a relatively lightweight isolation mechanism.

24.6 Conclusion

In consideration of inadequacy of existing isolation mechanisms, we propose an Intel VMX-based isolation mechanism called VIsolator. VIsolator can protect the integrity measurement mechanism, named Patos-RIP, in a Linux operating system. Meanwhile, VIsolator can avoid threats from a Linux operating system. VIsolator consists of a PTE management module and a hypervisor. The PTE management module is responsible for modifying the PTE of Patos-RIP's pages to control the access permission of Patos-RIP. The hypervisor is responsible for monitoring each modification of PTE to refuse untrusted software's modification of the PTE of Patos-RIP's pages.

At present, we can only provide write-protection for Patos-RIP, and our future work is to provide more protection functions: (1) implementing read-protection for Patos-RIP in order to prevent untrusted software from forging an illegal integrity measurement mechanism by reading Patos-RIP, (2) implementing execution-protection to make sure that Patos-RIP can jump to untrusted software only after security checking, and (3) implementing disk-protection for Patos-RIP.

References

1. Azab AM, Ning P, Sezer EC, Zhang X (2009) HIMA: a hypervisor based integrity measurement agent. In: Proceedings of the annual computer security applications conference
2. Shi W (2010) On methodology of modeling the trust base in operating systems. Comput Sci 37(6)
3. Swift MM, Bershad BN, Levy HM (2005) Improving the reliability of commodity operating systems. ACM Trans Comput Syst 23(1):77–110
4. Venema W (2009) Isolation mechanisms for commodity applications and platforms. Computer Science, RC24725(W0901-048)
5. Suh GE, Clarke D, Gassend B, Dijk M, Devadas S (2003) AEGIS: architecture for tamper-evident and tamper-resistant processing. In: Proceedings of the 17th annual international conference on supercomputing (ICS), ACM Press, New York, pp 160–171
6. McCune JM, Parno B, Perrig A, Reiter MK, Seshadri A (2008) Flicker: an execution infrastructure for TCB minimization. In: Proceedings of ACM European conference in computer systems (EuroSys)
7. Dyer J, Lindemanm M, Perez R, Sailer R, Doorn L V, Smith S W, Weingart S (2001) Building the IBM 4758 secure coprocessor. IEEE Comput 34(10):57–66
8. Rosenblum M, Garfinkel T (2005) Virtual machine monitors: current technology and future trends. IEEE Comput Soc 38(5):39–47
9. Sharif M, Lee W, Cui W (2009) Secure In-VM monitoring using hardware virtualization. In: Proceedings of 16th ACM conference on computer and communications security (CCS)

10. Garfinkel T, Rosenblum M (2005) When virtual is harder than real: security challenges in virtual machine based computing environments. In: Proceedings of USENIX 10th workshop on hot topics in operating systems
11. Drepper U (2008) The cost of virtualization. ACM Queue 6(1):28–35
12. VMware. Understanding full virtualization, paravirtualization, and hardware assist. http://www.vmware.com/files/pdf/VMware_paravirtualization.pdf
13. Intel Corporation (2009) Intel 64 and IA-32 architectures software developer's manual, vol 3A: system programming guide, part 1, number: 253668-031US, June 2009
14. Intel Corporation (2009) Intel 64 and IA-32 architectures software developer's manual, vol 3B: system programming guide, part 2, number: 253669-031US, June 2009
15. McCune JM, Qu N, Li Y, Datta A, Gligor VD, Perrig A (2010) TrustVisor: efficient TCB reduction and attestation. In: Proceedings of the IEEE symposium on security and privacy
16. Berger S, Caceres R, Goldman KA, Perez R, Sailer R, Doorn L (2006) vTPM: virtualizing the trusted platform module. In: Proceedings of 15th USENIX security symposium, pp 305–320
17. Seshadri A, Luk M, Qu N, Perrig A (2007) SecVisor: a tiny hypervisor to provide lifetime kernel code integrity for commodity OSes. In: Proceedings of the symposium on operating systems principles (SOSP)
18. Li X, Shi W, Liang Z, Liang B, Shan Z (2009) Operating system mechanisms for TPM-based lifetime measurement of process integrity. In: Proceedings of the IEEE 6th international conference on mobile adhoc and sensor systems (MASS), also (TSP 2009), IEEE computer society, pp 783–789
19. SecurAble. http://www.grc.com/securable.htm
20. Advanced Micro Devices (2005) AMD64 virtualization: secure virtual machine architecture reference manual. AMD Publication no. 33047 rev. 3.01, May 2005
21. Barham P, Dragovic B, Fraser K, Hand S, Harris T, Ho A, Neugebauer R, Pratt I, Warfield A (2003) Xen and the art of virtualization. In: Proceedings of the symposium on operating systems principles (SOSP)
22. Kernel based virtual machine. http://www.linux-kvm.org/page_Main

Chapter 25
Graph-Based Authorization Delegation Model

Jianyun Lei, Xiaohai Fang and Bingcai Zhang

Abstract There are still some shortcomings in the current trust management systems, for example, the evaluation of the trust, and the presentation of the trust among the entities. To address the issues above, a new authorization delegation model based on weighted directed graph is brought forward, uses weighted and directed graph to present the trust relationship between the entities. The model is simple and visual, and the algorithm of the compliance checking is analyzed and proved to be correct and efficient.

Keywords Access control · Trust management · Authorization · Delegation · Weighted directed graph

25.1 Introduction

With the popularity of the Internet applications and the rapid development of distributed systems, it becomes a basic requirement to exchange data and services through the Internet in open multi-domain environments. Therefore, it is a key factor in access control field that how to establish trust and do secure authorization and delegation between interoperability entities. There are many scholars that do a lot of in-depth research on this issue, and brought forward various solutions. Currently existing schemes that address security authorization can be divided into two categories: The first is to establish the map of the roles between the two interoperability administrative domains [1] and the other category is trust management [2].

J. Lei (✉) · X. Fang · B. Zhang
School of Computer Science, South-Central University for Nationalities,
430074 Wuhan, China
e-mail: leijianyun@mail.scuec.edu.cn

J. J. Park et al. (eds.), *Proceedings of the International Conference on Human-centric Computing 2011 and Embedded and Multimedia Computing 2011*, Lecture Notes in Electrical Engineering 102,
DOI: 10.1007/978-94-007-2105-0_25,

In recent years, trust management has gained extensive research and application, its main feature is supporting unknown entities to access the system resources through the authorization and delegation of the other entities, and using the appropriate trust delivery mechanism to support application scalability in a distributed systems. To support the application of distributed systems scalability, it is necessary to use distributed transitive authorization mechanism to achieve the corresponding requirements. Because of the distributed transitive authorization mechanisms, also some of the characteristics of trust itself, there is always some limitation in current trust management systems:

1. It is hard to achieve the quantification of trust and the expression of quantitative trust.
2. It is not easy to control the transfer of trust. Even if A trusts B, but A itself is not sure that B will or not abuse the privilege, such as B authorize and delegate to C who is not trusted by A.
3. Compared with centralized authorization, distributed authorization has great flexibility, and because of the dispersion of the authorization, it always leads to conflict authorization and ring authorization problems.

25.2 Model Design

25.2.1 Elements in Graph-Based Authorization Delegation Model

1. Entity	Entity is an object like system user, all entities can be used to uniquely represent a subject or object. Entity will be represented using nodes in the weighted directed graph.
2. Role	Role is the name of the given entity name space, it usually represent a set of access privilege.
3. Subject	Subject is an object who gives delegation to others in the process of authorization and delegation. Subject can not only be an entity, but also be a role. If the subject is a role, then the subject represents the set of all entities of the role. Subject will be represented using the beginning of a directed section in the weighted directed graph.
4. Object	Object is authorized object in the authorization and delegation. Object can not only be an entity, but also be a role. Similar to subject, if the object is a role, then the object represents the set of all entities of the role. Object will be represented using the end of a directed section in the weighted directed graph.
5. Trust value	Trust value is used to indicate the trust level of a subject to an object. A non-negative real number is used to

express the trust value. The greater the trust value is, the higher the trust level is. To facilitate the calculation below, the range of [0, 100] is prescribed to the trust value. In weighted directed graph, trust value can be expressed by the weight of the directed sections.

6. Authorization root: Authorization root is the origin point of the authorization, it is also the access control list (ACL) announced by the resource owners. ACL entry is a triad which contains privilege, subject and trust threshold. This trust threshold is a minimum that decide the access permission can be granted or not, that is to say, only when the direct or indirect trust value owned by the requester of the authority is greater than or equal to this value, requester can access the corresponding resources.

7. Delegation: Delegation is the process that authority passes from subject to object. Authorization certificate is used to represent a delegation. Authorization certificate and the issuer of authorization certificate must be digitally signed using its own private key. Weighted directed sections from one node to another node in the weighted directed graph will be used to represent delegation.

In this model, the concept of role follows Sandhu's definition in RBAC 96. Delegation is the most important concept of the model because it makes all the elements referred above related together, such as authorized source, trust threshold, subject, object, trust value, and so on.

25.2.2 Definition of Graph-Based Authorization Delegation Model

Definition 1 Authorization weighted directed graph:

$$G = \{G_r | r \in R\}$$

G_r is a weighted directed graph of authorization root r, $G_r = \{V, E, w\}$ represents all authorization which associated with r, in which, V is a node set, each node corresponds to a certain object's related authorization entities, E is a section set, each section is an ordered two-tuples that consisted of authorization publisher and authorization receiver, and each section is also the privilege that conferred from authorization publishers to authorization receiver. If $w(\langle g, s \rangle) \neq 0$, then there is a weighted directed section in G_r that direct from g to s and its weight is w.

Figure 25.1 is an example of authorization weighted directed graph.

The figure can be represented using triads:

$(\langle V_1,V_2\rangle, 90)$

$(\langle V_2,V_4\rangle, 80)$

$(\langle V_1,V_3\rangle, 95)$

……

It represents:

The trust value of node V_1 to node *V2* is 90

The trust value of node *V2* to node *V4* is 80

The trust value of node *V1* to node V_3 is 95

……

25.3 Compliance Checking Process in the Model

In authorization delegation models based on weighted directed graph, it is just as same as all authorization delegation models, resource access requests can be passed or not depends on "whether the certificate set *C* provided by the requester is able to demonstrate that the request set *r* is consistent with the local security policy *P*"[3]. It is the so-called compliance checking problem. Certificate chain searching and trust transmission controlling and judging the trust value from the authorization source to the end whether it is greater than the specified threshold are the key points of compliance checking problem. As long as the calculation of the trust path value from the authorization source to every other vertex is realized, we can judge the trust value whether it is greater than the required threshold, and the compliance checking process is completed.

In authorization delegation model based on weighted directed graph, the algorithm of calculating the trust path value from node V_0 to every other node can be described as below:

Directed graph can be expressed with adjacency matrix *adj* if $\langle V_i, V_j\rangle$ is a section, the value of *adj*$[i, j]$ equal to the directed section's weight, otherwise the value of *adj*$[i, j]$ will be 0. Initially *adj*$[i, i] = 0$, and *adj*$[i, i]$ is assigned with 1 to mark that *i* node has entered the first group in processing. Each element of array *dist*$[1..n]$ consists of two fields: *trust* field contains the trust value of node, and *pre* field is the serial number of the previous node on the path from V_0 to this node. When the algorithm ends, we can attain the maximum trust path from V_0 to V_i by tracing along with the *pre* field corresponding to the node, and the maximum trust value will be in the *trust* field.

25.3.1 Data Structure Definition

```
TYPE path = RECORD
   trust: real;
   pre: 0..n
   END;
```

VAR
adj: ARRAY[1..*n*,1..*n*] OF real;
dist: ARRAY[1..*n*] OF *path*;
max: real;
k,*i*,*u*:1..*n*;

25.3.2 Algorithm Description

25.3.2.1 Input Items

The serial number of the starting node V_O is given in the variable k, and the value of the *trust* field of the element in *dist* array indexed V_O is assigned with 100.

25.3.2.2 Output Items

The maximum trust path from node V_O to every other node.

25.3.2.3 Steps of the Algorithm

1. [The initial nodes are divided into two groups]

   (1) loop i step by 1, from 1 to n

   (i) $dist[i].trust \leftarrow adj[k, i]$
   (ii)if $dist[i].trust \neq 0$
   then $dist[i].\ pre \leftarrow k$
   else $dist[i].\ pre \leftarrow 0$

   (2) $adj[k, k] \leftarrow 0$

2. [add the nodes in the second group to the first group one by one]
   Loop until all nodes in the second group has been added into the first group

   (1) [Find the node with the maximum trust value in the second group]
   (i) $max \leftarrow 0; u \leftarrow 0$
   (ii) loop *i*step by 1, from 1 to n
   if $adj[i,i]=0$ and $dist[i].trust>max$
   then $u \leftarrow i; max \leftarrow dist[i].trust$
   (2) [add the found node into the first group]
   if $u = 0$
   then end algorithm [there is no node can be added to the first group]
   else $adj[u, u] \leftarrow 1$

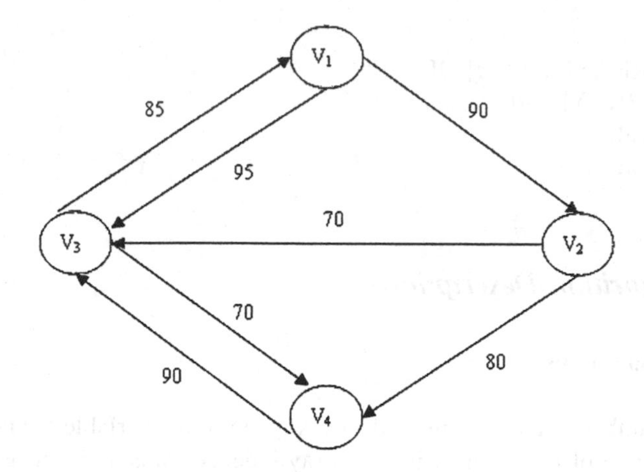

Fig. 25.1 Example of authorization weighted directed graph

Trust	pre
100	1
90	1
95	1
0	0

V_1 add to 1 group

Trust	pre
100	1
90	1
95	1
66.5	3

V_3 add to 1 group

Trust	pre
100	1
90	1
95	1
70	2

V_2 add to 1 group

Trust	pre
100	1
90	1
95	1
70	2

V_4 add to 1 group

Fig. 25.2 Algorithm simulated running instance

(3) [modify the trust value of node in the second group]
loop i step by 1, from 1 to n
if $adj[i, i] = 0$ and $dist[i].trust < dist[u].trust \times adj[u, i]/100$
then begin
$dist[i].trust \leftarrow dist[u].trust \times adj[u, i]/100;$
$dist[i].pre \leftarrow u$
end

25.3.3 Demonstration of the Algorithm

Take Fig. 25.1 for example to find the maximum trust value from node V_1 to every other nodes, values in array *dist* will be changed as the algorithm runs and the results are shown in Fig. 25.2.

25.4 Analysis of the Compliance Checking Algorithm

The time efficiency of the algorithm in calculating the trust path value from node V_O to every other node in the model is $O(n^2)$. Because the algorithm has to add n nodes into the first group, it is necessary to compare and modify each element in array *dist* with n elements one by one when a node is adding into the first group.

25.4.1 The Idea of the Algorithm

At first, there is only V_O in the first group, the second group contains all the other nodes. The trust value of the node V_O is 100, and the trust value of the nodes in the second group can be assigned in this way: if there is a section $\langle V_O, V_i \rangle$ in the graph, then the trust value of the node V_i is the weight of the section $\langle V_O, V_i \rangle$, otherwise the trust value of the node V_i is assigned with 0. Secondly, select a node V_m with maximum trust value in the nodes of the second group to the first group every time. There is an amendment to the trust value of every node in the second group when the node V_m is added to the first group. If added node V_m is a middle node which makes the trust path value from V_O to V_i greater than that before V_m is added as a middle node, then the modification of the trust value of the node V_i is necessary. After amendment, chose the maximum trust value node and add it into the first group, go on until all nodes of the graph have been added into the first group or there is no node to be added into the first group.

25.4.2 Proof of the Algorithm

Division of the two groups and judging the trust value of nodes in the algorithm are clearly in accordance with the basic ideas above. The only thing to prove the correctness of this approach is to prove the division of the two groups and the trust value still meet the requirements after adding a node into the first group. That is to prove the trust value of the node V_m is the maximum in the second group and the trust value is the greatest trust path value from the node V_O to V_m and V_m is the node of the greatest trust path value in the second group.

1. If the trust value of node V_m is not the maximum trust value from V_O to V_m and there is a another path from V_O through some nodes in the second group to V_m whose trust value is greater than that of node V_m and we assume the first passed node in the second group is node V_s, then the trust value of V_s must be greater than that from V_O through V_s to V_m and greater than that of V_m, this is conflict with the assumption that trust value of V_m is the greatest in the second group. So the trust value of node V_m is the maximum trust path value from node V_O to V_m.

2. Assume node V_x is any other node in the second group, and if only nodes in the first group can be the middle nodes in the maximum trust path to V_x, then its trust value canot be greater than any other trust path value from node V_O to V_m. This can be known by the definition of trust value. If the maximum trust path value from node V_O to V_x through middle nodes in the second group and assume that is a node named V_y, then the trust path value from V_O to V_y must be the trust value of V_y, it is less than or equal to the maximum trust path value from V_O to V_m, so it is obvious that the maximum trust path value from node V_O to node V_y canot be greater than the maximum trust path value from node V_O to node V_m, so the node V_m is the maximum trust path value in the second group.

25.5 Conclusion

Jun-Guo Liao et al. brought forward authorization delegation model based on trust-worthiness in 2007 [4], firstly proposed using trust-worthiness between entities and control the transmission of authority in authorization and delegation. But the existed problem is that there are no simple and intuitive tools to express the relationship between the entities and there is no specific algorithm in certificate chain searching. In this paper, authorization delegation model based on weighted directed graph overcomes the two shortcomings. The proposed model is simple and intuitive to understand. Algorithm of compliance checking is proved to be correct and efficient after analysis and demonstration.

References

1. Kapadia A, Al-Muhtadi J, Campbell R, Mickunas D (2000) IRBAC 2000: secure interoperability using dynamic role translation. Technical report UIUCDCS-R-2000-2162, University of Illinois, pp 245–267
2. Blaze M, Feigenbaum J, Lacy J (1996) Decentralized trust management. In: Proceedings of the 17th symposium on security and privacy. Oakland, California, USA. Los Alamitos: IEEE CS Press, pp 164–173
3. Blaze M, Feigenbaum J, Strauss M et al (1998) Compliance-checking in the policymaker trust management system. In: Hirschfeld R (ed) Proceedings of second international conference on financial cryptography. LNCS 1465. Anguila, British West Indies. 1998. Springer, Berlin, pp 254–274
4. Jun-Guo L, Fan H, Geng-Ming Z et al (2006) Trust worthiness-based authorization delegation model. Chin J Comput 29(8):1265–1270

Part VI
HumanCom 2011 Session 5: Ubiquitous Computing, Mobile Systems and Applications

Chapter 26
A Multiple Response Approach for Adaptive Learning Service with Context-Based Multimedia Contents

Xinyou Zhao, Qun Jin and Toshio Okamoto

Abstract Learners can benefit more from the emerging ubiquitous computing technology in more learning scenarios beyond traditional computer-based e-learning system. But due to the great diversity of device specification, learning contents, and mobile context existing today, learners may have a poor learning experience in the ubiquitous computing technology-enhanced learning (u-learning) environment. This paper proposes a multiple response approach for adaptive learning service with context-based multimedia contents, which recommends preferred media for learners according to the u-learning environment. Based on six learning statuses from SCORM, five learning responses of learning objects are used to reward or penalize the preferred media according to the learning context. In the evaluation experiment, the accessed object, time, location and mobile device are mainly used as context data. With the comparison between the controlled group and non-controlled group, the results show that the proposed method can improve the utilization rate of learning objects which implies that the learners are more interested in these recommended media. The results also show that the learning experience is improved.

Keywords Mobile learning · Pervasive computing · Context-aware contents · Multiple-response reward-penalize

---

X. Zhao (✉) · Q. Jin
Faculty of Human Sciences, Waseda University, Tokorozawa, Japan
e-mail: totoyou@ieee.org

Q. Jin
e-mail: jin@waseda.jp

T. Okamoto
Graduate School of Information Systems,
The University of Electro-Communications, Tokyo, Japan
e-mail: okamoto@ai.is.uec.ac.jp

J. J. Park et al. (eds.), *Proceedings of the International Conference on Human-centric Computing 2011 and Embedded and Multimedia Computing 2011*, Lecture Notes in Electrical Engineering 102,
DOI: 10.1007/978-94-007-2105-0_26,

26.1 Introduction

Information visualization is a well-established discipline [1]. The highly graphical, sophisticated approaches have been proposed to provide vast sets of information for users. These graphical schemes have been applied to the fields of information retrieval and exploration in an attempt to overcome search and access problems on conventional large-screen displays [2]. The users may conveniently access graphical information by personal computer, including multimedia contents [3].

With the development of communications and ubiquitous computing technologies, ubiquitous network society has emerged around us, e.g., u-Japan, u-Korea, and u-Home [4, 5]. Users may benefit more from new emerging ubiquitous computing and networking technology. Users may take pictures, record sounds, know their locations and moving speed, and even know the population density around them by these technologies.

Till now, ubiquitous computing technology has been increasingly integrated into many facets of our activities in daily living, including education or training [3, 6]. By the use of ubiquitous technology, anyone can access information and learning materials at any time and any place with mobile devices. It is easy to realize a seamless integration and provision of diverse learning services [7].

As a result, learners have a complete control over when they want to study and from which location they want to study. Learners will be empowered for their learning since they can learn whenever they want in ubiquitous learning environments [8]. Of course, the instructors may help learners, and they can edit and upload teaching multimedia contents. Ubiquitous device is an ideal tool to provide a seamless learning process at any place and any time for learners and instructors. In this paper, seamless learning process means that learning process may not be interrupted owing to different place, time, access device, et al. In other words, the learning system should provide best suitable learning contents for learners as long as they need.

Each content item in most learning systems is represented in multiple forms or versions in order to provide more efficient and more flexible tools for transforming digital content to satisfy the needs of end-users [9, 10]. The methodology has an extensive application on the context-based multimedia content delivery via ubiquitous channels, while the presentation and nature of information are restricted from learning environment [9]. The learning system ensures that a presentation will focus on the needs of the learner, rather than what the presenter finds interesting [11, 12]. Additionally, the separation of content and presentation can be also considered unavoidable in many cases [13].

In our previous study, we have proposed a three level model to create adaptive learning contents according to ubiquitous learning environment [3, 14]. This paper focuses on the presentation of learning objects in ubiquitous learning environment (ULE). We propose a multiple response approach for adaptive learning service in a way as close as to the presentation of learning objects to fit the learning situation.

The rest of this paper is organized as follows. Adaptive learning service is discussed in Sect. 26.2. Also, the issues on providing ubiquitous learning services are discussed. Section 26.3 proposes the method to recommend the preferred media based on learning contextual data. Section 26.4 discusses the evaluation and application scenario of the proposed method. Finally, Sect. 26.5 summarizes our present work and point to future research works.

26.2 Adaptive Learning Service

Based on ubiquitous computing technologies, many applications on personal computer, e.g., the successful e-Learning application [8, 15–17], have been altered to ubiquitous device. However, before they are applied in ubiquitous learning environments, there are important problems on how to precisely determine what learners want and need, and how to provide context-based content in a learner-friendly and effective way. Learners' needs are always conditioned by what they already get, or imagine they can get [9].

In addition, the specific theme shows the need for user-centered and context-based adaptive learning service development and personalized content delivery. Some visualization schemes may not be appropriate for small-screen devices. Although the display technology can deliver the high resolution required, the available screen space is not necessarily adequate for meaningful presentations and manipulation by the user [2]. When delivering multimedia contents to mobile learners, it is necessary for learning system to consider the context environment (such as, device features, network characteristics, location) and preferences of learners in question [14]. For example, rich contents can not be correctly displayed on mobile device [1, 15, 18]; different needs under different learning location and time [6, 9, 19], different learning activity under different navigation [20] or learning goals [21].

26.2.1 Adaptive Service in Learning System

The adaptivity and adaptability in e-learning systems are essential in order to adjust the changing of learner's situations dynamically. The main purpose of the adaptivity is to provide context awareness-based learning under learner's different skill and motivations. In general, the context data may be from learner, learning environment, instructor, or educational strategy and so on. A formal definition of the Law of Adaptation is described as follows [1, 3, 22].

Given a learning system, we say that a physical event is a stimulus for the system if and only if the probability that the system suffers a change or be perturbed (in its elements or in its processes) when the event occurs is strictly greater than the prior probability that suffers a change independently of E:

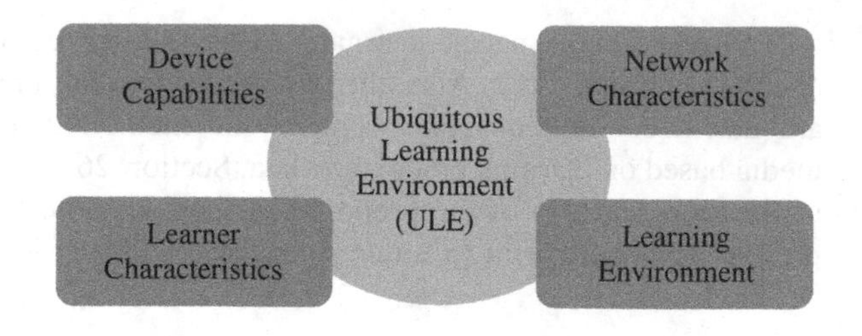

Fig. 26.1 Categories of Contexts in ULE

$$P(S \rightarrow S'|E) > P(S \rightarrow S') \tag{26.1}$$

Thus, from the technological point of view, the main challenge of these personalization techniques depends on correctly identifying characteristics of a particular learner [16]. The characteristics may be knowledge, skills, cognitive abilities, learning process/history and styles, learning preference, and so on.

In short, enhancing learning and performance is a function of adapting instruction and content to fit to the learner. In other words, the aim of adaptive e-learning system is coordinated with actual education and learning: right learning contents to the right learner at a right time in a right way—beyond any time, any place, any path, and any pace.

26.2.2 Issues for Providing Ubiquitous Learning Service

The dynamical and continual changing learning setting in u-learning environment gives more different learning contexts than those in e-learning. The task of context awareness in u-learning is to detect the ubiquitous situation and adapt to the changing context during learning process [3, 23]. Context awareness is the key in the adaptive u-learning system and must be integrated to a learning system seamlessly.

Figure 26.1 shows the context data used in u-learning environment, which are divided into four categories: device capabilities (DC, codec capabilities; input–output capabilities; device features), network characteristics (NC, static network features and dynamical network conditions), learning environment (LE, learner information and learning process; presentation preferences), and learner characteristics (LC, location; time; social awareness) [3, 24].

The system framework creates adaptive contents for u-learning, which is implemented based on adaptive behavior according to learning context awareness. The works are completed by an adaptation engine in general. The adaptation engine uses context data of u-learning as input data to produce the adaptation results [14].

Delivering learning resources designed for tabletop computers to u-learning environment is by no means a trivial task [17]. The learners get a poor learning experience in u-learning environment for the reason of great diversity mentioned and discussed above [3].

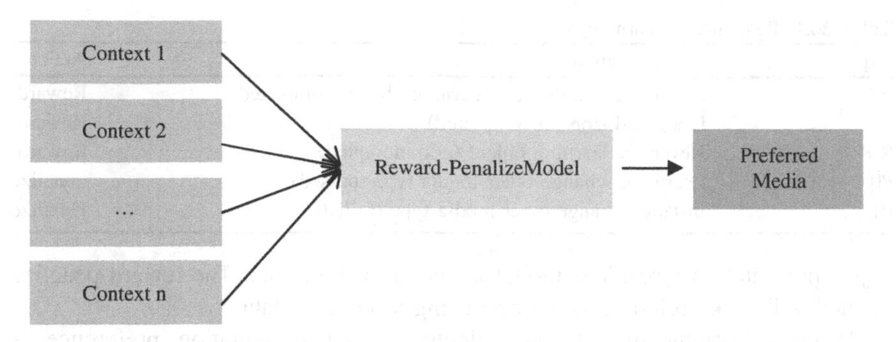

Fig. 26.2 Role of Reward-Penalize model

First of all, massive amount of contents, irrelevant to learners' preferences or contextual environment, will make learners feel frustrated and dissatisfied. Also, these make learners overloaded during learning. They also increase the learners' communication costs and channel burdens.

Secondly, most of learning contents, designed for desktop computers and high-speed network connections, are not suitable for network features with low bandwidth and handheld devices with limited resources and computing capabilities.

26.3 Multiple-Response Reward-Penalize Algorithm

In general, one learning response is used to reward or penalize learning objects based on students' reflections. If the learning response is favorable (success), the method rewards the selected learning object, while if the learning response is unfavorable (failure), the method penalizes the selected learning object. Economides calls the process reward-penalize method [22, 25], which can be represented by Formula (26.2).

$$\begin{cases} Reward & \leftarrow Success \\ Penalize & \leftarrow Failure \end{cases} \tag{26.2}$$

In this research, the system rewards (or penalizes) the selected contents preferred media type according to how much favorable (or unfavorable) the learning response was. This research introduces multiple responses algorithm for reward-penalize model. If the environment response is very favorable, the selected action is rewarded very much. On the other hand, if the learning response is marginally favorable, the selected action is rewarded little. Similarly, if the learning response is very unfavorable, the selected action is heavily penalized, and if the learning response is marginally unfavorable, the selected action is marginally penalized.

The learning context in the realization model is defined by u_i = [context 1; context 2; ... context n] and the learning action is defined by $r_i(t)$ = [learning

Table 26.1 Response of learning objects

Type	Events (status)	Action
R2	Learn, then go on next learning object (completed)	Reward
R1	Learn and stop (uncompleted)	Reward
R0	Never use learning object (not attempted)	Reward
P0	Learn, but change other media type (passed)	Penalize
P1	Instantly change other media type (failed)	Penalize

object presentation type]. The model is shown in Fig. 26.2. The reward-penalize method will rank preferred media according to context data.

At each learning time t, the selected media presentation preference is $r(u_i) = r_i, r_i \in \{r_1, r_2, \ldots, r_n\}$with probability$p_i(t) = p[r(u_i) = r_i]$. The selected preference $r(u_i)$ is taken as the context input to the learning environment.

If this results in a favorable environment response, the probability $p_i(t)$ is increased by $\Delta p_i(t)$ and $p_j(t), i \neq j$ is decreased by $\Delta p_i(t)$. Otherwise, if unfavorable environment response appears, $p_i(t)$ is decreased by $\Delta p_i'(t)$ and $p_j(t), i \neq j$ is increased by $\Delta p_i'(t)$. Here, $\Delta p_i'(t)$ and $\Delta p_i(t)$are the adaptation rates.

In the single response method, they are used by one same value. This behavior is shown in Formulas (26.3) and (26.4) in this research.

Preferred Media Type$_{\text{LO}}$ $(i \to j)$ received *positive response in Formula* (26.2).

$$\begin{cases} P_{ij}(t) = P_{ij}(t) + \Delta P_i(t)(1 - P_{ij}(t)) \\ P_{jk}(t) = P_{ik}(t) \cdot (1 - \Delta P_i(t)), k \neq j \end{cases} \tag{26.3}$$

Preferred Media Type$_{\text{LO}}$ $(i \to j)$ received *negative response in Formula* (26.3)

$$\begin{cases} P_{ij}(t) = P_{ij}(t) - \Delta P_i'(t)(1 - P_{ij}(t)) \\ P_{ik}(t) = P_{ik}(t) \cdot (1 + \Delta P_i'(t)), k \neq j \end{cases} \tag{26.4}$$

Our method is to use different adaptation rate for different learning response. When the learner's response is very good, the system heavily reward the selected preferred media type by increasing its probability rapidly. When the response is marginally good, the system correspondingly reward the selected preferred media type by increasing its probability slowly.

On the contrary, when the learning response is very bad, the system heavily penalizes the selected preferred media type by decreasing the probability very fast. When the learning response is a bit bad, the system penalizes the selected preferred media type less strictly by decreasing its probability slowly.

According to six status values of SCORM (2004) data model definition (Passed, Failed, Completed, Uncompleted, Not Attempted, Browsed) [26], the learning responses in this research are defined by five statuses in Table 26.1 (R: Reward, P: Penalize). The R0 is the threshold for a response to be considered as reward or penalty.

For example, the system will take the learning process as completed status (R2) if the learner goes on another learning object after he/she has accessed one learning object.

Table 26.2 Categories for preferred media type used

Sets	Description of sets
A	Learning objects under context u_i
B	Learning objects studied with media type of I
C	Learning objects studied with all media types used for I
D	Learning objects studied with one of media types (T) used for I

I Learning object, *T* Media type

In this research, a sequence for the reward-penalize response is a Fibonacci sequence (normalized to the [0, 1]). For the five levels of Fibonacci sequence, the value should be 0, 1/3, 1/3, 2/3, 1.

For different learning level, the selected preferred media type is different. So the system cannot take the recommended and self-selected contents as same weight during learning. This research divides all contents studied under learning context u_i into two categories: recommended contents and self-selected contents. The system assigns α to media type that is recommended, while $1 - \alpha$ is assigned to media type that is self-selected. They are represented in Formula (26.5).

$$\begin{cases} P_R = \alpha, & (\text{Recommended}) \\ P_S = 1 - \alpha, & (\textit{Self-selected}) \end{cases} \quad 0 \le \alpha \le 1 \tag{26.5}$$

Different α may be used in different learning system. For example, for primary learning or the invoice, the system should recommend suitable preferred media type for learners usually. In other words, the system assigns α with a higher weight for recommended contents.

On the contrary, for adult learning or expert, the learners can determine the learning objects by themselves under their knowledge, which means that self-selected contents are more important. In general, the Formula (26.6) gives the threshold of α.

$$\alpha = \begin{cases} \ge 0.5, & \text{High Priority} \\ = 0.5, & \text{Same Priority} \\ \le 0.5, & \text{Low Priority} \end{cases} \tag{26.6}$$

In order to compute the probability according to learning history, the learning objects used are categorized into four sets (*A*, *B*, *C*, and *D*) according to learning context u_i, which are described in Table 26.2.

The rank of preferred media type is used in this research to improve the Quality of Service (QoS) of adaptive contents. The probability$p_T(t)$ of the preferred rank value of MIME type (T) of learning object under context $u_i$is computed by Formula (26.7).

$$p_T(t) = \frac{\sum_{j \in D} p_{T,j}(t) / \sum_{j \in C} p_{T,j}(t)}{\sum_{j \in B} p_{T,j}(t) / \sum_{j \in A} p_{T,j}(t)} \tag{26.7}$$

Table 26.3 MR-RP algorithm (Type(L): Media Type of L)

Input
Context u_i
Output
Preferred Media Type;
Algorithm
Initialize the $p_i(t)$
forEach k in Response
If (k=favorable) then

$$p_i = p_i + R \cdot (1 - p_i)$$

$$p_j = p_j \cdot (1 - R), j \neq i$$

endFor
forEach k in Response
If (k=unfavorable) then

$$p_i = p_i - P \cdot (1 - p_i)$$

$$p_j = p_j \cdot (1 + P), j \neq i$$

endFor

$$p_i(t) = p_i(t) + p_i / \sum n$$

$$p_j(t) = p_j(t) + p_j / \sum n, j \neq i$$

$$\text{Preferred Media Type} = \begin{cases} Type(\text{Learning}Object) & p_i(t)=p_j(t), i \neq j; \\ Max(p(t)), & \text{others;} \end{cases}$$

Based on the discussion of multiple-response reward-penalize methods above, it is easy for us to formally define the multiple-response reward-penalize (MR-RP) algorithm to realize the preferred media type for ubiquitous learning environment, which is shown in Table 26.3.

26.4 Evaluation

According to the proposed method discussed above, let us suppose a u-learning scene (Fig. 26.3). Sam will attend a test at 9:30 today. During going to school, he wants to review the learning contents based on multimedia contents. At different location and time, the system should deliver suitable presentation format for Sam according to Sam's context environment.

In order to evaluate the adaptation mechanism, the developed realization model has been installed on a web system, which provides a general searching service for users (more 10,000 pages/day accessed by mobile users from the world). In the experimental evaluation system, the accessed items, time, location and device type are considered as learning context. The system provides types of contents such as text, image, document, and audio according to user's experience.

Figure 26.4 shows all accessing reports of users between April 19 and 27, 2010, which concludes the visitors, page views (PV), average page views, visiting time

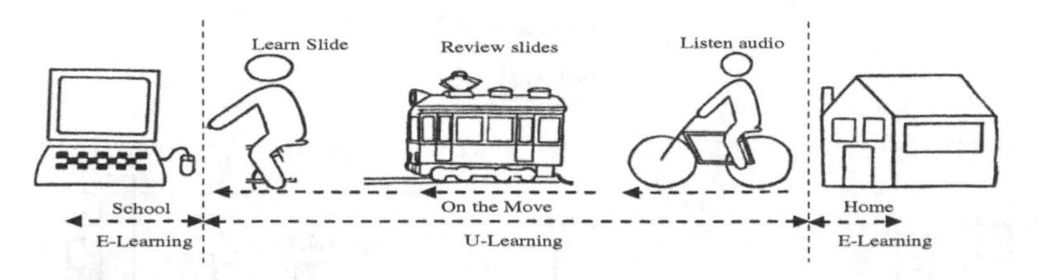

Fig. 26.3 Learning scene

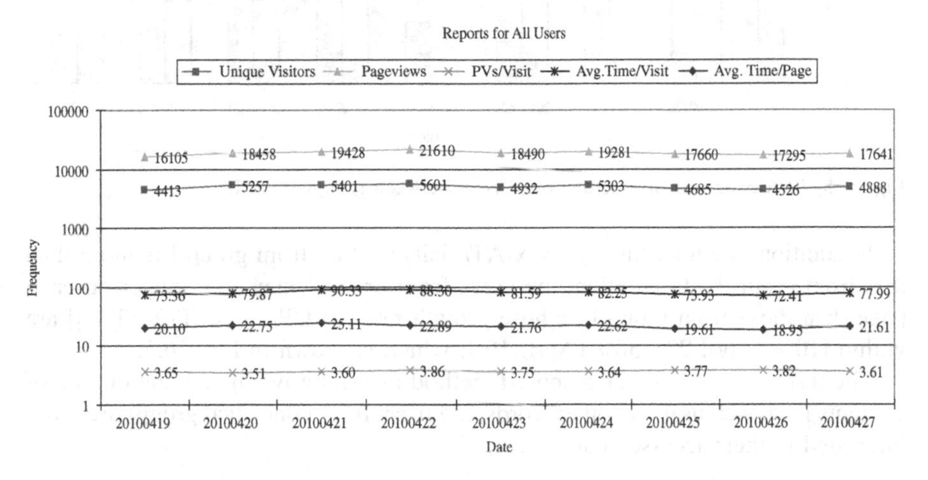

Fig. 26.4 Accessing reports for all users

per visit and average time per page (AT). Here the average time including the downloading and viewing/using time from each user.

During tracing, users are separated into two groups randomly: controlled group I for adaptation contents and non-controlled group II for general contents.

For controlled group I, the system creates adapted contents according to user's ubiquitous environment. For example, the system would provide audio type if there were more users to use audio under same contextual environment even though the contents were text message. For non-controlled group II, the system just sends originally accessing contents for users.

The increasing rate of page view and visiting time between group I and II is computed by (I–II)/II [I/II means the access time (T) or page (P) from group I or II].

Figure 26.5 shows that page view growth rate $[(PV_I - PV_{II})/PV_{II}]$ from group I to group II are within [12% (April 25), 63% (April 27)]. For example, there are 3.95 pages accessed by each user from group I on April 19, 2010. Corresponding to group I, each user from group II visited 3.28 pages, whose growth rate: (3.95−3.28)/3.28 = 20%.

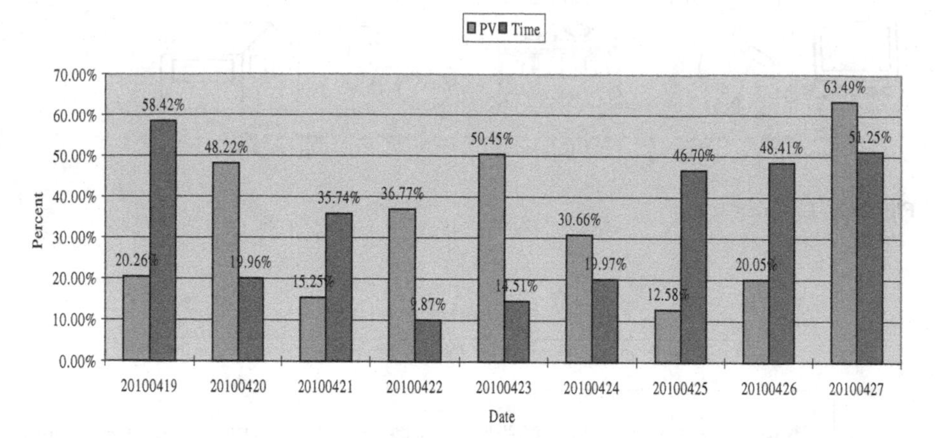

Fig. 26.5 Increasing rate of time and page view between groups I and II

In addition, the total time (=PV x AT/visit) visiting from group I is more than that from group II. In addition, the users from group I may stay longer on each page than those from group II, whose growth rate $[(AT/P_I - AT/P_{II})/AT/P_{II}]$ are within [10% (April 22), 58% (April 19)], which is shown in Fig. 26.5.

The results show that our proposed method may improve the utilization rate of contents from web users. In other words, the users from controlled groups are more interested in their accessed contents.

26.5 Conclusion

In this study, a recommendation method has been proposed for multimedia contents based on learning contextual data in ubiquitous learning environments. Based on the definition of learning response to accessed learning object, the multiple-response reward-penalize method has been proposed to suggest preferred media for learners. Finally, the learning objects, learning time, learning location and device type are evaluated. The results show that it has increased the utilization rate of learning objects. In other words, the learners are more interested in recommended learning contents.

As for future work, we will tackle the specific research question concerning "types of contextual information and sources" using in u-learning environment. It is a challenging work to define/reason different context (physical context, time context, learner context, resources context) and describe the learning environment (capabilities of ubiquitous device, characteristics network and learner) during moving. In addition, we realize it is important that content adaptation is required to provide a universal access from e-learning content provider or the Internet by ubiquitous device. The adapted contents should be designed to maximize the learner's Quality of Experience (QoE). The system should decide how to react while learning context is changing and select the appropriate adaptation and service parameters.

References

1. Card SK, Mackinlay JD, Shneiderman B (1999) Readings in information visualization-using vision to think. San Francisco, Morgan Kaufmann
2. Matt J, Gary M (2006) Mobile interaction design. Wiley, New York
3. Zhao XY, Okamoto T (2011) Adaptive multimedia content delivery for context-aware u-Learning. Int J Mobile Learn Org (IJMLO) 5(1):46–63
4. Yamazaki T (2007) The ubiquitous home. Int J Smart Home 1(1):17–22
5. Bell G, Dourish P (2007) Yesterday's tomorrows: notes on ubiquitous computing's dominant visión. Pers Ubiquitous Comput 11(2):133–143
6. Esra W, Niall W, Martin O (2008) Maintaining, changing and crossing contexts: an activity theoretic reinterpretation of mobile learning. Res Learn Technol (ALT-J) 16(1):41–57
7. Jin Q, Zhu Y, Shtykh RY, Yen NY, Shih TK (2011) Flowable services: a human-centric framework toward service assurance, Proceedings.of 10th international symposium on autonomous decentralized systems (ISADS2011), pp 653–658
8. Mohamed A (2009) Mobile learning: transforming the delivery of education and training. Athabasca University Press, Edmonton, pp 1–19
9. Germanakos P, Mourlas C (2008) Adaptation and personalization of web-based multimedia content. In: Multimedia transcoding in mobile and wireless networks, pp 1–29
10. Pavlov R, Paneva D (2007) Toward ubiquitous learning application of digital libraries with multimedia content. Cybernetics Info Technol 6(3):51–62
11. Longmire W A primer on learning objects. Learning, https://www.astd.org/LC/2000/0300_longmire.htm (Accessed May 3, 2011)
12. Parrish P (2004) The trouble with learning objects. Edu Technol Res Dev 52(1):49–67
13. Sicilia MA, García E (2003) On the concepts of usability and reusability of learning objects. Int Rev Res Open Distance Learn 4(2)
14. Zhao XY, Ninomiya T, Vilenius M, Anma F, Okamoto T (2008) Adaptive content delivery system for ubiquitous learning. J Info Sys Edu 7(1):95–102
15. Kojiri T, Tanaka Y, Watanabe T (2007) Device-independent learning contents management in ubiquitous learning environment. In: Proceeding of the world conference one-learning in corporate, government, healthcare, and higher education, pp 991–996
16. Shute V, Towle B (2003) Adaptive e-learning. Edu Psycholo 38(2):105–114
17. Sharples M (2007) Big issues in mobile learning, report of a workshop by the kaleidoscope network of excellence mobile learning initiative, pp 1–37
18. Yang SJH, Chen IYL, Chen R (2007) Applying content adaptation technique to enhance mobile learning on blackboard learning system. In: Proceedings of seventh ieee international conference on advanced learning technologies (ICALT 2007), pp 247–251
19. Hsiao JL, Hung HP, Chen MS (2008) Versatile transcoding proxy for internet content adaptation. IEEE Trans Multimedia 10(4):646–658
20. Chen J, Man H, Yen NY, Jin Q, Shih TK (2010) Dynamic navigation for personalized learning activities based on gradual adaption recommendation model. In: Proceedings of the 9th international conference on web-based learning(ICWL2010), December (2010), pp 31–40
21. Man H, Chen H, Jin Q (2010) Open learning: a framework for sharable learning activities. In: Proceedings of the 9th international conference on web-based learning(ICWL2010), December 2010, pp 387–391

22. Economides AA (1996) Multiple response learning automata. IEEE Trans Sys Man Cybern-Part B: cybernetics 26(1):153–156
23. Conlan O, Hockemeyer C, Wade V, Albert D (2002) Metadata driven approaches to facilitate adaptivity in personalized elearning systems. J Info Sys Edu l(1):38–44
24. Barbosa V, Carreras A, Arachchi HK, Dogan S, Andrade MT, Delgado J, Kondoz AM (2008) A scalable platform for context-aware and drm-enabled adaptation of multimedia content. In: Proceedings of ICT-mobilesummit 2008, Jun. 2008
25. Economides AA (2006) Adaptive mobile learning. In: Proceedings of the fourth IEEE international workshop on wireless, mobile and ubiquitous technology in education (WMTE'06), pp 26–28
26. SCORM (2004) Sharable content object reference model (SCORM). http://www.adlnet.gov/Technologies/scorm/default.aspx (Accessed May 3 2011)

Chapter 27
Enhancing the Experience and Efficiency at a Conference with Mobile Social Networking: Case Study with Find and Connect

Lele Chang, Alvin Chin, HaoWang, Lijun Zhu, Ke Zhang, Fangxi Yin, Hao Wang and Li Zhang

Abstract With mobile devices and wireless technology becoming pervasive at conferences, conference attendees can use location-based mobile social networking applications such as Foursquare and Gowalla to share their location and content with others. How to use the mobile devices and the indoor positioning technology to help the conference participants enhance the real-world interactions of people and improve efficiency during the conference? In this paper, we report our work in Nokia Find & Connect (NF&C) to solve this problem where we use location and encounters, together with the conference basic services, all through a mobile UI. To demonstrate the usefulness of the NF&C system, we conducted a field trial at the 7th International Conference on Ubiquitous Intelligence and

L. Chang (✉) · A. Chin · HaoWang · L. Zhu · K. Zhang · F. Yin · H. Wang
Nokia Research Center, 100176 Beijing, China
e-mail: ext-lele.chang@nokia.com

A. Chin
e-mail: alvin.chin@nokia.com

HaoWang
e-mail: ext-hao.10.wang@nokia.com

L. Zhu
e-mail: ext-lijun.2.zhu@nokia.com

K. Zhang
e-mail: ke.4.zhang@nokia.com

F. Yin
e-mail: ext-fangxi.yin@nokia.com

H. Wang
e-mail: Hao.ui.Wang@nokia.com

L. Chang · L. Zhu · L. Zhang
Department of Electronic Engineering, Tsinghua University, 100084 Beijing, China
e-mail: chinazhangli@mail.tsinghua.edu.cn

J. J. Park et al. (eds.), *Proceedings of the International Conference on Human-centric Computing 2011 and Embedded and Multimedia Computing 2011*, Lecture Notes in Electrical Engineering 102,
DOI: 10.1007/978-94-007-2105-0_27,

Computing and Autonomic and Trusted Computing (UIC/ATC 2010). Results show that NF&C can help the conference participants enhance the real-world interactions of people and improve efficiency at a conference effectively.

Keywords Mobile social networking · Indoor positioning · Conference-assisted system · Case study

27.1 Introduction

The number of mobile devices is growing at an explosive rate. Worldwide mobile device sales to end users totaled 1.6 billion units in 2010, a 31.8% increase from 2009. Smart phone sales to end users were up 72.1% from 2009 and accounted for 19% of total mobile communications device sales in 2010 [1]. With mobile phones having Internet and web capability, users can easily access and update their online social networks in real time and with real location using applications such as Foursquare and Gowalla. There are many applications that use GPS for location, but few that use indoor location positioning technologies.

However, Wi-Fi wireless networks are becoming increasingly pervasive in many of the indoor places like hotels, schools and offices. Wi-Fi can be used as indoor positioning technology for location tracking and navigation. Today, for example, Ekahau Real Time Location System [2] is one system that uses Wi-Fi to track tens of thousands of mobile objects, assets or people, in real time with only a few second update intervals, with down to 1–3 m accuracy. Wi-Fi is becoming a necessity at conferences and conference programs are readily available to read on mobile devices, however few conferences support a real time location-based conference system.

Much of the work on conference-based support systems can be divided into two parts: network and proximity-based systems and conference content-based systems. Network and proximity-based systems only focus on the interplay of networking and social contact at a conference, but do not provide other conference-supported services and are always dependent on specific device and network support. Conference content-based systems only provide services for the conference participants based on the conference content, but do not consider about the effect of user locations on enhancing the real-world interactions of people.

Considering the shortages of these two kinds of conference-based support systems, we designed Nokia Find & Connect (NF&C) that uses location as the basis for integration with the network and proximity-based system using encounters, together with the conference content-based services, all driven by a mobile UI. Our contribution is that we designed a system for enabling opportunistic social networking, that is, helping to connect with other people in the conference at specific opportunities based on recording encounters and social interactions, as well as providing a richer, enhanced social networking and conference experience.

To demonstrate the usefulness of our system, we conducted a field trial at the 7th International Conference on Ubiquitous Intelligence and Computing and Autonomic and Trusted Computing (UIC/ATC 2010) [3]. Through performing a user study with user behavior analysis and surveys, we show that NF&C can help the conference participants enhance the real-world interactions of people and improve efficiency at a conference.

Our paper is organized as follows. Section 27.2 summarizes related work. A full description of NF&C is given in Sect. 27.3. Section 27.4 covers the trial and outcomes of the application deployment at UIC/ATC 2010. Finally, we conclude the paper and provide avenues for future work in Sect. 27.5.

27.2 Related Work

Much of the work on conference-based support systems can be divided into two parts: network and proximity-based systems and conference content-based systems. Network and proximity-based systems such as [4–8] only focus on the interplay of networking and social contact at a conference, but do not provide other conference-supported services. For example, memetic spreading [4], opportunistic networking for mobile devices [5] and the live social semantics application [6] can only obtain the encounters of people, but not the user's position. IBM's RFID system [7] and Yuusuke's system [8], on the other hand, are only concerned about whether a person is in a session room, and also suffers from not being able to get the exact location of the user. All of these systems only focus on characterizing the statistical properties of human mobility and contact, such as reachability to connect indirectly with others. They lack other conference-supported services, such as building your schedule, and seeing a map of what's happening and how to get there. In addition, these systems are all dependent on specific device and network support, so they are complex to deploy.

On the other hand, there are conference content-based systems that provide services for the conference participants based on the conference content. For example, "Ask the Author" [9] helps audiences ask questions during a conference using mobile devices, such as a notebook and PDA. A conference information system in [10] allows users to query the progress of the conference and the current state of papers anytime and anywhere using a mobile client on a mobile phone and web server. Many conferences now provide their own conference applications such as SXSW GO [11] which was used for SXSW 2011. The application allows you to build your schedule, see a map of what's happening and how to get there, navigate the tradeshow and stay connected to the online social world like Twitter and Facebook. However, systems like [9, 10] only focus on one special service for the conference and do not consider about the interplay of networking and social contact at a conference, especially encounters. SXSW GO [11] and others do not consider about the locations of users or the social network in the real-world.

Our NF&C system differs from previous work in that we combine and integrate user location with a network and proximity-based system using encounters together with the conference schedule using a mobile UI. We do this by using readily available and popular technologies. We use mobile phones (instead of special hardware sensors), use the indoor Wi-Fi network, and specifically use the indoor location to record where attendees are, to provide location-based services such as finding where people are at the conference, where sessions are held, creating a personal conference schedule, and for establishing social networking connections such as exchanging personal information and communicating with others based on your encounters and similarity with others. In this way, we provide a much richer, integrated and social experience that provides opportunities for users to meet and connect with others.

27.3 Nokia Find & Connect

27.3.1 System Architecture

NF&C is a location proximity-based conference-supported system for conference participants to social network. The system architecture is a three-tier architecture, which includes the presentation layer, session layer and database layer as shown in Fig. 27.1.

There are two kinds of clients in the presentation layer. One is a mobile client for the conference participants which contains the Positioning Client and the Browser Client. The other one is a PC Client which shows the social network visualization and position in real time.

The Positioning Client and the Positioning Server compose the positioning subsystem. The Positioning Client collects Wi-Fi signal strengths from nearby WLAN access points at a user-specified interval, and sends them to the Positioning Server through the User Datagram Protocol. After the Positioning Server receives the Wi-Fi signal data, the Positioning Engine uses the Positioning Model and machine learning algorithms to approximate the positioning of the user. The Positioning Model is created by performing a site survey that involves recording the Wi-Fi signal strengths and access points of the conference place. For our implementation, we used an off-the-shelf commercial Wi-Fi positioning system, called the Ekahau Real Time Location System [2].

The Browser Client sends requests to and receives responses from the Web Server in JSON format through TCP. Then the server reparses the JSON message, and sends the result back to the Browser Client. The internal implementation of the Positioning Client, the Positioning Server, the mobile Browser Client and the Web Server is shown in Fig. 27.2.

The information stored in the database includes activity content, paper content, user content and message content. Activity content includes topic, time, location

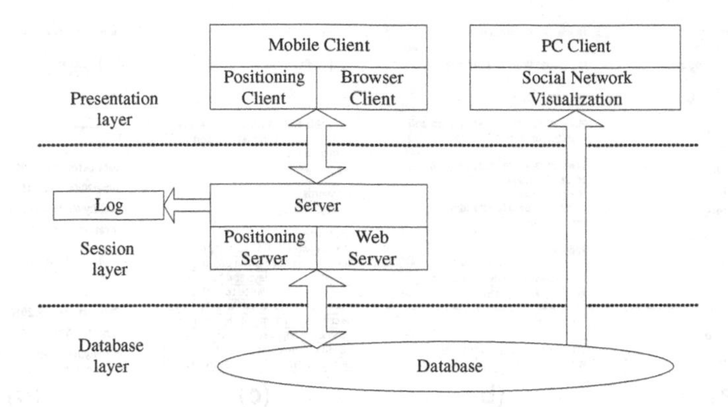

Fig. 27.1 System architecture of NF&C

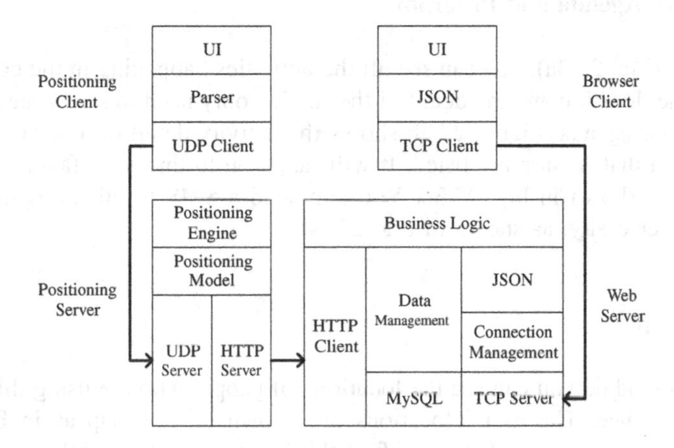

Fig. 27.2 Internal implementation of the mobile client and the server

and chair of each session, papers in that session and comments that users make during the conference. Paper content is title, keywords, abstract and author lists of all accepted papers. User content includes user profile, friends, friends recommended, agenda, favorite paper and encounter information of each user.

The server logs record every request from the clients including time, username and specific page accessed, parameters and location for further data analysis.

27.3.2 Features

Five functional modules are in the home screen: My Agenda, Program, Map, Social Network and Buzz.

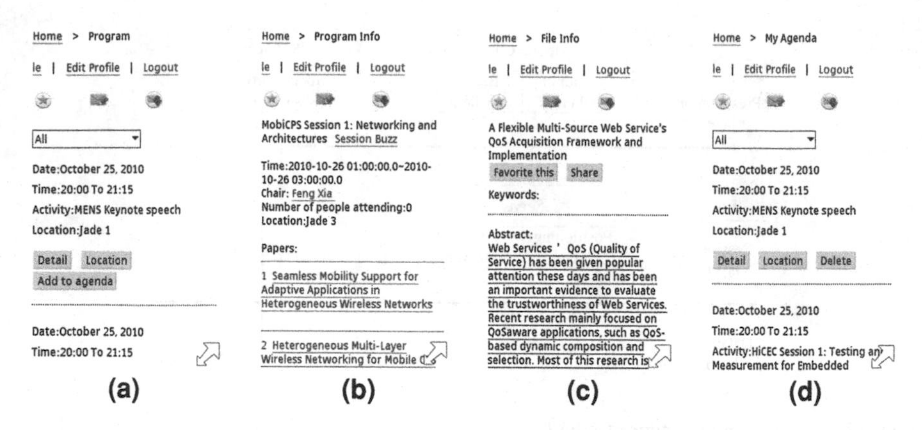

Fig. 27.3 Program & my agenda. **a** Program. **b** Program info. **c** Paper info. **d** My agenda

27.3.2.1 My Agenda and Program

In Program (Fig. 27.3a), you can see all the activities happening in the conference and the time, location and the detail of them. You only need to select the button to add it to your agenda. Figure 27.3b shows the activity detail of a session, and all the papers in that session are listed. It will help you to find your favorite session. Paper Info is shown in Fig. 27.3c. You can send a SMS to others to share your favorite paper easily, as shown in Fig. 27.6a.

27.3.2.2 Map

In the Map module you can see the locations of people who are using this system in the conference. The users' locations are shown on the map as in Fig. 27.4. Search Online People can help you find the location of the people you want to meet. You will never worry about missing the academic stars during the conference.

People on the map can be filtered by: all people, friends, sessions and only yourself. You can then select the link of the person to see his/her profile. If you select a specific user in the user list, you will see only that person on the map.

27.3.2.3 Profile and Social Network

The user's profile is similar to other social networking sites as shown in Fig. 27.5a. Users can manage their friends by adding and removing friends, and obtain the friend's details. When the user selects a friend or a contact that is nearby, that user can look at the contact details. These include downloading the contact's business card to the phone, and finding out the last encounter time and location.

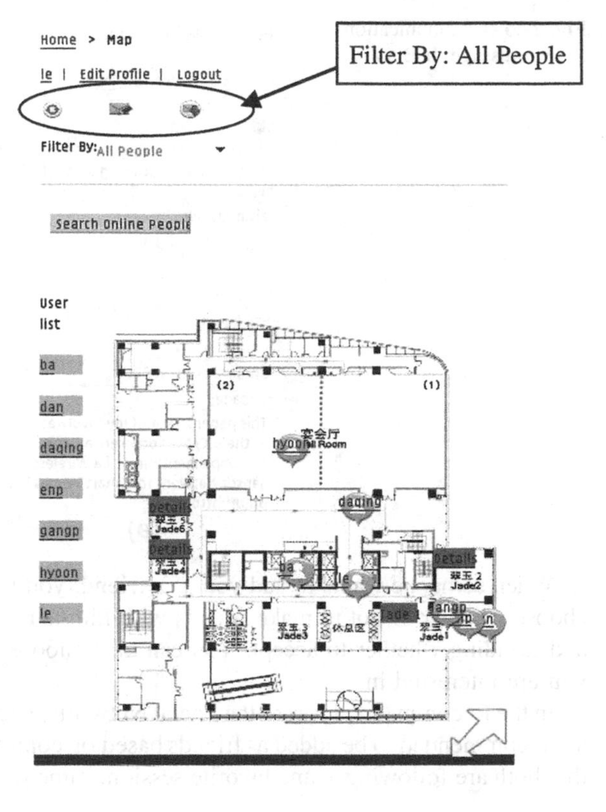

Fig. 27.4 Map

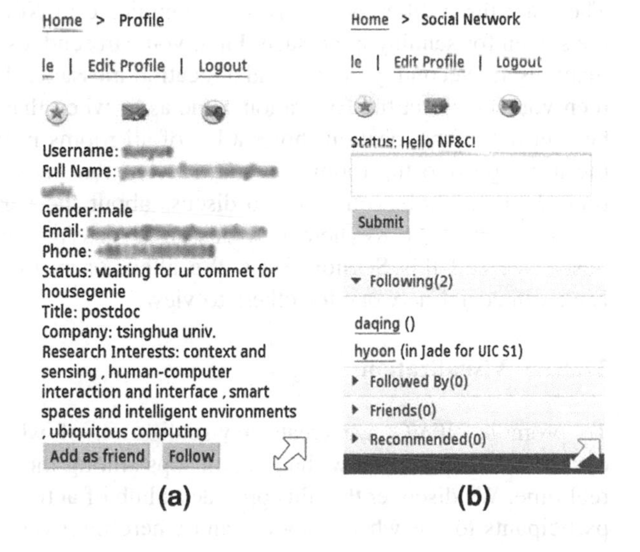

Fig. 27.5 Profile & social network. **a** Profile. **b** Social network

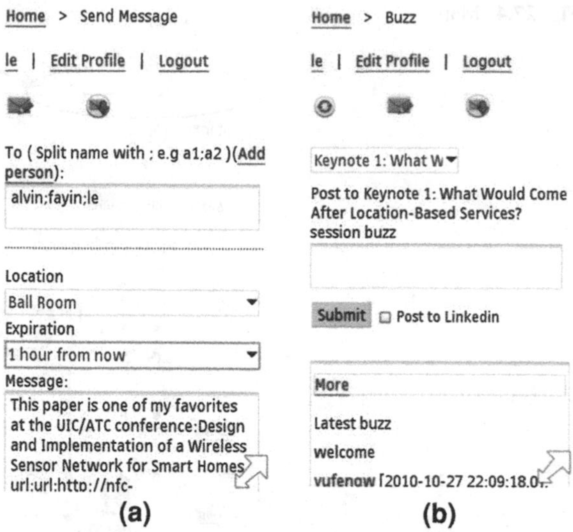

Fig. 27.6 Communication. **a** Send mssage. **b** Buzz

When someone wants to add you as a friend, you will receive a SMS. You can choose whether or not to make friends with him/her by reading the friend request and deciding whether to accept. You can also choose to follow the person whom you are interested in.

In the Recommended part of the Social Network page, there is a list of people that are recommended to be added as friends based on common friends, common people that both are following, same favorite session, same favorite paper and encounters.

27.3.2.4 Communication

There are three different means to communicate in NF&C. In Fig. 27.6a, we see the screen for sending a message. First, you can send a standard message to one or many users. Second, you can send a location message, if you choose the Location, then you can select the Expiration Time as to when that message will continue to be sent until. The location shows a list of all rooms in the conference. Whenever the users go into the room they can receive that message within the expiration time. In Fig. 27.6b, you can also discuss about the various sessions in the conference such as the Keynote 1 by posting a message and viewing other people's posts. We call this Session Buzz. We allow users to post the buzz also to the LinkedIn social network for others to view.

27.3.2.5 Visualization

To promote NF&C, we create a visualization of where all people are on the conference map and show the relationships among the people as a social graph in real time. We discover that this provides a hub of activity for participants and non-participants to see where they are and where their other friends are. The system

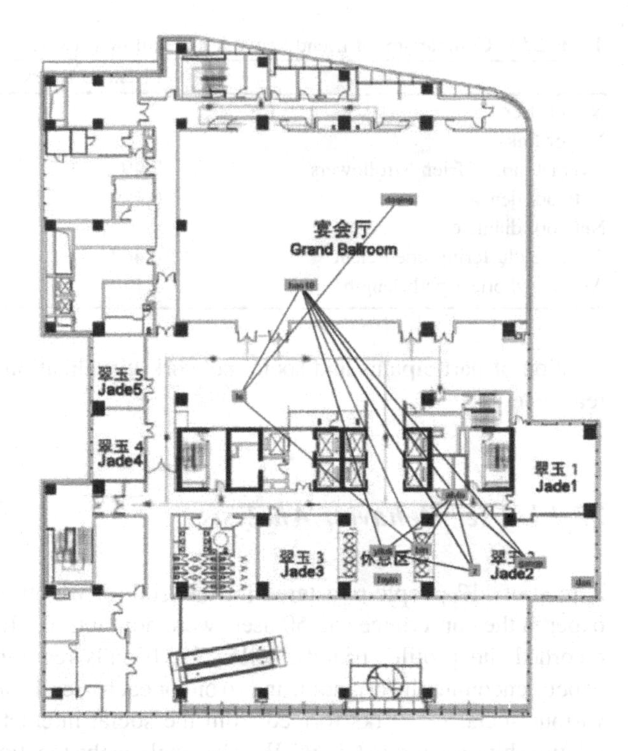

Fig. 27.7 Snapshot of the visualization of users at the conference and social graph in real time

uses the username to designate the user and if two users are friends then there will be a red line between them and a green line to indicate a follow. The visualization client can help the social scientists get the structure of the social network visually. Figure 27.7 shows one snapshot of the social network visualization captured at the conference.

27.3.2.6 Privacy

Privacy protection is an important part of the NF&C system. Permission was sought from all participants for collecting and using their data. A standard ethics participation agreement explained what the data was and how it was going to be used. We anonymized all the data in our analysis. Users can set whether to share location information to others or to friends or just private in their Profile.

27.4 Trial of NF&C System in UIC/ATC 2010

NF&C was developed and piloted in the UIC/ATC 2010 conference for four days during October 2010. During the trial, we provided 50 Nokia X6 and 50 Nokia 5800 phones for the participants. A big display was placed in the hall to show the

Table 27.1 Comparison of friend network and follow network

	Friend network	Follow network
No. of users	59	62
No. of links	221	184
Average no. of friends/followers	7.49	2.97
Network density	0.129	0.04865
Network diameter	4	6
Average clustering coefficient	0.462	0.387
Average shortest path length	2.120	2.679

location of participants and social network visualization as shown in Fig. 27.7, in real time.

27.4.1 User Behavior Analysis

A total of 112 people registered for the trial, of which 62 users were authors of a paper in the conference and 50 users were non-authors. In our database and log, we recorded the profile, friends, followers, friends recommended, agenda, favorite papers, encounter information and so on for each user. From this data, we analyze the various social networks formed from the social interactions of friend and follow relationships arising in NF&C. We also analyze the location distribution of the users.

27.4.1.1 Friend Network

From the acceptance of friend requests, we discover 59 users in total who have at least one friend in the friend network and 221 friend links generated. Each user has an average of 7.49 friends and 59% (257/436) of all friend requests are accepted.

A total of 2,829 friend recommendations are generated during the whole trial, and 274 of them are added as a friend request by 35 unique users. This means that about 10% of total friend recommendations are converted into friend requests.

We perform a network analysis of the friend network for all registered users and report the results in Table 27.1. Authors significantly dominate the friend network with 55 users.

Figure 27.8 shows the visualization of the friend network. In this figure, we have chosen the red dot as the selected person and the yellow dots are that person's friends.

27.4.1.2 Follow Network

A second form of social interaction network is the follow network which comprises of all users that follow others and others that follow that user.

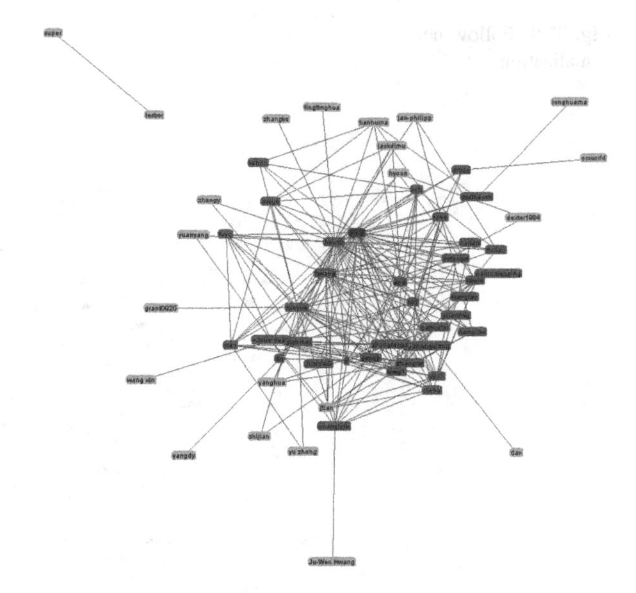

Fig. 27.8 Friend network visualization

A total of 62 users are in the follow network with 184 follow links generated, 57 unique users follow others and 37 unique users are followed. The reciprocal percentage is 0.1739, making it highly asymmetrical which makes sense because follow is a unidirectional relationship that does not require a response. Average degree is 2.96774 which means a user follows around 3 people. The comparison of social network properties of the friend network and the follow network is shown in Table 27.1.

Figure 27.9 shows the visualization of the follow network. In this figure, we have chosen the red dot as the selected person and the yellow dots as the people that person follows.

The numbers of people who are involved in the friend network and the follow network are almost the same. The number of links in the follow network is less than that of the friend network. The friend network is much denser than the follow network, has a smaller diameter, is more highly clustered with a larger clustering coefficient, and has similar average shortest path length. This means that the friend network is a tighter and intimate network than the follow network due to the reciprocity of friend requests and its social connotation, which is not a surprise.

27.4.1.3 Location Updates

From the trial, a total of 83 users uploaded 1,430,121 locations. Figure 27.10 shows the distribution of the location updates by every user which appears to follow an exponentially decreasing distribution. Aggregated positioning of users over the 4 days is shown in Fig. 27.11 and we can clearly see that by overlaying the scattered patterns to the conference map, we can clearly see which rooms in the conference had the most activity presence of users.

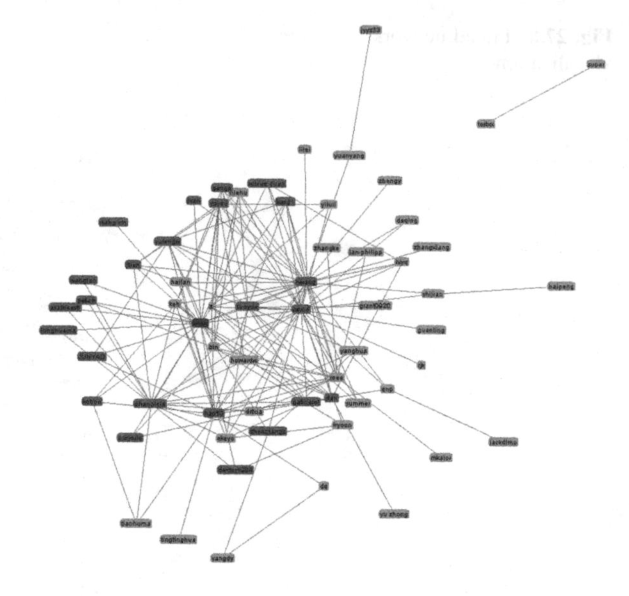

Fig. 27.9 Follow network visualization

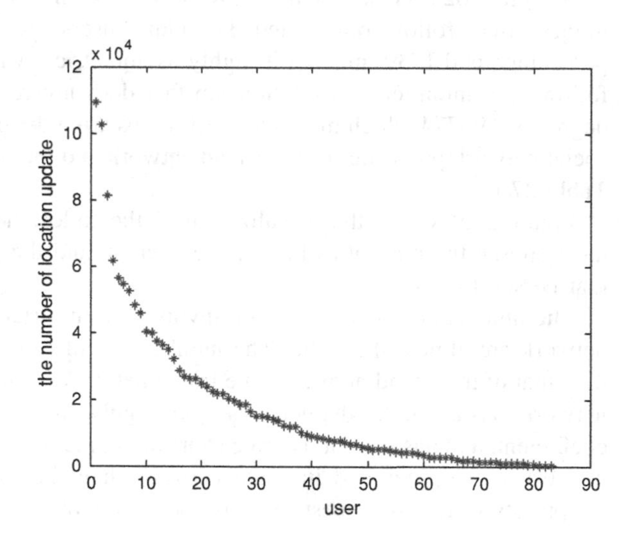

Fig. 27.10 Location updates by every user

27.4.2 Survey Results

We released two surveys in the conference. The pre-trial survey is the Social Network and Conference Survey which was provided to participants before the beginning of the conference. This was designed to get background knowledge of the social network behavior of the participants. The exit survey is the User Experience and Feedback Survey which was provided to participants at the end of the conference to obtain their feedback from using the software. We received 86

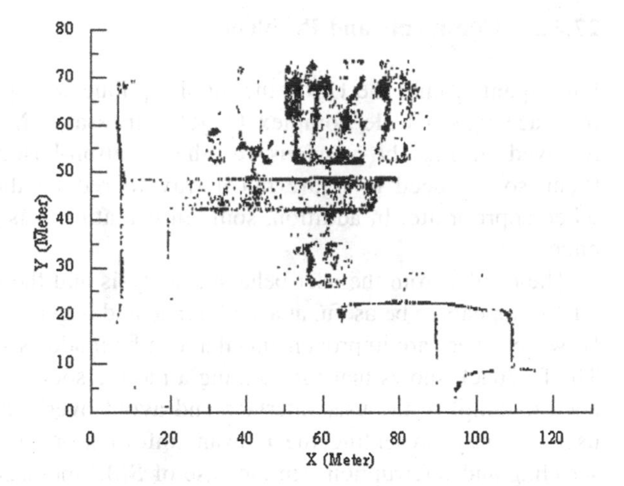

Fig. 27.11 Aggregated positioning of users

completed pre-trial surveys and 83 completed exit surveys. 49% of the participants were aged 25–35 years old, 79% were male and 91% were from academia.

27.4.2.1 User and Social Network Behavior Summary

From the Social Network and Conference Survey, we can reach the following conclusion. Most of the users use QQ as their online social network (OSN) and they use OSN at least once a day, and average 15 min to half an hour each session to keep in touch with friends. Some users use QQ on the phone at least once a day, while others never use OSN on phone. Most users never use proximity-based or LBS apps, some have used Google Latitude. All users have never used an indoor social networking application. Most do not use OSN for conferences, some have used Facebook and LinkedIn and over half of the users add people from the conference to Facebook. All users have never used a conference system designed for mobile devices.

Therefore, these users are prime to try out NF&C because they do have social networking experience, but not on the mobile and have not used any proximity-based location applications.

27.4.2.2 Usage and Usability

From the User Experience and Feedback Survey, we can reach the following conclusion. The user interface was average as to what they expected (62%), usable (easy to navigate) but difficult at times (67%), and overall ease of use was usable but difficult at times (60%). Location accuracy showed that they were in room when they entered the room. Social network visualization on the big display was also helpful. Most of the users made 0–5 new contacts.

27.4.2.3 Comments and Problems

Participants found the UI on the mobile phone was a little hard to use, and the interface was a little complex to get a message. Many SMS messages were received during the sessions for the location-based posts thus interrupting them, so we need to figure out a way to reduce the number and only send when appropriate. In addition, some information was not well organized in the pages.

The results from the user behavior analysis and the survey analysis prove that NF&C appears to be useful at a conference and is generally well accepted by users. However, there are improvements that can be made as shown from the comments. The feedback shows that for building a mobile social networking application, we need to simplify the user interface and avoid overloading all information to the user, by only presenting the relevant information at that time, especially with scrolling and interruptions (in the case of SMS messages).

27.5 Conclusion

In this paper, we presented a system called NF&C which is a location-based proximity system using encounters and integrates with the conference basic services using a mobile UI. We use mobile phones, indoor Wi-Fi network, and specifically use the indoor location to record where attendees are, to provide location-based services such as finding where people are at the conference, where sessions are held, creating a personal conference schedule, and for establishing social networking connections such as exchanging personal information and communicating with others based on your encounters and similarity with others. In this way, we provide a much richer, integrated and social experience that provides opportunities for users to meet and connect with others.

The system was deployed in the UIC/ATC 2010 conference. The results demonstrate that NF&C can help the conference participants enhance the real-world interactions of people and improve efficiency at a conference.

In the future we will improve the NF&C UI and build a better robust system to give users a better user experience. We would like to perform user studies in other conferences to improve the performance of NF&C, as well as gather more data for our research. Finally, we would like to deploy NF&C to other indoor environments like an office environment [12].

Acknowledgments We would like to thank all the attendees who participated in the Find & Connect trial at the UIC/ATC 2010 conference and provided much constructive feedback, as well as our colleagues who helped with the development and operations of the trial from the Find & Connect team.

References

1. Gartner says worldwide mobile device sales to end users reached 1.6 billion units in 2010. http://www.gartner.com/it/page.jsp?id=1543014
2. Ekahau incorporation. http://www.ekahau.com
3. UIC2010. http://www.nwpu.edu.cn/uic2010/
4. Borovoy R, Martin F, Vemuri S, Resnick M, Silverman B, Hancock C (1998) Meme tags and community mirrors: moving from conferences to collaboration. In: CSCW'98: Proceedings of the 1998 ACM conference on computer supported cooperative work, Seattle, pp 159–168
5. Hui P, Chaintreau A, Scott J, Gass R, Crowcroft J, Diot C (2005) Pocket switched networks and human mobility in conference environments. In: WDTN'05: Proceedings of the 2005 ACM SIGCOMM workshop on delay-tolerant networking, Philadelphia, pp 244–251
6. Van den Broeck W, Cattuto C, Barrat A, Szomsor M, Correndo G, Alani H (2010) The live social semantics application: a platform for integrating face-to-face presence with on-line social networking. In: PERCOM Workshops'10: Proceedings of the 8th annual IEEE international conference on pervasive computing and communications, Mannheim, pp 226–231
7. IBM uses RFID to track conference attendees. http://pcworld.about.com/od/businesscenter/IBM-uses-RFID-to-track-confere.htm
8. Kawakita Y, Nakamura O, Wakayama S, Murai J, Hada H (2004) Rendezvous enhancement for conference support system based on RFID. In: SAINT'04: Proceedings of the 2004 international symposium on applications and the internet workshops, pp 280–286
9. Deng Y, Chang H, Chang B, Liao H, Chiang M, Chan T (2005) "Ask the Author": an academic conference supported system using wireless and mobile devices. In: WMTE'05: Proceedings of the 2005 IEEE international workshop on wireless and mobile technologies in education, pp 29–33
10. Wan G, Yang Z (2010) Conference information system based on pervasive computing. In: CAIDCD'10: Proceedings of the 2010 IEEE 11th international conference on computer-aided industrial design & conceptual design, p 1254
11. SXSW GO. http://sxsw.com/node/6469/?
12. Zhu L, Chin A, Zhang K, Xu W, WangH, Zhang L (2010) Managing workplace resources in office environments through ephemeral social networks. In: UIC'10: Proceedings of the 7th international conference on ubiquitous intelligence and computing, pp 665–679

Chapter 28
Tools for Ubiquitous PACS System

Dinu Dragan and Dragan Ivetic

Abstract Ubiquitous computing in healthcare (ubiquitous healthcare) is a very appealing and beneficial goal. It is a mean to achieve patient-centric healthcare and a mean to improve healthcare in general. PACS system is important part of a modern day hospital. To achieve ubiquitous healthcare, it is necessary to build ubiquitous PACS system. Thus a traditionally closed system based on large and location-dependent workstations will be accessible to all users regardless of their location and devices they use. This is achievable only by improving PACS system functionality without sacrificing its interoperability.

Keywords Ubiquitous healthcare · PACS system · Medical image scalability · Data scalability

28.1 Introduction

Modern day hospitals represent a highly dynamic and interactive environment. To achieve patient-centric healthcare, it is necessary to adopt ubiquitous healthcare [1]. It improves healthcare because healthcare services are provided when requested and needed, not when they are offered [2].

Picture archive and communication system (PACS) is a backbone of medical imaging in modern day hospitals [3]. PACS system is exposed to the same dynamic conditions and interactive requirements as modern day hospitals. Quality

D. Dragan · D. Ivetic (✉)
Faculty of Technical Science/Computing and Control Department,
University of Novi Sad, Novi Sad, Serbia
e-mail: ivetic@uns.ac.rs

D. Dragan
e-mail: dinud@uns.ac.rs

J. J. Park et al. (eds.), *Proceedings of the International Conference on Human-centric Computing 2011 and Embedded and Multimedia Computing 2011*, Lecture Notes in Electrical Engineering 102,
DOI: 10.1007/978-94-007-2105-0_28,

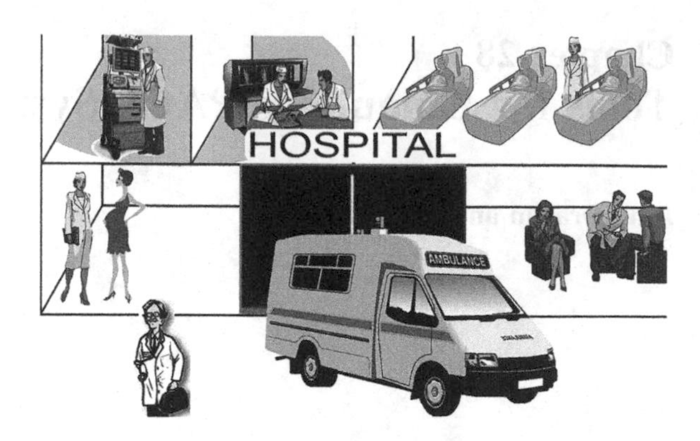

Fig. 28.1 Conventional PACS system—PACS data is accessible only in the upper two rooms containing desktop computers and PACS workstations

of medical service is directly influenced by PACS system [4]. The quality and availability of medical images influence the quality of medical diagnosis and patient treatment. Perceived quality of medical care is influenced by PACS system also. Patients are acquainted better with their disease and acquainted faster with the treatment. This improves patient-physician relationship.

Although it improved medical service and patient care greatly, conventional PACS system is not patient-centric in general (Fig. 28.1). Medical services of PACS system depend on large PACS workstations which are confined to fixed, predefined locations inside large hospitals. It is not possible to access PACS data remotely using random devices. As we can see in Fig. 28.1, only the upper two rooms (with gray background) containing PACS workstations and desktop computers have access to PACS data.

Due to the last decade development in the computing and network technologies, building ubiquitous PACS system became a reality [5], although the adoption is going slow. Beside conventional desktop computers and PACS workstations, new devices such as laptops, personal medical assistants, PDAs, and mobile phones can be used as integral part of PACS system [6]. PACS data should be accessible outside radiology reading room and physician's office and it should be possible to view medical images outside hospital centers. Ubiquitous PACS system enables equal patient treatment regardless of their location and devices on disposal.

The question is whether PACS system software infrastructure supports these new devices and what changes should be made to accommodate ubiquitous PACS system without sacrificing conventional PACS system interoperability. Communication in PACS system is based upon Digital Imaging and Communication in Medicine (DICOM) standard [8]. DICOM standard is more than just a communication protocol. It defines file format (header describing all the relevant medical data and image), data transmission, security, medical report management, and much more. DICOM standard has been defined more than 20 years ago. Building a ubiquitous PACS system was not a priority when the standard has been defined.

We will demonstrate that it is not necessary to disseminate and rebuild old PACS systems in order to create ubiquitous systems. It is possible to extend DICOM standard to support ubiquitous PACS system. Also some of the necessary components are already contained in the DICOM standard.

Ubiquitous healthcare computing and its connection to PACS system is described in Sect. 28.2. Backbone for building ubiquitous PACS system is described in Sect. 28.3. Section 28.4 concludes the paper.

28.2 Ubiquitous Healthcare and PACS

Digital medical data are useful for medical staff during many situations such as patient bad-side visits, emergency situations, out door consultations, etc. [6]. Conventional computing based on desktop computers is not a solution for these situations. The ubiquitous healthcare is computing paradigm in which computers are integrated in a seamless way in everyday medical practice and clinical work [1]. This means that computers are everywhere, at disposal to medical staff at any time. Beside conventional places, in ubiquitous healthcare computers are on hall walls, on cantina tables, beside patient bad-sides, in physician's hands, and in physicians pockets. Medical data flow is automated, context-aware, and instant [2].

A lot of time has been invested in ubiquitous healthcare. Most of the efforts have been focused on management aspects of medical systems. They cover customer service, patient and study management, staff management and scheduling, remote medical monitoring, teleconsultation, etc. All of these ubiquitous services are based on textual data and small sized binary data which are typical for Health Level 7 (HL7) medical standard [8].

In most part medical images have been completely ignored [9]. This is no wonder because ubiquitous computing is largely dependent on wireless technologies and mobile devices. The best way to support ubiquitous healthcare is through mobile devices carried by medical staff [2]. They are light and wearable. Most of them have the ability to connect on broadband mobile networks which makes them accessible almost everywhere. Also, they are quite common and widespread. Mobile devices are useful in many scenarios for ubiquitous healthcare because they can be carried in pocket. They can be manipulated with one hand while the other hand is available for other operations. They can be equipped with special readers (blue-ray, RFID, etc.) for context data update [10]. This would enable data displayed on mobile device to change as the physician moves from patient to patient.

Mobile devices are important part of telemedicine effort [11]. Medical images can be used on mobile devices in various medical disciplines [12–14]. They have been used for secondary consultation, in educational purposes, neurosurgery, radiology, emergency medicine, teleconsultation, etc. The most extensive use of handheld devices is in emergency medicine. In one case camera on mobile phones has been used for acquiring images of injuries on sight of accident or in medical vehicle [15]. These images are critical for achieving proper first aid to a patient and for the proper patient management at a hospital. In another case mobile phone to mobile phone transmission of medical images has been used to consult a senior specialist who was away from hospital [16]. Images have been taken from a film

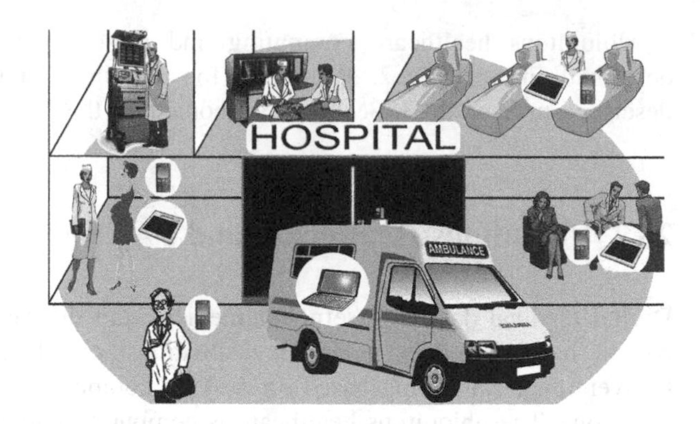

Fig. 28.2 Ubiquitous PACS system—PACS data is accessible in all parts of the hospital, even outside of the hospital

light box or a workstation display by camera on mobile phone. The consultations have been used successfully to decide patient treatment in emergency cases. However, DICOM communication protocol is not supported by the majority of the solutions and they have not been used for building ubiquitous PACS system.

Downside of mobile devices is their limited capacity, small display size, low resolution, and limited band-with of mobile networks. It is necessary to use additional services and techniques to achieve transmission of medical images to mobile devices. Several approaches have been proposed for medical image transmission to mobile devices [6]. They have been proposed as part of telemedicine effort, as part of PACS system extension, or both. Some of these services are suitable for ubiquitous PACS systems as we will describe latter in the paper.

28.2.1 Ubiquitous PACS System

As said in introduction, conventional PACS system is not ubiquitous. Ubiquitous PACS system is described in Fig. 28.2. PACS data are accessible in all parts of hospital complex, in corridors, waiting rooms, elevators, mass halls. Data is context sensitive. It is updating to accommodate physician's surrounding. For example, during patient visit, patient credentials are scanned using context reader of mobile device. The patient's latest PACS data are retrieved from PACS server and uploaded to physician's mobile device.

But ubiquitous PACS system is not limited to one building or hospital complex [9]. It encompasses outdoor staff, emergency vehicles, and remote sites. Thus medical staff and patients are part of PACS system even when they are not inside of a medical building.

Beside limitations in computing and network technologies (which can be overcome), great obstacle in building ubiquitous PACS system is presented in predating conventional PACS system [6]. Modern hospitals already possess PACS system. Usually these systems cost a lot of money. In many cases, redesigning or rebuilding PACS system to support ubiquitous computing is not an option. Also, there is always the question of back compatibility. Therefore, ideal solution would be the one which will only extend conventional PACS system to support ubiquitous computing [17]. Also, this should be achieved in DICOM manner as that is the standard way for communicating PACS data [6].

28.3 Ubiquitous PACS System Backbone

Computing and network resources needed for building ubiquitous PACS system already exist. There are mobile devices that can be used anywhere and wireless networks that can transmit data almost everywhere. Wherever possible, communication should be achieved using DICOM standard or at least in DICOM manner.

There are two main issues that should be addressed in ubiquitous PACS system [18, 19]:

- Security of transmitted data.
- Scalability of transmitted data.

Security issues are very important for ubiquitous PACS system because data access is not limited to fixed and secured entry points. Therefore, data sniffing and false authentication should be prevented. For a time, security issues have not been addressed in DICOM standard. This has change in previous 10 years with the growing interest in telemedicine [20]. DICOM has been expended to address the following security issues:

- Authentication—verifying the identity of entities involved in a DICOM operation.
- Confidentiality—guarding the data from disclosure to an entity which is not a party to the transaction.
- Integrity—verifying that data within an object has not been altered or removed during transport.
- Secure transport connection profiles using TLS 1.0 and ISCL protocols.

However, user log-in authentication is not supported by DICOM standard [20]. Therefore, before using DICOM based PACS system, user has to be identified using some other service of the global medical network.

Because there are many different devices and users involved in PACS communication, scalable data transmission should be supported in ubiquitous PACS systems. There are three contexts of data scalability:

- Usage context—only the data needed for the given context of use should be sent to the requesting device. For example, during patient visit only the last data for that patient are transmitted to the physician's device. It is not necessary to send data for all the patients in the same room or data related to some previous, unrelated illness.

- User context—only the data permitted to the user should be sent to the requesting device. Physicians should access only the data of their patients. Nurses should have only limited access and they should only see data relevant for the patient treatment.
- Device context—only the minimum amount of device supported data should be sent to the requesting device. Different devices have different capabilities. Mobile devices can handle and display only limited amount of PACS data. It is not necessary to send entire data scope to devices that cannot handle them.

All three contexts of data scalability should be supported in ubiquitous PACS system. They should be combined to achieve optimal informational and operational surrounding. Therefore, mechanisms for identifying usage, user, and device context should be provided.

Usage context could be implemented through mechanisms incorporated in working environment. Radio Frequency IDentification system (RFID) can be employed to automatically detect nearby patient [10]. RFID tag can be transformed to patient's ID which will be used to access relevant PACS data.

User context is achievable through user login mechanism and access rights [2]. This can be implemented at PACS server side. Only data matching the user credentials are sent back.

Device context, similar to user context, is achievable through user login mechanism. When user logs to the system, his/hers device information is sent to PACS server. Only data suitable to device capabilities are sent back. The entire process can be automated using device vendor information and identification mechanism such as we employed in [21]. We used WAP protocol tools to identify the type of mobile phone communicating to DICOM based PACS server.

All three contexts of data scalability can be implemented in DICOM standard with some limitations. Scalability of DICOM data has two dimensions [9], (Fig. 28.3):

- Scalability of relevant medical data.
- Scalability of relevant medical image.

Scalability of relevant medical data is natively supported in DICOM standard [22]. It is not necessary to transmit all the existing DICOM attributes. There is only minimum set required for successful communication. The rest of attributes is transmitted in accordance to user's request. PC user will receive more data than mobile phone user (Fig. 28.3). Relevant medical data is textual data of small size measured in kilobytes. It is adequate for all device types. The only limitation exists on mobile phones and PDAs [2]. They have small size display and there is limited amount of space for text and image. The most important medical data should be presented together with the medical image but only on user request. Data should be displayed on remote part of the screen independent of the image or as image overlay (Fig. 28.3b). Only as the last resort and only if there is too much data covering the entire image should the associated data be presented on multiple or

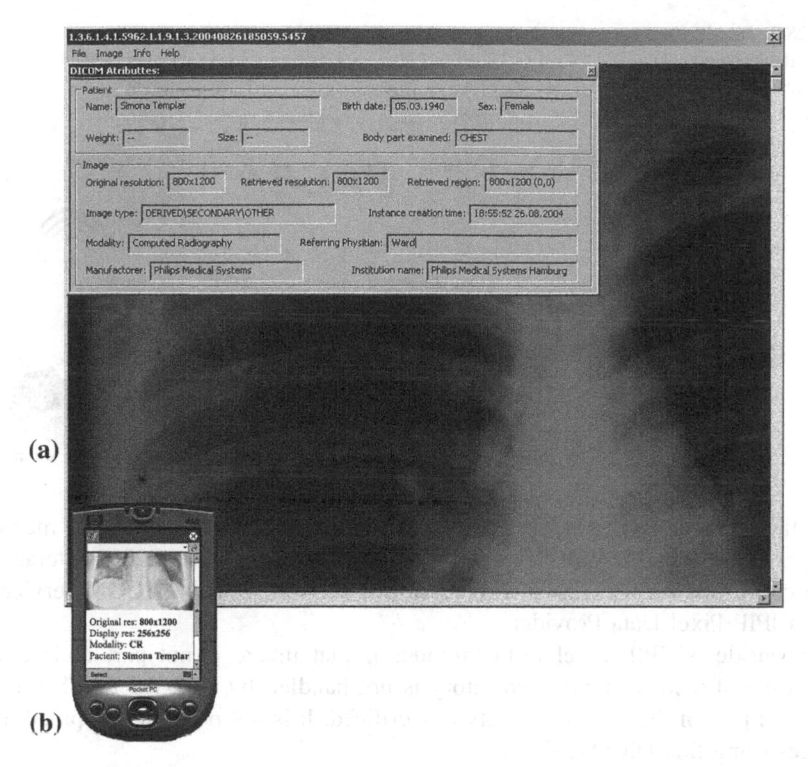

Fig. 28.3 Device context of data scalability: **a** PACS data transmitted to and presented on PC device (maximum amount of relevant medical data and highest image resolution); **b** PACS data transmitted to and presented on mobile phone (minimal amount of relevant medical data and highest image resolution)

scrollable screens. In that case the first screen should contain the image data and the associated data should be displayed on consecutive screens.

Scalability of medical image is very important as it minimizes image data transfer by "just enough quality delivered just in time" (Fig. 28.3). There is resolution and quality scalability [6]. Resolution scalability is important for ubiquitous PACS system because it enables transmission of just enough image data in quality suitable for devices capability and display size.

However, image scalability is not natively supported in DICOM standard [3]. Uncompressed image format is defined by the standard, but scalability is not supported. Yet, encapsulation of other compressions and image formats is supported in DICOM standard. Currently, DICOM standard supports run-length encoding, lossy and lossless JPEG compression, lossless and near lossless JPEG-LS compression, and lossy and lossless JPEG2000 image compression. None of these compression standards, except JPEG2000, supports image scalability.

Although the DICOM standard does not support it directly, it recognizes the importance of scalability. It is possible to access medical image in scalable way over communication protocol different than DICOM. The relevant medical data

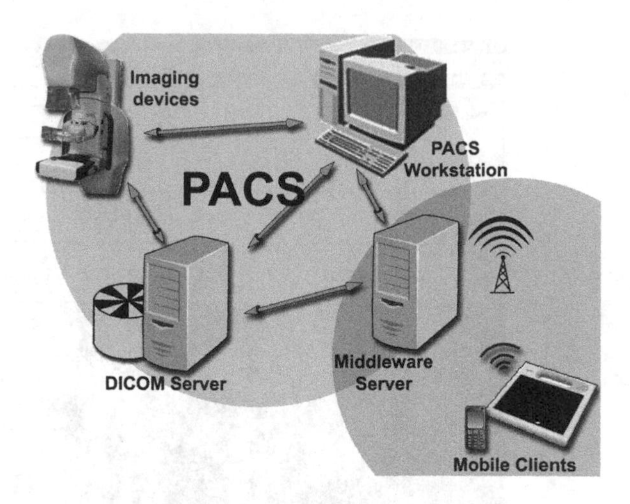

Fig. 28.4 Middleware server for scalable image transmission of ubiquitous PACS system

are still accessed over DICOM communication channel except for the medical image. At the moment only JPIP communication protocol (JPEG2000 Interactive Protocol) based on JPEG2000 compression is supported. This DICOM service is called JPIP Pixel Data Provider.

Downside of JPIP Pixel Data Provider is that image is not part of DICOM message and medical image repository is not handled by DICOM PACS server, therefore part of the interoperability is sacrificed. It is not possible to upload new images using this DICOM service.

There are two additional ways to enable scalable transmission of medical images in DICOM based PACS systems [6]. Both solutions do not impose any changes to the existing PACS systems.

28.3.1 Middleware Server for Scalable Image Transmission

In the first case, mobile devices communicate with DICOM based PACS server through middleware server (Fig. 28.4). Middleware server is in charge of requesting medical images from DICOM server. The received images are scaled on middleware server and transmitted to mobile clients. If necessary medical images are additionally processed and compressed at middleware server before they are transmitted. Images are transmitted to handheld devices over "more appropriate" communication protocol but DICOM server repository of medical images is used. Main downside of this approach is lack of medical image upload to DICOM server and additional processing. Therefore part of the interoperability is sacrificed. Medical image scalability is not supported inside PACS system but as addition to the system.

An example of solution for ubiquitous PACS system based on middleware server solution is a system for accessing medical image database on the go [23].

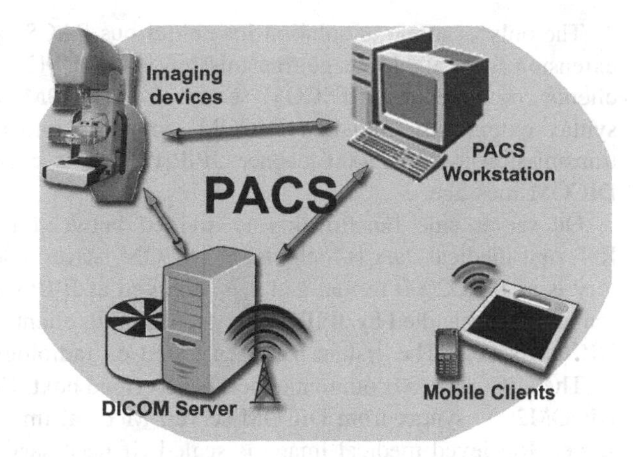

Fig. 28.5 DICOM extension for scalable image transmission of ubiquitous PACS system

The PDA client accesses a middleware server over wireless network. The server retrieves the medical image from the PACS and stores it locally. If necessary, the middleware server encodes the image into JPEG2000 format. Parts of the image (suitable for PDA display resolution) are transmitted back to the client device in a scalable way. Images are transmitted in preview resolution first.

Client request and relevant medical data are transmitted over Hypertext Transfer Protocol (HTTP). JPIP protocol is used to transmit medical image data. JPIP server acts as middleware server. A cache model is maintained on the server side for each session, so the JPIP server can avoid sending redundant data to the client. The system has been tested on radiology data.

28.3.2 DICOM Extension for Scalable Image Transmission

In the second case, mobile devices communicate with DICOM based PACS server directly (Fig. 28.5). To achieve this, it is necessary to extend DICOM communication and server with the support for scalable image transmission. It is possible to implement DICOM service for image scaling which will scale medical image according to client's request. DICOM syntax can be extended to support requests for scalable image transmission. Private Data Element syntax of DICOM standard is used to extend the standard syntax. Thus, it is completely transparent. This approach uses DICOM communication and medical image repository. It enables medical image upload for all devices involved in communication. Although it is a complex solution, DICOM extension for scalable image transmission is a complete DICOM solution for ubiquitous PACS system and it does not sacrifice the interoperability of conventional PACS system. Medical scalability is supported inside DICOM based PACS system and it is the integral part of the system. Even the conventional PACS workstations and desktop computers can request and receive scaled medical images. This is very useful for fast image browsing and preview.

The only example of solution for ubiquitous PACS systems based on DICOM extension for scalable image transmission is DICOM2000 system [6, 22]. Mobile clients communicate to DICOM server over DICOM2000 syntax. DICOM2000 syntax extends the standard DICOM syntax to support requests for scalable transmission in JPEG2000 meaner. JPIP request is encapsulated inside standard DICOM message.

On server side functionality is divided between DICOM and JPIP server. Relevant medical data is archived at DICOM server side. Medical image repository is in JPEG2000 format and it is archived at JPIP server side. Medical image scalability is handled by JPIP server according to client request conveyed through DICOM server. The system has been tested on radiology data.

The DICOM2000 communication is described next. Client request the image in DICOM2000 syntax from DICOM server. Medical image is requested from JPIP server. Retrieved medical image is scaled (if necessary) and transmitted back to DICOM server. Scaled medical image is encapsulated inside DICOM message with rest of relevant medical data and transmitted back to client.

28.4 Conclusion

Although the development of ubiquitous PACS system represents intensive research area, it has not yet reached the full-scale commercial implementation and use. Innovations in the computing and network technologies made ubiquitous PACS system a very real possibility. It is possible to make mobile devices integral part of PACS system thus bringing PACS system closer to ubiquitous computing system.

It is necessary is to achieve ubiquitous of PACS system without sacrificing its interoperability. DICOM standard is communication backbone of modern day PACS system. It supports secure transmission of PACS data and it partially supports scalable data transmission. With adequate tools, it is possible to upgrade DICOM based PACS to support scalable data transmission for user and usage contexts.

Full scalable data transmission in device context is what DICOM lacks and what keeps ubiquitous PACS system from becoming a reality. In standard DICOM syntax it is possible to transmit relevant textual medical data in scalable way. However, it is not possible to scalable transmit medical images. It is necessary to expend DICOM based PACS with the support for scalable image transmission. We described two of these solutions based on JPEG2000 image compression which natively supports resolution and quality scalability. Only the DICOM2000 extension is a complete DICOM based solution for ubiquitous PACS system which does not sacrifice the interoperability of conventional PACS system.

We developed DICOM2000 extension with ubiquitous computing in mind. The DICOM2000 PACS system has been tested in telemedicine. In the paper we identified additional tools necessary to build ubiquitous PACS system. These tools are compatible with existing conventional PACS systems. They can be included transparently and in simple way into existing PACS systems making them ubiquitous. This is intended for our future work.

Acknowledgments This work is financial supported by Ministry of Science and Technological Development, Republic of Serbia; under the project number TR32044 "The development of software tools for business process analysis and improvement", 2011–2014.

References

1. Bardram JE (2004) Applications of context-aware computing in hospital work: examples and design principles. In: Proceedings of the 2004 ACM symposium on applied computing—SAC '04, ACM, NY, pp 1574–1579
2. Ivetic D, Dragan D (2009) JPEG2000 aims to make medical image ubiquitous. Egypt Comput Sci J 31(5):1–13
3. Dragan D, Ivetic D (2009) Architectures of DICOM based PACS for JPEG2000 medical image streaming. Comput Sci Inf Syst J (ComSIS) 6(1):185–203
4. Chang BC et al (2010) Ubiquitous-severance hospital project: implementation and results. Healthc Inf Res 16(1):60–64
5. Seok-soo K, Young-hwan W (2006) Ubiquitous community system for medical information. Int J Multimedia Ubiquitous Eng 1(1):5
6. Ivetic D, Dragan D Medical image on the go. J Medical Sys (in press)
7. Dragan D, Ivetic D (2005) DICOM overview. In: Zebarca VM (ed) Works of the 7th international symposium "Young People and Multidisciplinary Research", vol 1, Intergraf Resita, pp 105–115
8. Elm H (2008) Do we really need standards in digital image management? Biomed Imaging Interv J 4(4):e20
9. Ivetic D, Dragan D Chapter 5: medical image streaming: dicom & JPEG2000 story. In: Kutais BG (ed) Internet policies and issues, vol 8, Nova Publisher (in press)
10. Jaejoon K et al (2006) Design and implementation of remote medical information system on ubiquitous environment. In: Dattatreya GR (ed) Proceedings of the 5th WSEAS International conference on circuits, systems, electronics, control & signal processing (CSECS'06), World Scientific and Engineering Academy and Society (WSEAS), Stevens Point, Wisconsin, USA, pp 186–191
11. Choong MK, Logeswaran R, Bister M (2007) Cost-effective handling of digital medical images in the telemedicine environment. Int J Med Inform 76(9):646–654
12. Reponen J (2006) Mobile image delivery and acquisition. Imaging Manag 6(5):16–17
13. Ahmad Z (2008) A picture paints a thousand words—the use of 3G camera mobile telephones in managing soft tissue injuries. Eur J Plast Surg 31(4):205–206
14. Pirris SM, Monaco EA III, Tyler-Kabara EC (2010) Telemedicine through the use of digital cell phone technology in pediatric neurosurgery: a case series. Neurosurgery 66(5):999–1004
15. Komnakos D et al (2008) Performance evaluation of an enhanced uplink 3.5G system for mobile healthcare applications. Int J Telemed Appl, p 11
16. Korim MT, Soobrah R, Hull P (2009) To admit or not: the use of camera mobile phone in trauma and orthopaedics at night. Eur J Orthop Surg Traumatol 19(3):217–221
17. Dragan D, Ivetic D (2009) Chapter 4: an approach to DICOM extension for medical image streaming. In: Katalinic B (ed) DAAAM international scientific book 2009. DAAAM International, Austria, pp 25–34
18. Doukas C, Maglogiannis I (2008) Article no. 25: adaptive transmission of medical image and video using scalable coding and context-aware wireless medical networks. In: Xiao Y (ed) EURASIP J Wireless Commun Netw. Hindawi Publishing Corporation, NY, pp 1–12

19. Le XH et al (2010) Activity-oriented access control to ubiquitous hospital information and services. Inf Sci 180(16):2979–2990
20. Tarbox L (2002) Security and DICOM. SPIE Medical Imaging 2002 http://medical.nema.org/dicom/spie2002/Security_by_Lawrence_Tarbox.ppt
21. Mirkovic J, Ivetic D, Dragan D (2007) Presentation of medical images extracted from DICOM objects on mobile devices. In: Proceedings of the 9th International Symposium of Interdisciplinary regional research "ISIRR 2007" Hungary–Serbia–Romania, Novi Sad, Serbia, 21–22 June 2007
22. Dragan D, Ivetic D (2009) Chapter 3: DICOM/JPEG2000 client/server implementation. In: Mihailovic D, Vojinovic Miloradov M (eds) Environmental health and humanity issues in down danubian region, multidisciplinary approaches. World Scientific Publishing, Singapore, pp 25–34
23. Tian Y et al (2008) Accessing medical image databases on the go. SPIE Newsroom, Electronic Imaging & Signal Processing [Serial on the Internet]. http://spie.org/x19505.xml

Chapter 29
Contact-Based Congestion Avoidance Mechanism in Delay Tolerant Networks

Chen Yu, Chuanming Liang, Xi Li and Hai Jin

Abstract Congestion in DTN is a prominent problem due to constrained resource, intermittent link and long propagation delay. Traditional congestion control mechanisms based on end-to-end feedback and rate control fail to work in such network environments. Considering the characteristics of DTN, we propose a new contact-based congestion avoidance mechanism (C2AM). The contact history and queuing information are both exploited to obtain the average sojourn time of each incoming message. Using this metric, we try to deter incoming messages with long sojourn time and large size. Motivated by Little's law, we introduce the average storage cost of each message. To our best knowledge, this is the first case where contact information is used to assist congestion control in DTNs. Using simulation, we evaluate the C2AM mechanism and show that it can avoid and alleviate congestion in DTN greatly and improves network performance.

Keywords Delay tolerant networks · Congestion control · Custody transfer

29.1 Introduction

The Internet has gained great success in the past decades of years since it uses a homogeneous set of communication protocols, called the TCP/IP protocol suite. But the usability of TCP/IP protocol is based on several important assumptions:

C. Yu (✉) · C. Liang · X. Li · H. Jin
Services Computing Technology and System Lab, Cluster and Grid Computing Lab,
School of Computer Science and Technology,
Huazhong University of Science and Technology,
430074 Wuhan, China
e-mail: yuchen@hust.edu.cn

J. J. Park et al. (eds.), *Proceedings of the International Conference on Human-centric Computing 2011 and Embedded and Multimedia Computing 2011*, Lecture Notes in Electrical Engineering 102,
DOI: 10.1007/978-94-007-2105-0_29,

an end-to-end path between the source and the destination should be established; round trip delay should be relatively small so that end-to-end feedback is viable; data rates in both directions should be relatively symmetric and loss or disruption of data on each link is slight. This is not the case in new emerging networks such as deep space networks, wireless military battlefield networks, terrestrial civilian networks. In these challenged networks, there may not be a contemporary end-to-end path from the source to the destination because of intermittent connectivity. In some scenarios such as deep space communication propagation delay is very large (minutes, days or even weeks). Because of the instability of wireless communication and node mobility, data rates maybe asymmetric and loss or disruption of data may be high. These kinds of networks are called *Delay/Disruption Tolerant Networks* or DTNs [1].

An overlay layer called the bundle layer which operates above the transport layer has been presented. The bundle layer employs persistent storage and Custody Transfer to help combat network interruption. Custody transfer is a mechanism to improve reliability by using hop-by-hop reliability in the absence of an end-to-end connectivity by transferring the responsibility of reliable delivery to intermediate nodes along a path from source to destination [2]. Custody transfer makes data retransmission caused by loss or corruption only occur within a hop thus improves storage usage of intermediate nodes and reduces end-to-end delay. As described in [2], a DTN node maintaining custody of a message is called a custodian. A custodian of a message will keep the message until it relinquishes custody of the message to another node or because of message expiration. Since messages have to be kept in DTN nodes (maybe for a very long time) until an outgoing link becomes available, buffer space in DTN nodes is much easier to be used up than in the internet. Custody transfer increases the likelihood greatly. With the exhaustion of buffer space in intermediate nodes, network congestion eventually occurs, which leads to the decreasing of network performance such as bundle delivery rate and end-to-end delay.

The link capacity between DTN nodes may vary between positive and zero capacity [3]. A contact refers to a period of time during which the link capacity is strictly positive and the delay and capacity can be considered to be constant. In this paper, we analyze the characteristics of DTN contact, using history learning to predict future contact and then figure out message sojourn time in the buffer space. After that, we study the congestion avoidance policies in the prospect of whether to decline a new incoming message when congestion is likely to occur. Motivated by Little's law, we derive the expected storage cost of each incoming message. When the buffer usage is above a certain threshold, incoming messages are rejected in an appropriate probability. This is somewhat similar to the Random Early Detection [4] mechanism in internet routers. Finally, we evaluate our approach with various network loads.

The rest of this paper is organized as follows. We survey related works in Sect. 29.2. In Sect. 29.3, we firstly analyze the long-term contact pattern in DTNs, then we derive expected storage cost for each incoming message and finally present the main idea of C2AM mechanism. We evaluate our approach in Sect. 29.4. Finally, we summarize this paper and discuss future work in Sect. 29.5.

29.2 Related Works

Although it has been many years since the concept of delay tolerant network was firstly presented, many previous studies are mainly focused on the routing problems [5, 6]. Many routing strategies assume that the buffer space in each node is unlimited since it is not their main focus. But realistic experiences have shown that many routing mechanism perform poorly when buffer space becomes a constraint because of congestion caused by resource exhaustion. For example, epidemic routing gains the optimal network performance as for message delivery rate and end-to-end delay when node buffer has infinite size. But its performance degrades significantly when buffer space is limited and even falls behind some forwarding-based routing strategies.

A great contribution was made by M. Seligman et al. [7] who presented the idea of migrating stored messages to adjacent nodes (usually neighbors) to avoid loss when custodian node becomes congested. Migrated messages will be retrieved back in appropriate time when the congestion is relieved. In some sense, this approach can be characterized as a form of congestion management implemented using distributed storage load balancing. The proposed solution consists of three basic steps: message selection, node selection, and retrieval selection. However, it works when congestion has already happened and thus is a reactive approach.

In [5], the authors use an economic pricing model and propose a rule-based congestion control mechanism where each router can autonomously decide whether to accept a bundle based on local information such as buffer usage. The value and risk of accepting the bundle are observed according to history statistics. If the bundle is non-custodial and the decision is to refuse it then the bundle is simply lost. But if that bundle is custodial, custody refusal caused by buffer oversubscribing or congestion will eventually propagate congestion back along the path to the original sender. The sender senses congestion and reduces the rate at which new bundles are issued. Finally the congestion is abated or disappears. But this approach does not take into consideration the link characteristics of DTNs, which is our main focus.

In [8], the problem of determining whether to accept a custody transfer is studied. On one hand, it is beneficial to accept a large number of messages to facilitate forwarding; on the other hand, if the receiving node overcommits itself by accepting too many messages, it may overuse its buffer space and thereby preventing itself from receiving further potentially important, high yield messages. The authors apply the concept of revenue management and employ dynamic programming to develop a congestion management strategy. We consider the congestion control problem from another point of view which takes into account the contact pattern and makes use of the idea of random early detection.

The authors in [9] firstly applied *Random Early Detection* (RED) and *Explicit Notification* (ECN) mechanism to delay tolerant networks. But they only consider the sizes of different priorities of queued messages, which is not enough to the intermittently congested network environment. There are also other papers studying the message dropping policy when congestion occurs at a DTN node. It maybe works effectively when most of the messages are non-custodial, but fails to work when most messages are custodial since dropping messages conflict with the basic idea of custody transfer. When custodial messages are unable to be forwarded in time, they can only be deleted because of TTL expiration. As for message dropping policies, ideas in the internet can be borrowed and it is beyond our topic.

The authors in [10] evaluated different scheduling and dropping policies in Vehicular Delay Tolerant Networks and mentioned that messages with longer remaining TTL should be sent first. They proposed three scheduling policies for VDTN: Priority Greedy, Round Robin, and Time Threshold. All the scheduling policies are on the traffic differentiation and do not consider the link characteristics.

Actually different congestion control mechanisms can work together to combat network congestion. Our approach can be seen as the previous stage of the message migration mechanism. When congestion happens, a message dropping mechanism can be applied to non-custodial messages. The integration of these mechanisms is an open issue and may require further study.

29.3 C2AM Mechanism

C2AM aims to solve the problem of whether to accept a new incoming message upon a contact opportunity based on local information and contact patterns. The new mechanism detects congestion when buffer usage reaches a certain degree and performs a contact-based buffer management.

29.3.1 Problem Description

C2AM has its focus on the problem of whether to accept a new incoming message when the node is congestion-adjacent or congested. It divides a node into three different states according to the degree of congestion. Based on these states, an RED-like mechanism is performed. We tune the amount of declined messages according to the degree of congestion and drop those messages that are almost unable to be delivered. Contact patterns are taken into account to derive the mark probability.

We consider the congestion control mechanism from the perspective of message cost. Generally speaking, in order to gain a high throughput rate, a DTN node tends to accept messages that would be forwarded sooner in the near future. The optimal solution should spend least time and forward most amount of messages. In traditional internet, this is only related to the incoming and outgoing data rates. But in delay tolerant networks, link characteristics have to be considered since it has a great impact on nodal congestion state. Messages that will wait for a very long time before its outgoing link become available will take up buffer space for a very long time and thus are not desirable. When buffer space becomes precious,

Fig. 29.1 Message sojourn time

message size is an important factor. Intuitively large message will occupy more buffer space and will drain the buffer space more quickly. Besides, message priority can be considered. When congestion is about to occur or has already happened, expedited message should be forwarded first and then turns to the normal and finally bulk messages.

29.3.2 Message Sojourn Time

Message sojourn time consists of contact waiting time and scheduling waiting time. Contact waiting time is the time a message has to wait before contact with its next hop is available, while scheduling waiting time refers to the time it needs to wait before all of its preceding messages are scheduled. A message may not be forwarded even if a contact with next hop is available because of link bandwidth limitations. This can be illustrated in Fig. 29.1. The white segment represent the time period when connection is down and forms the Contact Waiting Time. Black and following gray segment together form a contact in which small black segment represents a message transfer. A message has to wait until all preceding messages are transferred. Suppose a message *m* arrived at time instant t_{in}, it will wait until t_1 when a contact is up. In the time period between t_1 and t_{out}, the link is occupied by other message transferred.

There are mainly three types of contacts in DTNs: intermittent-scheduled contact, intermittent-opportunistic contact, and intermittent-predicted contact [3]. In realistic scenarios, the movements of mobile nodes are usually not completely random or opportunistic. In this case, the contact pattern can be learned based on historical information. This process may be unnecessary in intermittent-scheduled networks since it can be obtained directly.

The contact waiting time is mainly dependent on the inter-contact time T_{ic}. If a contact with the specific next hop is available, this value can be regarded as zero. Both nodes record the corresponding time instant TS_d (the time instant when a connection turns down) each time a connection between them turns down. When a connection to the same peer is established (say, at time instant TS_u), a new inter-contact time T_{ic_new} is sampled. That is:

$$T_{ic\_new} = TS_u - TS_d \tag{29.1}$$

To obtain the average inter-contact time, the exponential smoothing method can be used:

$$T_{ic} = \alpha T_{ic} + (1 - \alpha) T_{ic\_new}, \; 0 \leq \alpha \leq 1 \tag{29.2}$$

In this equation, α is the smoothing constant and can be adjusted according to the contact regularity.

Suppose the current time instant is TS_{now}, and a new message is upon entering but its next hop is not available. The predicted contact waiting T_{cw} is:

$$T_{\mathrm{cw}} = T_{\mathrm{ic}} + TS_{\mathrm{d}} - TS_{\mathrm{now}} \tag{29.3}$$

If the contact waiting time is unable to get due to lack of sufficient information or unknown destination, the residual *TTL* value *RTTL* of the message can be used.

Scheduling waiting time T_{sw} depends on the queued message size $\hat{L}_{\mathrm{q}}$ and link bandwidth *BS*.

$$T_{\mathrm{sw}} = \hat{L}_{\mathrm{q}}/BS \tag{29.4}$$

So the message sojourn time T_{ms} can be computed as following:

$$T_{\mathrm{ms}} = T_{\mathrm{ic}} + TS_{\mathrm{d}} - TS_{\mathrm{now}} + \hat{L}_{\mathrm{q}}/BS \tag{29.5}$$

When the message size is relatively small compared to the link capacity and the propagation delay is very large, the scheduling waiting time can be neglected. In this case the message sojourn time is replaced by the contact waiting time.

29.3.3 Storage Cost

We define projected message storage size $\overline{L}_{\mathrm{m}}$ as the product of message size and message priority P_{m}.

$$\overline{L}_{\mathrm{m}} = L_{\mathrm{m}} \times P_{\mathrm{m}} \tag{29.6}$$

Messages in DTNs have three kinds of priorities: expedited, normal, and bulk. A DTN node will always prefer to accept messages with higher priority. High priority messages are deemed to have high value. In turn, storage of lower priority messages will yield a higher price.

Storage cost C_{m} of a given message is:

$$C_{\mathrm{m}} = \left(T_{\mathrm{ic}} + T_{\mathrm{sd}} - TS_{\mathrm{now}} + \frac{\hat{L}_{\mathrm{q}}}{BS}\right)(L_{\mathrm{m}} \times P_{\mathrm{m}}) \tag{29.7}$$

To make the decision process easier, we can perform a per-flow control. Thus we replace L_{m} with incoming data rates R of corresponding flow. The rationale for this is Little's Law [11]. That is, the average number of queued messages is the product of incoming data rate and message sojourn time. Then we derive the per-flow waiting cost C:

$$C = \left(T_{\mathrm{ic}} + T_{\mathrm{sd}} - TS_{\mathrm{now}} + \frac{\hat{L}_{\mathrm{q}}}{BS}\right)(R \times P_{\mathrm{m}}) \tag{29.8}$$

Table 29.1 Relationship between queue size and node state

Queue size	Node state
$L_q < L_{max} \times TH_{min}$	NS
$L_{max} \times TH_{min} \le L_q \le L_{max} \times TH_{max}$	CAS
$L_{max} \times TH_{max} < L_q$	CS

29.3.4 Congestion State

We set a DTN node to three states according to its congestion level: *Normal State* (NS), *Congestion Adjacent State* (CAS), and *Congested State* (CS), which is presented in [12]. We map these states into the three stages of RED stage.

RED mechanism is a major approach for congestion control in internet routers. The main idea of RED is to respond to the buffer usage and to drop messages when buffer is about to be congested. In the NS state, the available buffer space is sufficient, and all incoming messages are accepted. As the buffer usage grows, the node turns into CAS state perceived by growing queuing delay. In this case, the probability for refusing a new incoming message increases too. If buffer usage keeps increasing, the buffer will finally be exhausted, that is, the node enters the CS state. In this case, messages are declined with no reason. Since node behavior is definite in the NS and CS state, we focus our study on the CAS state.

To reflect the long-term buffer usage, transient buffer increase brought by data burst transfer rate should be adapted. Therefore an average queue size should be maintained.

$$L_q = (1 - w)L_q + w\hat{L} \tag{29.9}$$

L_q is the average queue size, $\hat{L}$ is the current queue size and w is the smoothing constant. Because of intermittent link in DTNs, w is relatively smaller than in the internet case.

Suppose the minimum and maximum threshold of RED is TH_{min} and TH_{max}, the relationship between average queue size and node state is listed in Table 29.1.

C2AM focus on CAS state since it lets in all incoming messages when a node is in the NS state and refuses all of them in the CS state.

29.3.5 Congestion Avoidance Mechanism

We try to prevent unnecessary waste of buffer space at an early stage. Thus two rules are proposed.

Rule 1 If $RTTL < T_{cw}$, refuse the message.

This is because a message whose residual *TTL* is less than its contact waiting time is more likely to be dropped because of *TTL* expiration. Since it takes up buffer space but cannot be delivered, it causes a waste of buffer and other potential high yield messages may be declined because of buffer occupation.

Existing buffer management schemes in DTNs handle these kinds of messages until their *TTL* values expire. We call these kinds of messages dead messages and they should be cleared from the network as soon as possible.

Rule 2 If the destination of an incoming message is available right now and $L_m + L_q < L_{max}$, accept it.

This is the most desired case since this kind of message has no contact wait time and thus cause little storage cost.

After the two conditions of the above rules are neither satisfied, we perform queue management policy based on message storage cost and adjust it to the degree of congestion.

In RED, the basic mark probability of a given incoming message is:

$$P_b = \frac{L_q - TH_{min}}{TH_{max} - TH_{min}} \times P_{max} \tag{29.10}$$

That is, when value of buffer usage is between TH_{min} and TH_{max}, a new incoming message is rejected with probability P_b. We use this value to adjust the proportion of received messages.

Suppose the DTN node keeps track of the waiting cost of messages from every source and thus stores the average waiting cost for each source. Let N_s be the number of sources of all the encountered messages. We sort these sources according to their waiting cost. Suppose message *M* has a source *src*(*M*) and is in the *src*(*M*)-th place. Then the eventual decision is as following:

If $I_{src(M)} \leq P_b \times N$, accept it.

That is, C2AM sorts different data flows according to their storage cost and gives priority to messages with low storage cost. The proportion of accepted messages is determined by the degree of congestion.

29.4 Simulation and Analysis

29.4.1 Simulation Setup

To evaluate the efficiency of C2AM congestion control mechanism, we implement a simple simulation program. Since we only consider single node case, to bypass the routing details and for simplicity, we concentrate our focus only on one relay node. The network wide congestion may require network-wide nodes to cooperate in an effectively way, which usually works together with the multi-path routing algorithm to avoid the congested sub-area. The network topology comprises of 17 nodes. Eight of them keep sending bundles intermittently, and these messages are all relayed by node *N* and their destinations are another eight opportunistically

Table 29.2 Scheduling time of outing links

Link number	Up time	Down time
1	1,000	9,000
2	2,000	8,000
3	5,000	5,000
4	4,000	6,000
5	2,000	8,000
6	5,000	5,000
7	1,000	9,000
8	10,000	0

available DTN nodes. All links are intermittent but with varying up and down time.

To make the results more observable, we implement a simple discrete event generator and advance the simulation progress in a time-slicing fashion. Nodes keep records of contact up and down time, as well as bundle acceptation and forwarding or dropping time. By using these statistics, message sojourn time can be obtained and can be updated using the exponential smoothing average.

We select the link that can affect a significant percentage of network traffic so that congestion can easily occur. The intermittently connected links and their scheduling time are shown in Table 29.2.

29.4.2 Simulation Result Analysis

We have calculated the successful delivery rate (r) of bundles on the whole. In DTN where resource is constrained, the storage of each node has a significant effect on r. When node storage is large enough, congestion can never happen and there is no need to perform congestion control at all. But when node storage is below a certain threshold, since messages have to wait until an outgoing link becomes available, delivery rate is very sensitive to node storage and link scheduling. In the simulation case, node 0–7 keep generating bundles with speed varying from 10 to 100 messages per second, *TTL* value from 10 to 100. We have changed the node storage from 50 to 750 messages and compared the performance of C2AM with ACC in [7]. We can see from the result that when node storage is large enough, the difference between C2AM and ACC is not so obvious. That is because no congestion will occur in this case. With the decrease of buffer capacity, congestion is increasingly easier to occur and C2AM begins to take effect. C2AM can increase delivery rate by 20% at most.

The result is shown in Fig. 29.2 from which we can see that the successful delivery rate tends to be steady when storage approaches 600 messages. But for ACC, the rate tends to be steady when storage approaches 650 messages.

To evaluate the influence of intermittent link, we increase the link down time and observe the transferring performance of C2AM and ACC. Figure 29.3 shows the delivery rate of C2AM and ACC. Compared with Fig. 29.2, both C2AM and

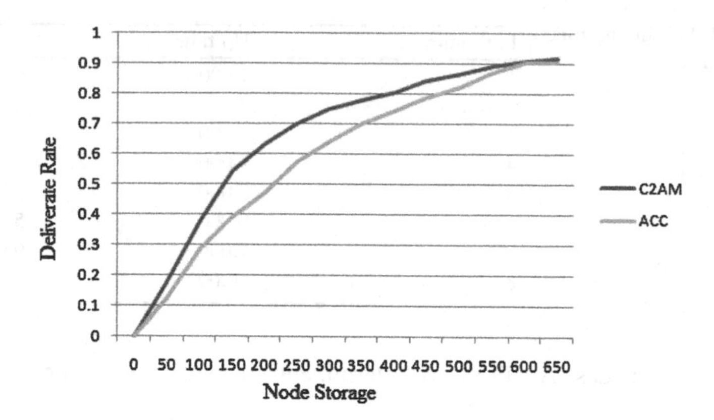

Fig. 29.2 Delivery rate comparison

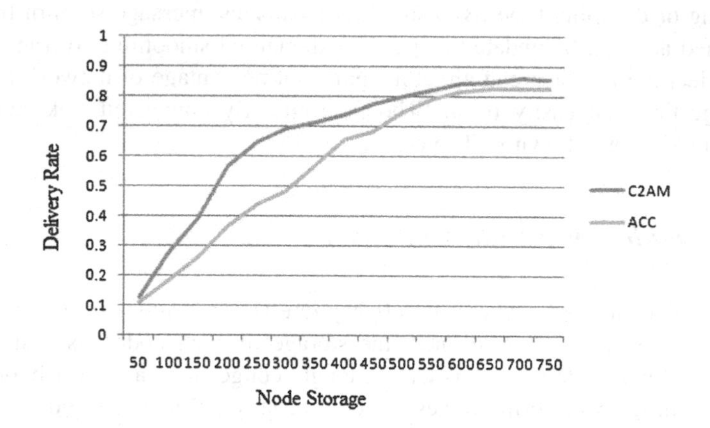

Fig. 29.3 Delivery rate of increased link down time

ACC have decreased delivery rate. But the difference between C2AM and ACC enlarges. We can see that the delivery rate of ACC decreases more quickly than C2AM and the performance of C2AM is superior.

We can also see from Fig. 29.3 that when buffer space is sufficiently large, delivery rate of C2AM is still a little higher than ACC. The reason for this is that C2AM discards messages which are probably unable to be forwarded in their remaining life. Clearing these kinds of messages in the early stage will prevent the node from unnecessary waste of buffer space.

C2AM differs with existing congestion control mechanisms in that it takes link characteristics as the main focus and thus is dedicated for delay tolerant networks. Since link down time comprises a major part of total message sojourn time, it should not be neglected. Promoting the forwarding of messages with low contact waiting time will improve the overall network performance especially in the congested nodes.

29.5 Conclusion

Against the characteristics of DTN that the resource is constrained and links are intermittently connected, we propose a new congestion control mechanism based on contact availability called C2AM. This mechanism exploits the contact scheduling to ease network congestion. In C2AM, bundle size, residual TTL, priority and specially, contact wait time are both considered. To reduce end-to-end delay, when a new contact is available, bundles with small RTTL value are transferred. C2AM differs with existing congestion control and buffer management schemes in that it takes into account the link characteristics and is dedicated to DTN environment. C2AM is a congestion avoidance mechanism and can work together with other congestion control methods such as migrating bundles to other adjacent nodes when congestion has already occurred.

We perform a simulation by implementing a simple routing scheduling topology. Simulation result has shown that C2AM outperforms ACC in delivery rate by 30% at the most.

In the future we will consider the details of cost function and look into the message scheduling policy when a new contact emerges. We believe that link characteristics of DTN will have a great impact on network performance. In our future work, we intend to evaluate C2AM in a further step and extend it to network-wide inter-node cooperation.

Acknowledgment The work is partly supported by National Science Foundation of China (No.61003220), Research Fund for the Doctoral Program of Higher Education of China (No.20090142120025), Fundamental Research Funds for the Central Universities (HUST:2010QN051) and Natural Science Foundation of Hubei Province (No.2010CDB02302).

References

1. Fall K (2003) A delay-tolerant networks architecture for challenged internets. In: Proceedings of ACM SIGCOMM'03, 25–29 August 2003, Karlsruhe, Germany
2. Seligman M (2006) Storage usage of custody transfer in delay tolerant networks with intermittent connectivity. In: Proceedings of ICWN'06, 26–29 June 2006, Las Vegas, Nevada, USA
3. RFC 4839. http://www.ietf.org/rfc/rfc4839.txt
4. Floyd S, Jacobson V (1993) Random early detection gateways for congestion avoidance. IEEE/ACM Trans Netw 1(4):397–413
5. Liu C, Wu J (2007) Scalable routing in delay tolerant networks. In: Proceedings of ACM Mobihoc'07, 9–14 September 2007, Montreal, Canada

6. Balasubramanian A, Levine B, Venkataramani A (2007) DTN routing as a resource allocation problem. In: SIGCOMM'07 Proceedings of the 2007 conference on applications, technologies, architectures, and protocols for computer communications, 27–31 August 2007, Kyoto, Japan, pp 373–384
7. Burleigh S, Jennings E, Schoolcraft J (2006) Autonomous congestion control in delay-tolerant networks. In: Proceedings of the 9th AIAA international conference on space operations (SpaceOps'06), June 2006, Rome, Italy
8. Zhang G, Liu Y (2008) Congestion management in delay tolerant networks. In: Proceedings of the 4th Annual international conference on wireless internet, 17–19 November 2008, Brussels, Belgium
9. Bisio I, Cola T, Lavagetto F, Marchese M (2009) Combined congestion control and link selection strategies for delay tolerant interplanetary networks. In: Proceedings of IEEE GLOBECOM 2009, 30 November–4 December 2009, Honolulu, Hawaii, U S A
10. Soares VNGJ, Farahmand F, Rodrigues JJPC (2009) Scheduling and drop policies for traffic differentiation on vehicular delay-tolerant networks. In: Proceedings of SoftCom'09, 24–26 September 2009
11. Little's Law. http://en.wikipedia.org/wiki/Little's_law
12. Hua D, Du X, Cao L, Xu G, Qian Y (2010) A DTN congestion avoidance strategy based on path avoidance. In: Proceedings of ICFCC'10, 3–5 June 2010, Iasi, Romania
13. Seligman M, Fall K, Mundur P (2006) Alternative custodians for congestion control in delay tolerant networks. In: Proceedings of SIGCOMM'06, 11–15 September 2006, Pisa, Italy

Part VII
HumanCom 2011 Session 6: Virtualization Technologies for Desktop Applications

Chapter 30
Visualization of Knowledge Map: A Focus and Context Approach

Jincheng Wang, Jun Liu and Qinghua Zheng

Abstract To address the low efficiency and understandability in traditional knowledge navigation visualization, a "Focus and Context" visualization approach for large-scale knowledge map (KM) is presented. It extracts "Backbone" of a KM based on its invulnerability and dynamically generates "Focus Area" by the focus knowledge unit and its adjacent areas. Experimental results show that the visualization approach highly speeds up the initialization time of KM, simultaneously displays both the local focus and global context, and effectively facilitates learner's navigate learning.

Keywords Knowledge map · Focus and context · Visualization

30.1 Introduction

Learning is a cumulative process where studying new knowledge relies on the old that the learner has ever accessed to [1, 2]. This kind of learning dependency refers to the necessity of completely mastering the premise knowledge before learning new KUs. Here, knowledge unit (KU) is defined as the smallest integral knowledge object in a given domain, such as definition, theorem, rule or algorithm, in which

J. Wang (✉) · J. Liu · Q. Zheng
MOE KLINNS Lab and SPKLSTNT Lab, Xi'an Jiaotong University, Xi'an, China
e-mail: wangjin.c520@gmail.com

J. Liu
e-mail: liukeen@mail.xjtu.edu.cn

Q. Zheng
e-mail: qhzheng@mail.xjtu.edu.cn

J. J. Park et al. (eds.), *Proceedings of the International Conference on Human-centric Computing 2011 and Embedded and Multimedia Computing 2011*, Lecture Notes in Electrical Engineering 102,
DOI: 10.1007/978-94-007-2105-0_30,

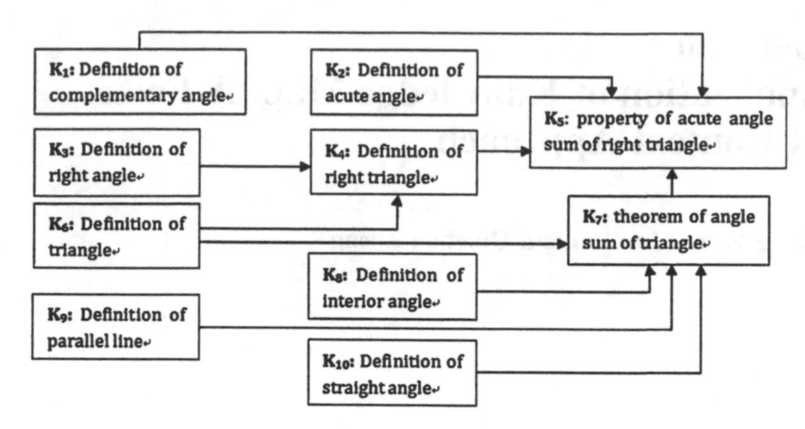

Fig. 30.1 Brief view of KM

definition, theorem, rule or algorithm also denotes the KUs semantic type [2]. For instance, in "Geometry", there exists a learning dependency from the KU "Definition of right angle" to the KU "Definition of right triangle", which indicates that a person should learn "Definition of right angle" before learning the "Definition of right triangle". A knowledge map (KM) is a network where the vertices are KUs and edges are learning dependencies which are in a same domain. As a kind of knowledge management method to visually represent knowledge [3], KMs help learners avoid the disorientation problem in learning by providing the visualization navigation learning path [2]. Figure 30.1 is a brief view of KM in Plane Geometry.

The visualization of KM is a key issue in navigation learning. According to the definition of KM, node-link diagrams are well suited to represent it, which can effectively show the global structure of a sparse or small size network [4]. However, the number of KU and learning dependency in different KMs is commonly $>10^3$. Therefore, KM visualization faces a major challenge: obtaining a representation which will not lead to overlap and confusion, but reduce the learner's visual burden and increase the navigation efficiency [5].

In this article, we proposed the "Focus and Context" visualization for KM based on the analysis of its network features. It extracts backbone based on invulnerability of KM to visualize the overall structure, within which focus area is generated by focus KU and its adjacent areas to show local detail. Experiment on the real KM data sets show: the representation greatly reduces the time complexity of KM loading, and takes into account of displaying the global structure and local details simultaneously, reduces the learner's visual burden and increases the navigation learning efficiency.

The rest of the paper is organized as follows. Section 30.2 discusses the related work. Section 30.3 analyzes the invulnerability of KM, according to which, Sect. 30.4 proposes backbone and focus area generating algorithm. Experimental results are discussed in Sect. 30.5. Section 30.6 presents the conclusions and our future research.

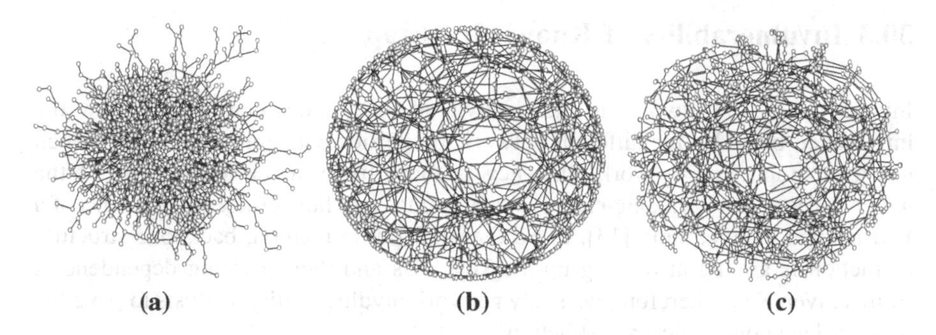

Fig. 30.2 Visualization of KM in Pajek with difference layout: **a** Kamada-Kawai, **b** 2DFruchterman-Reingold, **c** 3DFruchterman-Reingold

30.2 Related Work

To date, several kinds of network analysis tools can represent the KM. For example, Fig. 30.2 is the visualization in Pajek [6], which is an open source Windows program for analysis and visualization of large network having thousands or even millions of vertices.

As we have seen, Pajek excellently shows the overall structure of KM with different kinds of layout algorithms. But if the user wants to study the detail properties of each KUs or learning dependencies, several additional selecting and zooming operations are needed. So, these additional operations seriously limit the KMs readability and navigatability. To this end, two representations of networks based on node-linked and interaction mechanism is appropriate "Zoom and Pan (Z&P)" and "Focus and Context (F&C)" [7].

Z&P improves the readability and navigatability through zooming and panning and mainly focuses on displaying the local details of networks, which has been used in KM visualization [8, 9]. The representation well displays the detail properties of KUs and learning dependencies. However, as the network size increases, many KUs and learning dependencies will be folded and hided to keep the clarity display of focused KUs and its adjacent area. And the invisibility of folded and hided portion highly prevents the learners to master the KMs overall structure. Thus, it is not a good solution for navigation learning of KM.

Different from Z&P and Pajek, F&C integrates the global view and local details by nonlinearly displaying the focus details and roughly visualizing context simultaneously [10]. Recently, this method has been applied and performed excellently in many kinds of networks, such as RDF data, hierarchy structure of file system etc. In these networks, the creation of focus area and context depends on both their network structure features and application requirements [11, 12]. To the best of our knowledge, this paper is the first to study its application in KM.

30.3 Invulnerability of Knowledge Map

Invulnerability of networks to the removal of their nodes can always show the importance of nodes for different network properties. Generally, the node which has great impact on network efficiency is more likely to be a key node of the network. Here, network efficiency is a way to measure how efficiently the KUs of a KM navigate the learning [13]. In F&C visualization method, backbone structure extraction is aiming at picking up the key KUs and their learning dependencies from a given KM. Therefore, we study network invulnerability analysis to pave the way for backbone structure extraction.

To perform invulnerability analysis on KM, we introduced four network parameters and conducted experiments on four KMs (more than 9,300 KUs and 22,400 learning dependencies) from different domains.

30.3.1 Mainly Network Parameters

This section describes four network parameters which will be used in our invulnerability analysis subsequently: network efficiency, betweenness, closeness and degree centrality. In a given KM(KU, KA), let $KU = \{ku_i\}(1 \le i \le n)$ be the set of KUs and $KA \subseteq KU \times KU$ be the set of learning dependencies. The four parameters can be defined as follows:

Network Efficiency E_{globe} is used to measure the connectivity of KM [13, 14]:

$$E_{\text{globe}} = \frac{1}{n(n-1)} \sum_{0<i,j\le n, i\ne j} \frac{1}{d_{ij}} \tag{30.1}$$

d_{ij} is called the shortest path length between two knowledge units ku_i, $ku_j \in KU$.

Betweenness Centrality In KM, navigation learning path from one to the other non-adjacent KUs might depend on other units, especially on those on the shortest paths between the two. Betweenness indicates how the unit in the middle can have an influence on the others. In this paper, the units might be KUs and learning dependencies, which we call them node betweenness and edge betweenness:

Node Betweenness $b(ku)$ of $ku \in KU$ defines as:

$$b(ku) = \sum_{0<i,j\le n, i\ne j} \frac{\sigma_{ij}(ku)}{\sigma_{ij}} \tag{30.2}$$

in which σ_{ij} is the number of shortest path between ku_i, ku_j and $\sigma_{ij}(ku)$ is the number of path in σ_{ij} which includes knowledge unit ku.

Edge Betweenness $b(ka)$ of learning dependency $ka \in KA$ describes as [15]:

$$b(ka) = \sum_{0<i,j\le n, i\ne j} \frac{\sigma_{ij}(ka)}{\sigma_{ij}} \tag{30.3}$$

where σ_{ij} is the number of shortest path between ku_i, ku_j and $\sigma_{ij}(ka)$ is the number of path in σ_{ij} which includes learning dependency ka.

Closeness Centrality is based on how close a KU is to all the other KUs. In this paper, we use the simplest notion of closeness which is based on the concept of minimum distance of geodesic d_{ij}. The closeness centrality of ku_i is:

$$c(ku_i) = \frac{n-1}{\sum_{0<j<n} d_{ij}} \tag{30.4}$$

Degree Centrality is defined as the number of learning dependencies incident upon a KU:

$$d(ku_i) = \frac{k_i}{n-1} = \frac{1}{n-1} \sum_{0<j \le n} ka_{ij} \tag{30.5}$$

where k_i is the degree of point i. It is the truth that any of the KU in KM can at most have $n-1$ adjacent KUs, $n-1$ is the normalization factor introduced to make the definition independent of the size of the KM and to have $0 \le d(ku_i \le 1)$.

30.3.2 Invulnerability Analysis of Knowledge Map

To achieve F&C visualization of KM, the first step is to create the context which is constituted by numbers of KUs and learning dependencies among them. Considering KM as a network and centrality determines the relative importance of a unit within the network, we ranked the KUs on the basis of centrality, and found that different KU performs significant dissimilar in KM. Studying the impact on network efficiency by removing KUs based on centrality, we can find out which KUs are the most important in a KM. We also call this procedure invulnerability analysis, and different centrality indicates different strategy. In this paper, aiming at extracting the most suitable backbones for KMs, we employed several kinds of centrality to rank the KUs or learning dependencies, and studied the network efficiency by removing them according to the ranking.

Figure 30.3 presents the results of invulnerability analysis on "Plane Geometry", "Computer Network", "Computer Composition" and "C Programming Language" (shown in Table 30.1), in which x-axis denotes the proportion of removed KUs and y-axis indicates the network efficiency. In each domain, we employed six centralities to rank the KUs or learning dependencies: degree, node and edge betweenness, closeness, randomly edge and node removal.

It can be seen in Fig. 30.3, as the proportion of removed units increases, network efficiency decreases relatively smoothly by random removal strategies while a significant decline in first 10% by other centrality based removal strategies. It indicates that the key KUs and learning dependencies which greatly impact the navigation learning have been gathered by specific strategies.

Table 30.1 Experimental data sets

#ID	#Scale(KB)	#KU	#Learning-dependencies	Name
KM1	7,264	3,235	4,061	Computer Network
KM2	4,125	4,023	13,448	Computer Composition
KM3	3,530	1,618	4,843	C Programming Language
KM4	843	477	1,268	Plane Geometry

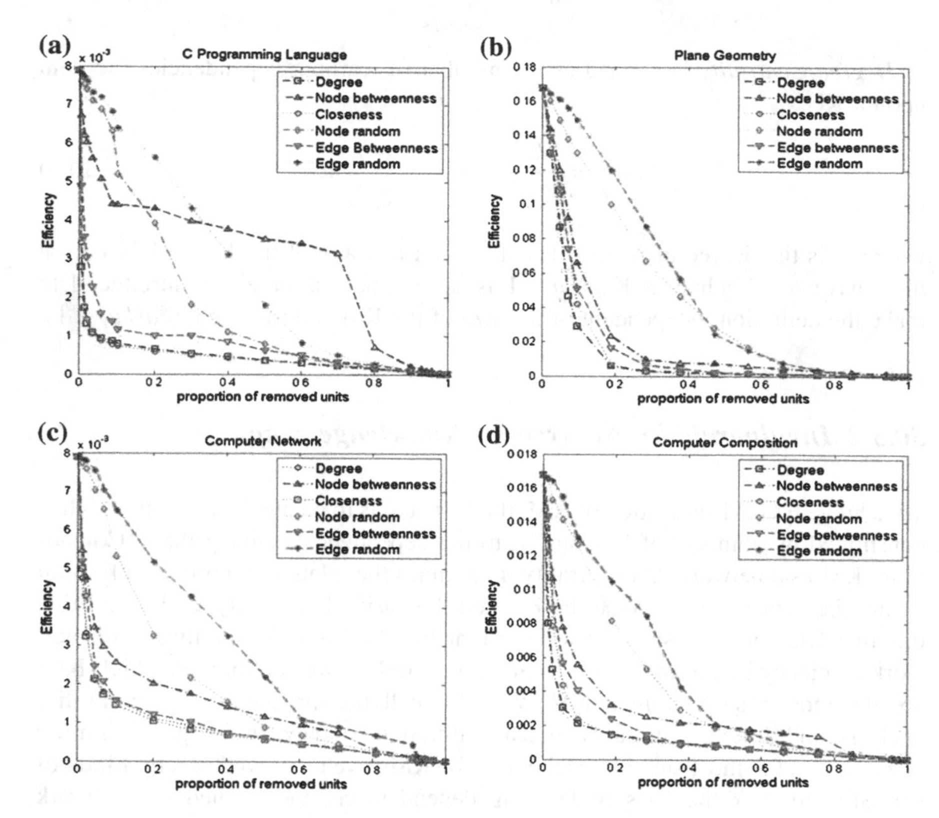

Fig. 30.3 The impact on network efficiency by different removing strategies on difference domain. **a** “C Programming Language”, **b** “Plane Geometry”, **c** “Computer Network”, **d** “Computer Composition”

30.4 Visualization of Knowledge Map

Based on invulnerability analysis on KM, we proposed a representation which extracts backbone structure by removing the unimportant KUs and learning dependencies as the context and creates focus area by the current focus KU and its adjacent area as the focus to achieve the F&C visualization for KM.

Input: *Knowledge Map*$(\mathcal{KU}, \mathcal{KA})$;
Output: *Knowledge Map* **backbone structure**$(\mathcal{KU}_S, \mathcal{KA}_S)$;
Parameter: Number of *knowledge unit* **in backbone**$\mathcal{N}_{bb}$;
Strategy to remove non backbone *knowledge unit*$\mathcal{AL}_g$;

Step1 : $\mathcal{KU}_S = \mathcal{KU}; \mathcal{KA}_S = \mathcal{KA}$;
Step2 : **Let** $\mathcal{KU}_k$ **to be the set of nun backbone KUs, applying** $\mathcal{AL}_g$ **to add the KUs into** $\mathcal{KU}_k$ **iteratively, until** $sum(|\mathcal{KU}| - |\mathcal{KU}_k|) <= \mathcal{N}_{bb}$.
Step3 : **For each** $ku \in \mathcal{KU}_k$:
$\mathcal{KU}_S = \mathcal{KU}_S - \{ku\}; \mathcal{KA}_S = \mathcal{KA}_S - \{ku\} \times \mathcal{KU} - \mathcal{KU} \times \{ku\}$.
If exists(ku_i, ku) **and**$(ku, ku_j) \in \mathcal{KA}$,
then$\mathcal{KA}_S = \mathcal{KA}_S + \{(ku_i, ku_j)\}$
Step4 : **Compute the edge betweenness of learning dependencies in set** $\mathcal{KA}_S$;
Step5 : **Find out the learning dependency** $ka \in \mathcal{KA}_S$ **which has the minimum edge betweenness and satisfies the weakly connectivity of** $(\mathcal{KU}_S, \mathcal{KA}_S - \{ka\})$;
If $ka \in \mathcal{KA}_S$ **exists, then** $\mathcal{KA}_S = \mathcal{KA}_S - \{ka\}$, **and turn to step 2;**
Otherwise, output $(\mathcal{KU}_S, \mathcal{KA}_S)$.

Fig. 30.4 Algorithm of backbone structure extraction: Steps 1–3 extract and remove the set of KU which is not belong to the backbone structure; Steps 4–6 filters learning dependencies according to the decreasing of edge betweenness. All steps must ensure the removal will not break the connectivity

30.4.1 Backbone Structure Extraction

The backbone structure of KM is constituted by the KUs and learning dependencies which are playing key roles in KM. With the invulnerability analysis of KM in Sect. 30.3, we proposed the methods to identify the key KUs and learning dependencies. Now we faced another challenge: how to ensure the weakly connected while extracting the backbone structure. A KM is weakly connected if its underlying undirected graph is connected, in which every KU is reachable from every other but not necessarily following the directions of the learning dependencies. In this section, we proposed a knowledge map backbone structure extraction algorithm based on invulnerability analysis and Edge betweenness filtering, as Fig. 30.4 shows.

In which parameter AL_g in Step 2 indicates the removal strategy, due to the difference computing process of each removal strategy, removal strategies which perform similar impact on network efficiency may also result in different contents of backbone. In Fig. 30.3, the specific removal strategies on KUs show the similar impact on network efficiency i.e. closeness, degree. To study which is more suitable for KM backbone extraction, we will conduct experiment on them.

30.4.2 Focus Area Generation

Focus area, constituted by the focus KU and its adjacent KUs and the learning dependencies among them, provides the detailed information of KUs and learning dependencies. KU is the basic unit of knowledge, and learning dependency is a necessary conditional relationship in KM. So, for a given learning dependency $(ku_i, ku_j) \in KA$, it does not mean that learners can study ku_i as soon as they finish the learning of ku_j, but all of the premise knowledge of ku_j have been completely mastered.

To deal with the necessary conditional relationship of learning dependency and extract the focus area of current focus, we proposed a method based on the maximum path length (MPL) to indicate the distance of two KUs. Here, MPL is opposite to the concept of shortest path length.

The computing procedure of focus area is: let the focus knowledge unit ku_0 be the start vertex, depth-first traversal the KM depending on learning dependencies without following their directions: for each transfer $(ku_p, ku_q) \in KA$, let the MPL from ku_p to ku_0 be l_p (the MPL from ku_0 to ku_0 is 0), if there exists a transfer $(ku_p, ku_q) \in KA$ and $l_p + 1 > l_q$, then $l_q = l_p + 1$. Choose m (the maximum number of KU will be displayed in focus area) of the KUs according to the decreasing of MPL, note them as ku_m. So the focus area is $(KU_m, KU_m \times KU_m \cap KA)$.

Depending on the selected knowledge unit ku_s, dynamically generates focus area $fa(ku_s)$, and then replaces ku_s with $fa(ku_s)$, transfers the learning dependencies between ku_s and other KUs in backbone to the adjacent KUs of ku_s which are in the focus area $fa(ku_s)$. With these procedures, F&C visualization of KM is achieved.

30.5 Experiments and Evaluation

To verify how the F&C representation performs in KM visualization, we implemented the F&C visualization tool based on the two algorithms which have been described before, and deployed it to the Yotta system [16].

Figure 30.5 is a screenshot from Yotta system. The left part of the screen is the visualized navigator of knowledge map with F&C method, and the right part is the resourcelist about the focus KU. In this figure, we can see the focus area of KM displays the local details of KM by presenting the name of KU and the specific learning dependencies among them as well as the backbone structure shows the rough structure of the KM as a global view. As the extraction of focus area highly relies on the size of display area, we just evaluate the algorithm of backbone structure extraction in this section.

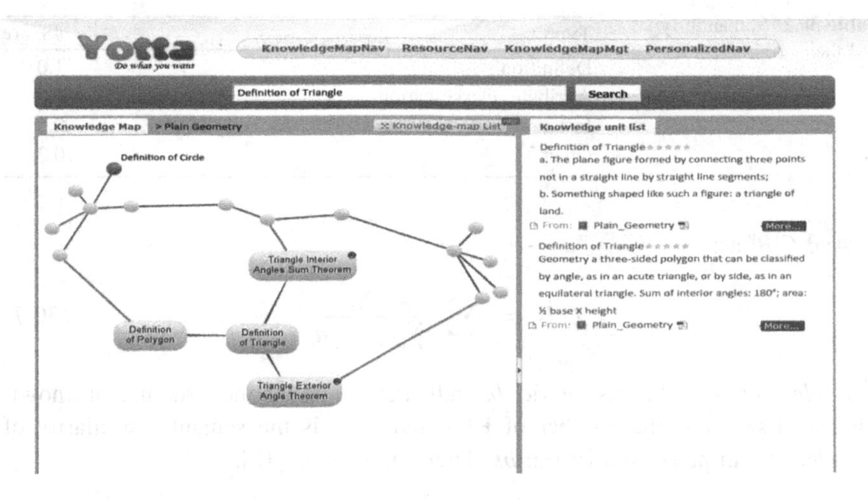

Fig. 30.5 The visualization tool deployed in Yotta

30.5.1 Data Sets

In order to validate the efficiency of our visualization method, we conducted experiments on a set of KMs that cover four domains: "Computer Network", "Computer System Structure", "C Programming Language" and "Plain Geometry", shown in Table 30.1. To the best of our knowledge, there are no public KM data sets. So we created the experimental data set through the process which generated KU and learning dependency in the given document set by using the method of [2, 17] and manually refined the extracted results.

30.5.2 Evaluation on Backbone Structure Extraction

To the best of our knowledge, there is no recognized standard for KMs backbone extraction evaluation. According to the efforts the visualization of KM expectation, we proposed the follow evaluation metrics (value ranges from 0 to 1):

Coverage $C[B]$: Note that *km* is the knowledge map of a given course created by *n* documents which are in the same domain. *d* is the number of chapter in one document, presume that there are totally *m* distinct chapters in *km*. Let BC_{set} be the chapter set in backbone and δ_i indicates whether the *i*th chapter is included in BC_{set}. Then δ_i can be defined as:

$$\delta_i = \begin{cases} 1 & \text{for chapter}_i \in BC_{\text{set}} \\ 0 & \text{otherwise} \end{cases} \tag{30.6}$$

Table 30.2 Semantic types and their original score

Type	#Score
Definition	1.0
Attribute, classification	0.8
Instance, case, method	0.5
Other	0.2

And $C[B]$ as:

$$C[B] = \sum_{0<i\leq m} \frac{\delta_i}{\sum_{0<j\leq n} d_j} \tag{30.7}$$

Irrelevance $I[B]$: In this metric, bs indicates the backbone structure of knowledge map km. n is the number of KUs and sim_{ij} is the semantic similarity of knowledge unit pairs (ku_i, ku_j) in bs. Then the sim_{ij} is [18]:

$$\mathrm{sim}_{ij} = 1 - \frac{1}{D} \times \mathrm{len}\,(ku_i,\ ku_j) \tag{30.8}$$

where D is the diameter of knowledge map, $\mathrm{len}\,(ku_i,\ ku_j)$ is the shortest path length of the two KUs in km and scaled by D despite of that any $\mathrm{len}\,(ku_i,\ ku_j)$ would not be greater than D. According to the definition of sim_{ij}, $I[B]$ can be defined as:

$$I[B] = \frac{2}{n(n-1)} \sum_{0<i<j\leq n} (1 - \mathrm{sim}_{ij}) \tag{30.9}$$

Original $O[B]$: In knowledge map, the KUs have their semantic types which usually indicate what kind of information it intends to deliver and may be classified into eight types. As the nature of KM, learning a new KU depend on the premise KUs according to the learning dependency on it. Obviously the premise KU is more original than the latter. According to the most frequency pattern of learning dependencies in KM which statistic in [2]. We allocate each type a score shows in Table 30.2.

With the extracted backbone structure bs of a knowledge map, let n be the number of KUs and os_i be the original score of $ku_i \in KU_{bs}$, we define the $O[B]$ in the following way:

$$O[B] = \frac{i}{n} \sum_{0<i\leq n} os_i \tag{30.10}$$

Based on the metrics described above, we conduct experiments to analyze the performance of backbone extraction by using three strategies: degree, closeness and node betweenness, and each strategies we extract 20 KUs for analysis. The result is shown in Table 30.3. It can be seen that degree centrality based strategy outperforms the other two on all of the metrics, demonstrating that the more the KU connect to others, the more important it is in KM.

Table 30.3 Result set on backbone structure evaluation

ID	Strategy	C[B]	I[B]	O[B]
KM1	Degree	0.94	0.75	0.82
	Node betweenness	0.85	0.65	0.74
	Closeness	0.91	0.72	0.77
KM2	Degree	0.95	0.78	0.84
	Node betweenness	0.81	0.63	0.68
	Closeness	0.9	0.71	0.79
KM3	Degree	0.89	0.73	0.8
	Node betweenness	0.77	0.62	0.71
	Closeness	0.88	0.7	0.73
KM4	Degree	0.79	0.75	0.73
	Node betweenness	0.68	0.65	0.64
	Closeness	0.77	0.72	0.67

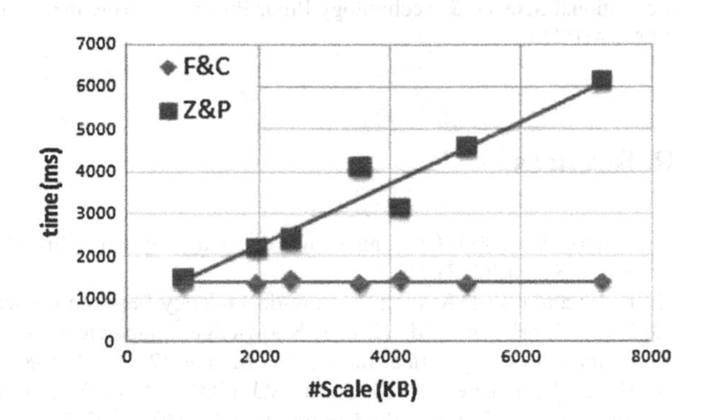

Fig. 30.6 Contrast on loading speed of two representations

30.5.3 *Efficiency of Visualization*

To evaluate the loading efficiency of F&C visualization, we also conducted the following experiments. The result is illustrated in Fig. 30.6, in which x-axis represents the scale of KM, and y-axis denotes the running time in milliseconds. The experiments were performed on the Intel Core 2 Duo CPU 3.00 GHz with 3 GB main memory and Microsoft Windows7. It can be seen while the running time of Z&P grows approximately linearly as the scale of KM increases, F&C reaches the zero loading speed growth. This evidences the F&C approach is more suitable for the large-scale KM.

30.6 Conclusion and Future Work

In this paper, we applied the F&C visualization method to KM by proposing an algorithm to constitute the backbone structure based on analysis of network invulnerability and presenting the algorithm of focus area extraction according to the MPL between KUs. Experimental results show that our method greatly improved the loading speed of large-scale KMs, and simultaneously represented the KMs global structure and local details in a relevantly small space, effectively reduced the users' navigation learning burden, thus improved the learning efficiency.

There are two specific directions in our future work. First, we will investigate the extraction and management strategy of focus area on server. In addition, we plan to study the request and transmit strategy of focus area.

Acknowledgments The research was supported in part by the National Science Foundation of China under Grant Nos. 60825202, 60803079, 61070072, the National Science and Technology Major Project (2010ZX01045-001-005), the Program for New Century Excellent Talents in University of China under Grant No. NECT-08-0433, the Doctoral Fund of Ministry of Education of China under Grant No. 20090201110060, Cheung Kong Scholar's Program, Key Projects in the National Science & Technology Pillar Program during the 11th 5-Year Plan Period Grant No. 2009BAH51B02.

References

1. Gordon JL (2000) Creating knowledge maps by exploiting dependent relationships. Knowl Based Syst 13(2–3):71–79
2. Liu J et al (2011) Mining learning-dependency between knowledge units from text. VLDB J
3. Lin FR, Hsueh CM (2006) Knowledge map creation and maintenance for virtual communities of practice. Inf Process Manag 42(2):551–568
4. Henry N, Fekete JD, McGuffin MJ (2007) NodeTrix: a hybrid visualization of social networks. IEEE Trans Vis Comput Graph 13(6):1302–1309
5. Ghoniem M, Fekete JD, Castagliola P (2004) A comparison of the readability of graphs using node-link and matrix-based representations. In: Proceedings of the IEEE symposium on information visualization, pp 17–24, 204
6. Networks/Pajek. http://vlado.fmf.uni-lj.si/pub/networks/pajek/
7. Herman I, Melancon G, Marshall MS (2000) Graph visualization and navigation in information visualization: a survey. IEEE Trans Vis Comput Graph 6(1):24–43
8. Lu J et al (2009) ETM toolkit: a development tool based on extended topic map. In: 13th international conference on computer supported cooperative work in design (CSCWD), pp 528–533
9. Mishra S, Ghosh H (2009) Effective visualization and navigation in a multimedia document collection using ontology. In: Chaudhury S et al (eds) Pattern recognition and machine intelligence, Springer, Berlin/Heidelberg, pp 501–506
10. Cohen M, Brodlie K (2004) Focus and context for volume visualization. In: Proceedings of the theory and practice of computer graphics 2004, pp 32–39, 233
11. Zhang K et al (2010) ZoomRDF: semantic fisheye zooming on RDF data. In: Proceedings of the 19th international conference on World Wide Web (WWW), ACM, Raleigh, pp 1329–1332
12. Lei R, Weixin W, Dongxing T (2008) Fisheye view for visualization of large tree by packing nested circles. J Comput Aided Des Comput Graph 20(3):298–303, 309
13. Marchiori M, Latora V (2007) A measure of centrality based on the network efficiency. New J Phys 9:188

14. Latora V, Marchiori M (2001) Efficient behavior of small-world networks. Phys Rev Lett 87(19):198701
15. Hashimoto TB et al (2009) BFL: a node and edge betweenness based fast layout algorithm for large scale networks. BMC Bioinformatics 10:19
16. Zheng Q, Qian Y, Liu J (2010) Yotta: A knowledge map centric e-learning system. In: IEEE 7th international conference on e-business engineering, Shanghai, pp 42–49
17. Chang X, Zheng Q (2008) Knowledge element extraction for knowledge-based learning resources organization. In: Leung H et al (eds) Advances in web based learning—ICWL 2007, Springer, Berlin/Heidelberg, pp 102–113
18. Hirst G (2001) Semantic distance in wordnet: an experimental, application-oriented evaluation of five measures. In: 2nd meeting North America chapter association for computational linguistics, 2001

Chapter 31
A Think on Security and Trusty Measurement for Virtualized Environment

Xiaolin Wang, Yan Sang, Yi Liu and Yingwei Luo

Abstract With the development of virtualization technology, some security problems have appeared. Researchers begin to suspect the security of virtualized environment and consider how to evaluate the security degree of virtualized environment. But there are not uniform virtualized environment security evaluation criteria at present. In this paper, we analyze the security problem in virtualized environment and present our virtualized environment security evaluation criteria. We take a further research on support technology and authentication method. First, we introduce the background of the problem. Next, we summarize the related work. After that we present our virtualized environment security evaluation criteria and introduce our work on support technology and authentication method. For support technology, we propose some safety guarantee technology for each security level. For authentication method, we give some ideas for the evaluation of each security level.

Keywords Virtualization · Security · Evaluation criteria · Authentication method

---

X. Wang · Y. Sang · Y. Liu · Y. Luo (✉)
Department of Computer Science and Technology, Peking University,
100871 Beijing, China
e-mail: lyw@pku.edu.cn

X. Wang
e-mail: wxl@pku.edu.cn

X. Wang · Y. Sang · Y. Liu · Y. Luo
The Shenzhen Key Lab for Cloud Computing Technology and Applications (SPCCTA),
Peking University Shenzhen Graduate School, Shenzhen, 518055 Guangdong, China

J. J. Park et al. (eds.), *Proceedings of the International Conference on Human-centric Computing 2011 and Embedded and Multimedia Computing 2011*, Lecture Notes in Electrical Engineering 102,
DOI: 10.1007/978-94-007-2105-0_31,

31.1 Introduction

With the development of multi-core system and cloud computing, virtualization technology has more and more advantages in practical application. Virtualization technology can benefit us in many ways. We can save computing resources with the help of virtualization technology, meanwhile, the stability and reliability of the system is improved greatly. With the development of virtualization technology, some security problems have appeared. Trusted computing in virtualized environment is facing great challenges.

In 2000, researchers from Air Force and Naval Postgraduate School addressed the problem of implementing secure VMMs on the Intel Pentium architecture [1]. They examined current virtualization techniques for the Intel Pentium architecture and identified several security problems. They mentioned that VMMs on the Intel Pentium architecture had following potential problems: First, resource sharing between virtual machines leads to a security problem. If two virtual machines have access to a floppy drive, information can flow from one VM to the other. Files could be copied from one VM to the floppy, thus giving the other VM access to the files. Second, a similar problem results from support of networking and file sharing. A VM could FTP to either a host OS or guest Linux OS and transfer files. Third, the ability to use virtual disks is also a security problem. A virtual disk is a single file that is created in the host OS. It is used to encapsulate an entire guest disk. Anyone with access to this file in the host OS could copy all information in the virtual disk to external media. Fourth, flaws in host OS design and implementation will render the virtual machine monitor and all virtual machines vulnerable. Fifth, implementation of serial and printer ports presents another security problem. When a user attempts to print, it will result in output to a printer file in host OS. Thus others could get information by reading the printer file.

In 2006, researchers from Microsoft and University of Michigan developed a virtual machine based rootkit named VMBR. The rootkit is based on $\times$ 86 CPU which supports virtualization technology, and inserts a malicious virtual machine monitor between operating system and hardware. The malicious VMM runs under operating system, threatens the security of the whole system by controlling operating system.

August 2006, in the Black Hat annual hacker conference held in Las Vegas, a researcher from Poland showed how to make malicious program attack Vista system successfully by using virtualization technology. This is the famous malicious program named Blue Pill. With the help of hardware virtualization technology used in AMD and Intel CPU, Blue Pill changes the state of operating system from normal to running in virtual machine, and a Blue Pill monitor is formed at the same time. The Blue Pill monitor can control and get any information that it is interested in, which affects the security of the system seriously.

Among the above three examples, the first one mainly focuses on the isolation between virtual machines, and the latter two examples mainly emphasize that the VMM has complete control over virtual machines, which will lead to serious consequences. How to ensure the security of virtualized environment and how to evaluate the security degree of virtualized environment have become current research focuses. In this paper, we analyze the security problem in virtualized environment and present our virtualized environment security evaluation criteria. We take a further research on support technology and authentication method. For support technology, we propose some safety guarantee technology for each security level. For authentication method, we give some ideas for the evaluation of each security level.

The rest of this paper is organized as follows. Section 31.2 introduces some related works about trusted computing in virtualized environment. Section 31.3 presents the virtualized environment security evaluation criteria. And in Sect. 31.4 we propose some safety guarantee technologies for each security level. Moreover, we give some ideas for the evaluation of each security level in Sect. 31.5. Finally the conclusions are drawn in Sect. 31.6.

31.2 Related Work

In the early research on virtualization technology, researchers begin to pay attention to security problems in virtualized environment.

IBM's PR/SM [2] (Processor Resource/System Manager) is a hardware facility that enables the resources of a single physical machine to be divided between distinct, predefined logical machines called "logical partitions". PR/SM provides separation of workloads, and prevents the flow of information between partitions. This trusted separation may be used where the separation is based on need to know, or where data at differing national security classifications must be isolated. PR/SM received an EAL5 evaluation under the Common Criteria, the highest assurance level for any hypervisor product on the market today.

In 1991, DEC developed a virtual machine monitor security kernel for the VAX architecture. The VAX Security Kernel supports multiple concurrent virtual machines on a single VAX system, providing isolation and controlled sharing of sensitive data. Rigorous engineering standards are applied during development to comply with the assurance requirements for verification and configuration management. The VAX VMM Security Kernel received an A1 rating from the National Computer Security Center.

In recent years, researchers keep exploring security problems in virtualized environment. Different research methods are used.

Jansen et al. [3] emphasize the architecture of virtual machine monitor. They look at ways of improving virtual machine (VM) security, specifically in the context of integrity of VMs, by adding scalable trusted computing concepts to a virtual machine infrastructure. They describe methods for doing integrity measurement recording, and reporting for VMs. And they place particular emphasis on how those methods can be made extensible and flexible with the goal of obtaining a generic attestation and sealing framework for VMs.

Matthews et al. [4] present the design of a performance isolation benchmark that quantifies the degree to which a virtualization system limits the impact of a misbehaving virtual machine on other well-behaving virtual machines running on the same physical machine. Their test suite includes six different stress tests—a CPU-intensive test, a memory-intensive test, a disk intensive test, two network-intensive tests (send and receive) and a fork bomb. And they test three flavors of virtualization systems—an example of full virtualization (VMware Workstation), an example of paravirtualization (Xen) and two examples of operating system level virtualization (Solaris Containers and OpenVZ). They find that the full virtualization system offers complete isolation in all cases and that the paravirtualization system offers nearly the same benefits—no degradation in many cases with at most 1.7% degradation in the disk intensive test. The results for operating system level virtualization systems are varied—illustrating the complexity of achieving isolation of all resources in a tightly coupled system.

Somani et al. [5] run different kind of benchmarks simultaneously in Xen environment to evaluate the isolation strategy provided by Xen. They emphasize different combinations—CPU and CPU, DISK and DISK, NET and NET, CPU and DISK, CPU and NET, NET and DISK. And they compare two different CPU scheduling algorithms—Credit and SEDF. They find that SEDF shows relatively good performance than credit scheduler.

Koh et al. [6] study the effects of performance interference by looking at system level workload characteristics. They allocate two VMs in a physical host, each of which runs a sample application chosen from a wide range of benchmark and real-world workloads. For each combination, they collect performance metrics and runtime characteristics using an instrumented Xen hypervisor. Through subsequent analysis of collected data, they identify clusters of applications that generate certain types of performance interference. Furthermore, they develop mathematical models to predict the performance of a new application from its workload characteristics.

Karger et al. [7] discuss trade-offs in complexity, security, and performance. They mainly explore three different modes—pure isolation, sharing hypervisors on a server, and sharing hypervisors on a client.

Weng et al. [8] present an access control mechanism for the virtual machine system, which is based on the BLP model. They prove that the virtual machine system with the access control mechanism and an initial secure state is a secure system. In addition, they implement a prototype of the access control mechanism for the virtual machine system based on Xen. They show that the access control mechanism is a feasible method to enforce isolation and limited sharing between VMs, and the added performance cost is acceptable.

Sailer et al. [9] design and implement a security architecture for virtualization environments that controls the sharing of resources among VMs according to formal security policies. The architecture is named sHype and designed to achieve medium assurance (Common Criteria EAL4). sHype has following goals—(i) near-zero overhead on the performance-critical path, (ii) non-intrusiveness with regard to existing VMM code, (iii) scalability of system management to many machines via simple policies, and (iv) support for VM migration via machine-independent policies.

Table 31.1 Virtualized environment security evaluation criteria

Level	Name
D1	Unprotected level
C1	Security isolation protection level
C2	Resource isolation protection level
C3	Performance isolation protection level
B1	Claim consistently protection level
A1	Active protection level

31.3 Virtualized Environment Security Evaluation Criteria

Recently researches on security problems in virtualized environment are mainly in the following areas—the architecture of virtual machine monitor, the evaluation of performance isolation, trade-offs in security and performance, formal methods, and the security policies in Xen. Researchers emphasize how to build a safe and trusted virtualized environment, and mainly concern about the performance isolation of virtual machine monitor. Although many researchers start to consider how to evaluate the security degree of virtualized environment. And many ways have been tried, such as designing a benchmark to evaluate performance isolation. But there are not uniform virtualized environment security evaluation criteria at present. So we analyze the security problem in virtualized environment from following aspects:

- Are virtual machines truly isolated? How is the degree of isolation? If a virtual machine is attacked by malicious program, can other virtual machines work normally as before? Will a virtual machine steal information from other virtual machines? Will the performance of a virtual machine influence that of other virtual machines?
- Is the virtual machine monitor trusted? Is the actual behavior of VMM the same as what it has claimed? For example, if a virtual machine has claimed that it can't be migrated, will the virtual machine monitor still migrate it?
- Can the virtual machine monitor protect the security of guest OS? Does the virtual machine monitor provide a new way for the attack of malicious program? When guest OS is attacked by malicious program, will the virtual machine monitor intercept the attack for guest OS?

Based on the above analysis, we try to construct our virtualized environment security evaluation criteria referring to that of the operating system. We divide the security degree of virtualized environment into following six levels: D1, C1, C2, C3, B1, A1 (as shown in Table 31.1).

Level D1 is unprotected level, which is the lowest level in virtualized environment security evaluation criteria. And there are not any security measures in Level D1. Level C1, Level C2, and Level C3 are lower security levels. Level B1 has medium security protection. The highest security level is Level A1. The detail definitions are as follows:

- *Level D1*: Unprotected level. The lowest level in virtualized environment security evaluation criteria. There are not any security measures in this level. It an't ensure isolation between virtual machines. If a virtual machine is attacked by malicious program, it can't ensure that other virtual machines work normally as before. When the performance of a virtual machine changes, other virtual machines' performance may be influenced. There is information disclosure between virtual machines. The actual behavior of VMM isn't the same as what it has claimed. It is possible that VMM will steal information from guest OS illegally. The virtual machine monitor can't intercept attacks for guest OS. Instead it provides a new way for the attack of malicious program.
- *Level C1*: Security isolation protection level. This level can ensure the security isolation between virtual machines. When a virtual machine is attacked by malicious program, other virtual machines can work normally as before.
- *Level C2*: Resource isolation protection level. The requirement of resource isolation is added in this level. It ensures not only security isolation but also resource isolation between virtual machines. There isn't information disclosure between virtual machines. A virtual machine won't steal information from other virtual machines.
- *Level C3*: Performance isolation protection level. It can ensure the security isolation, resource isolation, and performance isolation between virtual machines at the same time. When the performance of a virtual machine changes, other virtual machines' performance will not be influenced.
- *Level B1*: claim consistently protection level. It can ensure isolation between virtual machines. Moreover, the actual behavior of VMM is the same as what it has claimed in this level. If a virtual machine has claimed that it can't be migrated, the virtual machine monitor will not do that. And VMM will not steal information from guest OS illegally.
- *Level A1*: Active protection level. The highest level in virtualized environment security evaluation criteria. Except all the security measures mentioned above, VMM does not provide a new way for the attack of malicious program. Instead it will protect the security of guest OS actively and intercept the attack for guest OS.

The reasons why we divide the security degree of virtualized environment into the above six levels are as follows: First, we think the isolation between virtual machines is the most basic security requirement in virtualized environment. If we can not ensure the isolation between virtual machines, there must be significant security risks in this virtualized environment. So we consider the isolation between virtual machines at first. Among the three isolations, security isolation is the most important requirement. If a virtual machine can't work normally because of the fault of other virtual machines, let alone information disclosure between virtual machines. So we put security isolation in Level C1. Compared with performance isolation, resource isolation has a greater impact on the security of virtualized environment. Thus resource isolation is put in Level C2 and performance isolation is put in Level C3. After ensuring the isolation between virtual machines, we should consider the claim consistence of VMM. So Level B1 is about the claim

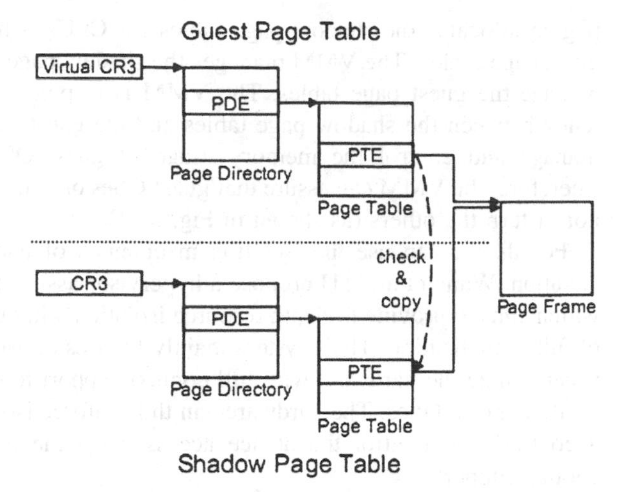

Fig. 31.1 Shadow page table

consistence of VMM. Furthermore, it is the highest level that VMM can protect the security of guest OS actively, so Level A1 is named active protection level.

After presenting the security classification of virtualized environment, we need to construct trusted computing environment from the following two aspects: First, provide some mechanisms in the VMM to achieve each security level from technical aspect. Second, authenticate each security level of virtualized environment.

31.4 Support Technologies

In this section, we are going to answer the following question: What mechanisms related to security are there in the VMM? How will these mechanisms support the security of virtualized environment?

In order to ensure security isolation from a technical level, we can consider the following aspects—first, because of attacks from malicious program, we must make sure of the isolation of network. Even if a virtual machine is attacked by malicious program, other virtual machines will be safe because the network is isolated. Second, we can realize a failure recovery mechanism. When a virtual machine breaks down and destroys shared resources, such as memory, disk, the failure recovery mechanism is started. We try to recover the shared resources, thus other virtual machines can work normally as before.

As for resource isolation, the following points are helpful—first, when a memory is freed by a virtual machine, the VMM should clear it before allocating it to other virtual machines. Second, the mechanism of shadow page table can be used to ensure resource isolation. Megumi Ito et al. [10] develop Gandalf, a lightweight VMM targeting embedded systems with multi-core CPUs. They focus on the shadow paging mechanism of Gandalf, which enables physical memory isolation among guest OSes by separating guest page tables from CPUs. Shadow paging allocates the shadow page tables for CPUs, which are separate from the guest page tables. The VMM manages the shadow page tables, and the guest OSes manage the guest page tables. The VMM is responsible for maintaining consistency between the shadow page tables and the guest page tables. The VMM can manage and control the memory usage of guest OSes using shadow paging. Therefore, the VMM can assure that guest OSes only use allocated memory and do not disturb the others (as shown in Fig. 31.1).

Fourth, we can use the isolation mechanism of hardware to ensure resource isolation. Wang et al. [11] propose a hypervisor-assisted, microkernel architecture which aims to provide in-depth resource isolation without the performance penalty of full virtualization. Their system mainly focuses on isolation of I/O devices, and it can utilize the hardware with full context support to directly assign a device to multiple guest OSes. The hardware can thus enforce isolation. Moreover, they use a controller to control the device access when the hardware does not provide enough support.

There are many ways to ensure performance isolation. First, static cache division can be used to ensure performance isolation. Virtual machines compete for shared caches, which leads to cache pollution and increases the frequency of cache invalidation. So the performance and isolation of virtual machines are seriously influenced because of competition for shared caches. Haogang et al. [12] present a way of static cache division to solve the above problem. Static cache division refers that allocate the cache when creating a virtual machine and the cache resource keeps unchanged during the whole life cycle of the virtual machine. Through experiments they find that compared with free competition for cache, static cache division is helpful for increasing the performance isolation of virtual machine on a multi-core platform. Second, we can use a mechanism of idle memory tax. Waldspurger et al. [13] present a mechanism of idle memory tax, which achieves efficient memory utilization while maintaining performance isolation guarantees. The basic idea is to charge a client more for an idle page than for one it is actively using. When memory is scarce, pages will be reclaimed preferentially from clients that are not actively using their full allocations. The tax rate specifies the maximum fraction of idle pages that may be reclaimed from a client.

In order to ensure the claim consistency, we can use the mechanism of migration lock. We set the migration flag in the configure file of the virtual machine when it is created. If the flag indicates that the virtual machine can't be migrated, when the VMM tries to migrate it, we will warn the VMM and stop migrating. Moreover, we allow that the virtual machine can set the migration flag actively during the runtime.

Our previous work—detecting memory leak via VMM [14], is an example of active protection technology. With the help of the VMM, we can record the running states of virtual machines, and find functions used to apply for/free memory. Based on this, some rules are used to detect memory leak. For example, memories which are not freed or accessed for a long time have the possibility of memory leak.

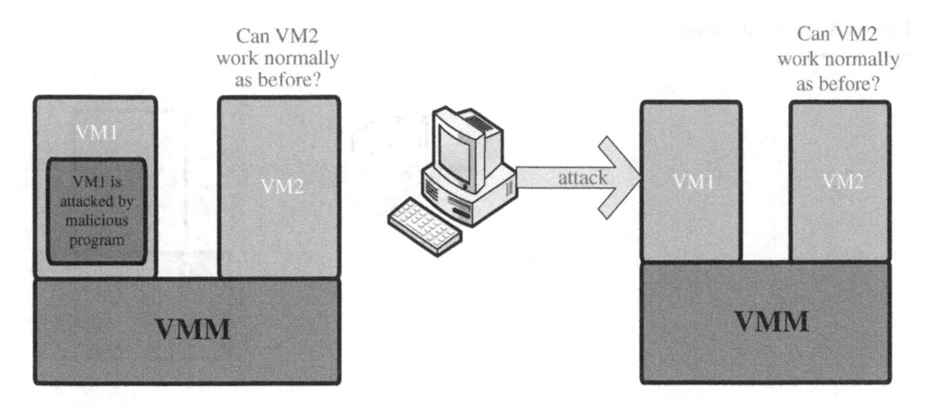

Fig. 31.2 Authentication method of level C1

31.5 Authentication Methods

Here we are going to explore the following three questions: First, how to authenticate the security of virtualized environment? Second, how to get evidences to prove that the virtualized environment is trusted? Third, what evidences should we need?

For authentication method of each security level, we give the following ideas:

- *Level D1*: Unprotected level. It is the lowest level in virtualized environment security evaluation criteria. So there is no need to authenticate this level.
- *Level C1*: Security isolation protection level. In order to assess whether the security degree of virtualized environment achieves Level C1, we can use the same way as performance isolation assessment. We can design a benchmark or an experiment to assess whether security isolation is ensured. To describe it in detail, we can make a virtual machine attacked by malicious program, and test whether other virtual machines can work normally as before (as shown in Fig. 31.2).
- *Level C2*: Resource isolation protection level. Recently in this aspect, only some researchers design a access control mechanism to ensure resource isolation between virtual machines. For the assessment of resource isolation, we can use the way like static code analysis. After analyzing the code mechanism of virtual machine monitor, we can find whether there is vulnerability in resource isolation. For example, whether the VMM clears the memory which is freed by a virtual machine before it allocates the memory to another virtual machine.
- *Level C3*: Performance isolation protection level. We can design different kinds of benchmarks to assess the performance isolation between virtual machines, such as memory-intensive, CPU-intensive, disk intensive, network-intensive. Perform stress testes in a virtual machine, and test whether other virtual machines' performance will be influenced. There are many researches on the assessment of performance isolation. Researchers have presented a lot of ways to assess performance isolation, see Sect. 31.2.

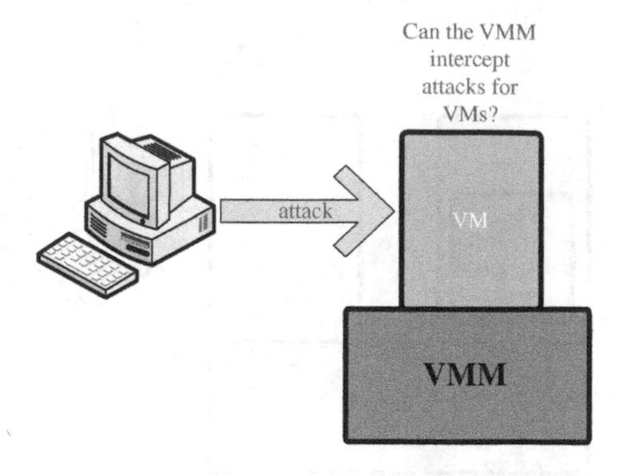

Fig. 31.3 Authentication method of level A1

- *Level B1*: Claim consistently protection level. In order to assess whether the security degree of virtualized environment achieves Level B1, we can use a way similar to the monitor mechanism. To describe it in detail, install a monitor for each virtual machine. When a virtual machine is migrated, there must be some characteristics recorded by the monitor. If the monitor finds that the VMM migrates a virtual machine illegally, the security degree can't achieve Level B1. In this aspect, the migration lock mentioned before can make sure that the VMM won't migrate a virtual machine illegally.
- *Level A1*: Active protection level. We can design some experiments to assess whether the security degree achieves Level A1. By using different kinds of attacks to virtual machines, we can test whether the VMM can intercept these attacks (as shown in Fig. 31.3). In this aspect, how to make the VMM intercept attacks for guest OS from a technical level is a more important question. We will research deeply in this aspect in the future.

31.6 Conclusion

In this paper, we analyze the security problem in virtualized environment and present our virtualized environment security evaluation criteria. We take a further research on support technology and authentication method. For support technology, we propose some safety guarantee technology for each security level. For authentication method, we give some ideas for the evaluation of each security level. We will research on implementation details in the future.

Acknowledgments This work is supported by the National Grand Fundamental Research 973 Program of China under Grant No. 2007CB310900, National Science Foundation of China under Grant No.90718028 and No. 60873052, National High Technology Research 863 Program of China under Grant No.2008AA01Z112, and MOE-Intel Information Technology Foundation under Grant No. MOE-INTEL-10-06.

References

1. Robin JS, Irvine CE (2000) Analysis of the Intel Pentium's ability to support a secure virtual machine monitor. In: Proceedings of the 9th USENIX security symposium
2. Certification report for processor resource/system manager (PR/SM) for the IBM system z10 EC GA1. BSI-DSZ-CC-0460-2008
3. Jansen B, Ramasamy HV, Schunter M Flexible integrity protection and verification architecture for virtual machine monitors. IBM Zurich Research Laboratory
4. Matthews PN, Hu W, Hapuarachchi M, Deshane T, Dimatos D, Hamilton G, McCabe M (2007) Quantifying the performance isolation properties of virtualization systems. In: Workshop on experimental computer science
5. Somani G, Chaudhary S (2009) Application performance isolation in virtualization. In: Proceedings of the 2009 IEEE international conference on cloud computing
6. Koh Y, Knauerhase R, Brett P, Bowman M, Wen Z, Pu C (2007) An analysis of performance interference effects in virtual environments. In: Performance analysis of systems and software
7. Karger PA, Safford DR (2008) I/O for virtual machine monitors: security and performance issues. IEEE Secur Priv 6(5):16–23
8. Weng C, Luo Y, Li M, Lu X (2008) A BLP-based access control mechanism for the virtual machine system. In: The 9th international conference for young computer scientists
9. Sailer R, Jaeger T, Valdez E et al (2005) Building a MAC-based security architecture for the Xen Open-Source Hypervisor. IBM Research Report RC23629
10. Ito M, Oikawa S (2008) Lightweight shadow paging for efficient memory isolation in Gandalf VMM. In: 11th IEEE symposium on object oriented real-time distributed computing
11. Wang J, Niphadkar S et al (2010) A virtualization architecture for in-depth Kernel isolation. In: Proceedings of the 43rd Hawaii international conference on system sciences
12. Jin X, Chen H, Wang X, Wang Z, Wen X, Luo Y, Li X (2009) A simple cache partitioning approach in a virtualized environment. In: The 2009 international workshop on architecture support on virtualization techniques (IWASVT2009), within IEEE ISPA 2009, August 10-12, 2009, Chengdu
13. Waldspurger CA (2002) Memory resouce management in VMware ESX server. In: 5th symposium on operating systems design and implementation
14. Wang X, Wang Z, Sun Y, Yi L, B Zhang, Luo Y (2010) Detecting memory leak via VMM. Chin J Comput 33(3)

Part VIII
EMC 2011 Session 1: Cyber-Physical Systems and Real-Time Systems

Chapter 32
An Efficient Deadline and Period Assignment Scheme for Maintaining Temporal Consistency Under EDF

F. Zhu, J. Li and G. Li

Abstract In order to maintain real-time data freshness (or temporal consistency) while minimizing the imposed processor workload, much work has been done to address the period and deadline assignment problem for real-time update transactions, such as the More-Less scheme [18] for fixed priority scheduling, and the $\mathcal{ML}_{\mathrm{EDF}}$ and $\mathcal{HS}_{\mathrm{EDF}}$ algorithms [19] for dynamic priority scheduling. This paper studies the period and deadline assignment problem for update transactions scheduled by the earliest deadline first (EDF) scheme. Based on a sufficient feasibility test for EDF [3], we formulate the assignment problem as a Linear Programming problem and propose the Improved More-Less ($\mathcal{IML}_{\mathrm{EDF}}$) algorithm. $\mathcal{IML}_{\mathrm{EDF}}$ can achieve a considerably low workload in a time-efficient manner. Extensive experiments are conducted and the results verify the efficiency of $\mathcal{IML}_{\mathrm{EDF}}$, when compared with existing methods with comparable quality.

Keywords Real-time databases · Temporal consistency · Period and deadline · Earliest deadline first

---

This work was partially supported by the National Science Foundation of China [Award No. 60873030] and the Research Fund for the Doctoral Program of the Ministry of Education of China [Award No. 20090142110023].

F. Zhu · J. Li (✉) · G. Li
School of Computer Science and Technology,
Huazhong University of Science and Technology, Luoyu Road 1037#, Wuhan, China
e-mail: jianjunli@smail.hust.edu.cn

G. Li
e-mail: guohuili@mail.hust.edu.cn

J. J. Park et al. (eds.), *Proceedings of the International Conference on Human-centric Computing 2011 and Embedded and Multimedia Computing 2011*, Lecture Notes in Electrical Engineering 102,
DOI: 10.1007/978-94-007-2105-0_32,

32.1 Introduction

Real-time database systems (RTDBS) have been widely used in many applications that require timely processing of large amounts of real-time data, such as aerospace and defense systems, industrial automation and air traffic control systems, etc. Typically, a real-time database is composed of a set of real-time data objects, each of which models the current status of a real world entity in the external environment. Different from data stored in traditional databases, a real-time data object is only valid in a given time period, which is defined as its temporal validity interval. In order to maintain temporal validity (a.k.a. temporal consistency), each real-time data object must be refreshed by a sensor update transaction before its temporal validity interval expires, or else the RTDBS cannot respond to environmental changes timely. Given the temporal consistency requirement, one important issue in designing RTDBSs is to determine the sampling periods and deadlines for sensor update transactions so that the temporal consistency can be maintained while the resulting processor workload can be minimized. During the past decade, many schemes and algorithms, such as Half-Half (HH) [7, 14], More-Less (ML) [18], and deferrable scheduling algorithm for fixed priority transactions (DS-FP) [6, 17], have been proposed to address this problem.

Although there has been much work devoted to the period and deadline assignment problem, most of them are based on *fixed priority* scheduling, while the same problem under *dynamic priority* scheduling receives comparatively less attention. An exception is the work conducted by Xiong et al. in [19]. Xiong et al. examined deriving periods and deadlines for update transactions scheduled under earliest deadline first (EDF). Firstly, based on a sufficient feasibility condition of EDF scheduling with deadlines no greater than their corresponding periods, a *linear time* algorithm, $\mathcal{ML}_{\mathrm{EDF}}$, was proposed to solve the assignment problem. Although $\mathcal{ML}_{\mathrm{EDF}}$ is time efficient, it is based on a sufficient but not necessary feasibility condition, which means it cannot result in an optimal solution in terms of minimizing CPU workload. Secondly, by relaxing the deadline constraint to be arbitrary and assuming the time of the system to be discrete, a *branch and bound* based search algorithm, $\mathcal{OS}_{\mathrm{EDF}}$, was proposed with the objective to obtain the optimal solution. The problem with $\mathcal{OS}_{\mathrm{EDF}}$ is that it does not scale well with the problem size. Hence thirdly, a heuristic search-based algorithm, namely $\mathcal{HS}_{\mathrm{EDF}}$, which is more efficient than $\mathcal{OS}_{\mathrm{EDF}}$ and is capable of finding a solution if one really exists, is further proposed. Compared to $\mathcal{OS}_{\mathrm{EDF}}, \mathcal{HS}'_{\mathrm{EDF}}$s efficiency is achieved at the expense of increased processor workload.

In this work, we address the same problem as [19], i.e., determining periods and deadlines for firm periodic real-time update transactions scheduled under EDF. Since the relaxation of task deadlines to be arbitrary has little impact on the resulted CPU workload of update transactions, as indicated in [19], we focus on the case where transaction deadlines are not greater than their corresponding periods. We intend to derive deadlines and periods for transactions in a time-efficient manner but can result in a relatively low processor workload. Based on a sufficient feasibility test for EDF scheduling [3], we propose the $\mathcal{IML}_{\text{EDF}}$ algorithm, within which we formulate the assignment problem as a linear programming (LP) problem. $\mathcal{IML}_{\text{EDF}}$ can result in considerably lower workloads than $\mathcal{ML}_{\text{EDF}}$, due to that $\mathcal{IML}_{\text{EDF}}$ is based on processor demand analysis, which is more accurate than the utilization-based feasibility test applied in $\mathcal{ML}_{\text{EDF}}$. Compared with $\mathcal{HS}_{\text{EDF}}$, the resulted workload from $\mathcal{IML}_{\text{EDF}}$ is slightly higher, since $\mathcal{IML}_{\text{EDF}}$ is based on a sufficient feasibility condition while $\mathcal{HS}_{\text{EDF}}$ is based on a necessary and sufficient one. But $\mathcal{IML}_{\text{EDF}}$ is more time-efficient, as it has a polynomial time complexity.

The remainder of the paper is organized as follows: Sect. 32.2 reviews the definition of temporal validity and presents some assumptions. Section 32.3 details the $\mathcal{IML}_{\text{EDF}}$ algorithm. Section 32.4 presents our performance studies. Section 32.5 briefly describes the related work, and finally, conclusions are given in Sect. 32.6.

32.2 Background and Assumptions

32.2.1 Temporal Validity for Data Freshness

In a real-time database a data object is a logic image of a real-world entity. As the state of a real-world entity changes continuously, to monitor the entity's state faithfully, real-time data objects must be refreshed by update transactions, which are generated periodically by intelligent sensors, before they become invalid. The actual length of the temporal validity interval of a real-time data object is application dependent [12, 14, 15]. We assume that a sensor always samples the value of a real-time data at the beginning of its update period.

Definition 1 [14] A real-time data object (X_i) at time t is temporally valid if, for its jth update finished last before t, the sampling time ($r_{i,j}$) plus the validity interval (α_i) of the data object is not less than t, i.e., $r_{i,j} + \alpha_i \geq t$.

A data value for real-time object X_i sampled at any time t will be valid from t up to $(t + \alpha_i)$.

32.2.2 Notations and Assumptions

In this paper, we use $\mathcal{T} = \{\tau_i\}_{i=1}^N$ and $\mathcal{X} = \{X_i\}_{i=1}^N$ to denote a set of periodic sensor update transactions and a set of temporal data, respectively. All temporal data are assumed to be kept in main memory. For each data $X_i (1 \leq i \leq N)$, there is a validity interval length α_i associated with it. Transaction τ_i is responsible for updating the corresponding data X_i periodically. Since each sensor transaction updates different data, no concurrency control is considered. Each update

Table 32.1 Symbols and definitions

Symbol	Definition
X_i	Real-time data object i
τ_i	Sensor update transaction updating X_i
C_i	Computation time of τ_i
α_i	Validity interval length of X_i
T_i	Period of τ_i
D_i	Relative deadline of τ_i
U_i	Processor workload of τ_i
$U_{i_{\min}}$	Lower bound on processor workload of τ_i
$U_{i_{\max}}$	Upper bound on processor workload of τ_i
$\mathcal{U}$	Processor workload of $\{\tau_i\}_{i=1}^{N}$

transaction τ_i is periodic and is characterized by the following five-tuple: $\{C_i, D_i, T_i, T_{i_{\min}}, T_{i_{\max}}\}$, where C_i is the execution time, D_i is the relative deadline and T_i is the period. Furthermore, $T_{i_{\min}}$ and $T_{i_{\max}}$ denote the lower and upper bound of τ_i's period, respectively. The utilization of τ_i is $U_i = \frac{C_i}{T_i}$. Similarly, the minimum and maximum utilizations of τ_i are $U_{i_{\min}} = \frac{C_i}{T_{i_{\max}}}$ and $U_{i_{\max}} = \frac{C_i}{T_{i_{\min}}}$, respectively. We use $\mathcal{U}$ to denote the total processor utilization of $\mathcal{T}$, i.e., $\mathcal{U} = \sum_{i=1}^{N} \frac{C_i}{T_i}$. The same as [19], we assume the system is synchronous, i.e., all the first instances of sensor transactions are generated at the same time, and the jitter between sampling time and release time of a job is zero. But different from [19], we assume the time of the system to be continuous. Table 32.1 presents the formal definitions of the frequently used symbols. Throughout the paper we assume the scheduling algorithms are preemptive and ignore all preemption overhead. For presentation convenience, we use terms transaction and task, workload and utilization interchangeably in the following discussion.

32.3 Improved More-Less Scheme Under EDF

In this section, we detail the Improved More-Less scheme for temporal consistency maintenance under EDF. Section 32.3.1 formulates a constrained optimization problem for period and deadline assignment. Section 32.3.2 gives the design of $\mathcal{IML}_{\text{EDF}}$ by solving the optimization problem under a sufficient feasibility condition, which was first introduced in [3].

32.3.1 Constrained Optimization Problem Under EDF

Given a set of update transactions, the optimal solution to the deadline and period assignment problem has to minimize processor workload $\mathcal{U}$ while maintaining the temporal consistency under EDF. In order to maintain real-time data freshness and comply with the More-Less scheme adopted in our approach, the same as in [19], we have the following three constraints for T_i and D_i:

- Validity constraint: $T_i + D_i \leq \alpha_i$.
- Deadline constraint: $C_i \leq D_i \leq T_i$.
- Feasibility constraint: $\mathcal{T}$ with derived deadlines and periods is schedulable by using EDF.

Now the period and deadline assignment problem for real-time update transactions can be described as: given a set of transactions $\mathcal{T} = \{\tau_i\}_{i=1}^{N}$ with C_i and α_i specified for each τ_i, determine T_i and D_i for τ_i so that the set of transactions can be scheduled under EDF to maintain real-time data freshness, i.e., the validity and deadline constraint can be satisfied, and the processor workload $\mathcal{U}$ can be minimized.

In terms of EDF scheduling, Baruah et al. considered the case where task deadlines are no greater than periods and proposed an exact, albeit complex, schedulability test [1], which is later improved in [2]. The test is restated in the following theorem.

Theorem 1 *Given a periodic task set with $D_i \leq T_i$, the task set is schedulable if and only if the following constraint is satisfied: $\forall L \in \{kT_i + D_i | kT_i + D_i \leq \min(B_p, H)\}$,*

$$L \geq \sum_{i=1}^{N} \left(\lfloor \frac{L - D_i}{T_i} \rfloor + 1 \right) C_i \tag{32.1}$$

where $k \in \mathbb{N}$ (the set of natural numbers including 0), B_p and H denote the busy period and hyperperiod, respectively.

Based on Theorem 1, the period and deadline assignment problem can be formulated as,

$$\min : \mathcal{U} = E(U_1, U_2, \ldots, U_N) = \sum_{i=1}^{N} U_i \tag{32.2}$$

$$\text{s.t.} : L \geq \sum_{i=1}^{N} \left(\lfloor \frac{L - D_i}{T_i} \rfloor + 1 \right) C_i \tag{32.3}$$

$$L \in \{kT_i + D_i \leq \min(B_p, H)\}, k \in \mathbb{N} \tag{32.4}$$

$$U_i \geq U_{i_{\min}} \quad \text{for } i = 1, 2, \ldots, N \tag{32.5}$$

$$U_i \leq U_{i_{\max}} \quad \text{for } i = 1, 2, \ldots, N \tag{32.6}$$

The first two constraints state the schedulability condition under EDF, while the rest two represent the validity and deadline constraint. Note that $\mathcal{U}$ is minimized only if $T_i + D_i = \alpha_i$. Otherwise, T_i can always be increased to reduce processor workload as long as $T_i + D_i < \alpha_i$ [19]. Then according to the deadline constraint, the minimum period of $\tau_i, T_{i_{\min}}$, can be determined as $\frac{\alpha_i}{2}$, which further means $U_{i_{\max}} = \frac{C_i}{T_{i_{\min}}} = \frac{2C_i}{\alpha_i}$. But by now we cannot decide $U_{i_{\min}}$, we will show how to determine $U_{i_{\min}}$ in the following subsection.

32.3.2 Designing $\mathcal{IML}_{\mathrm{EDF}}$ Using a Sufficient Feasibility Condition

Solving the constrained optimization problem defined in (32.2)–(32.6) can be extremely time-consuming, since verifying the constraint in (32.1) for all L values leads to high computation complexity. Hence, we investigate solving the problem approximately, i.e., deriving a solution which satisfies the constraint in (32.3)–(32.6) quickly but can lead to a processor workload near the optimal one. Now consider the following relaxed schedulability condition [3],

$$L \geq \sum_{i=1}^{N} \left(\frac{L - D_i}{T_i} + 1 \right) C_i \tag{32.7}$$

It is not difficult to see that if (32.7) is satisfied then (32.1) must also be satisfied. We consider utilizing (32.7) to simplify the feasibility test in the above constraint optimization problem. The reason is that with (32.7), the schedulability of a task set can be determined based on a single L value, L^*, as indicated by Chantem et al. [3]. Before presenting their result, we first introduce two useful lemmas on which their work is based. The two lemmas are also very useful in our work.

Lemma 1 *Given a set $\mathcal{T}$ of N tasks with $D_i \leq T_i (1 \leq i \leq N)$, let $L_1 < L_2$. If the constraint in (32.7) is satisfied for L_1, then it is also satisfied for L_2.*

Lemma 1 can be trivially proved by rewriting (32.7) as $L \geq \frac{\sum_{i=1}^{N} C_i - \sum_{i=1}^{N} U_i D_i}{1 - \sum_{i=1}^{N} U_i}$.

Lemma 2 [3] *Given a task set $\mathcal{T}$, let the tasks in $\mathcal{T}$ be ordered in a non-decreasing order of deadlines (D_i) and suppose that the minimum task deadline, $D_{\min}$, is unique. Regardless of the choices of periods, any task set that is schedulable must satisfy the following property:*

$$\sum_{i=1}^{j} C_i \leq D_j, \ \forall j = 1, \ldots, N \tag{32.8}$$

Note that Lemma 2 states a necessary condition for any task set to be schedulable under EDF. Now we present Chantem et al.'s result as the following theorem shows.

Theorem 2 [3] *Consider a set $\mathcal{T}$ of N tasks that satisfy the condition in Lemma 2. Let the tasks in $\mathcal{T}$ be sorted in a non-decreasing order of task deadlines. The task set $\mathcal{T}$ is schedulable if*

$$L^* \geq \sum_{i=1}^{N} \left(\frac{L^* - D_i}{T_i} + 1 \right) C_i \tag{32.9}$$

where

$$L^* = \begin{cases} \min_{i=1}^{N}(T_i + D_i) : & D_1 + T_1 > D_2 \\ D_2 : & \text{otherwise} \end{cases}$$

Theorem 2 paves the way to finding a simpler constrained optimization problem formulation for the purpose of period and deadline calculation. Since the complexity of the assignment problem defined in (32.2)–(32.6) mainly comes from that (32.3) must be satisfied for all possible values of L, we consider utilizing the schedulability test stated in Theorem 2 to solve the assignment problem approximately but with less complexity. First, with $D_i + T_i = \alpha_i$ and $U_i = C_i/T_i$, (32.9) can be simply transformed to,

$$\sum_{i=1}^{N} (L^* - \alpha_i) U_i \leq L^* - 2 \sum_{i=1}^{N} C_i \tag{32.10}$$

Then by Theorem 2, the problem in (32.2)–(32.6) can be relaxed as,

$$\min : \mathcal{U} = E(U_1, U_2, \ldots, U_N) = \sum_{i=1}^{N} U_i \tag{32.11}$$

$$\text{s.t.} : \sum_{i=1}^{N} (L^* - \alpha_i) U_i \leq L^* - 2 \sum_{i=1}^{N} C_i \tag{32.12}$$

$$L^* = \begin{cases} \min_{i=1}^{N} \alpha_i : & D_1 + T_1 > D_2 \\ D_2 : & \text{otherwise} \end{cases} \tag{32.13}$$

$$U_i \geq U_{i_{\min}} \quad \text{for } i = 1, 2, \ldots, N \tag{32.14}$$

$$U_i \leq U_{i_{\max}} \quad \text{for } i = 1, 2, \ldots, N \tag{32.15}$$

Now we try to solve the constraint optimization problem in (32.11)–(32.15). In what follows, we formulate the constraint optimization problem as a LP problem.

It is important to note that Theorem 2 requires that a task set be sorted according to task deadlines and satisfy inequality (32.8). But task deadlines are what we need to calculate, i.e., we have no information about any task deadlines initially. We first consider sorting the task set by validity interval length (α_i). After computing a deadline for each task, we will resort the tasks based on their deadlines to make sure the conditions in Lemma 2 and Theorem 2 are satisfied.

There are two possible values for L^*, as shown in (32.13). Since we do not know any D_i, we cannot decide whether L^* is D_2 or $\min_{i=1}^N \alpha_i$ by now. We first assume $D_1 + T_1 > D_2$, then according to Theorem 2, we only need to check $L^* = \min_{i=1}^N \alpha_i = \alpha_1$ for schedulability, which indeed leads to a constant L^* value in (32.12). For τ_1, the task with the minimum validity interval length, we have $L^* - \alpha_1 = 0$, which means U_1 does not affect the establishment of (32.12). In order to minimize the processor workload and satisfy the constraint in (32.8), it is obvious that we can have $D_1 = C_1, T_1 = \alpha_1 - C_1$ and $U_1 = \frac{C_1}{\alpha_1 - C_1}$. Note that by setting $D_1 = C_1$, we implicitly mean τ_1 is the task with the minimum deadline, since for other tasks $\tau_j (2 \leq j \leq N)$ to satisfy the inequality condition in Lemma 2, there must be $D_j \geq \sum_{i=1}^j C_i > C_1 = D_1$.

Now we need to guarantee that D_1 is the smallest deadline when determining deadline for other tasks. We take into account the computation time of τ_1 (C_1) to determine the utilization lower bound ($U_{i_{\min}}$) for other tasks, i.e., $U_{i_{\min}} = \frac{C_i}{\alpha_i - C_1 - C_i} (2 \leq i \leq N)$. This lower bound guarantees that all task deadlines are larger than D_1. But note that the lower bound determined in this way can not guarantee other tasks $\tau_i (2 \leq i \leq N)$ satisfy inequality (32.8). We need to rectify the solution to guarantee that (32.8) can be satisfied for all tasks and we will detail this issue in the design of our $\mathcal{IML}_{\mathrm{EDF}}$ algorithm. Given the utilization upper bound defined in Sect. 32.3.1 and the lower bound defined above, the assignment problem can be stated as,

$$\min : \mathcal{U} = E(U_1, U_2, \ldots, U_N) = \sum_{i=1}^{N} U_i \tag{32.16}$$

$$\text{s.t.} : \sum_{i=1}^{N} (\alpha_1 - \alpha_i) U_i \leq \alpha_1 - 2 \sum_{i=1}^{N} C_i \tag{32.17}$$

$$U_i \geq \frac{C_i}{\alpha_i - C_1 - C_i} \quad i = 2, \ldots, N \tag{32.18}$$

$$U_i \leq 2C_i / \alpha_i \quad i = 2, \ldots, N \tag{32.19}$$

It is clear to see that (32.16)–(32.19) is a LP problem and a linear solver, such as Lindo and Matlab, can be used to solve it. In the following discussion, a solution which satisfies the constraint stated in (32.17)–(32.19) but may not necessarily minimize the CPU workload $\mathcal{U}$ is defined as a *candidate* solution to the LP problem. A similar definition can be applied to the problem in (32.11)–(32.15).

By solving the above LP problem, we can obtain a deadline and a period for each task, then we find transaction τ_m with the minimum deadline from the new calculated transaction deadlines, i.e., $D_m = \min\{D_i\}_{i=2}^N$. We interchange the order of τ_m with τ_2, then $D_2 = D_m$. Now there are two possible cases for D_2, one is $D_1 + T_1 > D_2$ and the other is $D_1 + T_1 \leq D_2$.

For the first case, obviously as indicated by (32.13), after resorting the task set in a non-decreasing order of deadline, if all tasks in the task set satisfy the constraint in (32.8), then we have identified a feasible solution. Otherwise, a rectification operation is required to guarantee the schedulability of the derived solution. We will discuss the rectification issue later.

Let us now consider the second case where $D_1 + T_1 \leq D_2$. According to Theorem 2, we need to check whether $L^* = D_2$ satisfy (32.12) to determine feasibility, but actually we use $L^* = \alpha_1$. Note that since $D_2 \geq D_1 + T_1 = \alpha_1$, if (32.12) holds at $L^* = \alpha_1$, it should also be satisfied at $L^* = D_2$ following Lemma 1, which means the solution calculated at $L^* = \alpha_1$ is also a *candidate* one for the problem in (32.11)–(32.15).

Algorithm 1 $\mathcal{IML}_{EDF}$

---

input: A set of update transactions $\mathcal{T} = \{\tau_i\}_{i=1}^N$ with $\{C_i\}_{i=1}^N$ and $\{\alpha_i\}_{i=1}^N$, sorted in a non-decreasing order of α_i.
output: Deadlines $\{D_i\}_{i=1}^N$ and periods $\{T_i\}_{i=1}^N$.

```
1  begin
2      D_1 = C_1;   T_1 = α_1 − D_1;
3      Compute {U_i}_{i=2}^N by solving the LP problem stated in (16)–(19);
4      for 2 ≤ i ≤ N do
5          T_i = C_i/U_i;   D_i = α_i − T_i;
6      Find transaction τ_m with the minimum deadline from the new caculated
       deadlines and interchange τ_2 with τ_m;
7      Sort T in a non-decreasing order of deadline;
8      for 3 ≤ i ≤ N do
9          if D_i < Σ_{j=1}^{i} C_j then
10             D_i = Σ_{j=1}^{i} C_j;   T_i = α_i − D_i;
11 end
```

For the solution obtained by solving the LP problem, by now we cannot guarantee that inequality (32.8) in Lemma 2 can be satisfied for all tasks, i.e., we cannot guarantee the obtained solution is a feasible one, since we only take C_1 into consideration when determining the utilization lower bound for task $\tau_i (2 \leq i \leq N)$. We need to check whether the necessary condition stated in Lemma 2 is satisfied for $\tau_i (2 \leq i \leq N)$. If there exists some task τ_j such that inequality (32.8) is not satisfied, i.e., $D_j < \sum_{i=1}^{j} C_i$, then a rectification operation is required to guarantee the schedulability of the task set. In order to satisfy inequality (32.8) and minimize the processor workload, it is natural to increase the deadline of τ_j to be $D_j = \sum_{i=1}^{j} C_i$, and adjust the period of τ_j accordingly, i.e., $T_j = \alpha_j - D_j$. But we should ensure that after conducting the rectification operation, the solution becomes a feasible one, as the following theorem shows.

Theorem 3 *Given a solution obtained by solving the LP problem, after sorting the task set in a non-decreasing order of deadline, if* $D_j (1 < j \leq n)$ *does not satisfy inequality* (32.8), *then by setting* $D_j = \sum_{i=1}^{j} C_i$, *the solution becomes a feasible one.*

proof After conducting the rectification, all tasks satisfy inequality (32.8), therefore, we only need to prove that (32.12) still holds for the rectified task set.

Given $D_j \leq \sum_{i=1}^{j} C_i (2 \leq j \leq N)$, increasing the deadline of τ_j, which inversely decreasing T_j leads to high workload of τ_j. We use U_j and U_j' to denote the workload before and after increasing D_j, respectively. With $U_j < U_j'$ and $L^* = \alpha_1 < \alpha_j$, it is obvious that if (32.12) can be satisfied for U_j, it should also be satisfied for U_j'. It is easy to see that the task order with non-decreasing deadlines can still be maintained by increasing D_j to be $\sum_{i=1}^{j} C_i$ when necessary. Therefore, the theorem follows.

Detail of $\mathcal{IML}_{\text{EDF}}$ is shown in Algorithm 1. Our algorithm aims to find a feasible solution to the optimization problem defined in (32.2)–(32.6) in a time-efficient manner. The deadline and period for τ_1 are determined to be C_1 and $\alpha_1 - C_1$, respectively, as shown in line 2. We formulate the assignment problem as a LP problem (stated in (32.16)–(32.19)) which can be solved by a linear solver. Then we find task τ_k with the minimum deadline and interchange the order of τ_k with τ_2, we further check whether $\alpha_1 < D_2$ holds. Finally, we sort the task set in a non-decreasing order of derived deadlines and conduct a rectification operation (lines 8–10) to guarantee a feasible solution. Since the LP problem can be solved in polynomial time, it is not difficult to see that our approach is also in polynomial time complexity.

Since $\mathcal{IML}_{\text{EDF}}$ utilizes a more accurate sufficient feasibility condition than $\mathcal{ML}_{\text{EDF}}$, it can result in a lower workload. Compared to $\mathcal{HS}_{\text{EDF}}$, the workload by $\mathcal{IML}_{\text{EDF}}$ is slightly higher, but the workload difference between them is not large. Moreover, $\mathcal{IML}_{\text{EDF}}$ is more time-efficient than $\mathcal{ML}_{\text{EDF}}$ due to it has a polynomial time complexity.

32.4 Performance Evaluation

32.4.1 Simulation Model and Assumptions

We have conducted experiments to compare the performance of *HH*, $\mathcal{ML}_{\text{EDF}}$, $\mathcal{HS}_{\text{EDF}}$[1] and $\mathcal{IML}_{\text{EDF}}$. The update transaction workload produced by *HH*, $\mathcal{ML}_{\text{EDF}}$, $\mathcal{HS}_{EDF}$ and $\mathcal{IML}_{\text{EDF}}$, as well as the execution time under the four

[1] Since $\mathcal{OS}_{\text{EDF}}$ doesnot scale well with problem size, we do not consider it in our experiments.

Table 32.2 Experimental parameters and settings

Parameter class	Parameters	Meaning	Value
System	N_{CPU}	No. of CPU	1
	N_T	No. of data objects	[50, 300]
	α_i(ms)	Validity interval of X_i	[4,000, 8,000]
Update	C_i(ms)	Time for updating X_i	[5, 15]
Transactions	Trans. length	No. of data to update	1

schemes have been compared. It is demonstrated that $\mathcal{IML}_{\text{EDF}}$ can result in a relatively low CPU workload, in a time-efficient manner.

Table 32.2 shows a summary of the parameters and default settings used in our experiments. For convenience of comparison, we use the same baseline values for the parameters as [19], which are originally from air traffic control applications [12]. Two categories of parameters are defined: system and update transaction. For system configurations, a single CPU, main memory based RTDBS is considered. The number of real-time data objects is uniformly varied from 50 to 300 to change the workload in the system. The validity interval length of each real-time data object is assumed to be uniformly distributed in [4,000, 8,000] ms. For update transactions, it is assumed that each update transaction updates one data object, and the execution time for each transaction is uniformly distributed in [5, 15] ms. The proposed algorithms are implemented in C++, while the constraint optimization problem from (32.16)–(32.19) is solved by using Matlab.

32.4.2 Experimental Results

32.4.2.1 Comparison of CPU Workload

In our experiments, the CPU workload of update transactions produced by HH, $\mathcal{ML}_{\text{EDF}}$, $\mathcal{HS}_{\text{EDF}}$ and $\mathcal{IML}_{\text{EDF}}$ are quantitatively compared. Update transactions are generated randomly according to the parameter settings in Table 32.2. The resulting processor workloads from the four schemes are depicted in Fig. 32.1, in which the x-axis denotes the number of update transactions and the y-axis denotes the resulted CPU workload.

From the experimental results, we observe that $\mathcal{IML}_{\text{EDF}}$ consistently outperforms HH and $\mathcal{ML}_{\text{EDF}}$. The discrepancy widens as the number of transactions increases. When the number of transactions grows up to 300, the difference reaches to 30%. The major reason that $\mathcal{IML}_{\text{EDF}}$ outperforms HH and $\mathcal{ML}_{\text{EDF}}$ in terms of CPU workload lies in the more accurate feasibility condition stated in Theorem 2.

Compared with $\mathcal{HS}_{\text{EDF}}$, $\mathcal{IML}_{\text{EDF}}$ derives solutions with slightly larger workload. As can be seen from Fig. 32.1, the workload produced by $\mathcal{IML}_{\text{EDF}}$ is close to that of $\mathcal{HS}_{\text{EDF}}$. The discrepancy is small and widens as the transaction

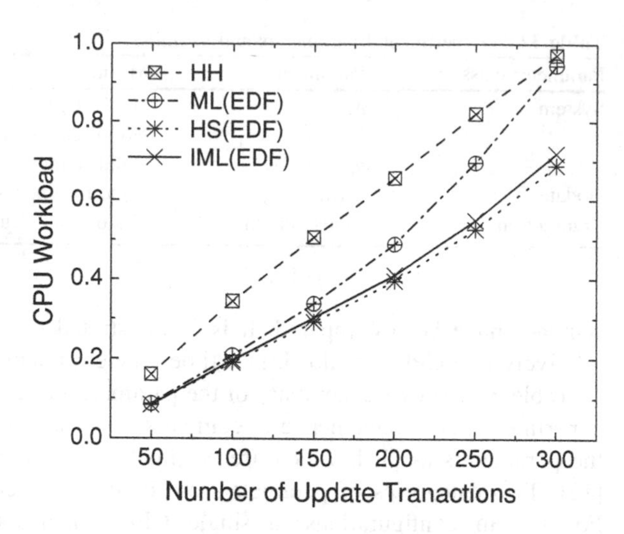

Fig. 32.1 Resulted workload

Table 32.3 Execution time

#	Execution time (s)			
	Half-Half	$\mathcal{ML}_{EDF}$	$\mathcal{HS}_{EDF}$	$\mathcal{IML}_{EDF}$
50	<0.01	<0.01	66.4	<0.01
100	<0.01	<0.01	704.5	<0.01
150	<0.01	<0.01	2202.1	<0.01
200	<0.01	<0.01	4032.8	<0.01
250	<0.01	<0.01	8299.9	<0.01
300	<0.01	<0.01	16524.2	<0.01

number increases. When N_T grows up to more than 200, we observe that $\mathcal{IML}_{\text{EDF}}$ results in about 2% larger workload than $\mathcal{HS}_{\text{EDF}}$. The reason is that both $\mathcal{IML}_{\text{EDF}}$ and $\mathcal{HS}_{\text{EDF}}$ are built on the processor demand analysis based feasibility condition for EDF scheduling, but $\mathcal{IML}_{\text{EDF}}$ is based on a sufficient one, while $\mathcal{HS}_{\text{EDF}}$ is based on the exact (necessary and sufficient) one. Compared to $\mathcal{IML}_{\text{EDF}}$, $\mathcal{HS}'_{\text{EDF}}$s lower workload is achieved at the expense of larger computation overhead, due to that it is a search-based algorithm. Although the workload derived from $\mathcal{IML}_{\text{EDF}}$ is slightly higher than $\mathcal{HS}_{\text{EDF}}$, $\mathcal{IML}_{\text{EDF}}$ is more time-efficient, as illustrated next.

32.4.2.2 Comparison of Execution Time

The execution times of the four schemes under varied transaction number are shown in Table 32.3. As one can see, HH, $\mathcal{ML}_{\text{EDF}}$ and $\mathcal{IML}_{\text{EDF}}$ spend very little time to obtain a solution. With the growth of task set size, the execution times of these three approaches are consistently less than 0.01s and actually, the time difference of them is negligible. The reason is that both HH and $\mathcal{ML}_{\text{EDF}}$ are linear-time algorithms, while $\mathcal{IML}_{\text{EDF}}$ has polynomial-time complexity.

Compared with the aforementioned three algorithms, $\mathcal{HS}_{\text{EDF}}$ takes significantly longer time to obtain a solution, and the execution time increases exponentially with the growth of transaction number. When $N_T = 50$, the execution time of $\mathcal{HS}_{\text{EDF}}$ is about one minute. But when N_T increases to 300, the time increases to almost 4.6 h. The main reason that $\mathcal{HS}_{EDF}$ takes such a long time to obtain a solution lies in that it involves in solving the 0–1 knapsack problem iteratively, which is very time-consuming. Note that in our experiments, we utilize the Bintprog() function in MatLab to solve the knapsack problem.

32.5 Related Work

There has been a lot of work on RTDBSs for maintaining real-time data freshness [4, 5, 8–11, 13, 15–17]. [16] studies the performance of two well known concurrency control algorithms, two-phase locking and optimistic, in maintaining temporal consistency of shared data in a hard real-time systems. [10] investigates real-time data-semantics and proposes a class of real-time access protocol called similarity stack protocol (SSP). The trade-off between data consistency and system workload is exploited in [7], where similarity-based principles are combined with the HH scheme to reduce workload by skipping the execution of task instances.

All the work mentioned above assumes the deadlines and periods of update transactions are given, hence gives no answer to the period and deadline assignment problem for maintaining temporal consistency. [18] is the first work to address the period and deadline assignment problem for real-time update transactions. The More-Less scheme is proposed in [18] to solve the assignment problem with *deadline monotonic* scheduling, a *fixed priority* scheduling algorithm. While More-Less is based on periodic task model, the DS-FP proposed in [17] follows a sporadic task model. DS-FP reduces processor workload by adaptively adjusting the separation of two consecutive instances of update transactions while satisfying the validity constraint. [6] addresses how to improve the schedulability test condition of DS-FP, a necessary and sufficient schedulability condition for DS-FP, along with a new schedulability test algorithm are presented. [8] investigates how to maintain the mutual temporal consistency of real-time data objects.

32.6 Conclusions

Temporal consistency maintenance of real-time data is an important problem in real-time database systems. We have addressed the period and deadline assignment problem from the perspective of EDF scheduling. Based on a sufficient schedulability test of EDF scheduling, we formulate the assignment problem as a LP problem and propose a novel efficient algorithm, $\mathcal{IML}_{\text{EDF}}$, to reduce the imposed workload while maintaining temporal validity of real-time data. Our experimental results verify the efficiency of $\mathcal{IML}_{\mathrm{EDF}}$.

References

1. Baruah S, Rosier L, Howell R (1990) Algorithms and complexity concerning the preemptive scheduling of periodic, real-time tasks on one processor. Real-Time Syst 2(4):301–324
2. Buttazzo G (2005) Hard real-time computing systems: predictable scheduling algorithms and applications. Springer, Heidelburg
3. Chantem T, Hu X, Lemmon M (2006) Generalized elastic scheduling. In: Proceedings of RTSS, pp 236–245
4. Gerber R, Hong S, Saksena M (1994) Guaranteeing end-to-end timing constraints by calibrating intermediate processes. In: Proceedings of RTSS, pp 192–203
5. Gustafsson T, Hansson J (2004) Dynamic on-demand updating of data in realtime database systems. In: Proceedings of ACM symposium on applied computing, New York, USA, pp 846–853
6. Han S, Chen D, Xiong M, Mok A (2008) A schedulability analysis of deferrable scheduling using patterns. In: Proceedings of ECRTS, pp 47–56
7. Ho S, Kuo T, Mok A (1997) Similarity-based load adjustment for real-time dataintensive applications. In: Proceedings of RTSS, pp 144–154
8. Jha A, Xiong M, Ramamritham K (2006) Mutual consistency in real-time databases. In: Proceedings of RTSS, pp 335–343
9. Kang K, Son S, Stankovic J, Abdelzaher T (2002) A QoS-sensitive approach for timeliness and freshness guarantees in real-time databases. In: Proceedings of ECRTS
10. Kuo T, Mok A (1993) SSP: a semantics-based protocol for real-time data access. In: Proceedings of RTSS, pp 76–86
11. Lam K, Xiong M, Liang B, Guo Y (2004) Statistical quality of service guarantee for temporal consistency of real-time data objects. In: Proceedings of RTSS
12. Locke D (1997) Real-time databases: real-world requirements. Kluwer international series in engineering and computer science, pp 83–92
13. Nyström D, Tešanovic A, Nolin M, Norström C, Hansson J (2004) COMET: a component-based real-time database for automotive systems. In: Proceedings of the workshop on software engineering for automotive systems
14. Ramamritham K (1993) Real-time databases. Distrib Parallel Databases 1(2):199–226
15. Ramamritham K (1996) Where do time constraints come from? Where do they go?. J Database Manag 7:4–11
16. Song X, Liu J (1995) Maintaining temporal consistency: pessimistic vs. optimistic concurrency control. IEEE Trans Knowl Data Eng 7(5):786–796
17. Xiong M, Han S, Lam K. (2005) A deferrable scheduling algorithm for real-time transactions maintaining data freshness. In: Proceedings of RTSS, pp 27–37
18. Xiong M, Ramamritham K (2004) Deriving deadlines and periods for real-time update transactions. IEEE Trans Comput 53(5):567–583
19. Xiong M, Wang Q, Ramamritham K (2008) On earliest deadline first scheduling for temporal consistency maintenance. Real-Time Syst 40(2):208–237

Chapter 33
Scheduling Performance of Real-Time Tasks on MapReduce Cluster

Fei Teng, Lei Yu and Frédéric Magoulès

Abstract This paper addresses the scheduling problem of real-time tasks on MapReduce cluster. Since MapReduce consists of two sequential stages, segmentation execution enables cluster scheduling to be more flexible. In this paper, we analyze how the segmentation between Map and Reduce influences cluster utilization. We find out the worst pattern for schedulable task set, based on which the upper bound of cluster utilization is deduced. This bound function, as a general expression of the classic Liu's result, can be used for on-line schedulability test in time complexity O(1). Theoretical results show proper segmentation between Map and Reduce is beneficial to MapReduce cluster with a higher utilization bound.

Keywords MapReduce · Real-time scheduling · Periodic task · Utilization bound

---

This work was done in part while Lei Yu was visiting Ecole Centrale Paris under the Top Academic Network for Developing Exchange and Mobility program

---

F. Teng (✉) · F. Magoulès
Ecole Centrale Paris, Chatenay-Malabry, France
e-mail: fei.teng@ecp.fr

F. Magoulès
e-mail: frederic.magoules@hotmail.com

L. Yu
Ecole Centrale de Pekin, Beihang University, Beijing, China
e-mail: yulei@buaa.edu.cn

J. J. Park et al. (eds.), *Proceedings of the International Conference on Human-centric Computing 2011 and Embedded and Multimedia Computing 2011*, Lecture Notes in Electrical Engineering 102,
DOI: 10.1007/978-94-007-2105-0_33,

33.1 Introduction

MapReduce is a popular framework to process distributable computation on a collective cluster [1]. MapReduce consists of two individual steps, inspired by Map and Reduce functions. Firstly, input data is chopped into smaller sub-problems, each of which runs a Map operation. After all Maps finish, the intermediate answers of all sub-problems are assemble and then reassigned to Reduces according to different keys generated by Maps. Reduces can only start after all Maps are completed, which illustrates a special feature, that is, sequential segmentation of task execution.

Dealing with different real-time tasks on a MapReduce cluster can benefit users from sharing a common large data set [2, 3]. However, the traditional scheduling schemes need to be revised in terms of particular characteristics such as segmentation and interdependence between Map and Reduce. It provides the primary motivation of our study on scheduling of real-time tasks on MapReduce cluster. In this paper, we shall assume the computation ability of cluster as a whole by hiding assignment detail of every Map/Reduce task in the interior of cluster.

This paper is organized as follows. Section 33.2 introduces the related research of real-time scheduling schemes. In Sect. 33.3, a MapReduce scheduling model and a less pessimistic utilization bound are presented. Next Sect. 33.4 discusses scheduling performance of our mathematical model. Finally, Sect. 33.5 concludes the paper.

33.2 Related Works

Task scheduling research is one of the hottest topics in the real-time system. Liu [4] proved rate monotonic (RM) algorithm as the optimal strategy for periodic tasks scheduling on uniprocessor. With RM algorithm, the lowest upper utilization bound is 0.69 when n approaches infinity. Although Liu's bound is only sufficient, not necessary, it is widely used in schedulability test. Because it is easily implemented and fast enough for on-line test. As long as the utilization of a given task set is beneath this bound, schedulability is guaranteed.

To raise system utilization bound, strict constrains of RM algorithm are relaxed by subsequent researchers. Shih [5] improved the upper bound approaching 0.84 by deferring the deadline slightly. Burchard [6] increased the processor utilization under the condition where all periods in the task set have values that are close to each other. Sha [7] studied a special case in which task periods are harmonic. The schedulability bound then reaches near one. Han [8] modified the task set with smaller but harmonic periods, and deduced a less pessimistic bound at the cost of higher computational complexity. Bini [9] proposed a new hyperbolic bound, so that the schedulability analysis has the same complexity as Liu's bound, but with better accuracy. In addition, Lehoczky [10] generalized deadline monotonic (DM) scheduling scheme as the optimal policy for systems with arbitrary deadline. Peng [11] proposed system hazard to check whether assigned tasks miss their deadlines. In the specific case where periodic tasks are executed on MapReduce cluster, the combination of sequential computing and parallel computing exerts new challenges on real-time schedulability tests. We shall make efforts on analysis of scheduling performance on MapReduce cluster.

33.3 The Generalized Bound on MapReduce Cluster

33.3.1 System Definition

Assume that there is a task set $\Gamma = (\tau_1, \tau_2, \ldots, \tau_n)$ including n independent periodic tasks on MapReduce cluster. Any task τ_i consists of a periodic sequence of requests. When a request arrives, a new instance is created. The interval between two successive instances is called period T_i. Without losing generality, we let $T_1 < T_2 < \cdots < T_n$.

In rate monotonic scheduling, task with higher request rate has higher priority, so the priority of task is in inverse proportion to its period. When a running task encounters a new request of a higher priority task, it has to hand over cluster to the new one. This behavior is called preemptive. When all tasks with high priorities are completed, the suspending task is reloaded. The overhead of preemptive and reload is negligible in our model.

MapReduce solves distributable problem using a large number of computers, collectively referred to as a cluster with certain computing capability. One task is partitioned into n_m Maps and n_r Reduces. The numbers of n_m and n_r depend on particular cases, varying from one task to another. All Maps performed in parallel will finish in a certain time M_i, which means total time required to complete n_m Map operations. Total time spent on n_r Reduces is reduce execution time R_i. For simplification, we assume Reduce R_i is in proportion to Map M_i, and $\alpha = R_i/M_i$ is introduced to express the ratio between the two operations. Here we simply let all tasks use the same α. The whole computation time for task τ_i is $C_i = M_i + R_i = M_i + \alpha M_i = \frac{1}{\alpha}R_i + R_i$.

One remarkable character of MapReduce is that any of n_r Reduce operations can not be submitted till n_m Map operations are all finished. In the following context, we use Map/Reduce to signify the whole executing process of Map/ Reduce operations.

As former assumption, request of each instance occurs when a new period begins, so the Map request is consistent with the request of the whole task. The time when Reduce request is submitted makes a huge impact on cluster utilization. If Reduce always executes as soon as Map finishes, two stages of Map and Reduce are continuous. Hence the task can be considered as a normal case without segmentation, the bound of which is the famous Liu's bound. If Reduce makes its request not in hurry, this tradeoff will be beneficial to cluster utilization by making better use of spare time. We introduce parameter $\beta = T_{R_i}/T_{M_i}$ to reveal the segmentation ratio. The same β is applied for all tasks in task set Γ. Clearly, $T_i = T_{M_i} + T_{R_i} = T_{M_i} + \beta T_{M_i} = \frac{1}{\beta} T_{R_i} + T_{R_i}$.

Utilization of τ_i is the ratio of computation time to its period $u_i = C_i/T_i$. System utilization U is the sum of utilization for all the tasks in task set, that is, $U = \sum_{i=1}^{n} u_i$.

33.3.2 Optimal Utilization Bound

In order to get the lowest utilization, we firstly find out the worst pattern for schedulable task set on MapReduce cluster. In this section, T_{M_i} and T_{R_i} are treated as relative deadlines for Map and Reduce, respectively. Map M_i instantiates at the beginning of a new period, and must be finished before T_{M_i}. At the moment T_{M_i}, Reduce R_i makes its request, and the execution lasts for T_{R_i} at most.

Lemma 1 *For a task set* $\Gamma = (\tau_1, \tau_2, \ldots, \tau_n)$ *with fixed priority assignment where* $T_n > T_{n-1} > \cdots > T_2 > T_1$*, if the relative deadline of Reduce is not longer than Map* ($\beta \le 1$)*, the worst pattern which allows all tasks to be scheduled is*

$$\begin{aligned} M_1 &= T_2 - T_1 \\ M_2 &= T_3 - T_2 \\ &\vdots \\ M_{n-1} &= \tfrac{1}{1+\beta} T_n - T_{n-1} \\ M_n &= (2+\alpha) T_1 - \frac{1+\alpha}{1+\beta} T_n \end{aligned}$$

Proof Suppose a task set fully utilizing MapReduce cluster, in which $M_1 = T_2 - \left\lfloor \frac{T_2}{T_1} \right\rfloor \cdot T_1 + \epsilon$. When $\epsilon > 0$, we reduce Map runtime M_1 with ϵ, that is $M_1 = T_2 - \left\lfloor \frac{T_2}{T_1} \right\rfloor \cdot T_1$. Meanwhile, M_2^a need to be improved to the amount of $M_2^a = M_2 + \epsilon$ to maintain the full processor utilization. Through this adjustment cluster utilization U^a is consequently smaller than original utilization U, because

$$U - U^a = \epsilon(1+\alpha)\left(\frac{1}{T_1} - \frac{1}{T_2}\right) > 0 \tag{33.1}$$

On the contrary, when $\epsilon < 0$, we let M_2 get longer to fully use the cluster as $M_2^b = M_2 + \left\lceil \frac{T_2}{T_1} \right\rceil \cdot \epsilon$. The corresponding utilization U^b decreases again, owing to

$$U - U^b = \epsilon(1+\alpha)\left(\frac{1}{T_1} - \left\lceil\frac{T_2}{T_1}\right\rceil\frac{1}{T_2}\right) > 0 \quad (33.2)$$

We therefore make a conclusion that cluster utilization could reach minimum, as long as ϵ approaches zero. When $\epsilon = 0$, the period T_1 enlarges $\left\lfloor\frac{T_2}{T_1}\right\rfloor$ times as $T_1^c = \left\lfloor\frac{T_2}{T_1}\right\rfloor \cdot T_1$. Compare new utilization U^c with U

$$U - U^c = (1+\alpha)\left(1 - 1/\left\lfloor\frac{T_2}{T_1}\right\rfloor\right)\left(\frac{T_2}{T_1}\right) \geq 0 \quad (33.3)$$

This revise further pulls down the cluster utilization, which gives us inspirations that closer periods degrade the system utilization. If we try to search for the worst pattern, the smallest value of $\left\lfloor\frac{T_2^c}{T_1^c}\right\rfloor = 1$ should be taken. To sum up, we have

$$M_1 = T_2 - T_1 \quad (33.4)$$

Using similar methods, we shall get more results

$$M_i = T_{i+1} - T_i, \qquad i = 2, 3, \ldots, n-2 \quad (33.5)$$

Next, we construct a new task set by keeping the same $T_1, T_2, \ldots T_{n-2}$ and halving the period $T_{n-1}^d = T_{n-1}/2$. In order to fully utilize cluster, Map execution time M_{n-1} is transferred from τ_{n-1} to τ_n, so $M_n^d = M_n + M_{n-1}$. A lower utilization U^d is achieved than old U.

$$U - U^d = M_{n-1}\frac{1}{T_{n-1}} - \frac{1}{T_n} > 0 \quad (33.6)$$

We then resort the task set according to the length of period assuring $T_n > T_{n-1} > \cdots > T_2 > T_1$. Because $T_{n-1}^d \leq \frac{1}{1+\beta} \cdot 2T_{n-1}^d = \frac{1}{1+\beta}T_{n-1} < \frac{1}{1+\beta}T_n$, another condition $\frac{1}{1+\beta}T_n > T_{n-1}$ appears where smaller utilization exists. So Map M_{n-1} is obtained

$$M_{n-1} = \frac{1}{1+\beta}T_n - T_{n-1} \quad (33.7)$$

Time left for Map execution M_n is

$$M_n = \frac{1}{1+\beta}T_n - \sum_{i=1}^{n-1} C_i - \sum_{i=1}^{n-1} M_i = (2+\alpha)T_1 - \frac{1+\alpha}{1+\beta}T_n \quad (33.8)$$

The above worst pattern stands for the most pessimistic situation where the least utilization can be calculated. Under the condition given by Lemma 1, we derive schedulable upper bound on MapReduce cluster.

Theroem 1 *For a task set* $\Gamma = (\tau_1, \tau_2, \ldots, \tau_n)$ *with fixed priority assignment where* $T_n > T_{n-1} > \cdots > T_2 > T_1$, *if the length of reduce is not longer than map* ($\beta \leq 1$), *the schedulable upper bound in terms of cluster utilization is* $U = (1+\alpha)\ \left\{n[(\frac{2+\alpha}{1+\beta})^{1/n} - 1] + \frac{\beta-\alpha}{1+\beta}\right\}$.

Proof To simplify the notation, we introduce parameters $\gamma_1, \gamma_2, \ldots, \gamma_n$

$$T_i = \gamma_i T_n, \qquad i = 1, 2, \ldots, n-1 \tag{33.9}$$

Computation time of n tasks is expressed as

$$\begin{aligned} C_i &= (1+\alpha)(\gamma_{i+1}T_n - \gamma_i T_n), \qquad i = 1, 2, \ldots, n-2 \\ C_{n-1} &= (1+\alpha)(\frac{1}{1+\beta}T_n - \gamma_{n-1}T_n) \\ C_n &= (1+\alpha)[(2+\alpha)\gamma_1 T_n - \frac{1+\alpha}{1+\beta}T_n] \end{aligned} \tag{33.10}$$

The cluster utilization U is

$$U = (1+\alpha)\left[\sum_{i=1}^{n-2}\frac{\gamma_{i+1}-\gamma_i}{\gamma_i} + \frac{\frac{1}{1+\beta} - \gamma_{n-1}}{\gamma_{n-1}} + (2+\alpha)\gamma_1 - \frac{1+\alpha}{1+\beta}\right] \tag{33.11}$$

In order to compute the minimum value of U, we set the first order partial derivative of function U with respect to variable γ_i to zero

$$\frac{\partial U}{\partial \gamma_i} = 0, \qquad i = 1, 2, \ldots, n-1 \tag{33.12}$$

For variable γ_i, we get the equation

$$\begin{aligned} \gamma_1^2 &= \frac{1}{2+\alpha}\gamma_2 \\ \gamma_i^2 &= \gamma_{i-1}\gamma_{i+1}, \qquad i = 2, \ldots, n-1 \end{aligned} \tag{33.13}$$

The general expression of γ_i is

$$\gamma_i = \frac{1}{2+\alpha}\left(\frac{2+\alpha}{1+\beta}\right)^{i/n}, \quad i = 1, 2, \ldots, n-1 \tag{33.14}$$

By substituting general value of γ_i into U, we obtain the least cluster utilization

$$U = (1+\alpha)\left\{n\left[\left(\frac{2+\alpha}{1+\beta}\right)^{1/n} - 1\right] + \frac{\beta-\alpha}{1+\beta}\right\} \tag{33.15}$$

Moreover, a symmetric utilization bound is easily deduced using similar method as Theorem 2. If the length of reduce is longer than map ($\beta > 1$), the schedulable upper bound of cluster utilization is

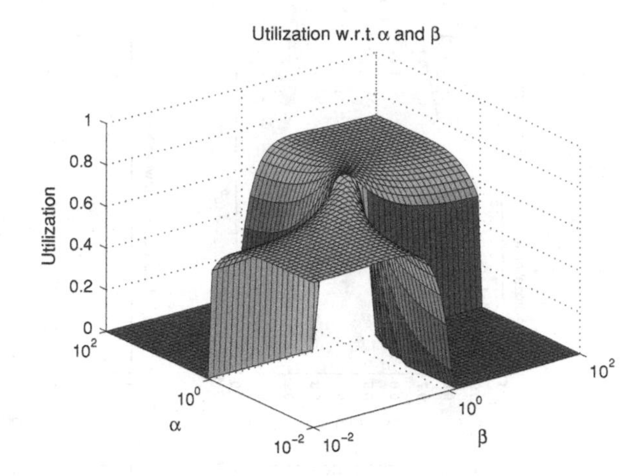

Fig. 33.1 Utilization bound

$$U = (1 + \frac{1}{\alpha})\left\{ n\left[\left(\frac{\beta + 2\alpha\beta}{\alpha + \alpha\beta} \right)^{1/n} - 1 \right] + \frac{\alpha - \beta}{\alpha + \alpha\beta} \right\} \tag{33.16}$$

On a real MapReduce cluster, numerous tasks are executed concurrently, so the number n is typically very large. Therefore, for all practical purposes, we are most interested in the cluster utilization as $n \to \infty$. When n is infinite, the limit of U is

$$U_\infty = \lim_{n \to \infty} U = \begin{cases} (1+\alpha)[\ln(\frac{2+\alpha}{1+\beta}) + \frac{\beta - \alpha}{1+\beta}] & \beta < 1 \\ (1+\frac{1}{\alpha})[\ln(\frac{\beta + 2\alpha\beta}{\alpha + \alpha\beta}) + \frac{\alpha - \beta}{\alpha + \alpha\beta}] & \beta \geq 1 \end{cases} \tag{33.17}$$

33.4 Results Analysis

Figure 33.1 outlines the fluctuation of utilization bound with respect to α and β, where α shows the proportion of execution time between Map and Reduce and β illustrates the ratio between two relative deadlines.

Seen from Fig. 33.1, the bound is a symmetrical plane on the axis $\alpha = \beta$. It implies that the value of α and β should be harmonious, that is to say, difference between α and β can not be too dramatic. Easily understood, if a long Map ($\beta < 1$) is just given a short relative deadline ($\alpha > 1$), it is impossible to schedule all the tasks in time before periods expire. That is why the cluster utilization dips to zero when assignment of the two variables goes to opposite directions.

We take cross sections of different α and β, in the Fig. 33.2. In three subfigures of row one, α is given $0.5, 1, 2$, respectively. The same values of β is fixed in row two. No matter the value of α or β is greater, less or equal to 1, the bound first ascends, and then descends gradually. Cluster utilization reaches the highest point at $\alpha = \beta$.

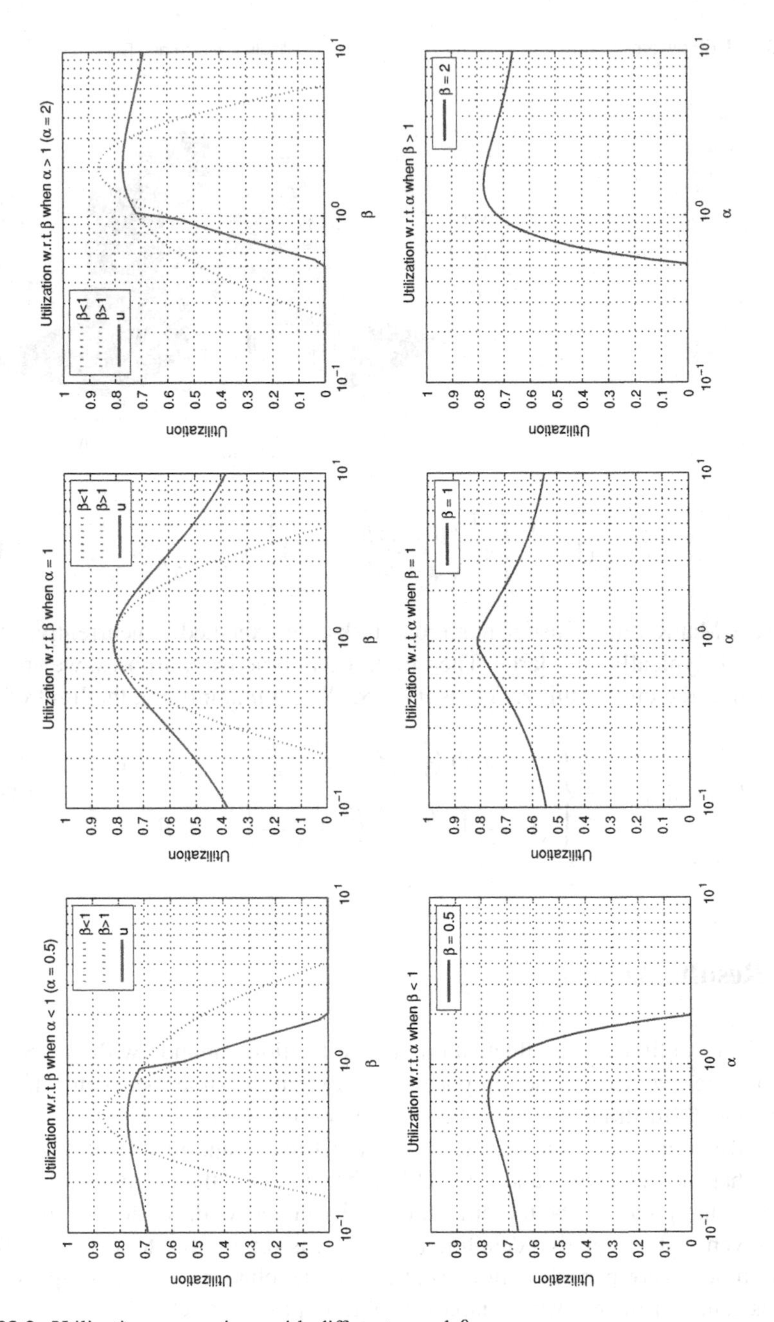

Fig. 33.2 Utilization comparison with different α and β

However, curves in the second row are no longer jumpy, but turns to be smooth. Because the pessimistic bound is piecewise function with respect to β, not to α.

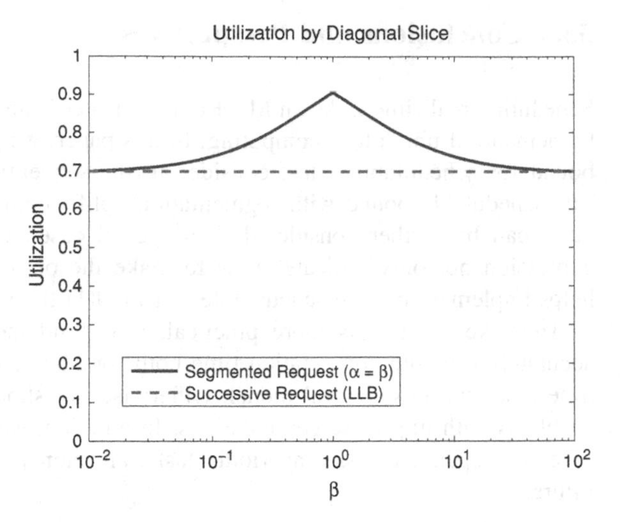

Fig. 33.3 Optimal utilization

We set the first order partial derivative of function U with respect to variable β to zero

$$\begin{aligned} &(1+\alpha)(\beta-\alpha)=0 && \beta\leq 1 \\ &(1+\tfrac{1}{\alpha})\left(\frac{\beta+1}{\beta}-\frac{\alpha+1}{\alpha}\right)=0 && \beta>1 \end{aligned} \tag{33.18}$$

Both equations are true under the same condition $\alpha=\beta$. It is fair and in accord with reality, that is, Map and Reduce share one period exactly according to their proportion. Figure 33.3 is drawn when $\alpha=\beta$. The bound uplifts steadily due to segmentation of Map and Reduce. When β approaches zero, the least cluster utilization is near 0.7. The amount of utilization rises as β goes up till $\beta=1$, peaking at 0.81, which is also a global maximum point. After that, cluster bound declines slowly to 0.7 again, when β increased to infinity. Notice that Liu's utilization bound $U=n(2^{\frac{1}{n}}-1)$ can be represented in a task set with $\beta=0$ or $\beta=\infty$. $\beta=0$ is treated as an extreme case where the time spend on Reduce is negligible, so Map execution time stands for the whole computation time. The case of $\beta=\infty$ implies that Reduce execution occupies the whole computation time. Therefore, our new bound can be considered as a general expression of Liu's bound.

Consequently, our result raises the original Liu's work. This augmentation comes from the flexibility of MapReduce itself. The execution of Map operations should be first promised, because Reduces need to collect all the output of Maps. However, the moment when Reduce makes a request influences over the final cluster utilization. If Reduce can wait in a reasonable period and can hand over cluster to more pressing task with lower priority, it is possible to achieve more dynamic allocation and higher utilization bound than the case in which no segmentation exists between Map and Reduce.

33.5 Conclusions and Perspectives

Scheduling real-time tasks on MapReduce cluster is an important issue developed by demand of ubiquitous computing. In this paper, we proposed a new utilization bound for schedulability test considering the sequential feature of MapReduce. The schedulable bound with segmentation uplifts compared to Liu's bound. The latter can be further considered as a special case of the former. This general expression not only indicates how to make the optimal segmentation, but also helps implement on-line schedulable test in O(1) time complexity.

To make the results more practical, it is fundamental to find a method to accurately estimate computation time both for Map and Reduce, since execution time is known in our ideal model. Otherwise, we should address the scheduling problems with imprecise computations. In addition, we shall extend our result to cases of dependent tasks, aperiodic tasks and non preemptive execution in the future.

Acknowledgments The authors acknowledge financial support from the Chinese Science Council. The second author also acknowledge partial financial support form the Chinese Universities Scientific Fund of BUAA (Grant No. YWF1003020).

References

1. Dean J, Ghemawat S (2008) Mapreduce: simplified data processing on large clusters. Commun ACM 51(1):107–113
2. Pan J, Le Biannic Y, Magoulés F (2010) Parallelizing multiple group-by query in sharenothing environment: a mapreduce study case. In: Proceedings of the 19th ACM international symposium on high performance distributed computing, ser. HPDC '10. ACM, New York, pp 856–863
3. Yu L, Magoulés F (2009) Service scheduling and rescheduling in an applications integration framework. Adv Eng Softw 40(9):941–946
4. Liu CL, Layland JW (1973) Scheduling algorithms for multiprogramming in a hard-realtime environment. J Assoc Comput Mach 20(1):46–61
5. Shih WK, Liu JW-S, Liu CL (1993) Modified rate-monotonic algorithm for scheduling periodic jobs with deferred deadlines. IEEE Trans Softw Eng 19(12):1171–1179
6. Burchard A, Liebeherr J, Oh Y, Son SH (1994) Assigning real-time tasks to homogeneous multiprocessor systems. University of Virginia, Charlottesville, VA, USA, Tech. Rep.
7. Sha L, Goodenough JB (1990) Real-time scheduling theory and ada. IEEE Comput 23(4): 53–62
8. Han C-CJ (1998) A better polynomial-time schedulability test for real-time multiframe tasks. In: IEEE real-time systems symposium, pp 104–113
9. Bini E, Buttazzo GC, Buttazzo G (2003) Rate monotonic analysis: the hyperbolic bound. IEEE Trans Comput 52(7):933–942
10. Lehoczky JP (1990) Fixed priority scheduling of periodic task sets with arbitrary deadlines. In: Proceedings of the 11th real-time systems symposium, pp 201–209
11. Peng D-T, Shin KG (1993) A new performance measure for scheduling independent realtime tasks. J Parallel Distrib Comput 19:11–26

Chapter 34
An Efficient Metadata Management Method in Large Distributed Storage Systems

Longbo Ran and Hai Jin

Abstract Efficient metadata management is a critical aspect of overall system performance in large distributed storage systems. Directory subtree partitioning and traditional hashing are two common techniques used for managing metadata in such systems, but both have some shortcomings, for example, in directory subtree partitioning method, the root node easily becomes a bottleneck, and for traditional hashing, a large directory hierarchy must be maintained and traversed. In this paper, we present a new method called directory-level-based metadata management that has high efficiency while avoiding the shortcomings of traditional methods.

Keywords Metadata management · Metadata server · Storage system

34.1 Introduction

In traditional file storage systems, metadata and data are normally managed by single machine and stored on single device [1], but in many large distributed storage systems the data and metadata management are separated and managed by metadata servers (MDS) cluster. Although the size of metadata is generally small compared to the overall storage capacity of a system, from 50 to 80% of all file system accesses are to metadata [2], so efficient metadata management in large distributed storage is very important.

There are two popular approaches on metadata management. The first is directory subtree partitioning, which partitions the namespace according to

L. Ran · H. Jin (✉)
Cluster and Grid Computing Lab, Services Computing Technology and System Laboratory, Huazhong University of Science and Technology, 430074 Wuhan, China
e-mail: hjin@hust.edu.cn

J. J. Park et al. (eds.), *Proceedings of the International Conference on Human-centric Computing 2011 and Embedded and Multimedia Computing 2011*, Lecture Notes in Electrical Engineering 102,
DOI: 10.1007/978-94-007-2105-0_34,

directory subtrees [3]. In directory subtree partitioning, the metadata is managed by individual metadata servers. In order to get any metadata from the directory tree, a lookup operation has to be executed from the root node to the right node. When the tree is big enough the root node will be a bottleneck because all other nodes are attached to the root node. The second is pure hashing, which uses hashing to widely and evenly distribute the namespace among the metadata servers [4]. Pure hashing assigns metadata to metadata servers based on the hash of file identifier, file name, or other related values, which results in more balanced workloads than directory subtree partitioning. However, a directory hierarchy must be maintained and traversed in order to provide standard hierarchical directory semantics. If the hash is based on the full pathname, it may be the only information the client has about the file, and a large amount of metadata might have to be moved when a directory name is changed. If the hash uses only the filename such as Lustre [5], files with the same name will hash into the same location, even if they are in different directories, which would lead to a bottleneck during large parallel accesses to different files with the same name in different directories.

We present directory-level-based hash (DLBH) metadata management, a new metadata management architecture, which is designed to provide high-performance, scalable metadata management.

DLBH is designed to meet the following general goals: (1) fast *ls* response; (2) fast builds standard hierarchical directory semantics; (3) fast rebuilds hierarchical directory when metadata is changed; (4) scalable; (5) flexible.

The rest of this paper is organized as follows. Section 34.2 provides some related works. In Sect. 34.3, the design and implementation of DLBH is presented. Performance evaluations of DLBH are given in Sect. 34.4. Finally, we provide a conclusion in Sect. 34.5.

34.2 Related Work

Directory subtree partitioning and hashing are two common methods in metadata management.

Some systems, such as NFS [6], AFS [2], Coda [7], Sprite [8], and Farsite [9], use static tree partition. They divide the namespace tree into several non-overlapped subtrees and assign them statically to multiple MDSes. This approach allows fast directory operations without causing any data migration. However, due to the lack of efficient mechanisms for load balancing, static tree partition usually leads to imbalanced workloads especially when access traffic becomes highly skewed [10].

To provide a good load balancing across metadata servers, many recent systems distribute files across servers based on the hash of some unique file identifier, such as Lustre, Vesta [4] and InterMezzo [11]. All hash the file pathname and/or some other unique identifier to determine the location of metadata and/or data. As long as such a mapping is well defined, the simple strategy has a number of advantages.

Clients can locate and contact the responsible MDS directly and, for average workloads and well-behaved hash functions, requests are evenly distributed across the cluster.

Lazy hybrid (LH) [12] provides a scalable metadata management mechanism based on pathname hashing with hierarchical directory management. LH avoids the bottleneck of directory subtree partitioning and pure hashing by combining their best aspects and by propagating expensive directory name and permission changes lazily and distributing the overhead of these potentially costly operations. LH provides high performance by avoiding hot spots in the metadata server cluster and minimizing the overhead of disk access and server communication. However, in doing so the locality benefits are lost while the system remains vulnerable to individually popular files. More importantly, the low update overhead essential to LH performance is predicated on the low prevalence of specific metadata operations, which may not hold for all workloads.

Dynamic sub-tree partition [13] utilizes a dynamic subtree partitioning strategy to distribute workload while maximizing overall scalability. It utilizes embedded inodes to exploit the locality of reference presented in both scientific computing and general-purpose workloads. Metadata is stored using a two-tiered strategy, initially writing updates to a log for fast commits to stable storage for quick recovery and cache warming, and later committing directory contents to an object storage pool. It leverages the dynamic metadata distribution and collaborative caching framework to avoid flash crowds by preemptively replicating popular metadata and distributing client requests to file system hot spots.

Ceph [14] maximizes the separation between data and metadata management by using a pseudo-random data distribution function (CRUSH) [15], which is derived from RUSH (Replication under scalable hashing) [16] and aims to support a scalable and decentralized placement of replicated data. This approach works at a smaller level of granularity than the static tree partition scheme and thus might cause the overheads of slower metadata lookup operations. When an MDS joins or leaves, all directories need to be re-computed to reconstruct the tree-based directory structure, potentially generating a very high overhead in large-scale file systems.

Bloom filter-based approaches provide probabilistic lookup. A bloom filter is a fast and space-efficient data structure to represent a set. Due to the high space efficiency and fast query response, bloom filters have been widely utilized in storage systems, such as Summary Cache [17], Globus-RLS [10] and HBA [18].

34.3 DLBH

There are some common facts for metadata management: (1) directories Conjunction. It means that whenever we are visiting a file, the other files or directories in the same directory have more possibilities to be visited; (2) *ls* operation is the most frequency directory operation.

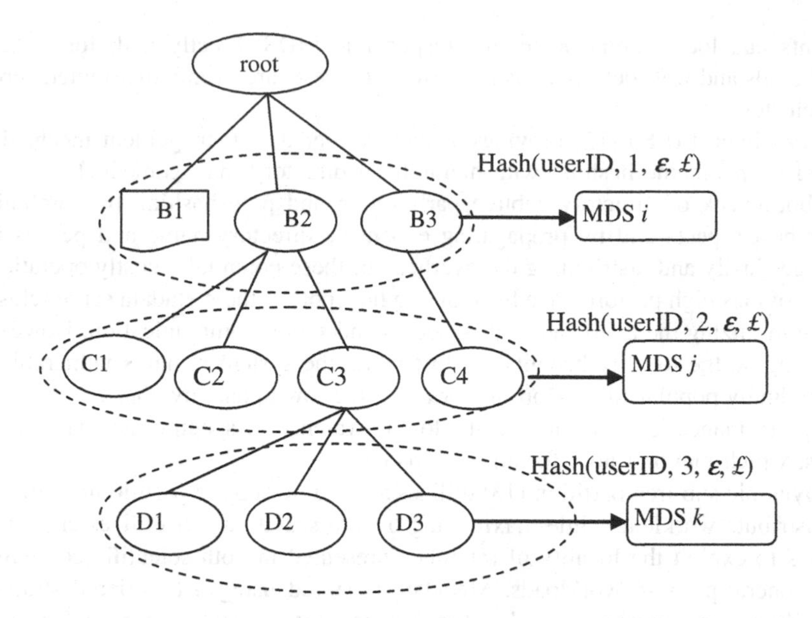

Fig. 34.1 DLBH schemes. It shows a directory tree, and the metadata is mapped to different MDSs by hashing

It is a good choice putting the metadata at the same level in a directory tree in an MDS. Unlike full path or file name hashing, we use directory level hash called DLBH.

Directory level shows the distance from the root to the directory. For example, file root\movie\classic\2009\a.mov, the level is 4.

The hash function in DLBH is described as:

$$\mathrm{Hash}(userID, d, \varepsilon, \pounds). \tag{1}$$

d is denoted as directory level. ε is denoted as metadata backup factor that control the number of the backups. £ is denoted as MDS scale size list and is initialized when (1) executes at the first time. *userID* and £ are stored in Global information server (GIS) that stores the user information, including userID, password, £ etc. The information will be downloaded when login.

Figure 34.1 shows a directory tree, and the metadata is mapped to different MDSs by hashing. It looks very simple but it will be very highly efficient.

34.3.1 Metadata on MDS

In order to store the metadata, there is 4D big table shown in Fig. 34.2.

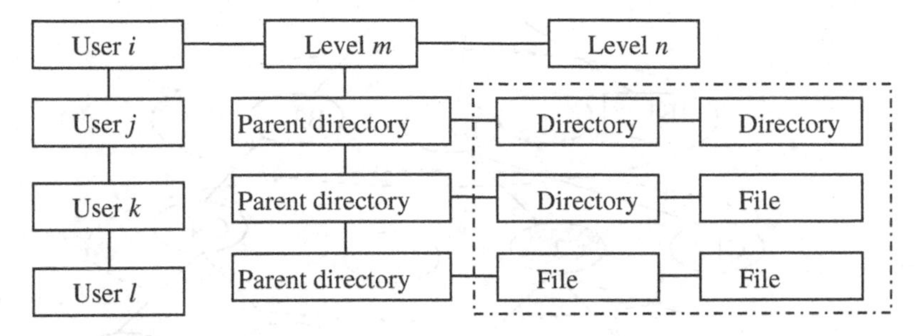

Fig. 34.2 Metadata structure on MDS. This shows a metadata view on metadata server. At first it is a user list, each user has a directory list, each directory level has a parent directory list, and then each parent has a directory or file list

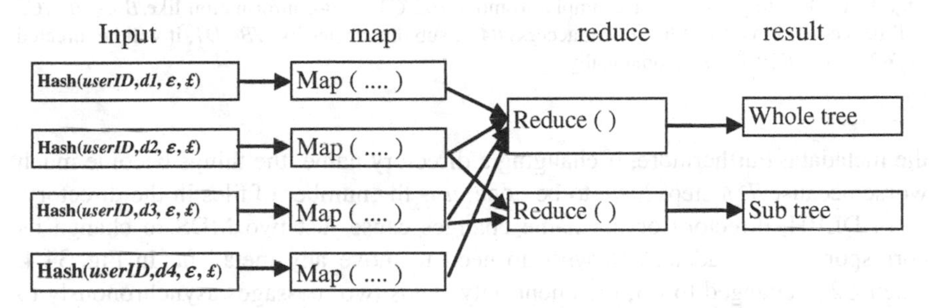

Fig. 34.3 Directory hierarchy reconstruction through Map/Reduce. This shows how to get the whole directory tree or sub directory tree by google Map/Reduce

34.3.2 Directory Hierarchy Reconstruction

In pure hash, no matter it is full path hash or filename hash, it is very difficult to reconstruct the completed directory tree, sometimes much redundant information is used to help to reconstruct the directory tree. But it is still not an easy thing and causes other problems such as data consistency.

In traditional hash schemes, in order to reconstruct directory the hierarchy of all the directories maintain the sub-directories location information. If the clients want to get the directory hierarchy recursive procedure needs to be executed from root to the current directory level.

In DLBH it is highly efficient to reconstruct directory hierarchy. It is very fast to complete directory tree though parallel algorithm instead of recursive procedure. Using Map/Reduce mode to describe the algorithm is shown in Fig. 34.3.

34.3.3 Directory or File Name Change

Directory or file name changes will cause metadata movement and burst out network communication. In lazy hybrid (LH), it will cause at least 7 steps to move

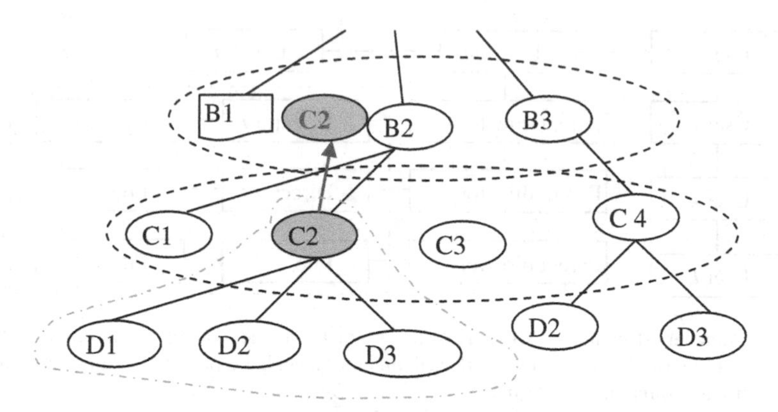

Fig. 34.4 Directory move. For example, to move */B2/C2* to */B4*, information like*/B4* →*/B2/C2* will be recorded in GIS. When client access */B4* or sub directories like */B4/D1*, it will be directed to */B2/C2* or */B2/C2/D1* automatically

the metadata. Furthermore, if changing a directory name, the things become much worse because 7**n* steps have to be done, *n* is the number of files in the directory.

In DLBH, directory or file name changes cause just two MDS to change its corresponding metadata, and with no need to move any metadata. In Fig. 34.4, when *C2* is changed to *C3*, the client only sends two messages asynchronously to finish the directory name change.

In details, if *C2* is stored on DMS *j*, and the directory level is *d*, the client sends two message to MDS *j* [Hash(*userID*, *i*, ε, £)] and top level MDS *k* [Hash(*userID*, *i-1*, ε, £)] to update the metadata.

34.3.4 Metadata Moving

It is very terrible to move a directory to another directory for the hashing scheme. It may cause all the sub directory and files to change the metadata.

In DLBH this kind of operation changes the level of the directories and files so it will cause large metadata movement. In order to reduce metadata movement the GIS records the directory moving information.

34.4 Load Balancing, Scalability, and Failover

34.4.1 Load Balancing

DLBH inherits the advantage of pure hash but overcomes its shortcoming, that is, the directory tree is hard to be reconstructed. But when the directories or files in

Table 34.1 Mapping information on GIS

Directory level	Hash scale
di, dn,dm	*£i*
dj	*£j*
dk	*£k*
dl	*£l*

the same level become too many a single MDS will become a bottleneck. So enhanced load balancing mechanism Virtual sub level (VSL) is needed.

DLBH provides the flexible scheme to add VSLs. If files or directories at some level *d* are too many, a sub level $d + 0.5$ will be added, the system sorts the files and directories by name, and finds the middle file name. The file metadata that bigger than middle file name will move to MDS Hash(*userID, d + 0.5*, ε, *£*).

In order to implement VSL, the VSL information needs to be stored in a big table on GIS. When the VSL information becomes large enough, it will cause the system to remove the metadata on MDS.

34.4.2 Scalability

In mass storage system, the number of MDSs will grow with the increase of data. If the scale of MDSs changes the hash function will usually be changed and thus cause large amount of metadata to be migrated, which is really a disaster for MDS.

DLBH provides a method called Multi-level Hash. When the scale of MDSes changes the DLBH makes the new levels and chooses a part of old levels to adapt the new scale. In order to support this function, the GIS records the mapping information of the directory level and the hash scale shown in Table 34.1.

For example when the scale changes from *£i* to *£j*, DLBH will use Hash(*userID*, *dj*, ε, *£j*) to locate the MDS which stores the new level *dj* files, but the old level will still use Hash(*userID*, *dj*, ε, *£i*). So, if the scale changes the old metadata will not be moved. Only if the MDS is overloaded will DLBH trigger the metadata moving.

34.4.3 Failover

In large scale storage systems, failover mechanism is strongly needed especially for the storage systems built up by low-cost servers. The basic method is to provide enough backups. In DLBH parameter ε define the basic backup number and the backups are dynamically added with the scale increase of MDSes. All the MDSes send heartbeat to GIS and the GIS will choose another MDS to replace it by giving it the same MDS id if some MDS fails.

34.5 Conclusions

We present DLBH metadata management, a scalable metadata management mechanism based on directory name hashing with hierarchical directory management. DLBH avoids the bottlenecks of directory subtree partitioning and mass metadata movement in pure hashing in case of the MDS scale changes or directory name changes. DLBH provides high performance in directory hierarchy reconstruction and high scalability by Multi-level Hash. DLBH also provides enhanced load balancing mechanism VSL to balance the workloads of MDSes. The preliminary results show high performance and scalability of DLBH.

References

1. Ousterhout JK, Costa HD, Harrison D, Kunze JA, Kupfer M and Thompson JG (1985) A trace-driven analysis of the Unix 4.2 BSD file system. In: Proceedings of the 10th ACM symposium on operating systems principles (SOSP'85), December, pp 15–24
2. Morris JH, Satyanarayanan M, Conner MH, Howard JH, Rosenthal DSH, Smith FD (1986) Andrew: a distributed personal computing environment. Commun ACM 29(3):184–201
3. Levy E, Silberschatz A (1990) Distributed file systems: concepts and examples. ACM Comput Surv 22(4)
4. Corbett PF, Feitelso DG (1996) The Vesta parallel file system. ACM Trans Comput Syst 14(3):225–264
5. Braam PJ Lustre whitepaper. http://wenku.baidu.com/view/72b0fb1cfad6195f312ba642.html
6. Pawlowski B, Juszczak C, Staubach P, Smith C, Lebel D and Hitz D (1994) Nfs version3: design and implementation. In: Proceedings of the summer 1994 USENIX technical conference, pp 137–151
7. Satyanarayanan M, Kistler JJ, Kumar P, Okasaki ME, Siegel EH, Steere DC (1990) Coda: a highly available file system for a distributed workstation environment. IEEE Trans Comput 39(4):447–459
8. Nelson MN, Welch BB, Ousterhout JK (1988) Caching in the sprite network file system. ACM Trans Comput Syst 6(1):134–154
9. Adya A, Bolosky W, Castro M, Cermak G, Chaiken R, Douceur J, Howell J, Lorch J, Theimer M, Wattenhofer RP (2002) Farsite: federated, available and reliable storage for an incompletely trusted environment. ACM SIGOPS Oper Syst Rev 36:1–14
10. Chervenak A, Palavalli N, Bharathi S, Kesselman C, Schwartzkopf R (2004) Performance and scalability of a replica location service. In: Proceedings of HPDC
11. Braam PJ, Nelson PA (1999) Removing bottlenecks in distributed file systems: Coda & intermezzo as examples. In: Proceedings of Linux expo
12. Brandt SA, Miller EL, Long DDE, Xue L (2003) Efficient metadata management in large distributed storage systems. In: Proceedings of MSST
13. Weil S, Pollack K, Brandt SA, Miller EL (2004) Dynamic metadata management for petabyte-scale file systems. In: Proceedings of the 2004 ACM/IEEE conference on supercomputing (SC'04)
14. Weil S, Brandt SA, Miller EL, Long DDE, Maltzahn C (2006) Ceph: a scalable, high-performance distributed file system. In: Proceedings of the 7th conference on operating systems design and implementation (OSDI'06)
15. Weil S, Brandt SA, Miller EL, Maltzahn C (2006) Crush: controlled, scalable, decentralized placement of replicated data. In: Proceedings of the 2006 ACM/IEEE conference on supercomputing (SC'06)

16. Honicky RJ, Miller EL (2004) Replication under scalable hashing: a family of algorithms for scalable decentralized data distribution. In: Proceedings of IEEE IPDPS, April 2004
17. Fan L, Cao P, Almeida J and Broder AZ (2000) Summary cache: a scalable wide area web cache sharing protocol. IEEE/ACM Trans Netw 8(3)
18. Zhu Y, Jiang H and Wang J (2004) Hierarchical Bloom filter arrays (HBA): a novel, scalable metadata management system for large cluster-based storage. In: Proceedings of IEEE cluster computing

Chapter 35
Study on Methods for Extracting Comprehensive Information Based on the Sequential Dynamic Weighted Fusion

Bo Sun, Wenqiang Luo and Jigang Li

Abstract In view of the optimal weighting fusion for mean of multi-sensor observation data which needs to obtain accurately observation variance of each sensor and the distribution of weights which has a significant effect on the results of fusion, a method of weighted fusion is proposed which estimates the variance in real time and allocates weights dynamically. In order to improve the accuracy of optimal dynamic weighted fusion, a new method of sequential optimal dynamic weighted fusion is proposed. Comparative analysis shows that the accuracy of sequential optimal dynamic weighted fusion is one times higher than that of the optimal dynamic weighted fusion. As an example, we studied the monitoring data of landslide with the method of sequential optimal dynamic weighted fusion, results shows that this method is not only feasible in theory but also more effective than the method of optimal dynamic weighted fusion in practical applications.

Keywords Multi-sensor data fusion · Weighting factors · Landslides

35.1 Introduction

In recent years, the technique of multi-sensor data fusion has become a hot issue in domestic and international. Currently, study on the technique has been expanded to engineering field besides the military field and these show great application value. Many scholars had made corresponding research from different angles [1–4]. The method of optimal weighted fusion is often used in engineering practice. Two

B. Sun (✉) · W. Luo · J. Li
School of Mathematics and Physics, China University of Geoscienes,
430074 Wuhan, China
e-mail: 531554158@qq.com

J. J. Park et al. (eds.), *Proceedings of the International Conference on Human-centric Computing 2011 and Embedded and Multimedia Computing 2011*, Lecture Notes in Electrical Engineering 102,
DOI: 10.1007/978-94-007-2105-0_35,

questions have been discovered in studying of the method. Firstly, we need to know observation variance of each sensor. However, observation variance of each sensor is difficult to obtain in practical applications. Secondly, the distribution of weights has a significant effect on the results of fusion. If weights are not distributed appropriately, the fusion precision and reliability may not be significantly improved. Therefore, a method of weighted fusion is proposed which estimates the variance in real time and allocates weights dynamically based on the above two problems. The proposed method can generate optimal fusion results with least mean square error directly from multi-sensor observation data, without requirement of any prior knowledge on system and observation noises. Further, in order to improve the accuracy of optimal dynamic weighted fusion, we propose the method of sequential optimal dynamic weighted fusion.

35.2 Paper Preparation Optimal Weighting Fusion for Mean of Multi-sensor Observation Data

We consider a multi-sensor system consisting of n sensors. The results of optimal weighted fusion of multi-sensor observation data is expressed by the following equation [5]:

$$X = \sum_{i=1}^{n} w_i x_i \tag{35.1}$$

$$\sum_{i=1}^{n} (0 \leq w_i \leq 1) \tag{35.2}$$

where i represents sensor index, X is the really value need to estimate, x_i is the observation value of the sensor i, $w_i$is the weights of the sensor i.

We assume that σ_i is the observation mean square error of the sensor i. The optimal weight distribution is expressed by the following equation [6]:

$$w_i = \frac{1/\sigma_i^2}{\sum_{i=1}^{n} 1/\sigma_i^2} \tag{35.3}$$

By (35.3), the overall variance and the fusion accuracy can be represented as

$$\begin{aligned} \sigma^2 &= \sum_{i=1}^{n} w_i^2 \sigma_i^2 \\ &= \frac{1}{\sum_{i=1}^{n} 1/\sigma_i^2} \end{aligned} \tag{35.4}$$

$$\sigma = \sqrt{\frac{1}{\sum_{i=1}^{n} 1/\sigma_i^2}} \tag{35.5}$$

where σ^2 is the overall variance of observation error of n sensors, σ denotes as fusion accuracy. Obviously, the smaller σ value is, the higher fusion accuracy is. In fact, the method is theoretically optimum and there are many defects in practical application.

35.3 Optimal Dynamic Weighting Fusion of Multi-sensor Observation Data

Equation (35.3) shows that the determination of weighting factors is related to the estimation of observation variance of each sensor. In order to obtain the optimal dynamic weighting factors, the observation variance of each sensor should be obtained. However, it is difficult to obtain the variance of each sensor in practical applications. Therefore, the method of dynamic weighted fusion is proposed which estimates the variance in real time and allocates weights dynamically based on the above problem [7].

Since the true value of the observation parameter is unknown, assume that the average value of n sensors is a reference value of the current observation parameter's actual value. Therefore, the deviation between sensor i observation value and the average value of n sensors is used as the deviation between sensor i observation value and the observation parameter true value. The observation variance of each sensor is calculated based on the above assumptions. After obtaining the observation variance of each sensor, we can calculate the optimal weighting factor of each sensor by (35.3). The steps involved can be summarized as follows:

$$\bar{x}(k) = \frac{1}{n}\sum_{i=1}^{n} x_i(k) \quad (k = 1, 2, \ldots, N) \tag{35.6}$$

$$\Delta x_i(k) = x_i(k) - \bar{x}(k) \quad (k = 1, 2, \ldots, N) \tag{35.7}$$

$$\Delta x_i = \frac{1}{N}\sum_{k=1}^{N} \Delta x_i(k) \quad (i = 1, 2, \ldots, n) \tag{35.8}$$

$$\bar{\sigma}_x(i) = \sqrt{\frac{\sum_{k=1}^{N} \left(\Delta x_i(k) - \Delta x_i\right)^2}{N-1}} \quad (i = 1, 2, \ldots, n) \tag{35.9}$$

$$\bar{w}_i = \frac{\frac{1}{\bar{\sigma}_x^2(i)}}{\sum_{i=1}^{n} \frac{1}{\bar{\sigma}_x^2(i)}} \quad (i = 1, 2, \ldots, n) \tag{35.10}$$

$$x(k) = \sum_{i=1}^{n} \bar{w}_i * x_i(k) \quad (k = 1, 2, \ldots, N) \tag{35.11}$$

where k represents time index, $\bar{x}(k)$ is the average value of n sensors in k moment, $\Delta x_i(k)$ is the deviation between sensor i observation value and average value in k

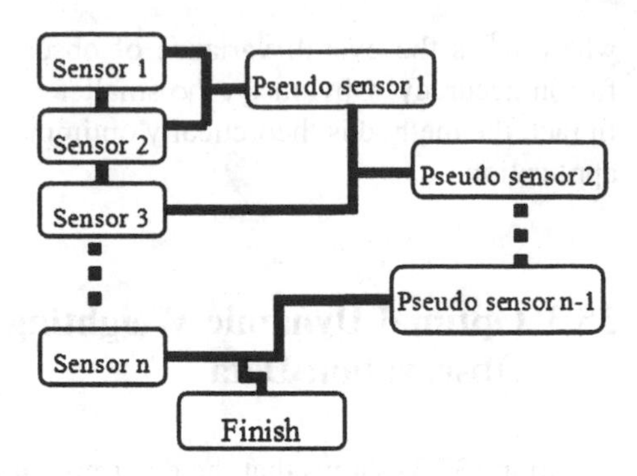

Fig. 35.1 The structure of sequential optimal dynamic weighted fusion

moment, Δx_i is the mean value of sensor i deviation, $\bar{\sigma}_x(i)$ is the standard deviation of sensor i deviation, $\bar{w}_i$ is the weight of sensor i, $x(k)$ is optimal estimate of the observation parameter.

35.4 Sequential Optimal Dynamic Weighting Fusion of Multi-sensor Observation Data

In order to improve the accuracy of optimal dynamic weighted fusion, the method of sequential optimal dynamic weighted fusion is proposed on the basis of studying of optimal dynamic weighting method. The basic idea of the method is to change the structure of the original fusion. Firstly, two of the sensors fuse with dynamic optimal weighted fusion and then the fused data is used as a pseudo sensor monitoring data. Secondly, the pseudo sensor and the next sensor fuse with optimal dynamic weighted fusion. According to the above method, fusion will stop until last sensor take part in fusion.

Since sensors have different sampling rate and the system have different transport delay in actual multi-sensor dynamic system, we often meet asynchronous condition. According to the basic idea of the sequential optimal dynamic weighted fusion, we will take a principle of the first come first fusion in practice to shorten the waiting time and improve fusion efficiency. Therefore, the new method not only inherits the advantage of the method of optimal dynamic weighted fusion, but also improves the efficiency of integration. Figure 35.1 is the structure of the sequential optimal dynamic weighted fusion.

35.5 The Example Analysis

The monitoring data of four monitoring station of a landslide is expressed by the following Table 35.1.

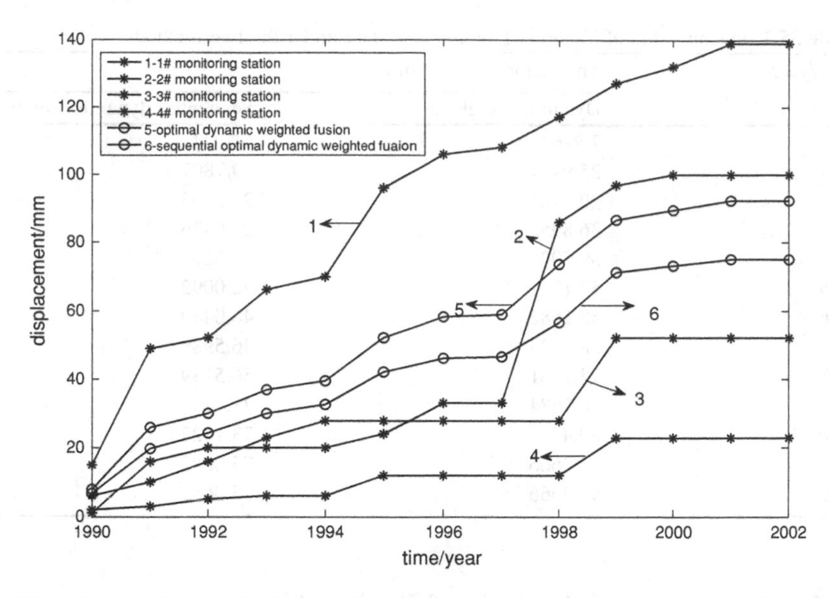

Fig. 35.2 The result of optimal dynamic weighted fusion and sequential dynamic weighted fusion

Table 35.1 The displacement values of the respective monitoring station

Time/ year	The displacement values (mm)			
	1# monitoring station	2# monitoring station	3# monitoring station	4# monitoring station
1990	15	1	6	2
1991	49	16	10	3
1992	52	20	16	5
1993	66	20	23	6
1994	70	20	28	6
1995	96	24	28	12
1996	106	33	28	12
1997	108	33	28	12
1998	117	86	28	12
1999	127	97	52	23
2000	132	100	52	23
2001	139	100	52	23
2002	139	100	52	23

Table 35.2 Fusion accuracy of two methods

Fusion method	Fusion accuracy value (mm)
Optimal dynamic weighted	183470
Sequential optimal dynamic weighted	72966

Table 35.3 Fusion value of Table 35.1 monitoring data under the two methods

Time/year	The fusion values (mm)	
	Dynamic weighted	Sequential dynamic weighted
1990	7.9164	6.7224
1991	25.9997	19.7807
1992	29.9787	24.2203
1993	36.8952	29.9426
1994	39.5028	32.6211
1995	52.1771	42.0092
1996	58.1682	46.0440
1997	58.9923	46.5989
1998	73.7161	56.5149
1999	86.7984	71.3153
2000	89.4822	73.1227
2001	92.3666	75.0653
2002	92.3666	75.0653

35.5.1 Compare and Analysis of Fusion Accuracy

The optimal dynamic weighted fusion and the sequential optimal dynamic weighted fusion are applied to fuse the data of Table 35.1. Fusion accuracy of two methods can be obtained by (35.5). The specific results are represented by the following Table 35.2.

Since the fusion data of every moment is used as a pseudo sensor monitoring data, the number of the sensor is virtually increased. The accuracy of fusion can be also improved by the sensor of poor accuracy when it enters fusion.

35.5.2 Compare of the Effects of Data Fusion

The optimal dynamic weighted fusion and the sequential optimal dynamic weighted fusion are applied to fuse the data of Table 35.1. Fusion values and the effects of two methods are represented by Table 35.3 and Fig. 35.2.

Table 35.3 data shows that the gap of fusion value of two methods has expanded unceasingly. The gap increased almost 17 times from 1990 to 2002. Because of applying sequential optimal dynamic weighting method virtually increases the number of sensors to enhance the fusion accuracy, so reliability of sequential optimal dynamic weighting is higher in practical applications. Figure 35.2 also shows that the displacement curve after fusion reflects the overall trend of landslide displacement. So the sequential optimal dynamic weighting method is not only feasible in theory but also effective in practical applications.

35.6 Conclusion

Firstly, the method of sequential optimal dynamic weighted can directly generate optimal fusion results with least mean square error and the fusion efficiency is higher. Secondly, we can see that the fusion accuracy of sequential dynamic weighted is higher through Table 35.2. Finally, a practical example of landslide explains the sequential dynamic weights weighting method is not only feasible in theory but also effective in practical applications.

Acknowledgments It is project supported by a grant from the National Basic Research Program of China (973 program) (2011CB710605).

References

1. Mirabadi A, Mort N, Schmid f (1996) Application of sensor fusion to railway systems. In: IEEE/SICE/RSJ International conference on multisensor fusion and integration systems. IEEE Press, Washington, pp 185–192
2. Ma XM (2008) Application of data fusion theory in coal gas fire prediction system. In: International conference on integration computation technology and automation. IEEE Press, Hunan, pp 572–575
3. Wijk O, Christensen HI (2000) Triangulation-based fusion of sonar data with application in robot pose tracking. IEEE Robot Autom Soc 16:740–752
4. Wang J, Zhang S, Zhang S, Bao F (2009) The application of multi-sensor data fusion technology in partial discharge detector. In: International conference on information engineering and computer science. IEEE Press, Wuhan, pp 1–4
5. Zhu P, Xu B, Xu B (2009) Research on Multisensor state fusion estimation based on weighted least squares. In: IEEE International symposium on intelligent information technology application. IEEE Press, Wuhan, pp 746–750
6. Ling L, Li Z, Chen C, Gu Y, Li C (2000) Optimal weight distribution principle used in the fusion of multi-sensor. Chin J Geol Hazard Control 8:36–39
7. Hao Y, Liu Z (2010) Fusion algorithm of angular velocity weighted averages for GFSINS based on dynamically allocating weights. J Chin Inert Technol 18:16–21

Part IX
EMC 2011 Session 2: Distributed Multimedia Systems

Chapter 36
A Scalable Multithreaded BGP Architecture for Next Generation Router

Kefei Wang, Lei Gao and Mingche Lai

Abstract To satisfy the requirement for the large-scale and high-speed Internet, the critical interdomain routing standard BGP on core routers must address the challenges of performance and scalability. The introduction of multicore and multithreading technique could effectively resolve the computing bottleneck in singlecore processor and the view consistency problem in cluster system, thus becoming an important approach to further improve the performance of core routers. In this paper, a multithreaded BGP architecture named Threaded BGP (TBGP) is proposed based on neighbor session division, achieving the parallel processing of different sessions on multiple threads by employing data parallelism. Experimental results show that TBGP yields good performance improvement compared to traditional BGP, where the average speedups of route learning time for TBGP under iBGP and eBGP sessions reach 1.79 and 2.73, and its peer switch time is decreased greatly by 45% on average with session number increasing to 100, 200 and 300. Meanwhile, TBGP scales better than BGP by improving the CPU efficiency so that it processes more routes when affording 300 peer sessions under iBGP and eBGP sessions.

Keywords Multithreaded · BGP · Performance · Scalability

36.1 Introduction

With the scaling of Internet, the routing software on core routers needs to compute a large number of routing reachable information, which increases non-linearly and has exceeded the abilities of computing and storage on current routers. And with

---

K. Wang · L. Gao (✉) · M. Lai
Department of Computer, National University of Defense Technology,
410073 Changsha, China
e-mail: gaolei@nudt.edu.cn

J. J. Park et al. (eds.), *Proceedings of the International Conference on Human-centric Computing 2011 and Embedded and Multimedia Computing 2011*, Lecture Notes in Electrical Engineering 102,
DOI: 10.1007/978-94-007-2105-0_36,

the interconnection denseness of the interdomain routing system, the increasing implementation complexity and policy diversity of routing software, the interdomain routing standard Border Gateway Protocol (BGP) has faced the performance and scalability challenges for applying on the next generation internet. Agarwal et al. [1] examined the BGP data from 196 Cisco routers in Sprint IP network, and the analysis result revealed that BGP processes contributed over 60% CPU load. The fast consumption of computing ability attributes to the routing dynamics [2], for example the core routers might receive hundreds of thousands of route updates per min. In this case, some update messages might be delayed or even dropped if not processed in good time. On the other hand, the increasing connection density of autonomous system (AS) has also introduced a great challenge for better scalability to satisfy the performance requirement by the incremental peer sessions. The black-box tests [3] on commercial routers had shown that they could not afford to maintain the BGP peer sessions even without extra message traffic on the control plane when session number went up to 250. In this context, the pass-through time was increased by several orders and thus BGP could not provide the sufficient efficiency to guarantee the quality of delay and loss-sensitive applications, such as streaming multimedia, gaming, and telecommuting/video conferencing applications. In addition, some other instability events, like BGP outburst update 4], route oscillations [5, 6], slow convergence [7, 8], configuration errors [9], malicious attacks [9], etc., which further deteriorate BGP efficiency, also present a higher performance demand on BGP.

The mainstream routers usually adopt the multi-processor structure of centralized control plane or the cluster one of the fully distributed control plane, but they still cannot satisfy the increasing requirements for performance and scalability. With the centralized control plane, the performance of routers is restricted by the single core processor, which could easily become the bottleneck of routers without sufficient computing ability for processing all kinds of routing protocols. And with the fully distributed control plane in cluster, routers could achieve high performance and be scaled up by the distributed processing of various routing protocols, but should maintain the consistent view for all nodes within the cluster, thus introducing much communication overhead among cluster nodes via network, degrading the factual efficiency of routing software on control plane.

Applying multicore technique to routers provides alternatives to address the above problems in current routers. Multicore processor is composed of multiple tile-like computing modules to support higher computing ability, larger communication width and faster synchronization operations, and it also exhibits better scalability with nature consistency by sharing the system memory among different cores. However, it's not sufficient only to use multicore in routers, where the routing software cannot make full use of the abundant computing resources. Thus, the parallel architecture exploration for routing protocols by multithreading becomes a critical issue to improve router performance on multicores.

In this paper, a scalable multi-threaded BGP architecture named TBGP is presented to meet the increasing performance and scalability demand for BGP. We develop the data parallelism in BGP and a neighbor-based data division method is employed in our architecture to empower multiple threads processing in parallel with the decoupled executions among different peer sessions. Experimental results show that compared with traditional BGP, the average speedups of route learning time under iBGP and eBGP sessions in TBGP reach 1.79 and 2.73 respectively, the peer switch time of TBGP is decreased greatly by 45% on average with session number increasing to 300. Meanwhile, TBGP scales better in comparison with BGP by affording 300 peer sessions under iBGP and eBGP sessions.

36.2 Related Works

Many technologies [10–18] on distributed processing and parallel computing have been put forward for improving the performance of transport protocols and routing ones. Bjorkman et al. [10] analyzed the parallelism of TCP/IP on the message level, connection level, protocol level and task level, and proposed a queuing network model to predict the performance when capturing the effects of lock and memory contentions. Xiao et al. [11] presented the parallel routing table computation approach for OSPF, which divided the OSPF area into several regions and calculated the routing table by different nodes in parallel. For BGP, Klockar et al. [12] proposed a distributed router based on modularized BGP, where multiple BGP processes worked in parallel and each node selected path according to its local route information, but the forwarding table was divided into several parts and updated by BGP proxy. Nguyen et al. [13] presented a distributed BGP architecture to increase the scalability and resiliency of routers, in which a set of BGP processes run on multiple control cards of a router and cooperate with each other to maintain the behavior of a single routing protocol process communicating with its peer in the network. Hamzeh et al. [14] proposed a distributed architecture for BGP with the way of master–slave task separation, and they mainly focused on the research of the consistency algorithm for the routing table replications of different tasks, to decrease the overhead of the communication latency and scalability to a large number of peer sessions. Zhang et al. [15] proposed a fully distributed BGP model by distributing route computation on multiple agents. It extended BGP and put forward CBGP for the synchronizations of routes and status information. A tree-based distributed model of BGP was devised in [16], which gave the corresponding algorithms for dispatching sub-tasks under two typical scalable router systems. And afterward, Xu et al. [17] considered the task balance on different nodes based on two kinds of iteration trees, decreasing the communication cost among nodes and increasing the computing efficiency by 91.8% in comparison with neighbor-based division scheme. The above studies have enhanced the BGP performance effectively, but they are constrained to apply to clusters or multi-processors, where long communication latency and high synchronization overhead by distributed memory always bottleneck the further performance improvement of BGP. Our previous work [18] tried to exploit the speculative parallelism in BGP, but the high complexity of functional verification and a great deal of overhead for lock operations also limited its speedups.

36.3 Architecture of TBGP

The router connecting to multiple peer neighbors is always exchanging information through BGP, which represents the decoupled characteristic that the messages from different neighbors can be processed in parallel for their weaker correlations and the ones from the same neighbor have to be processed orderly due to their close correlations. Thus, as the routers shift to multicore systems, we firstly propose the TBGP protocol to exploit the potential parallelism by dispatching multiple neighbor sessions on different parallelized threads, thereby improving the protocol efficiency.

In general, the TBGP is composed of one master thread and multiple slave threads just as shown in Fig. 36.1. The master thread has the responsibility of initializing process, creating slave threads, answering and evenly distributing peer sessions among different slave threads. After initialization of startup module, the event process module is running for scheduling all kinds of events to invoke other modules' work. The peer manager module monitors and answers the connection requests from neighbor sessions, dispatches a slave thread for the new session, and assigns the socket address to the specified slave thread. In order to balance the session overheads of different threads, the master thread adopts a round-robin dispatch scheme in its implementation. The route scanner function is performed in route detection module of master thread to detect the reachability of all the routes in BGP routing information base (RIB). In addition, master thread also processes all the interactive requests in the interaction module, providing a unified access interfaces with which users could configure protocol, lookup various state information, and so on.

Then, the slave thread is the actual execution entity for a cluster of sessions. For each slave thread, it is responsible for maintaining the finite state machine (FSM) operations, keeping session connectivity, processing update messages and managing protocol behaviors by itself. Similar with master thread, the event process module is used to push slave thread running, so that the multiple sessions triggered by events of different threads may work in parallel. In Fig. 36.1, with the session process module, each slave thread receives the routes from its corresponding neighbor peer sessions, and filters them according to the input/output policies from BGP RIB. When routes have been updated by local slave thread in route selection module, they also need to be propagated to other slave threads by route advertise module for announcing to all neighbor sessions, maintaining the behavior consistency with BGP. Then, the TBGP deploys a shared routing table to keep the consistent route view for all the threads, ensuring to make the correct route decisions and advertise the globally optimal routes. With multicore systems, the access to the shared routing table supports for the high-bandwidth communication and fast synchronization, thereby addressing the synchronization problem of

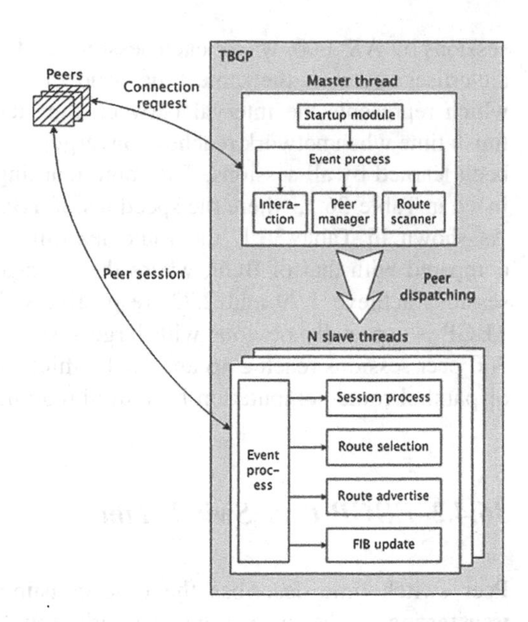

Fig. 36.1 Architecture of TBGP

distributed route storage [18] and providing a better aggregated performance. Besides, each slave thread writes updated routes into forwarding information base (FIB) by FIB update module just like in traditional BGP.

36.4 Experiments

Let us consider the TBGP implementation based on Quagga 0.99.9 [19] which provides with a high quality, multi-server routing engine and has interactive user interfaces for routing protocols. For Quagga software package, its bgp daemon is responsible for handling the projected TBGP protocol and interacting with the zebra daemon that is used for routing table management as well as route redistribution.

Through adjusting the number of threads, peer sessions and advertised routes, the performance of TBGP is further evaluated by Spirent AX4000 in terms of route learning time, eBGP peer switch time, and CPU usage. Then, we run the TBGP protocol with the prefix database of routing table snapshots extracted from route views [20], and compare its performance with traditional BGP. All the experiments are performed on Intel dual quad core Xeon server with Linux 2.6.18-8AX.

36.4.1 Route Learning Time

Under the configuration with four slave threads, we select the number of routes ranging from 250 thousands to 3 millions and establish a number of eBGP or iBGP sessions by AX4000, where each session sends data flow to the TBGP system and advertises routes at the same time. And then we capture the route learning time which represents the interval between the start time of packet sending and the finish time when network reaches convergence in which the advertised routes have been learned by all sessions. The route learning time of BGP and TBGP is partly listed in Table 36.1, where the speedups of TBGP beyond BGP are also displayed. As shown in Table 36.1, the route learning time of TBGP is greatly reduced compared with that of BGP, where the average speedups under iBGP and eBGP sessions achieve 1.79 and 2.73, respectively. The performance improvement of TBGP is especially obvious with large session numbers, and the speedups under 300 peer sessions reach 2.46 and 3.64, which attributes to the enhanced efficiency of parallel route computation by multithreading in TBGP.

36.4.2 EBGP Peer Switch Time

Peer switch time describes the time consuming by switching packets to new transferring paths when the network topology changes or routes become unreachable. We use two ports of AX4000 to advertise two sets of EBGP routes, each set of which has same route reachable information but different path attributes compared with the other set. By withdrawing one set of routes on which packets are transferred initially, data packets then have to be turned to the other set of routes as substitute paths for maintaining the continuous packet transmission. And then we attain the EBGP peer switch time by AX4000 after the system reaches convergence as shown in Fig. 36.2.

The statistical peer switch time of TBGP and BGP with larger session numbers are depicted in Fig. 36.2. When session number increasing to 100, 200 and 300, the peer switch time of TBGP is decreased greatly by 45% averagely in comparison with BGP, in which the connection of neighbor sessions cannot be maintained with the increasing route number. That's because the performance of BGP tends to be saturated when further increasing both the number of peer sessions and routes, resulting that BGP cannot process the keepalive packets when data packets are switched to new paths. And TBGP succeeds in completing all the tests and presents better computing performance and scalability to support more peer sessions with processing more routes.

36.4.3 CPU Usage

Finally, we employ the Top system command supported by Linux operating system to attain the CPU usages by TBGP process as well as BGP one, which is bound on a certain core of the multicore processor. The statistical CPU usages with the increased route numbers under the configurations of 25, 50, 200 and 300 peer

Table 36.1 Route learning time of TBGP and BGP (s)

Session type	Session number	Route number	BGP	TBGP	Speedup	Session type	Session Number	Route Number	BGP	TBGP	Speedup
IBGP	25	12.5	9	7	1.3	EBGP	25	25	18.5	11	1.68
		25	12	8.9	1.35			50	40	23	1.74
		50	24.5	16.7	1.47			75	65	35	1.86
		100	53	32.5	1.63			100	104	45	2.31
		200	110	61.5	1.79			200	236	104	2.27
	50	25	17.5	11.2	1.56		50	25	9.3	4	2.33
		50	35	22	1.59			50	57	21	2.71
		100	75	47.8	1.57			100	188	65	2.89
		200	150	84	1.79			200	390	145	2.69
	100	50	40	22	1.82		100	20	45	15.7	2.87
		100	81	40	2.03			30	72	24.2	2.98
	200	20	10.5	5.7	1.84			50	180	60.6	2.97
		100	50	26	1.92			100	400	123.8	3.23
		200	135	63.8	2.12		200	10	44	12.5	3.52
	300	30	23	9.8	2.35			20	310	88.5	3.5
		150	112	45.5	2.46		300	12	172	47.3	3.64

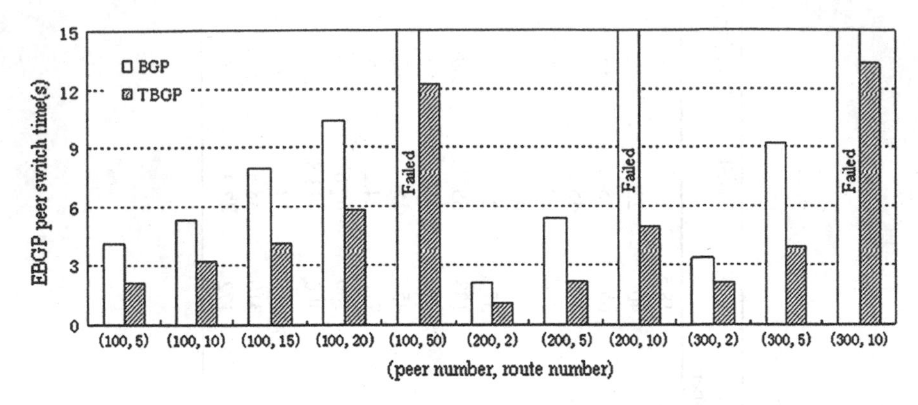

Fig. 36.2 EBGP peer switch time of TBGP and BGP

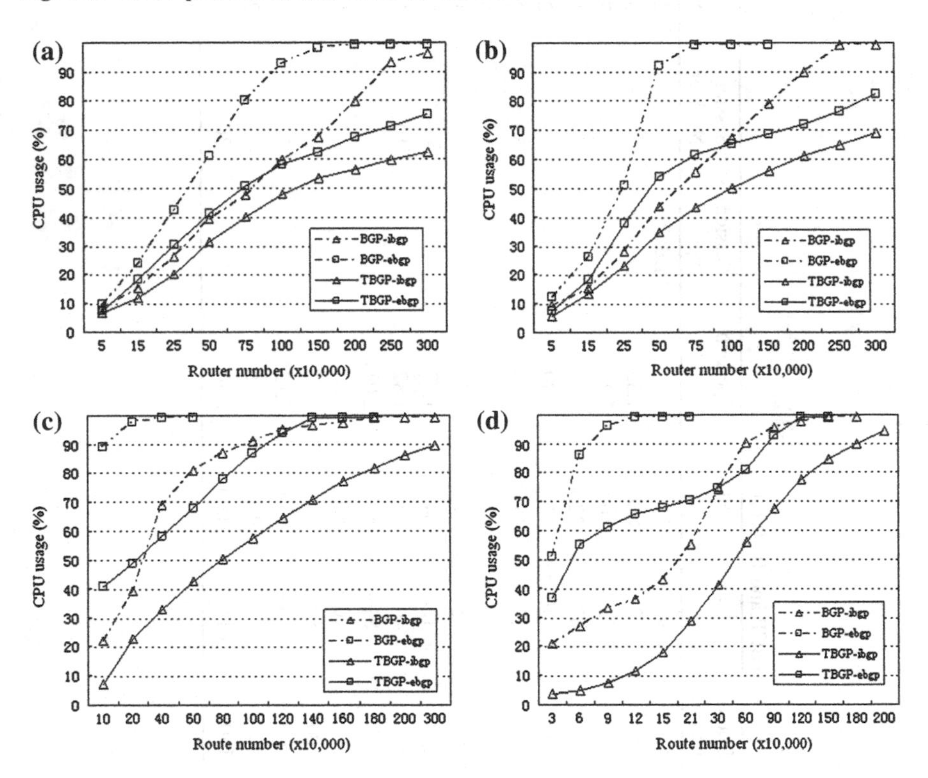

Fig. 36.3 CPU usages of TBGP and BGP under eBGP or iBGP sessions **a** peer number is 25 **b** peer number is 50 **c** peer number is 200 **d** peer number is 300

sessions are shown in Fig. 36.3. Under both type of iBGP sessions and eBGP ones, the CPU usages of BGP quickly reach full load in all tests, especially with larger peer session number, the number of routes processed by BGP is reduced sharply, e.g., BGP only computes 200 thousands and 120 thousands routes when configuring 200 and 300 peer sessions respectively. The reason for the decreased route processing efficiency is that BGP only makes use of limited computing resources with other cores idleness on multicore processor. And TBGP makes full use of the computing resources and its saturated rate for occupying CPU is relatively smaller. With 50 eBGP sessions, TBGP consumes 83% CPU resources for processing 3 million routes, and when connecting 300 peer sessions, it still could process 1.2 million routes at most, providing better scalability to suit the future network scaling.

36.5 Conclusions

A scalable multithreaded BGP architecture named as TBGP is proposed in this paper to address the challenges of BGP performance and scalability for applying on the next generation internet. The architecture of TBGP is composed of one master thread and multiple slave threads based on neighbor session division, achieving the parallel processing of different sessions on multiple threads by using data parallelism. We implement TBGP on Quagga, and the experimental results show that compared with traditional BGP the average speedups of route learning time under iBGP and eBGP sessions reach 1.79 and 2.73 respectively, the eBGP peer switch time of TBGP is decreased greatly by 45 percents averagely with session number increasing to 300. Meanwhile, TBGP scales better in comparison with BGP by affording 300 peer sessions under iBGP and eBGP sessions.

References

1. Agarwal S, Chuah CN et al (2004) Impact of bgp dynamics on router cpu utilization. In: Proceedings of passive and active measurement workshop, LNCS France, vol 3015, pp 278–288
2. Feldmann A, Maennel O, Mao ZM (2004) Locating internet routing instabilities. In: Proceedings of SIGCOMM, pp 205–218
3. Feldmann A, Kong HW, Maennel O, Tudor A (2004) Measuring bgp pass-through times. In: Passive active measurement workshop, vol 3015, pp 267–277
4. Labovitz C, Malan GR et al (1998) Internet routing instability. IEEE/ACM Trans Netw 6(5):515–558
5. Griffin T, Shepherd FB et al (2002) The stable paths problem and inter-domain routing. IEEE/ACM Trans Netw 10(1):232–243
6. Varadhan K, Govindan R, Estrin D (2000) Persistent route oscillations in inter-domain routing. Comput Netw 32(1):1–16
7. Labovitz C, Ahuja A, Bose A et al (2000) Delayed internet routing convergence [C], In: ACM SIGCOMM, vol 9, pp 293–306
8. Sun W, Mao ZM, Shin KG (2006) Differentiated bgp update processing for improved routing convergence. In: Proceedings of ICNP, pp 280–289
9. Karlin J, Forrest S, Rexford J (2006) Pretty good bgp: improving bgp by cautiously adopting routes. In: Proceedings of ICNP, pp 290–299

10. Bjorkman M, Gunningberg P (1998) Performance modeling of multi-processor implementations of protocols. IEEE/ACM Trans Netw 6(3):262–273
11. Xiao X, Ni LM (1998) Parallel routing table computation for scalable IP routers [C]. International workshop on network based parallel computing: communication, architecture, and applications
12. Klockar T, Hidell M et al (2005) Modularized bgp for decentralization in a distributed router, Winternet grand finale workshop, Sweden
13. Nguyen KK, Jaumard B (2009) A scalable and distributed architecture for BGP in next generation routers. ICC workshops
14. Hamzeh W, Hafid A (2009) A scalable cluster distributed BGP architecture for next generation routers. In: Proceedings of LCN
15. Zhang X, Zhu P, Lu X (2005) Fully-distributed and highly-parallelized implementation model of bgp4 based on clustered routers. In: ICN 2005, LNCS 3421:433–441
16. Wu K, Wu J, Xu K (2006) A tree-based distributed model for BGP route processing. In: Proceedings of HPCC LNCS. Springer, Berlin, vol 4208, pp 119–128
17. Xu K, He H (2008) BGP parallel computing model based on the iteration tree. J China Universities Posts Telecommun 15(Suppl):1–8
18. Gao L, Gong Z et al (2008) A TLP approach for BGP based on local speculation. Ser F Sci China 51(11):1772–1784
18. Qugga project. http://ww.quagga.net/download/
20. RouteViews project. http://archive.routeviews.org/bgpdata

Chapter 37
Meta Level Component-Based Framework for Distributed Computing Systems

Andy Shui-Yu Lai and Anthony Beaumont

Abstract Adaptability for distributed object-oriented enterprise frameworks in multimedia technology is a critical mission for system evolution. Today, building adaptive services is a complex task due to lack of adequate framework support in the distributed computing systems. In this paper, we propose a Metalevel Component-Based Framework which uses distributed computing design patterns as components to develop an adaptable pattern-oriented framework for distributed computing applications. We describe our approach of combining a meta-architecture with a pattern-oriented framework, resulting in an adaptable framework which provides a mechanism to facilitate system evolution. This approach resolves the problem of dynamic adaptation in the framework, which is encountered in most distributed multimedia applications. The proposed architecture of the pattern-oriented framework has the abilities to dynamically adapt new design patterns to address issues in the domain of distributed computing and they can be woven together to shape the framework in future.

Keywords Distributed computing · Meta architecture · Design patterns · Component-based framework

---

A. S.-Y. Lai (✉)
Department of Information and Communications Technology,
Institute of Vocational Education, Hong Kong, China
e-mail: andylai@vtc.edu.hk

A. Beaumont
Department of Computer Science, Aston University,
Aston Triangle, Birmingham, UK
e-mail: a.j.beaumont@aston.ac.uk

J. J. Park et al. (eds.), *Proceedings of the International Conference on Human-centric Computing 2011 and Embedded and Multimedia Computing 2011*, Lecture Notes in Electrical Engineering 102,
DOI: 10.1007/978-94-007-2105-0_37,

37.1 Introduction

Adaptability for distributed Object-Oriented (OO) enterprise frameworks in multimedia technology is a critical mission for system evolution. This paper presents an approach for constructing an adaptive OO design framework using distributed computing design patterns as components under a meta-level architecture, which we call Metalevel Component-Based Framework (MELC). MELC can be used for constructing adaptive distributed computing applications. It is based on modelling layers of architecture to meet the challenges of customization, reusability, and extendibility in multimedia and distributed computing technology. Design patterns and frameworks both facilitate reuse by capturing successful software development strategies. From the software engineering point of view, frameworks focus on reuse of concrete designs, algorithms, and implementations in a particular programming language. In contrast, design patterns focus on reuse of abstract design and software micro-architecture [1]. Frameworks can be viewed as a concrete realization of families of design patterns that target a particular application-domain [2]. Suzuki and Yamamoto pioneered the application of meta-architecture to design a web server which supports dynamic adaptation of flexible design decisions in the web server design space [3, 4]. Their framework is merely for web servers. However, it demonstrates the possibility of applying a meta-architecture to attack the problem of framework adaptation.

37.2 MELC: Adaptive Layer and its Kernel

We introduce an adaptive layer into pattern-oriented frameworks by adding a higher level of abstraction with a meta-architecture. The meta-level will be designed with a pattern-oriented approach by using system components that are instances of design patterns. The base level will contain the application components. The framework architecture separates system functional concerns. By adding a meta-architecture layer on top of the pattern-oriented framework, we will be able to provide an *adaptive level* in addition to the pattern level and class level, as shown in Fig. 37.1 [5, 6].

MELC supports dynamic adaptation of feasible design decisions in the framework design space by specifying and coordinating meta-objects that represent various functional components within the distributed environment. The framework has the adaptability to enable system evolution. This approach resolves the problem of how to allow for dynamic adaptation which is encountered in most distributed applications. The concept of using a meta-architecture to produce an adaptable pattern-oriented framework for distributed computing is new and has not so far been explored elsewhere.

A meta-architecture provides a mechanism for changing the structure and behavior of software systems dynamically. An application can be split into two parts: system behavior and business application. The meta-level provides

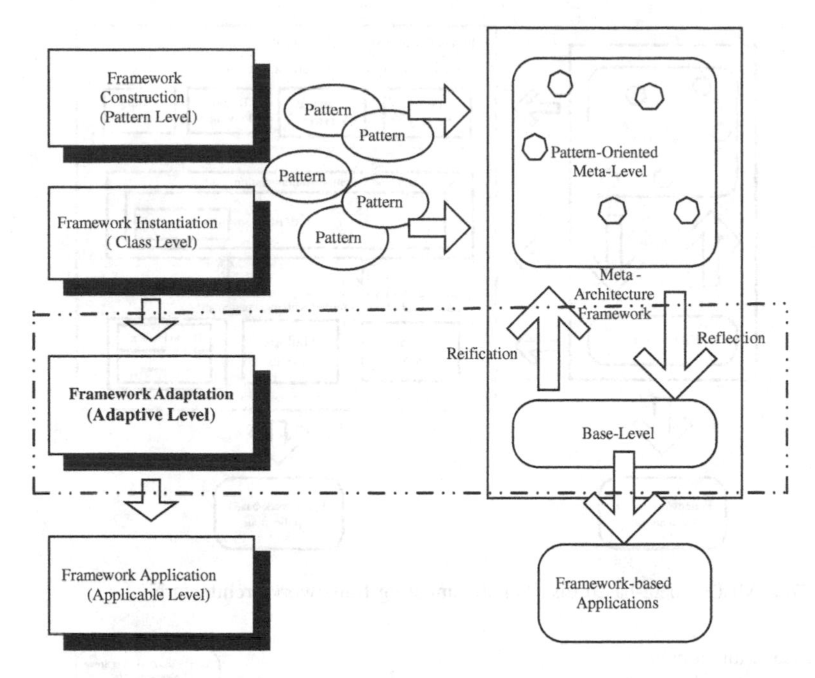

Fig. 37.1 Adaptive level introduced in meta-based pattern-oriented framework

information about selected system properties and behaviors, and the base level consists of application logic. The implementation of system functionality is built at the meta-level. Any changes in information kept at the meta-level affect subsequent base-level behavior. Figure 37.2 shows the framework architecture where the meta-level contains distributed computing patterns and the base level contains base objects, such as multimedia application servers.

The Kernel of MELC, shown in more detail in Fig. 37.3 provides the core functions for the adaptation between metalevel and base level. It includes: *Meta objects* and *Meta Space Management* which handles metalevel configuration, *Reflection Management* which provides dynamic reflection from the metalevel to the base level, and *Reification Management* which provides dynamic reification from the base level to the metalevel.

37.3 Implementation of Meta-Based Kernel in MELC Framework

37.3.1 Meta Objects and Meta Space Management

In MELC, meta objects are distributed computing patterns which are used to perform distributed computing operations. In Fig. 37.4, the class called `Meta-Space` acts as a director which manages a pattern repository to store the

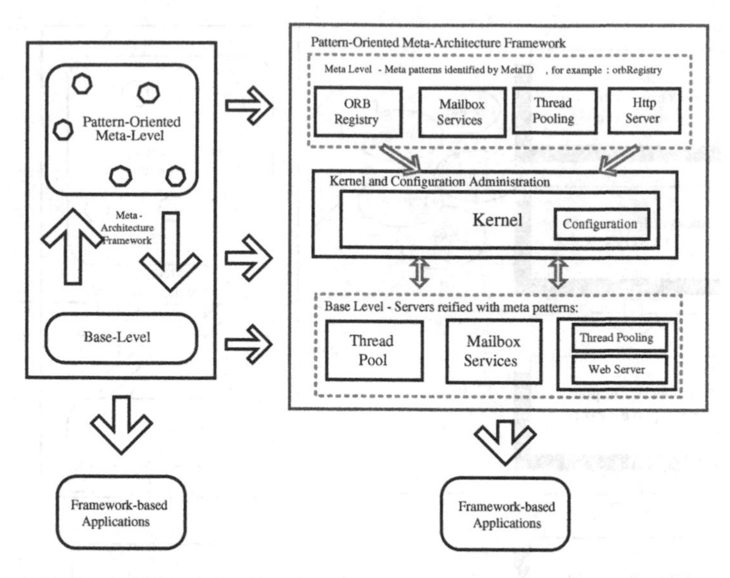

Fig. 37.2 MELC meta-based distributed computing framework architecture

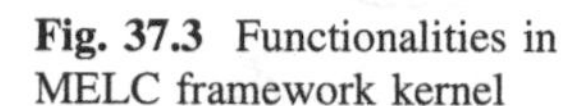

Fig. 37.3 Functionalities in MELC framework kernel

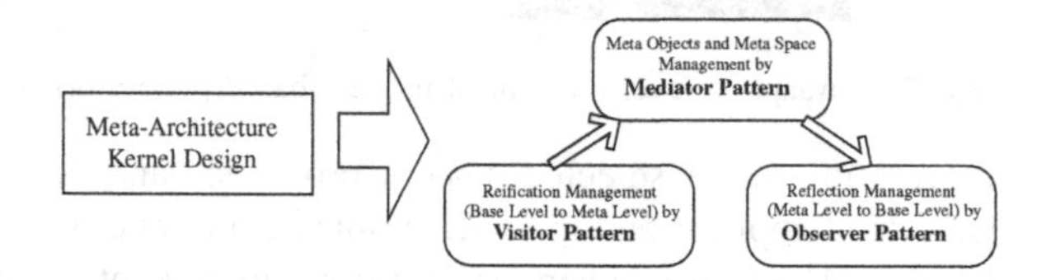

Fig. 37.4 Meta objects and meta space management with mediator design pattern

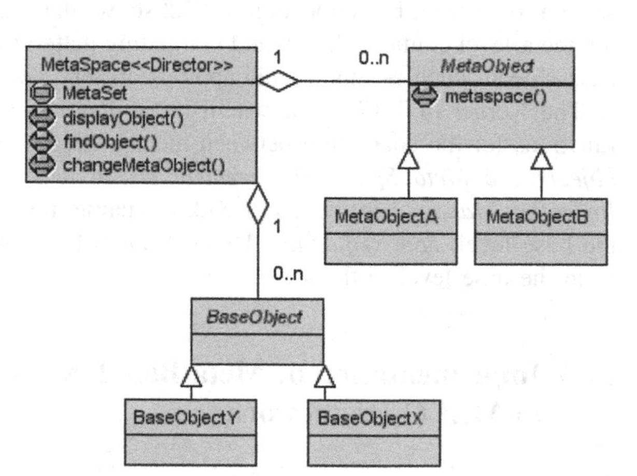

distributed computing patterns, and coordinates the interaction, reflection and reification between the metalevel and the base level. The mediator pattern is employed in such a way that the distributed computing patterns in the meta space can be manipulated by adding, removing and replacing meta objects.

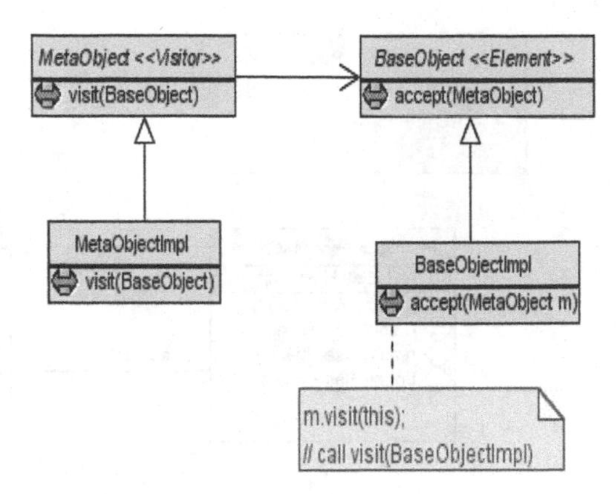

Fig. 37.5 Reification management in MELC framework with visitor design pattern

A mediator is responsible for controlling and coordinating the interactions of group objects. The MetaSpace class functions as a mediator and serves as an intermediary, promoting loose coupling by keeping meta objects in the group from referring to each other explicitly. Meta objects only know about the mediator, thereby reducing the number of interconnections at the metalevel. The mediator lets the administrators vary the interaction independently.

37.3.2 Reification Management in MELC

Reification is the process of making meta objects accessible to the base-level. The meta objects are the operations and represent the functionality provided in a distributed computing environment. They can be added, changed or removed from the metalevel. The manipulation of meta objects has an immediate effect on the behavior of base objects. We use the visitor design pattern to implement reification between the metalevel and the base level.

With the visitor pattern, we can apply the two class hierarchies: one for the elements being operated on (base level)—Element Classes Hierarchy, and one for the visitors that define operations on the elements (metalevel)—Visitor Classes Hierarchy. The visitor pattern shown in Fig. 37.5 represents a distributed computing operation to be performed on the elements at the base level. Visitor lets you define a new distributed component without changing the base classes of the elements on which it operates. We can create a new operation by adding a new subclass to the visitor class hierarchy. In other words, we can add new distributed computing functionality simply by defining new visitor subclasses (Meta Object Implementation).

37.3.3 Reflection Management in MELC

The key objects in the observer pattern shown in Fig. 37.6 are *Subject* and *Observer*. A subject is a description in reflection and observers are the

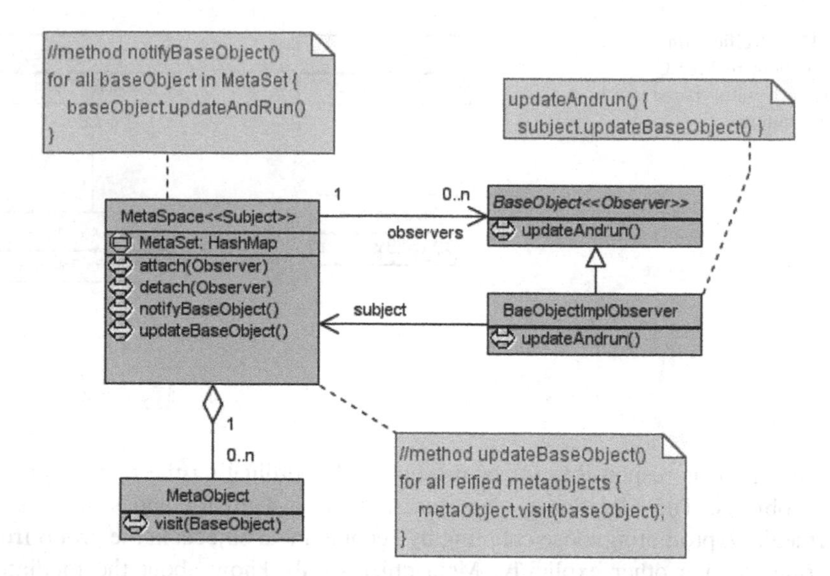

Fig. 37.6 Reflection management in MELC framework with observer design pattern

described objects in reflection. A subject may have any number of dependent observers.

In the reflective system of our meta-architecture, the observers are the base objects and the subject is the meta space director. All observers (base objects) are notified whenever the subject (meta space) undergoes a change in state. MetaSpace can be considered as a change manager.

37.4 Implementation of Meta Components

37.4.1 Installation of Meta Components

The management and deployment of meta components in the meta space are simple and uniform. In Fig. 37.7, we use Thread Pool design pattern as an example to illustrate how a distributed computing design pattern can be easily conformed to a meta components in MELC meta space. Kernel classes are the core meta objects. The kernel classes shown in Fig. 37.8 manipulate the interaction between meta objects and are generalized for conformation from distributed computing patterns to meta objects.

On the class level, distributed computing pattern components simply extend an abstract class called `MetaObjectImpl` from the kernel package to form meta components, and, on the object level, the instantiation of meta components means that they are ready to deploy to meta space.

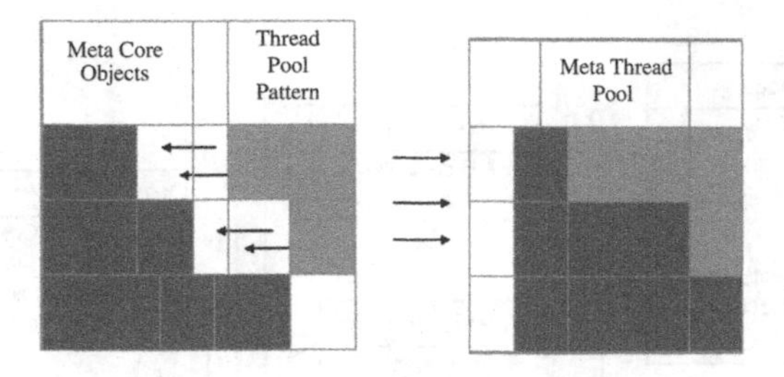

Fig. 37.7 Meta component installation at meta level

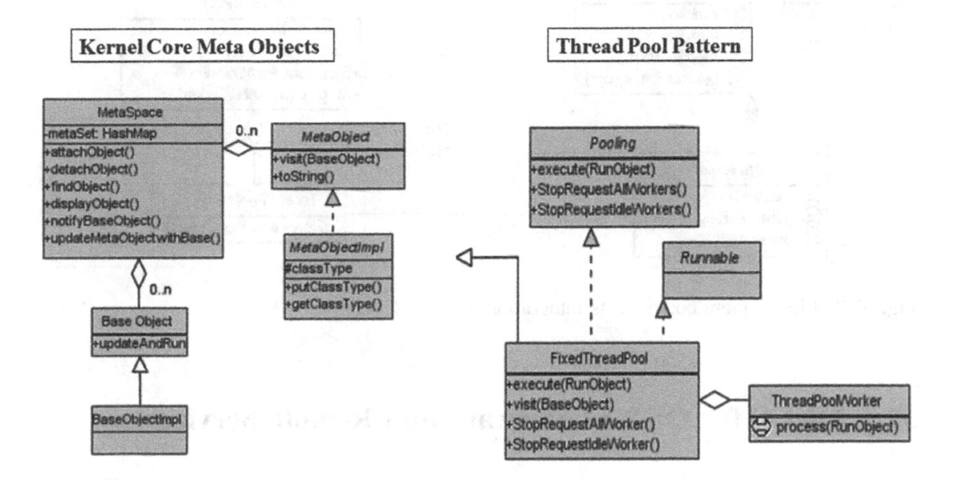

Fig. 37.8 MELC meta components installation at meta level

37.4.2 Integration of Meta Components

MELC supports dynamic integration of meta components at run-time. For example, in a multimedia bookshop application, the Thread Pooling Pattern and Http Server Pattern are two meta components and they are deployed to the meta space and that situation is illustrated in Fig. 37.9.

It shows that the integration between meta components, `ThreadPool` and `HttpServer`, at the metalevel, which is used to provide a *Thread Pooling HTTP Server* to the base level. Every time a HTTP request comes to the base object, the meta space will handle the request by checking whether the base object is reified with the `ThreadPool` at the metalevel. If the `ThreadPool` found is reified with the base object, `MetaSpace` will let `HTTPServer` pass its `HTTPWorker` to the thread pool workers to process.

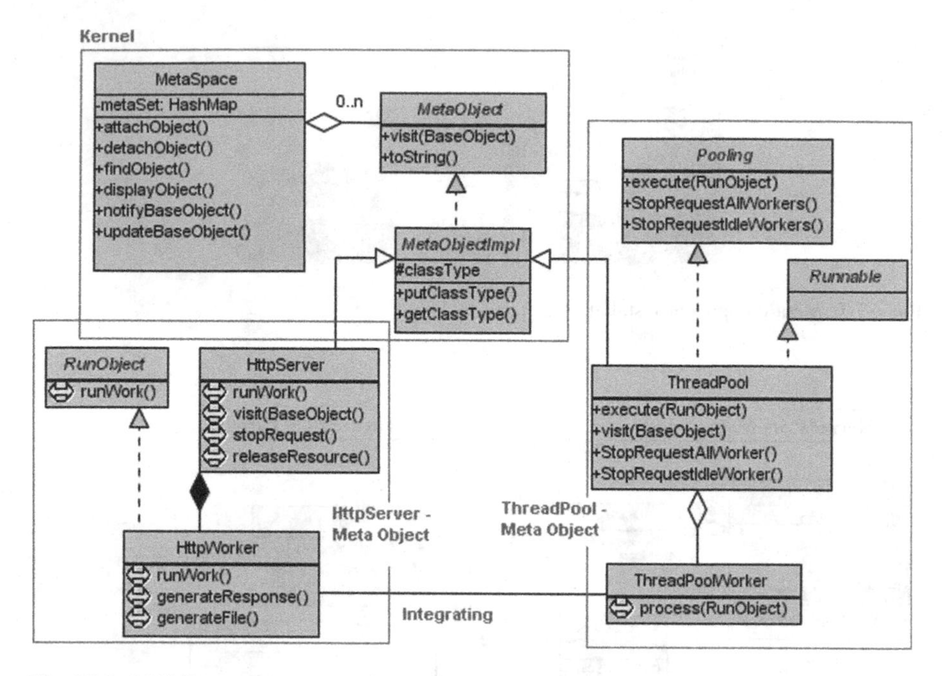

Fig. 37.9 MELC meta components integration at meta level

37.5 MELC for ORB Middleware and Remote Services

At the heart of distributed object computing are Object Request Brokers (ORBs), which automate many tedious and error-prone distributed programming tasks [5]. Fig. 37.10 presents the class diagram of the meta level ORB component which is used for remote operation invocation [2].

The `TCPConnection` class is responsible for the transport of messages between the environment of a remote caller and the environment of the callee. The ORB proxy of the base object E-Bookshop listens for requests. The role of ORB proxy checks that the calling meta object in caller's request has reified and started at the base level and also that the required remote business objects in MELC have been installed ready for access. ORB proxy at the base level then passes the requests to `ORBRegister` at the meta level.

The ORB proxy at the base level acts as an ORB server that receives ORB requests from the remote multimedia bookshop clients. The sequence diagram for the process is illustrated in Fig. 37.11. Note the invocation `Operation` object of ORB, is embedded within the target object. It uses the `ORBDispatcher` to dispatch the `Operation` to the relevant meta object for processing.

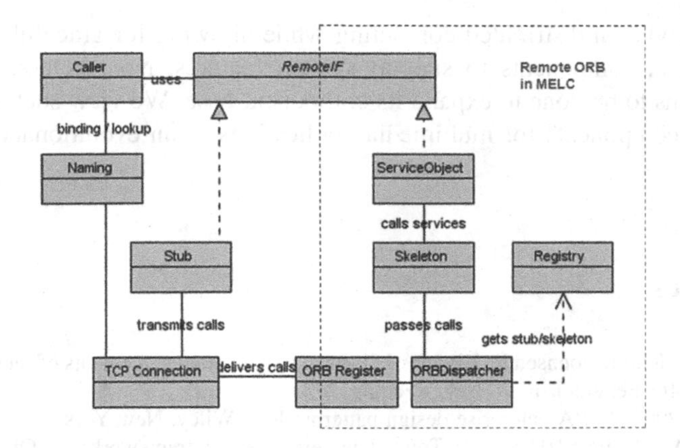

Fig. 37.10 Meta level component ORB in MELC

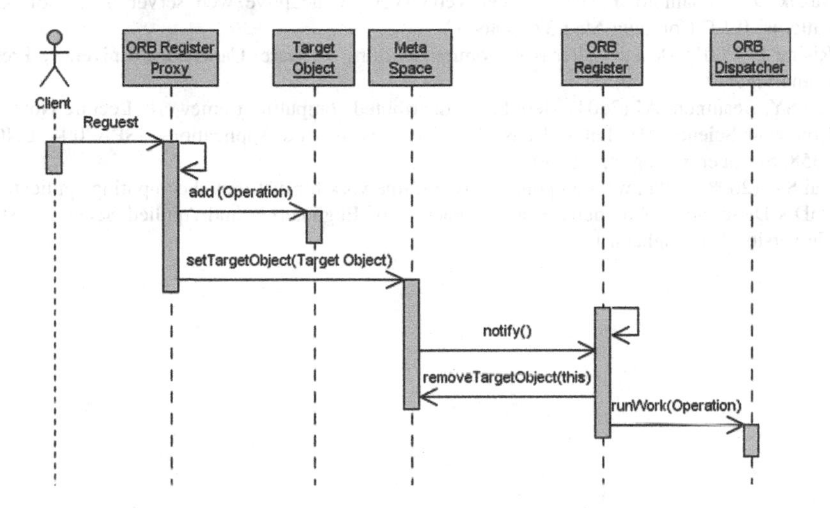

Fig. 37.11 Sequence diagram for the ORB Register to remove target object in MELC repository

37.6 Conclusion

We have presented a new approach to the design of a framework for distributed business and multimedia applications. We have described a MELC based on meta-architecture employing distributed computing patterns as components for meta objects. This paper describes how MELC can meet diverse requirements in distributed computing systems, and how its meta-architecture creates the kind of adaptive and configurable run time system that is critical for most distributed computing frameworks in multimedia technology today. We believe that MELC makes a significant contribution towards flexible architectural interfaces for distributed computing systems. In particular, the notion of a modularity preserving meta-architecture affords the protection of current abstraction boundaries of system behavior in distributed computing while allowing for graceful integration between meta components to support system features. Nonetheless, significant work remains to be done to expand its component base. We view such distributed computing components for multimedia applications as an evolutionary extension of MELC.

References

1. Gamma E, Helm R, Johnson R, Vissides J (1995) Design patterns: elements of reusable object-oriented software. Addison-Welsey, Reading, MA
2. Grand M (2002) JAVA enterprise design patterns. John Wiley, New York
3. Yacoub SM, Ammar HH (2000) Toward pattern-oriented frameworks. J. Object-Oriented Program 12(8):25–35
4. Suzuki J, Yamamoto Y (1999) OpenWebServer: an adaptive web server using software patterns. IEEE Commun Mag 37(4):46–52
5. Rising L (2001) Design patterns in communications software. Cambridge University Press, Cambridge
6. Lai SY, Beaumont AJ (2004) Meta-based distributed computing framework. Lecture Notes of Computer Science, Parallel and Distributed Processing and Applications, ISPA2004, LNCS 3358, Springer Verlag, pp 85–90
7. Lai SY. (2008) Meta level component-based framework for distributed computing application. PhD's Dissertation, Computer Science, Faculty of Engineering and Applied Science, Aston University, Birmingham, UK

Part X
EMC 2011 Session 3: Embedded Systems, Software and Applications

Chapter 38
Design of Remote Engine Room Monitoring System Based on Niche Stack TCP/IP

Sanjun Liu, Huihua Zhou and Chao Li

Abstract A distributed remote engine room monitoring system with multi-processors is designed based on SOPC technology in this paper. In the system, an embedded Web server is developed with the help of Niche Stack TCP/IP stack on a DE2 development board, while more than one DE0 development boards are used to monitor by integrating μC/OS-II. The Web server and the monitoring module communicate with each other by ZigBee. There are three schemes to protect network security. It is demonstrated that the system is stable, reliable, and flexible, and also convenient for upgrading and maintenance.

Keywords SOPC · Niche Stack TCP/IP · μC/OS-II · ZigBee · Embedded system

38.1 Introduction

This paper proposes a system that uses SOPC technology [1], sensor technology, ZigBee [2, 3] wireless communication technology, multi-CPU technology [4], modern controlling technology and so on, to implement the main function of remote monitoring by embedding μC/OS-II [5, 6] and Niche Stack [7, 8] TCP/IP Stack on DE0 and DE2 [9] (short names of series of development boards

S. Liu (✉) · H. Zhou · C. Li
School of Information Engineering, Hubei University for Nationalities,
Hubei 445000 Enshii, China
e-mail: liusanjunbox1@126.com

H. Zhou
e-mail: mhzheng3i@gmail.com

C. Li
e-mail: springxun@163.com

J. J. Park et al. (eds.), *Proceedings of the International Conference on Human-centric Computing 2011 and Embedded and Multimedia Computing 2011*, Lecture Notes in Electrical Engineering 102,
DOI: 10.1007/978-94-007-2105-0_38,

developed by Altera and Terasic for SOPC, such as DE0, DE2, and DE4, etc.) developing platform.

In the system, NicheStack TCP/IP stack, μC/OS-II and zip file system [10] are used to create the Web server on DE2. More than one DE0 are used as monitoring modules which are integrated with varies sensors and controlling equipments flexible to be added or removed to meet custom's requirements. The real-time data can be passed from DE0 to DE2 by ZigBee wireless module, in turn, DE2 can control every DE0, enabling the customers to monitor engine room by internet anywhere away.

Here, the term "engine room" consists of four meanings: first, it refers to common living homes, the second refers to general public facilities, and the third refers to substations that are used to supply power, especially, the worldwide known "IoT" [11] (Internet of Things) accords with the concept. There are two Nios II microprocessors implemented on DE0, one of which is used to monitor, and the other is used as a sound player. The sound wave files are stored in flash memory via zip file system [10].

Some of the IP Cores [12] (also known as hardware components) together with their software drivers are developed by ourselves with SOPC Builder component editor [13]. There are three schemes to protect network security, and the system can be also controlled by touch screen offline.

38.2 Hardware Design

Considering the system may be mounted with many sensors, may process real-time video and audio data, may be flexible for adding and removing instruments, need real-time operating system and NicheStack TCP/IP Stack to implement Web server, need an extra processor to be used as sound player, etc., so the system is done by using large capacity FPGA-Cyclone II (mounted on DE2) and high performance microprocessor-Nios II of Altera corporation based on the SOPC technology. For advanced information on SOPC technology and Nios II microprocessor associated with their development tools, such as Quartus II, IDE, and SOPC Builder, refer [13, 14].

38.2.1 Hardware Schematic of System

The system hardware is mainly composed of four parts which are DE2 module, DE0 module, ZigBee, and sensors module, as are shown in Fig. 38.1.

The DE2 module which is integrated with DM9000A [15] is mainly used to fetch audio and video data and to implement Web server. The DE0 module is used to process data gained by sensors, and then to transmit the data to DE2 by ZigBee. Accordingly, DE0 responses as the commands obtained from DE2. Sensors

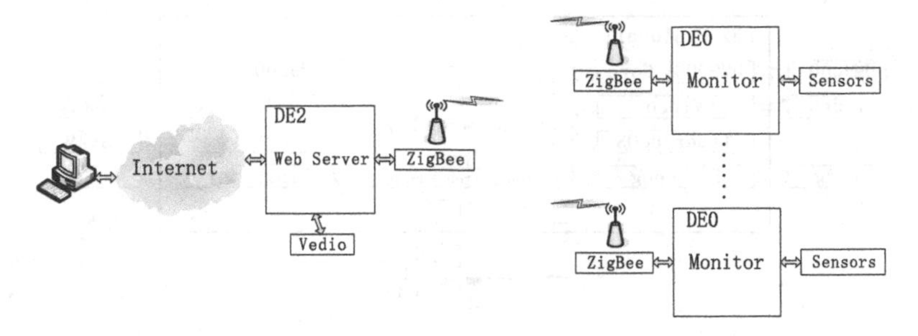

Fig. 38.1 Hardware schematic of the whole system

module includes many kinds of sensors which refer to those not only commonly used but also unusually or specially used. On DE0, there is also a second Nios II microprocessor that is used as sound player.

Why do we make the system distributed, rather than integrate the functions of Web server and monitoring together on one development board [16]? In fact, most of the embedded systems for remote monitoring work in this way. The possible reasons may be as follows:

1. *Save IP address resources*: Distributed system occupies only one IP address on Web server module, which is less than the system in which every monitoring board occupies an IP address that is very limited now.
2. *Reduce system cost*: Both monitoring module and Web server module require a lot of RAM, ROM and resources of FPGA. If we integrate the two module together in one FPGA, advanced types must be used, so the system cost will be high. If the two functions are implemented separately in different FPGAs, common types are enough to meet the requirements [17].
3. *Reduce the influence of environment*: Monitoring modules may need to work in harsh environment, but for Web server modules it is not so. What the Web server modules need is stability. If the two kinds of modules are implemented respectively, Web server modules may not work in harsh environment, and can avoid the influence of environment and other modules

38.2.2 Hardware Structure of DE2 and DE0

The hardware structure of DE2 and DE0 are respectively shown in Figs. 38.2 and 38.3. It is clearly seen that there is one Nios II processor on DE2, and there are two Nios II processors on DE0, every processor is connected with some IP cores [12]. Some of the important IP cores are detailed as follows.

In Fig. 38.2, PLL refers to phase lock loop [17], it is used to generate 100 M Hz clock from system crystal of 50 MHz, supplying to Nios II processor which must be of high speed because of complexity of the Web server routines. The timers

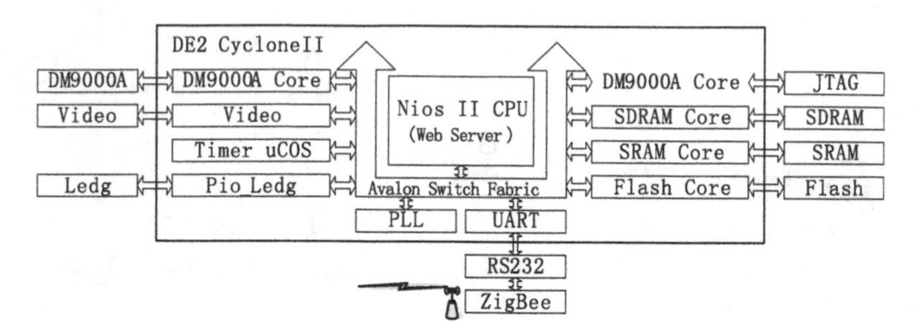

Fig. 38.2 Schematic of system on DE2 development board

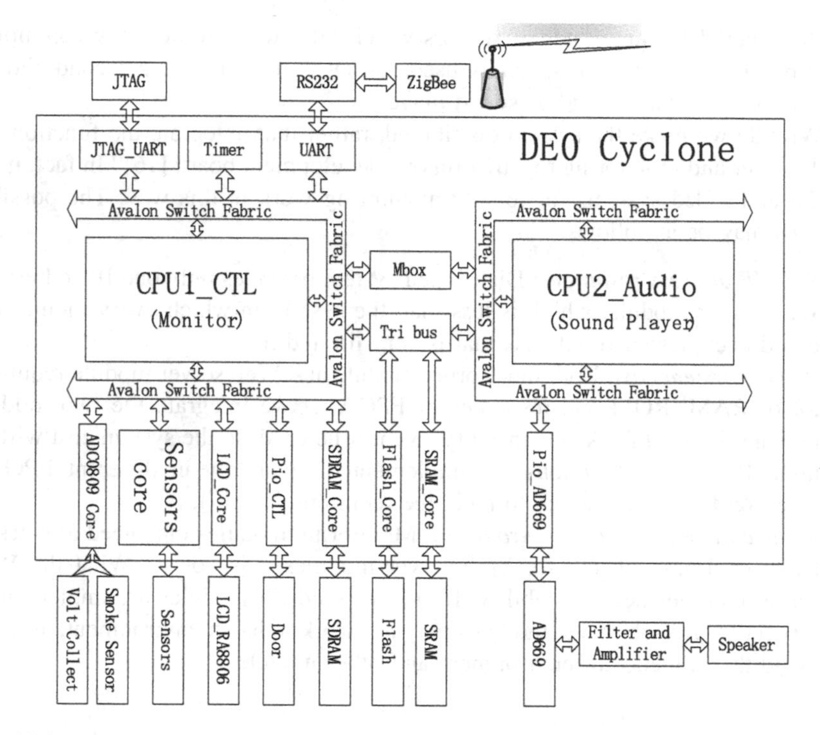

Fig. 38.3 Schematic of system on DE0 development board

[17] in both Figs. 38.2 and 38.3 are used to generate periodic interrupt for operation system μC/OS-II.

SRAM, SDRAM and Flash are system memories, when the system is powered up, programs and data (including the format of .jpg, .gif, .wav, etc, that are needed by Web server) will be copied automatically from Flash to SRAM or SDRAM by a routine called Boot_Loader [17]. ZigBee module has an interface of RS232, so there are UART Cores [13, 17] in the system, enabling to communicate between processors.

The DM9000A Core represent the hardware driver of the chip DM9000A [15], aiming to connect its data bus, address bus, RD, WR and chip select wires to Avalon bus of Nios II processor. The DM9000A [15] is a fully integrated and cost-effective low pin count single chip Fast Ethernet controller. We can implement embedded Web server based on NicheStack TCP/IP Stack with DM9000A's hardware IP Core [12] and its software drivers [14] both of which are developed by ourselves.

In Fig. 38.3, the processor on named "CPU1_CTL" aims to monitor and control sensors, and to communicate with the processor on DE2 by ZigBee, etc. The second processor named "CPU2_Audio" is used to play audio, just like a sound card.

The two processors communicate with each other by mailbox core [13, 17] Mbox. Once the CPU1_CTL receives the information needed to play a certain audio by CPU2_Audio, it will send a message to CPU2_Audio by Mbox. The CPU2_Audio always keeps waiting for the messages, once it receives a message, it will play the corresponding audio whose number is found from the message. For more details on mailbox core and how to create multi-processor system, refer [4, 13].

38.2.3 Custom Built IP Cores of the System

An SOPC Builder IP Core [12] is a hardware design block available in SOPC Builder component library that can be instantiated in an SOPC Builder system. One of the most significant features of SOPC technology is users can create and edit any component freely with SOPC Builder. If your components are created and integrated into the SOPC Builder component library, you can reuse them easily by just clicking the mouse and share them with others [12, 17]. Nowadays, there are a lot of corporations and companies that develop IP Cores for sale in the world, such as Altera, Actel,etc. [17].

There are 4 custom built IP Cores [12] in the system, which are DM9000A Core, ADC0809 Core, temperature-humidity sensor Core and LCD_RA8806 Core. The ADC0809 Core is developed by containing buses (such as data bus and address bus) and wires (such as RD, WR and CS) that are interfaced with Avalon bus of CPU1_CTL. Sensors Cores represent not only the common used sensors' cores (such as temperature core, humidity core, and so on), but also those cores that are unusually used (such as biosensor[18], water level sensor and so on). For details on custom built IP Cores, refer [12].

38.3 Software Design of the System

38.3.1 Software Design of DE0

On DE0 development board, the program of CPU1_CTL is implemented by μC/OS-II [5], while the program of CPU2_Audio is implemented by a single-threaded super-loop routine [14] without operating system. There are 4 tasks and an interrupt service routine in DE0's program, which are T1_SensorTest,

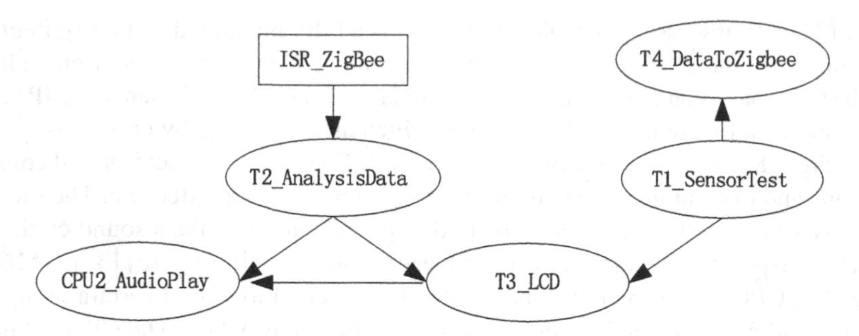

Fig. 38.4 Information intercommunion of routines and tasks on DE0

T2_AnalysisData, T3_LCD, T4_DataToZigbee and ISR_Zigbee respectively. Assume CPU2_Audio's program name is "CPU2_AudioPlay", the relationship among the 4 tasks, ISR_Zigbee and CPU2_AudioPlay can be shown in Fig. 38.4. Where the arrows represent the data flows.

T1_SensorTest aims to obtain the data from sensors, and then refresh the corresponding global variables. T2_AnalysisData is used to analyze the data received from DE2, so as to make the related equipments response. T3_LCD is used to display the data of sensors on touch screen, and to give according commands to the controlled equipments. The objective of T4_DataToZigbee is to send the sensors' data to Web server. ISR_Zigbee is ZigBee's interrupt service routine with the aim to arrange the received data into an array, and then the array will be analyzed by T2_AnalysisData.

38.3.2 Schemes of Web Server

Nios II IDE [14] contains the μC/OS-II and NicheStack TCP/IP Stack software components [7, 8], providing designers with the ability to build networked embedded systems for the Nios II processor quickly. NicheStack TCP/IP Stack is a small-footprint implementation of the TCP/IP suite. The focus of NicheStack TCP/IP Stack is to reduce resource usage while providing a full featured TCP/IP stack [8]. The NicheStack TCP/IP Stack is designed for use in embedded systems with small memory footprints, making it suitable for Nios II processor systems.

Considering the advantages of NicheStack TCP/IP Stack above, we implemented the Web server with μC/OS-II and NicheStack TCP/IP Stack on DE2. And we use the read-only zip file system [14] to store and access varies data of files (such as .jpg, .gif, etc.). It is shown that the zip file system is the most convenient tool in imbedded system that facilitates users to use the files [19].

To develop a web server with NicheStack TCP/IP Stack, the procedure is almost the same as other TCP/IP suites. First, we use the function bind() and listen() to bind and then listen to a socket, then wait for the clients to log on. If there is a client that emerges, we invoke the function FD_SET() to launch the

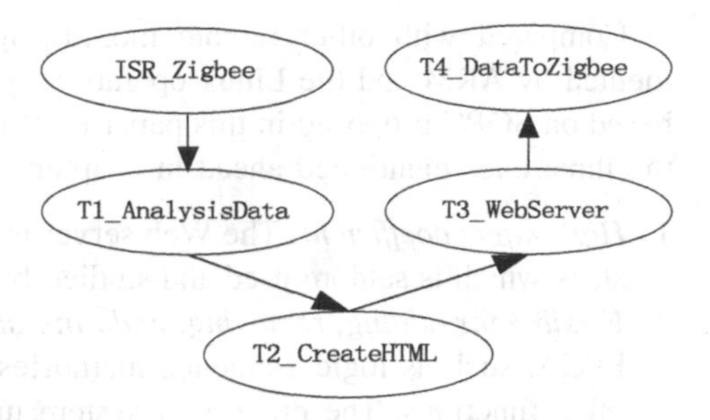

Fig. 38.5 Information intercommunion of routines on Web server

according socket, and at last we send or receive data by the function send() or rcv(). For details on how to use NicheStack TCP/IP stack, refer [7, 8].

38.3.3 Relationship among Tasks of the Web Server

There are three tasks and an interrupt service routine except the Web server task T3_WebServer. The three super-loop tasks are T1_AnalysisData, T2_Create-HTML and T4_DataToZigbee. The interrupt service routine is ISR_Zigbee with the responsibility to write the data received from DE0 into an array whenever an interrupt of ZigBee comes.

T1_AnalysisData is used to analyze the data in the array periodically in order to obtain useful information which will be needed to form the HTML file by T2_CreateHTML. T3_WebServer can obtain commands from the Web, and then pass them to T4_DataToZigbee sequentially, so as to make the DE0 response accordingly. The information intercommunion of routines and tasks is shown in Fig. 38.5.

38.3.4 Schemes of Network Security

There are three schemes of network security in the system. Firstly, users must enter the right password when log on the system. Secondly, you can suspend the task of Web server by push a button on DE2 development board with the other routines and tasks operating normally. And the third scheme is that the system can keep a record of the clients' IP addresses, which take great advantages for network security investigation.

38.4 Conclusions

This paper introduces a new scheme to create remote monitoring systems and Web servers by SOPC technology. It is demonstrated that system operates perfectly. All the data can be shown and refreshed on time on the Web, which can be accessed by a common IE browser.

Compared with other remote monitoring systems which are always implemented by ARM and the Linux operation system [20, 21], the distributed system based on SOPC proposed in this paper have quite a lot of other advantages except the three ones mentioned ahead in chapter 1.2, as are summarized as follows:

1. *High safety coefficient*: The Web server is implemented by Niche Stack TCP/IP stack which is seldom used and studied by hackers [22], so the invasion is low.
2. *Flexible for adding, removing, updating and maintenance*: The left resources of FPGA, such as logic elements, memories and pins can be used to implement other functions. The process of system updating is just reconfiguration of the data in EPCS [17]. Especially, it is easy for an FPGA to add new processors, forming a multi-processor system.
3. *Stable*: The hardware configuration data is stored in EPCS, and the software is stored in flash. Once the system fails, the hardware will be configured again and the software will reloaded either when powered up again, which means that they system is completely renewed. It is fortunate that the overall time of renewing is less than a second [17], as if the system did not fail!

Acknowledgments This work was partially supported by the Natural Science Foundation of Huhei Province under Grant No. 2009CDA143 and D20111901.

References

1. Jun K, Guanzhong D, Pengju H (2007) Design of intelligent node of distributed measuring and controlling network based on SOPC. J Appl Res Comp 24:282–285
2. Ye X, Miao X (2010) The intelligent smart home content networking system based on ZigBee. J Mod Building Electron 1:25–27
3. Microchip ZigBee Protocol Stack TM [EB/OL], http://www.microchip.com
4. Altera Corporation (2009) Creating Multi-processor Nios II Systems Tutorial
5. Altera Corporation (2009) Using μC/OS-II RTOS with the Nios II Processor Tutorial
6. Jean JL (2002) MicroC/OS-II The Real-Time Kernel, 2nd edn. CMP Book Company Press, New York
7. Altera Corporation (2009) Using the Niche Stack TCP/IP Stack
8. Altera Corporation (2009) Ethernet and the Niche Stack TCP/IP Stack -Nios II Edition
9. Zhigang Z (2007) The design tutorial of FPGA and SOPC—ON DE2. M. Xi An University of Electronic Science and Technology Press, Xi An, China
10. Altera Corporation (2009) NiosII Software Developer‘s Handbook: Read-Only Zip File System
11. Lintao J (2010) Internet and Internet of Things. J Telecommun Eng Technol Standard 2:1–4
12. Altera Corporation (2009) Quartus II Version 9.1 Handbook Volume 4: Component Editor
13. Altera Corporation (2009) Quartus II Version 9.1 Handbook Volume 4, SOPC Builder
14. Altera Corporation (2009) NiosII Software Developer’s Handbook: Altera-Provided Development Tools
15. Davicom Semiconductor Inc (2005) DM9000A Ethernet Controller with General Processor Interface
16. Zheng Q, Zheng Y, Du C (2007) The remote internet controlling system based on SOPC. J China Water Transp 5:159–160

17. Ligong Z (2006) The foundation tutorial of embedded system SOPC. M. Beijing Aerospace University Press, Beijing
18. Pingji X, Ruili X (2010) Attachment of Tyrosinase on mixed self-assembled momolayers for the construction of electrochemical biosensor. J Chin Chem Lett 21:1239–1242
19. Peng Y, Xiuli Z (2007) Research of transformation from inner code to dot matrix with read-only zip files system. J Exp Technol Manag 6:62–66
20. Anfu Y, Shengfeng X (2011) Design and implementation of the DM9000 network device driver based on ARM and linux. J Comp Eng Sci 2:27–31
21. Zidao M, Zhengbing Z (2011) Software realization of video surveillance terminal based on ARM-linux. J Comp Meas Control 2:456–459
22. Xianyong O (2010) Computer network information security research. J Netw Security Technol Appl 3:63–67

Chapter 39
Design of an Efficient Low-Power Asynchronous Bus for Embedded System

Guangda Zhang, Zhiying Wang, Hongyi Lu, Yourui Wang and Fangyuan Chen

Abstract With the widespread use of embedded system, efficient and low-power designs have become crucial. The increasing scale of circuits leads to many problems in traditional synchronous designs. Asynchronous circuits use handshake signals to control communications between different modules, resolving problems caused by global clock. In order to implement full-asynchronous communications between asynchronous devices, this paper studies key techniques about asynchronous bus and implements an asynchronous bus, Pipeline-based Asynchronous bus for low energy (PABLE). We propose an asynchronous pipeline structure to improve bus's performance and research the asynchronous arbitration circuit to provide a stable and efficient asynchronous mechanism. Experimental results show that, for a single transfer, the read or write latency of PABLE would be lower than synchronous bus in the case of more than 60%. The average power consumption of PABLE decreases by 41% compared with synchronous bus.

Keywords Asynchronous bus · Pipeline · Arbiter · Low power

G. Zhang (✉) · Z. Wang · H. Lu · Y. Wang · F. Chen
School of Computer, National University of Defense Technology,
Changsha, Hunan, People's Republic of China
e-mail: zhanggd@nudt.edu.cn

Z. Wang
e-mail: zywang@nudt.edu.cn

H. Lu
e-mail: hylu@nudt.edu.cn

Y. Wang
e-mail: wyr_nudt@nudt.edu.cn

F. Chen
e-mail: fycsky@nudt.edu.cn

J. J. Park et al. (eds.), *Proceedings of the International Conference on Human-centric Computing 2011 and Embedded and Multimedia Computing 2011*, Lecture Notes in Electrical Engineering 102,
DOI: 10.1007/978-94-007-2105-0_39,

39.1 Introduction

The embedded system has special requirements in efficient and low-power circuit designs. Synchronous design method is still the mainstream in embedded designs. However, as the semiconductor technology proceeds to deep sub-micron, transistors on a single chip become more and more. Chip size increases largely. Problems of clock skew and high power consumption have become the main bottlenecks in synchronous circuit designs. Asynchronous circuits use handshake signals to control communications between different modules on a chip, resolving problems in traditional synchronous circuits which are caused by global clock, such as clock skew, crosstalk effect and high power consumption, achieving an average-case performance with perfect reusability and robustness. In recent years, asynchronous circuit designs have caught widespread attentions.

In order to extend the use of asynchronous modules more widely in embedded system, we must provide a full-asynchronous interconnecting strategy for asynchronous circuits so that the circuits could take full advantage of asynchronous features. Recent studies on asynchronous bus are few and most focus on the designing of asynchronous functional units [1, 2] and asynchronous circuits' characters [3]. Besides, asynchronous NoC attracts more attentions due to the rapid development of multi-core systems [4, 5]. As the main interconnecting structure of embedded system, bus is used to transmit information between various components on an embedded chip. The typical research of asynchronous bus is the Manchester Asynchronous Bus for Low Energy (MARBLE) designed by Manchester University [6]. MARBLE splits transfers of data and address, uses arbitration tree to implement a centralized arbitration, places FIFO between the bus and devices to buffer data and address. However, the multi-level arbitration tree makes the circuit more complex and increases its area cost. The simple pipeline in MARBLE is not sufficient to improve the transmission efficiency of the bus. To some extent, FIFO can be used to improve bus's throughput, but also brings area cost. CHAIN [7] is an asynchronous point-to-point interconnection method which supports multiple masters and slaves, and splits transactions like MARBLE. It uses many pipeline latches to increase throughput. The throughput of CHAIN increases by 30% more than that of MARBLE, while the latency increases by 40% [8]. Kaiming Yang et al. have designed an asynchronous ring bus to support multi-core systems recently [9]. The transfer time of the shortest distance in the ring bus is 1.5 ns approximately in TSMC 0.18 μm process, but the performance differs in different arbitration strategies. In addition, different data encoding methods, such as 1-of-4 and "return-to-zero" have been studied to improve asynchronous bus's performance and both obtain good results [8, 10]. A number of existing asynchronous arbiters for multiple bus systems have been discussed in [11, 12] indicating that, except for fairness, the stability and efficiency of arbitration mechanism should also be concerned in the process of asynchronous circuit designs.

In order to provide a mature and efficient full-asynchronous interconnecting strategy for asynchronous circuits, this paper studies key techniques about

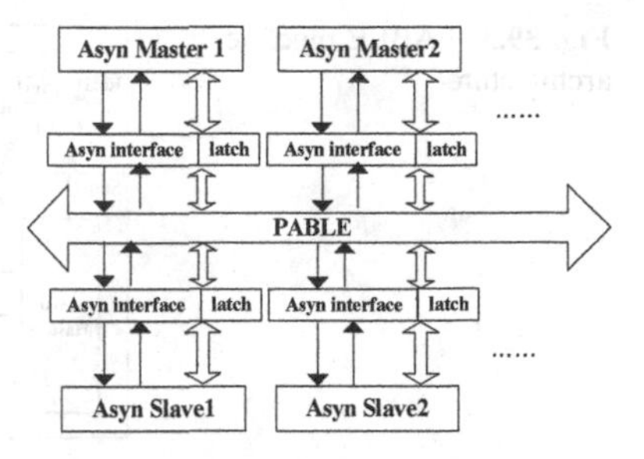

Fig. 39.1 PABLE-based environment

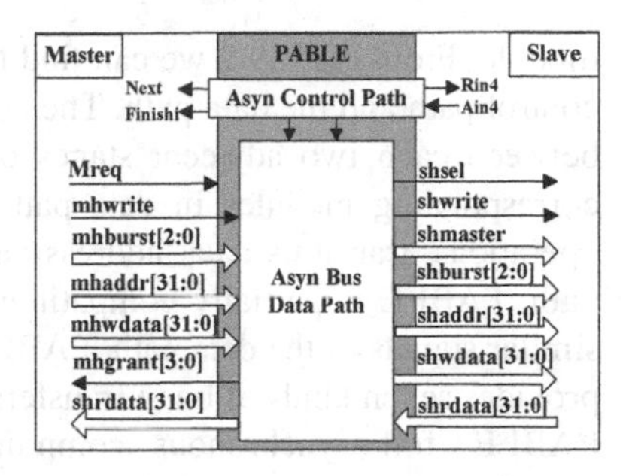

Fig. 39.2 PABLE interface module

asynchronous bus and implements a full-asynchronous bus named Pipeline-based Asynchronous bus for low energy (PABLE) which is partially compatible with the AMBA AHB protocol. We define an asynchronous interface specification for PABLE and present a new asynchronous pipeline structure. A stable and efficient asynchronous arbitration circuit is proposed. This paper lays foundation for the design and implementation of asynchronous bus from architecture and circuit level, and explores main techniques in asynchronous circuit designs.

39.2 The Architecture and Interface of PABLE

Figure 39.1 shows a typical target environment of PABLE, illustrating how support can be provided for asynchronous communications. The environment is full-asynchronous without global clock, in which communications between bus and devices are controlled by handshake signals. PABLE supports multiple asynchronous masters and slaves. The introduction of a latch into the bus interface at each port (see Fig. 39.1) is used to support the pipeline. With an asynchronous wrapper, PABLE could also support GALS (Globally Asynchronous Locally Synchronous) system [13].

As an important part of bus protocols, interface protocol specifies the format and type of interface signals. Through the interface, external signals are transformed into internal signals that PABLE could use directly. The interface module of PABLE is divided into two sides due to different input and output signals from masters and slaves. Figure 39.2 gives a detailed description of the interface

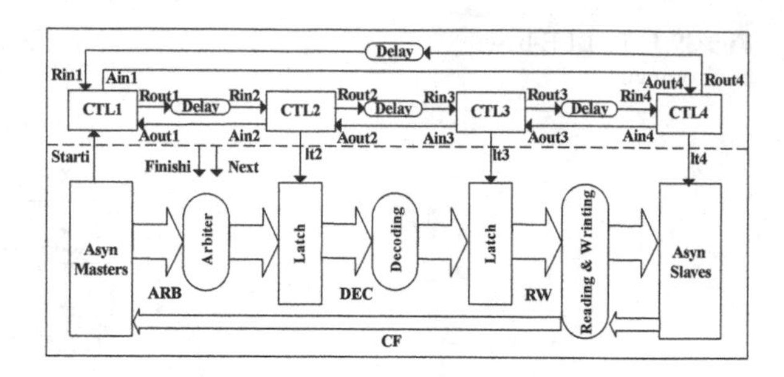

Fig. 39.3 PABLE pipeline architecture

module. From Fig. 39.2 we can find that PABLE is partitioned into two parts, the control path and the data path. The control path implements handshake operations between each two adjacent stages of pipeline and sends control signals to the corresponding modules in data path. The data path mainly implements logical operations, transmits data, address and other signals in the control of the control path. PABLE is partially compatible with the AMBA AHB protocol due to the similar signals of the data path. PABLE supports 32-bit address and data transfers, provides seven kinds of burst transfers and supports pipelined operations. Through PABLE, full-asynchronous communications could be implemented between asynchronous masters and slaves with a high performance and low-power.

39.3 The Asynchronous Pipeline of PABLE

Pipeline has been widely used in synchronous circuit designs, which could significantly improve the performance. In this paper, based on the de-synchronization [14] method, taking into account the practical applications of PABLE and reducing unnecessary area and power consumption, we choose the full-decoupled 4-phase bundled-data protocol [15] to design asynchronous pipeline architecture.

39.3.1 Pipeline Architecture

The overall structure of PABLE pipeline is shown in Fig. 39.3. PABLE has been divided into four pipeline stages. All stages are connected to each other to form a ring. The four stages are Arbitration stage (ARB), Decoding stage (DEC), Read-write stage (RW) and Confirming stage (CF). Communication between each two adjacent stages is controlled by handshake signals. The asynchronous controller CTL implemented with a full-decoupled method [15] is used to control handshake signals between adjacent stages. For the whole pipeline, it needs four handshakes to complete an independent data read or write transfer. The four handshakes are corresponding to four stages of the pipeline respectively. Before PABLE works, masters should prepare all the data, address and control signals. Then they issue requests to request the bus, which would make the bus work. The process of every pipeline stage is introduced below.

(1) *ARB stage*: This stage implements the first handshake. Requests from masters trigger the control path of PABLE to work. Before arbitration operation starts, the arbitration circuit needs to decide whether the request signal Reqi of master i is "valid". If master i has been granted the bus and it is occupying one of the pipeline stages now, its request Reqi is "invalid" and would be ignored by the arbitration circuit. Otherwise, if the master isn't occupying any pipeline stages, its request signal is set to "valid" state. The arbitration circuit only deals with valid requests. In this way, we could ensure that before a master's last transfer is completed, its new request would not take any effect. This would avoid mistakes made by repeated or de-asserted requests. If one of the requests is valid and the signal Next generated by DEC stage is high, the ARB stage would run arbitration operations. Then the granted master's data, address and control signals are transmitted to the next stage.

(2) *DEC stage*: The second handshake mainly implements decoding operation and deals with burst transfers. It is controlled by signal lt2 sent from the asynchronous controller CTL2. Under the control of lt2, data, address and control signals are transmitted to the third pipeline stage. Simultaneously, PABLE sends out the signal Next to tell the arbitration circuit in ARB stage whether the ARB stage could continue the next arbitration operation. The signal Next is generated by a control component in DEC stage, which detects the burst transfers and determine the level of Next.

For a burst transfer, the signal Next should be kept low to lock the arbitration circuit so that all requests are blocked before the ARB stage and no masters could be granted any more. The master that starts the burst transfer would get an exclusive access to the bus from the first passing of DEC stage. The DEC stage has a dedicated address calculation component to calculate address for burst transfers. When a master is granted the bus and performing a fixed length burst, it is not necessary to continue requesting the bus and pass the ARB stage. Therefore, the master performing burst transfer would only pass three stages except for the ARB stage in every beat. When burst operation comes to its last beat and enters DEC stage, the signal Next would be set to high level by the control component. Thus the arbitration circuit could return to work and deal with requests from other masters.

(3) *RW stage*: The third handshake implements the reading and writing operations. If it is a write transfer, the reading-writing module would write data into right asynchronous slave's address. If it is a read transfer, the slave would get address from the corresponding masters.

(4) *CF stage*: The fourth handshake implements the confirming operation. For a read transfer, the reading-writing module would read data from the right slave's address and send data to the corresponding master. Then PABLE would issue signal Finishi to tell the corresponding master i that data has been read from slaves successfully and the transfer is completed. For a write transfer, PABLE will send

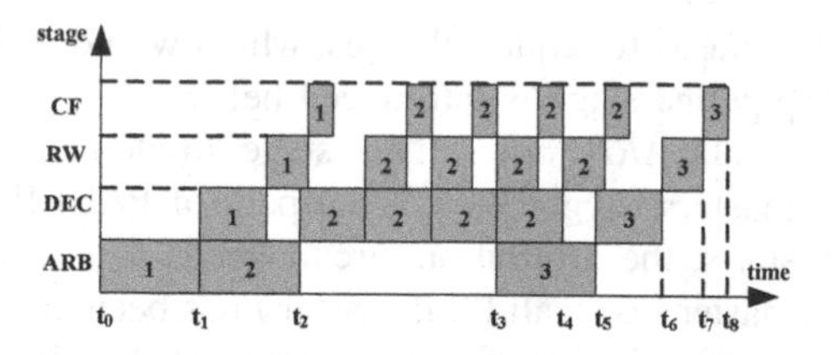

Fig. 39.4 An example of asynchronous pipeline

Finishi immediately to confirm this transfer. When ARB stage detects the signal Finishi, it will change the state of request signal of master i into "valid".

39.3.2 Example and Analysis

Here we give an example of how the asynchronous pipeline works, as shown in Fig. 39.4. We suppose that there are four masters M0, M1, M2 and M3 attached to PABLE. M1 is granted the bus at t0. At time t1, ARB stage detects that both M1 and M2 are requesting the bus (the request of M1 has not been de-asserted in time). However, M1 was granted at t0 so that the request of M1 is "invalid". Simultaneously, PABLE finds the transfer of M1 is in single mode in the DEC stage. Then the signal Next is set high to unlock the arbitration circuit. Thus, the arbitration circuit could grant M2's request and M2 starts occupying bus. At t2, the control component in DEC stage detects that the transfer of M2 is in a 4-beat increments burst mode (INCR4) so that the signal Next is set low to block the arbitration circuit. Although ARB stage receives a request from M3, the arbitration module does not work and Req3 is blocked. At t3, the control component in DEC stage detects that it is the last beat of the INCR4 burst of master M2. The signal Next is set high to allow arbitration circuit to continue to work. Then M3 is granted the bus.

For the whole pipeline, both incrementing and wrapping burst transfers only need to pass the ARB stage once when master is granted. We use P_{ARB} to indicate the percentage that a burst transfer passes the ARB stage in all beats. Thus, an INCR4 burst transfer needs to pass the ARB stage once in all the 4 beats. The $P_{ARB\text{-}INCR4}$ is $1/4 = 25\%$. For an INCR16 burst transfer, $P_{ARB\text{-}INCR16}$ is $1/16 = 6.25\%$. Therefore, with the increasing beats of a burst transfer, the value of P_{ARB} becomes smaller. The throughput of PABLE is more determined by the DEC stage, rather than the longest ARB stage.

Then we analyze the pipeline's performance for a burst transfer theoretically to reflect the total performance of the pipeline. For a burst transfer, we use n to represent the number of beats in a burst (n = 1, 4, 8 or 16). Each value of n is corresponding to a value of P_{ARB} ($P_{ARB} = 1/n$). T_{ARB} and T_{DEC} represent latencies of the ARB stage and the DEC stage respectively. From the experimental result we know that, T_{ARB} is longer than T_{DEC}, and both T_{ARB} and T_{DEC} are longer than latencies of the other two stages. We use t_{RC} to represent the sum of latencies of

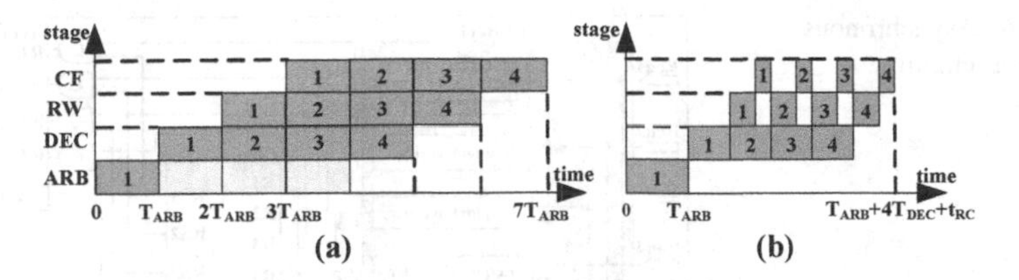

Fig. 39.5 Comparation between the synchronous pipeline and asynchronous pipeline

RW stage and CF stage. Then we could calculate the speedup ratio S_{asyn} of the asynchronous pipeline as (1) below. For synchronous bus of the same function, the speedup ratio S_{syn} is limited by the longest stage, i.e. the ARB stage. The speedup ratio S_{syn} of the synchronous bus is calculated as (2).

$$S_{\text{asyn}} = \frac{4nT_{\text{ARB}}}{(T_{\text{ARB}} + T_{\text{DEC}} + t_{\text{RC}}) + (n-1)T_{\text{DEC}}}. \quad (39.1)$$

$$S_{\text{syn}} = \frac{4nT_{\text{ARB}}}{4T_{\text{ARB}} + (n-1)T_{\text{ARB}}} = \frac{4n}{n+3}. \quad (39.2)$$

We assume that $a = T_{DEC}/T_{ARB}$, $b = t_{RC}/T_{ARB}$ $(a < 1, b < 2)$, in which "a" and "b" are both constants determined by the practical circuit. Then (39.1) could be transformed to (39.3). From (39.3) we get that, if $a = 1$ and $b = 2$, S_{asyn} and S_{syn} would have the same value for a burst transfer. If $a < 1$ and $b < 2$, S_{asyn} would be larger than S_{syn}. Apparently, the asynchronous bus obtains a better speedup ratio than the synchronous bus. With the increasing of n, i.e. the decreasing of P_{ARB}, the speedup ratio S_{asyn} of the pipeline would increase. Figure 39.5 presents a contrast between asynchronous pipeline and synchronous pipeline for an INCR4 burst. For an INCR4 burst, the asynchronous pipeline needs a period of (T_{ARB} + $4T_{DEC}$ + t_{RC}) to deal with, while the synchronous bus needs $7T_{ARB}$, which is longer than the asynchronous.

$$S_{\text{asyn}} = \frac{4n}{an + b + 1}. \quad (39.3)$$

In conclusion, with the asynchronous pipeline architecture implemented in PABLE, for a single transfer, it needs four handshakes to complete a read or write operation. Asynchronous pipeline provides parallelism mechanism for data transmissions on bus, allows that when some masters are occupying the bus, the other masters could also request and be granted. For burst transfers, after the arbitration operation in the first beat, it only needs three handshakes to complete a data read or write transfer in the following beats, which accelerates the flow of data on bus and improve the transmission efficiency of PABLE. With the asynchronous pipeline structure, the performance of PABLE could be largely improved.

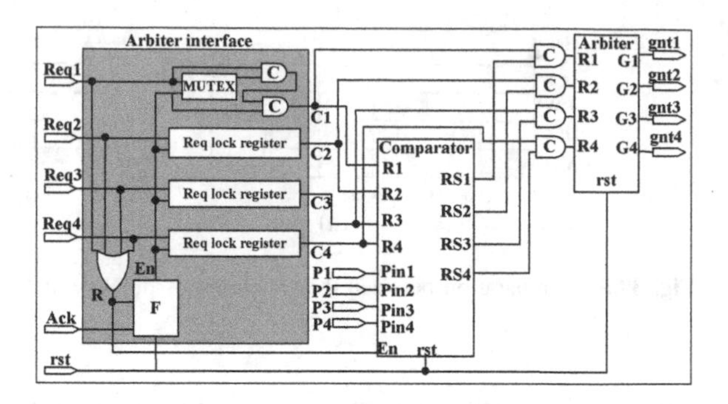

Fig. 39.6 Asynchronous arbitration circuit

39.4 Asynchronous Arbitration Circuit

Synchronous bus uses uniform clock edges to control all requests' coming. The arbitration circuit just needs simple combinational logic to realize different arbitration algorithms. Asynchronous bus doesn't have a global clock. Different arriving time of requests from asynchronous masters could easily cause the meta-stability problem. Therefore, we must redesign the asynchronous arbitration circuit to provide a stable and efficient arbitration mechanism.

PABLE uses a centralized scheme with an arbitration circuit composed of three modules, including the Interface module, Comparator module and Arbiter module, to implement the asynchronous arbitration, as shown in Fig. 39.6.

39.4.1 Arbiter Interface

Arbiter interface module is used to sample request signals from all masters, resolving the meta-stability problem caused by different arriving time of requests. It consists of an F module and n Request Lock Registers (n represents the number of masters). Figure 39.7 shows the truth table and one of the implementations of the F module. F module is connected to request signals through an OR-gate. Any requests will set the input port R of F module high. The other two input ports are Ack and rst, which are respectively used to receive acknowledge signal and reset the output.

Each Request Lock Register is composed of a mutual exclusion element (MUTEX) and two Muller C-elements. A possible implementation of MUTEX [16], as shown in Fig. 39.8, is used to pass two inputs R1 and R2 to the corresponding outputs G1 and G2 in such a way that at most one output is active at the given time. If one input arrives well before the other, the latter is blocked until the first input is de-asserted. When both input signals are asserted at the same time, the circuit becomes meta-stable with both signals x1 and x2 halfway between the supply and ground. The meta-stability filter prevents these undefined values from propagating to the outputs. G1 and G2 are both kept low until x1 and x2 differ by more than a transistor threshold voltage.

Fig. 39.7 The truth table and circuit of F module

Rst	R	Ack	Q_n	Q_{n+1}
0	-	-	-	0
1	0	-	-	0
1	1	0	0	1
1	1	1	0	1
1	1	0	1	1
1	1	1	1	0

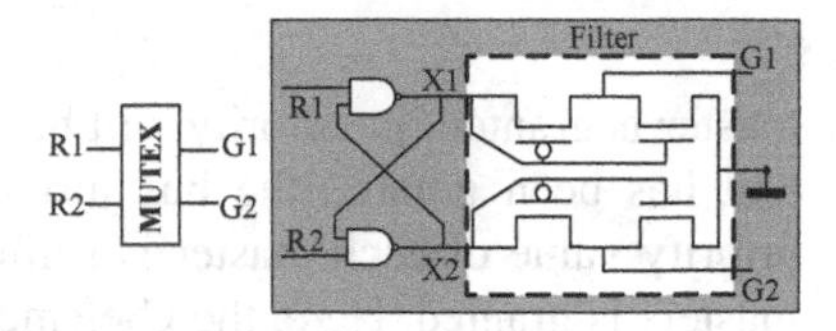

Fig. 39.8 The mutual exclusion element: symbol and possible implementation [16]

Muller C-element [16] is one of the main building blocks of SI (Speed-independent) asynchronous circuits. It has two input ports and one output. When two inputs are both high, the output is high. When both of inputs are low, the output is low. For the other situations the output will keep to the last state.

When at least one request is sensed, the input port R of F module will go high, which makes the output En high (see Fig. 39.6). Then request signals are latched into Request lock Registers. The principle of Request lock Register is that the MUTEX will give priority to the early coming input. Before enable signal En goes high, the early coming requests would be granted by MUTEX. After En's going high, the later coming requests would be blocked and would not be granted. If the request Reqi succeeds the MUTEX, through two C-elements, the output signal Ci will go high. C-element is used to keep the output signal Ci high until the request Reqi is de-asserted and acknowledge signal Ack turns high. When one of the masters is finally granted the bus by ARB stage, the Ack signal is set high to release the registers so that the arbitration interface module could continue to receive new requests. The cooperation of MUTEX and C-elements in interface module would eliminate the meta-stability problem and ensure that the arbitration circuit could sample requests correctly.

39.4.2 Dynamic Priority

PABLE uses dynamic priority discipline as the arbitration strategy. The designed circuit based on dynamic priority algorithm is composed of a priority comparator and a round-robin arbiter connected with C-elements (see Fig. 39.6).

The comparator uses dedicated priority ports to receive priority information from different masters. Master that has the highest priority will gain the access. When there are many masters that have the same highest priority, the corresponding outputs RSi of comparator will be set high. Afterwards, in a round-robin scheme, the arbiter module will grant one master. The arbitration circuit needs to keep a priority list which stores the priority information of all masters. If one

Fig. 39.9 The transition diagram of priority list

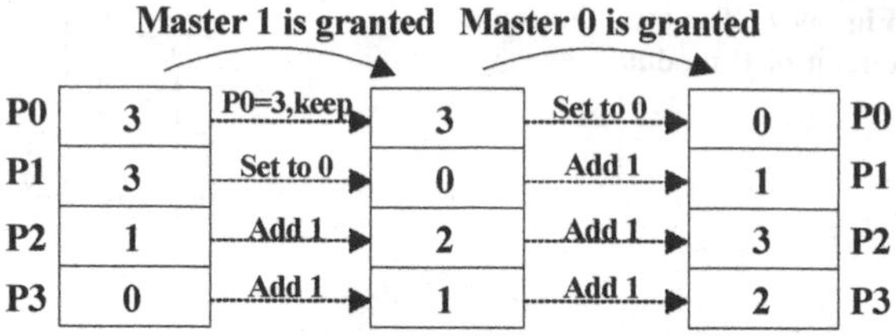

Master 1 is granted Master 0 is granted

P0	3	P0=3,keep	3	Set to 0	0	P0
P1	3	Set to 0	0	Add 1	1	P1
P2	1	Add 1	2	Add 1	3	P2
P3	0	Add 1	1	Add 1	2	P3

master is granted, its priority will be set to the lowest value. Therefore, the master that has been granted the bus just before would have the lowest priority. The priority value of each master is shifted in a circular way each time when one of masters is granted. Here, the C-elements between comparator and arbiter are used to ensure that only the requesters that have requested the bus and have been selected by the comparator module will reach to the arbiter module. Figure 39.9 gives an example of how the priority list changes. With such a dynamic priority arbitration strategy, the whole arbitration circuit would be efficient and fair.

39.5 Analysis and Simulation of PABLE

39.5.1 Performance

The clock frequency of synchronous bus depends on the critical path's length of data path, leading to a worst-case performance. For asynchronous bus, the pipeline stage does not need to wait for clock edge. As long as the combinational logic is completed, the data would be transferred to the next stage. Asynchronous bus provides an average-case performance.

In order to analyze the performance of PABLE, we use Design Compiler to synthesize the four pipeline stages of PABLE respectively under the UMC 0.18 μm CMOS technology. The transfer latency of each stage is 1.37, 1.25, 0.52 and 0.41 ns, corresponding to ARB, DEC, RW and CF stage respectively. The control overhead of each stage of the control path is 0.77 ns. Thus, the delay elements in ARB and DEC stage should be set to more than 0.60 and 0.48 ns for each. For RW and CF stage, the transfer latency is less than the corresponding control overhead so that no delay element is needed. The latency of RW and CF stage is both 0.77 ns. Therefore, for a single read or write transfer on PABLE, the total latency could reach to 1.37 + 1.25 + 0.77 + 0.77 = 4.16 ns.

For a synchronous bus that has the same function, the circuit is simple relatively. Under the same technology, for a single read or write transfer, the total latency of synchronous bus is 4.62 ns. Therefore, for a single transfer, the read or write latency of PABLE would decrease by 9.96% compared with the synchronous bus.

In reality, as a result of the 4-phase bundled-data pipeline of PABLE, there must be a "return-to-zero" part in the control path which surely leads to

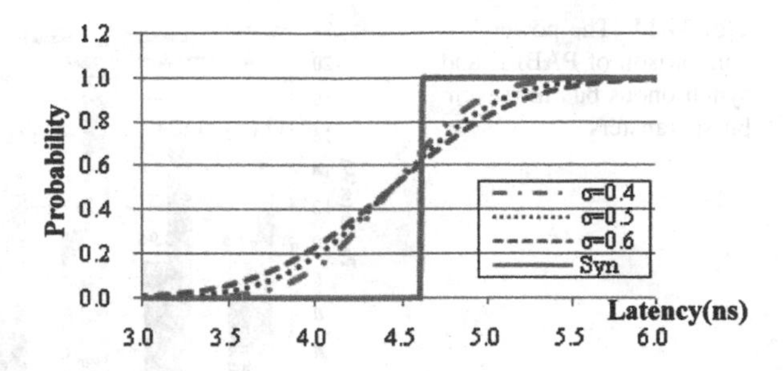

Fig. 39.10 The performance comparison of PABLE and synchronous bus

performance losses. The actual overhead of each control module is greater than 0.77 ns. In order to reduce the unnecessary performance losses caused by "return-to-zero", we use asymmetric delay elements [17] for delay matching. In asymmetric delay elements, the rise delay and fall delay are inconsistent. Therefore, by adjusting the asymmetric delay units, we could make the handshake operation return-to-zero quickly to improve the pipeline's performance. After adding the asymmetric delay elements we re-synthesize PABLE, the read (or write) latency becomes 4.72 ns, larger than the initial 4.16 ns. According to [18], the actual performance of asynchronous pipeline follows normal distribution between the worst and the best case performance. Thus, we could assume that under the best and the worst conditions, the read (or write) latencies are 4.16 and 4.72 ns respectively. When the standard deviation σ is 0.4, 0.5 or 0.6, for a single transfer the probability that the PABLE's latency is less than the synchronous bus is 67.4, 64.1 ans 61.8% respectively (see Fig. 39.10). We could conclude that, in the case of more than 60%, for a single transfer the latency of PABLE is less than synchronous bus.

39.5.2 Power Consumption

Power consumption of synchronous bus consists of three parts, including the clock power, leakage power and dynamic power. Asynchronous circuit eliminates the clock power, but increases the power caused by control path. For a synchronization module i, its total power consumption is calculated as (39.4) below. The asynchronous version of the power equation is calculated as (39.5) below.

$$\mathrm{Pbus(i)\_syn} = \left(\sum \mathrm{E(i)\_j} \times \mathrm{tj}\right)/\mathrm{te} + \mathrm{p(i)\_s} + \mathrm{Pclk}. \tag{39.4}$$

$$\mathrm{Pbus(i)\_asyn} = \left(\sum \mathrm{E(i)\_j} \times \mathrm{tj}\right)/\mathrm{te} + \mathrm{P(i)\_s} + \mathrm{Pctl\_path}. \tag{39.5}$$

E(i)_j represents the produced dynamic power when module i of the data path is running the j-class operations. tj represents the times of j-class operation. te is the running time. P(i)_s represents the leakage power of module i. Pclk represents the power caused by the clock turning. Pctl_path represents the power produced by

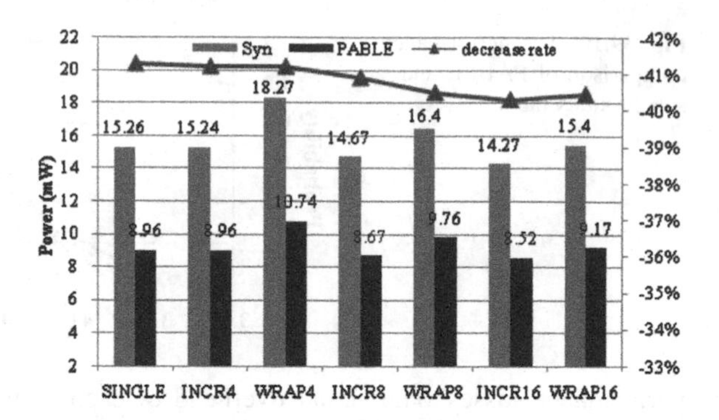

Fig. 39.11 The power comparison of PABLE and synchronous bus for seven burst transfers

control path. From (39.4) and (39.5) we find that, the main difference between synchronous and asynchronous bus is the clock power and power produced by the control path. For synchronous bus, a large part of power is generated by clock, regardless of whether the circuit is working. For asynchronous bus, the power of control path Pctl_path is only generated when there are logical operations in data path.

In order to analyze the actual power consumption of PABLE, under the UMC 0.18 μm CMOS technology, we use Design Compiler to synthesize PABLE and obtain VCD (Value Change Dump) files of 7 bursts. With the gate-level netlist, VCD files and associated library files, we use PrimePower to analyze the power consumption of PABLE and the synchronous bus of the same functions. The synchronous bus works at a frequency of 200 MHz. Request signals are sent to PABLE and synchronous bus every 10 ns from four simulated masters. Results are shown in Fig. 39.11.

In all the 7 bursts the power of PABLE keeps between 8.52 and 10.74 mW (see Fig. 39.11). The power of 4-beat wrapping burst (WRAP4) reaches to 10.74 mW which is the highest. The lowest is the 16-beat incrementing burst (INCR16) which is only 8.52 mW. Compared to the synchronous bus we could find that, in all the seven bursts the power of PABLE decreases by an average rate of 41% (see the curve of decreasing rate in Fig. 39.11). Therefore, we could conclude that, PABLE has an obvious advantage in power consumption compared with synchronous bus.

From Fig. 39.11 we also find that, for a single transfer of PABLE, the pipeline is not fully-filled which leads to a low-power consumption of 8.96 mW. For the other burst transfers, the pipeline can be well filled and the power is relatively high. When pipeline is busy enough or nearly filled, the power consumption of INCR16 burst is lower than INCR4's. It is because that the proportion of arbitration operation in INCR16 is smaller than that in INCR4 in the whole pipeline ($P_{ARB\text{-}INCR16} < P_{ARB\text{-}INCR4}$), which causes less signal turnings. The power of INCR16 nearly decreases by 4.9% compared with INCR4. The wrapping bursts have the similar conditions to the incrementing bursts. In conclusion, when the pipeline is in the same "busy" state, with the increasing beats of burst transfer, the power consumption of PABLE will be gradually reduced.

In addition, for burst transfers of the same beats, the wrapping bursts causes more power than the incrementing bursts. It is because the wrapping bursts require more logic elements to calculate address, leading to higher power consumption.

39.5.3 Area

For area, the data path of PABLE is redesigned for asynchronous operations and becomes more complex. PABLE also adds additional control path which largely increases its area. The result of synthesis demonstrates that PABLE has an area of 0.063 mm^2, increasing by 14.5% than synchronous bus.

39.6 Conclusion

This paper mainly studies key techniques about asynchronous bus, designs and implements an efficient low-power asynchronous bus PABLE, which is partially compatible with the AMBA AHB protocol. The Final simulating results show that compared to synchronous bus of the same function, PABLE adds a complex control path, which increases its area. However, PABLE has obvious advantages in performance and power consumption. For a single read or write operation, in the case of more than 60%, the latency of PABLE is lower than synchronous bus. The average power consumption of PABLE approximately decreases by 41% compared with synchronous bus. In conclusion, the overall performance of PABLE is better than synchronous bus. As to power consumption, PABLE has an obvious advantage.

This paper lays the foundation for further studies on asynchronous circuit designs in embedded system. For lacking of efficient EDA and synthesis tools and mature methodology, asynchronous designs confront with many problems and difficulties. Asynchronous designs couldn't satisfy complicated circuit designs and the circuits usually couldn't have a really high performance. Our future work would mainly focus on the studying of methodology of asynchronous designs and improving asynchronous circuits' performance. With further exploration in asynchronous designs, the embedded system will benefit more from asynchronous features and have expectations of obtaining a much better performance and lower power consumption.

Acknowledgments This work is supported by National Natural Science Foundation of China under Grant No.60873015 and Innovation Fund of Graduate School of National University of Defense Technology under Grant Nos. S100605, B100601 and CX2010B026.

References

1. Toms WB, Edwards DA (2010) M-of-N code decomposition for indicating combinational logic. In: 2010 IEEE symposium on asynchronous circuits and systems. IEEE CS Press, Grenoble, pp 15–25
2. Sheikh BR, Manohar R (2010) An operand-optimized asynchronous IEEE 754 double-precision floating-point adder. In: 2010 IEEE symposium on asynchronous circuits and systems. IEEE CS Press, Grenoble, pp 151–162
3. Joon Chang IK, Park SP, Roy K (2010) Exploring asynchronous design techniques for process-tolerant and energy-efficient subthreshold operation. IEEE J Solid-State Circuits 45:401–410
4. Badrouchi S, Zitouni A, Torki K, Tourki R (2005) Asynchronous NoC router design. J Comput Sci 1:429–436
5. Song W, Edwards D (2011) Asynchronous spatial division multiplexing router. J Microprocess Microsyst 35:85–97
6. Bainbridge WJ, Furber SB (2000) MARBLE: an asynchronous on-chip macrocell bus. J Microprocess Microsyst 24:213–222
7. Bainbridge J, Furber S (2002) Chain: a delay-insensitive chip area interconnect. IEEE Micro 22:16–23
8. Jung EG, Choi BS, Lee DI (2003) High performance asynchronous bus for soc. In: Proceedings of the 2003 international symposium on circuits and systems (ISCAS '03). IEEE Press, Bangkok, pp 505–508
9. Yang K, Lei K, Chiu J (2010) Design of an asynchronous ring bus architecture for multi-core systems. In: 2010 international computer symposium (ICS 2010). IEEE CS Press, Tainan, pp 682–687
10. Bainbridge WJ, Furber SB (2001) Delay insensitive system-on-chip interconnect using 1-of-4 data encoding. In: Proceedings of the 7th international symposium on asynchronous circuits and systems. IEEE CS Press, Salt Lake City, pp 118–126
11. Hasasneh N, Bell I, Jesshope C (2007) Scalable and partitionable asynchronous arbiter for micro-threaded chip multiprocessors. J Systems Archit: The EUROMICRO J 53:253–262
12. Yan C, Greenstreet M, Eisinger J (2010) Formal verification of an arbiter circuit. In: 2010 IEEE symposium on asynchronous circuits and systems. IEEE CS Press, Grenoble, pp 165–175
13. Zhiyi Y, Baas BM (2010) A low-area multi-link interconnect architecture for GALS chip multiprocessors. IEEE Trans VLSI 18:750–762
14. Shi W, Shen L, Ren H, Su B, Wang Z (2010) The approach of power optimization for de-synchronized circuits. J Comput Aided Design Comput Graphics 22:2155–2161
15. Furber SB, Day P (1996) Four-phase micropipeline latch control circuits. IEEE Trans Very Large Scale Integration Syst 4:247–253
16. Sparsø J, Furber S (2001) Principles of asynchronous circuit design—A systems perspective. Hardbound, Boston
17. Christos PS, Luciano L (2003) De-synchronization: asynchronous circuits from synchronous specifications. In: Proceedings of symposium on System-on-Chip (SoC). IEEE CS Press, Washington, pp 165–168
18. Andrikos N, Lavagno L, Pandini D, Sotiriou CP (2007) A fully-automated desynchronization flow for synchronous circuits. In Proceedings of IEEE design automation conference. ACM Press, New York, pp. 982–985

Chapter 40
Exploring Options for Modeling of Real-Time Network Communication in an Industrial Component Model for Distributed Embedded Systems

Saad Mubeen, Jukka Mäki-Turja and Mikael Sjödin

Abstract In this paper we investigate various options for modeling real-time network communication in an existing industrial component model, the rubus component model (RCM). RCM is used to develop resource-constrained real-time and embedded systems in many domains, especially automotive. Our goal is to extend RCM for the development of distributed embedded and real-time systems that employ real-time networks for communication among nodes (processors). The aim of exploring modeling options is to develop generic component types for RCM capable of modeling real-time networks used in the industry today. The selection of new component types is based on many factors including compliance with the industrial modeling standards, compatibility with the existing modeling objects in RCM, capability of modeling legacy systems and legacy communications, ability to model and specify timing related information (properties, requirements and constraints), ease of implementation and automatic generation of new components, and ability of the modeled application to render itself to early timing analysis.

Keywords Distributed embedded systems · Real-time systems · Real-time network · Component-based development · Automotive embedded systems

---

S. Mubeen (✉) · J. Mäki-Turja · M. Sjödin
Mälardalen Real-Time Research Centre (MRTC), Mälardalen University,
Västerås, Sweden
e-mail: saad.mubeen@mdh.se

J. Mäki-Turja
e-mail: jukka.maki-turja@mdh.se

M. Sjödin
e-mail: mikael.sjodin@mdh.se

J. Mäki-Turja
Arcticus Systems, Järfälla, Sweden

J. J. Park et al. (eds.), *Proceedings of the International Conference on Human-centric Computing 2011 and Embedded and Multimedia Computing 2011*, Lecture Notes in Electrical Engineering 102,
DOI: 10.1007/978-94-007-2105-0_40,

40.1 Introduction

Embedded systems are found in many domains and their applications range from simple consumer products to sophisticated robotic systems. Approximately 98% of the processors produced today are embedded processors [1]. The software that runs on these processors (embedded software) has drastically increased in size and complexity in the recent years. For example, a modern premium car contains approximately 100 embedded processors and the embedded software may consist of nearly 100 million lines of code [2]. In order to deal with such complexity, the research community proposed model- and component-based development of embedded systems [3, 4].

In distributed real-time and embedded systems, the functionality of an application is distributed over many nodes (processors) and the nodes communicate with each other by sending and receiving messages over a real-time network. A component model for the development of these systems is required to support resource efficient development, abstraction of the application software from communication infrastructure, modeling of real-time network communication, development of nodes to be deployed in legacy systems with predefined rules of communication, early analysis during the development, and efficient development tools.

In this paper, we explore various options for the development of special purpose components capable of modeling real-time network communication in distributed embedded systems. These components are to be added in the rubus component model (RCM) which is an industrial model used for component-based development of resource-constrained real-time and embedded systems in many domains. In this paper, we focus on component-based development of embedded systems in automotive domain. RCM has evolved over the years based on the industrial needs and implementation of the state-of-the-art research results. At present, RCM is able to develop only single-node real-time and embedded systems. Our aim is to enable RCM to also model distributed embedded systems that employ real-time networks for inter-node (inter-processor) communication.

The purpose of the new component types is to encapsulate and abstract the communications protocol and configuration in a component-based and model-based software engineering setting. The new components should support state-of-the-practice development processes of distributed embedded systems where communications rules are defined early in the development process. The components should allow the newly developed nodes to be deployed in legacy systems (with predefined communication rules) and enable the nodes to adapt themselves to changes in the communication rules (e.g. due to re-deployment in a new system or due to upgrades in the communication system) without affecting the internal component design. Moreover, the components should enable the modeled application to render itself to early timing analysis during the development.

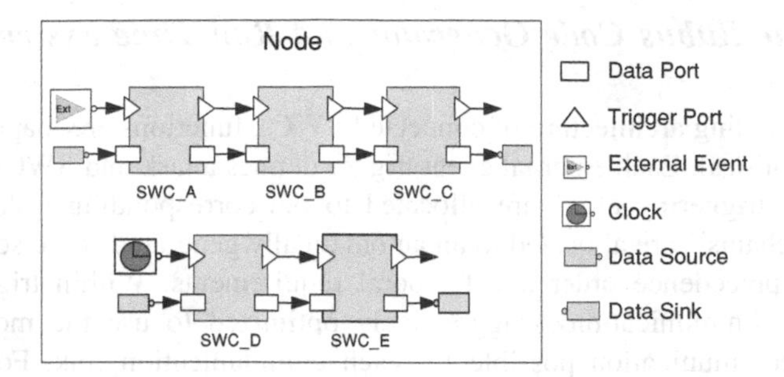

Fig. 40.1 An example system modeled in RCM

40.2 The Rubus Concept

The Rubus concept is based around the Rubus Component Model [5] and its development environment Rubus-ICE (Integrated Component development Environment) [6] providing modeling tools, code generators (also generating glue-code), analysis tools and run-time infrastructure. The overall goal of Rubus is to be aggressively resource efficient and to provide means for developing predictable and analyzable control functions in resource-constrained real-time embedded systems.

40.2.1 The Rubus Component Model

The purpose of RCM is to express the infrastructure for software functions, i.e., the interaction between the software functions in terms of data and control flow. One important principle in RCM is to separate code and infrastructure, i.e., explicit synchronization or data access should all be visible at the modeling level. In RCM, the basic component is called a software circuit (SWC). By separating code and infrastructure RCM facilitates analysis and reuse of components in different contexts (an SWC has no knowledge how it connects to other components). The execution semantics of software components (functions) is simply: upon triggering, read data on data in-ports; execute the function; write data on data out-ports; and activate the output trigger. Figure 40.1 shows how components interact with external events and sinks with respect to both data and triggering in an example real-time system modeled with RCM. The triggering events can be interrupts, internal periodic clocks, internal and external events. RCM has the possibility to encapsulate SWCs into software assemblies enabling the designer to construct the system at various abstraction levels.

40.2.2 *The Rubus Code Generator and Run-Time System*

From the resulting architecture of connected SWCs, functions are mapped to tasks (run-time entities). Each external event trigger defines a task and SWCs connected through the triggering chain are allocated to the corresponding task. All clock triggered "chains'' are allocated to an automatically generated static schedule that fulfills the precedence order and temporal requirements. Within trigger-chains, inter SWC communication is aggressively optimized to use the most efficient means of communication possible for each communication link. For example, there is no use of semaphores in point-to-point communications within a trigger chain. Another example is the sharing of memory-buffers between ports when there are no overlapping activation periods. The Rubus tool suite facilitates the optimization of the construction of schedule and allocation of SWCs to tasks, e.g., optimization of response times for different types of tasks or memory usage. The run-time system executes all tasks on a shared stack, thus eliminating the need for static allocation of stack memory to each individual task.

40.2.3 *The Rubus Analysis Framework*

RCM supports the end-to-end timing analysis and resource requirement estimations. It facilitates the expression of real-time requirements and properties on the architectural level, e.g., it is possible to declare real-time requirements from a generated event and an arbitrary output trigger along the trigger chain. For this purpose, the designer has to express real-time properties of SWCs, such as worst-case execution times and stack usage. The scheduler will take these real-time constraints into consideration when producing a schedule. For event-triggered tasks, response-time analysis is performed and the results are compared to the requirements.

40.3 Related Component Models

There are a number of models that support the model-based and component-based development of distributed systems. For example, distributed component object model (DCOM) [7], Common object request broker architecture (CORBA) [8], Enterprise JavaBeans (EJB) [9], etc. These models in their original form are not suitable for the development of distributed embedded systems which are often resource constrained and have real-time requirements. This is because these models require excessive amount of computing resources, have large memory foot print and have inadequate support for modeling of real-time network communication. There are very few commercial component models for the development of distributed embedded and real-time systems in the automotive domain. In the past few years, the research community and industry focused more on the component-based development of automotive embedded systems which led to the development of various models.

40.3.1 AUTOSAR

AUTOSAR [10] is standardized software architecture for the development of software in automotive domain. It can be considered as a distributed component model [11] in which the application software is defined by means of software components (SWCs). At design time, the virtual function bus (VFB) is responsible for the distribution of SWCs in a single node or on several nodes (according to the requirements of the application). VFB also provides virtual integration of SWCs and communication among them. The run-time representation of VFB for each electronic control unit (ECU) is defined by the run-time environment (RTE). The communication services for SWCs are provided by the basic software (BSW) via RTE.

When AUTOSAR was developed, there was no focus put on the ability to specify and handle timing related information (real-time requirements, properties and constraints) during the process of system development. On the other hand, such requirements and capabilities were strictly taken into account right from the beginning during the development of RCM. AUTOSAR enables the development of embedded software at a relatively higher level of abstraction as compared to RCM. A SWC in RCM resembles more to a runnable entity in AUTOSAR instead of AUTOSAR SWC. A runnable entity is a schedulable part of AUTOSAR SWC. As compared to AUTOSAR, RCM clearly distinguishes between the control flow and the data flow among SWCs in a node. AUTOSAR hides the modeling of execution environment. On the other hand, RCM explicitly allows the modeling of execution requirements, e.g., jitter, deadlines, etc., at an abstraction level that is very close to the functional modeling and at the same time, abstracts the implementation details.

In RCM, special purpose objects and components are used if SWCs require inter-ECU communication otherwise SWCs communicate via data and trigger ports by using connectors. On the other hand, AUTOSAR does not differentiate between intra-node and inter-node communication at modeling level. Unlike RCM, there are no special components in AUTOSAR for inter-node communication. AUTOSAR SWCs use interfaces for all types of communications which can be of two types, i.e. Sender–Receiver and Client–Server. The Sender–Receiver communication mechanism in AUTOSAR is very similar to the pipe-and-filter communication mechanism in RCM.

40.3.2 TIMMO and TADL

TIMing Model (TIMMO) [12] is an initiative to provide AUTOSAR with a timing model. TIMMO provides a predictable methodology and a language called Timing Augmented Description Language (TADL) [13]. TADL can express timing requirements and timing constraints during all design phases in the development of automotive embedded systems. TADL is inspired by Modeling and Analysis of Real Time and Embedded systems (MARTE) [14] which is a UML profile for model driven development of real-time and embedded systems. TIMMO uses the structural modeling provided by EAST-ADL [15] (a standard domain-specific architecture description language for automotive embedded systems). TIMMO methodology and its model structure abstract the modeling of communication at implementation level where they propose to use AUTOSAR. Both TIMMO methodology and TADL have been evaluated on prototype validators. To the best of our knowledge there is no concrete industrial implementation of TIMMO. In TIMMO-2-USE project [16], the results of TIMMO will be brought to the industry.

40.3.3 ProCom

ProCom [17] is a component model for the development of distributed embedded systems. It is composed of two layers. At the upper layer, called ProSys, it models the system by means of concurrent subsystems that communicate with each other by passing messages via explicit message channels. Unlike an SWC in RCM, a subsystem is active (it has its own thread of execution). At the lower layer, called ProSave, a subsystem is internally modeled in terms of functional components. These components are implemented as a piece of code, e.g., a C function. Like SWCs in RCM, ProSave components are passive (they cannot trigger themselves and require an external trigger for activation). ProCom gets inspiration from RCM. There are a number of similarities between the ProSave modeling layer and RCM. For example, components in both ProSave and RCM are passive. Similarly, both the models separate the data flow from the control flow. Moreover, both the models use pipe-and-filter mechanism for communication. At modeling level, ProCom does not differentiate between inter-node and intra-node communication. On the other hand, RCM clearly distinguishes modeling of inter-node and intra-node communication. ProCom uses two step deployment modeling, i.e., virtual node modeling and physical node modeling [18]. At present, physical node modeling is a work in progress. The validation of a distributed embedded system is yet to be done with ProCom. The development environment and tools accompanying ProCom are still evolving.

40.3.4 COMDES-II

COMDES-II [19] provides a component-based framework for the development of distributed embedded control systems. To some extent, it is similar to ProCom as it models the architecture of a system at two levels. At upper level, an application is modeled as a network of actors (active components) which communicate with each other by sending labeled messages. At the lower level, the functionality of an actor is modeled by means of function blocks (FBs). Similar to the SWCs in RCM, FBs are passive. The operating system (OS) employed by COMDES-II implements fixed-priority timed multitasking scheduling. On the other hand, Rubus-OS implements hybrid scheduling that combines both static–cyclic scheduling and fixed-priority preemptive scheduling [20]. COMDES-II is a relatively new research project and the support for development tools and run-time environment was provided recently [21]. On the other hand, RCM and its tool suite are relatively mature as they are being used in the industry for the development of embedded systems for more than 10 years [6].

40.3.5 RT-CORBA, Minimum CORBA and CORBA Lightweight Services

There are a number of middleware technologies introduced by the object management group (OMG) such as real-time CORBA, minimum CORBA and CORBA lightweight services. These middleware technologies can be used for the development of real-time and distributed embedded systems [22]. In some projects [23, 24], Real-time CORBA has been used to develop distributed embedded and real-time systems. Because of higher resource requirements, these models may not be suitable for the development of distributed embedded systems that are resource constrained and have hard real-time requirements.

40.3.6 Discussion

By comparing the models discussed in this section with RCM, we propose that RCM can be considered a suitable choice for the development of resource-constraint distributed embedded systems for many reasons. Some of the reasons are: its ability to completely handle and specify the timing related information (i.e., real-time require-ments, properties and constraints during all the stages of system development); it has a small run-time footprint (timing and memory overhead); it implements state-of-the-art research results; it has a strong support for development and analysis tools, etc.

40.4 Modeling of Real-Time Network Communication

A component model for the development of distributed real-time and embedded systems is required to automatically generate network communication for any type of protocol. However, this is often not the practice in the industry because of many factors including the dependencies due to existing communications and network protocols, early design decisions about the network communication, requirement for early analysis, deployment of the developed system in a legacy system having predefined rules for communication, etc. In this section, we will introduce the model representation of a physical network. We will also explore different options to build new components in RCM capable of modeling real-time network communication.

40.4.1 Network Specification: Modeling Object for a Physical Network

In order to represent the model of communication in a physical network, we propose the addition of a new object type, i.e., the Network Specification (NS) in RCM. There will be one NS for each network protocol. It is composed of two parts, i.e., a protocol independent part and a protocol dependent part. The protocol independent part defines a message and its properties. A message is an entity that is used to send information from one node to another via a network. The properties of a message that are defined by NS include message ID, sender node ID, list of receiver nodes IDs, list of RCM signals included in the message, etc. A signal in RCM has a name, RCM data-type, resolution, real-time properties, etc.

The protocol dependent part of NS is uniquely defined for each protocol. For example, it will be different for different protocols such as CAN (Controller Area Network) [26], CANopen [27], HCAN (Hägglunds Controller Area Network) [28], MilCAN (CAN for Military Lands System domain) [29], Flexray, etc. Therefore, there is one NS per bus. The protocol dependent part of NS contains the complete information of all the frames which are sent to or received from the bus. Moreover, it describes the frame properties. A frame is a formatted sequence of bits that is actually transmitted on the physical bus. In RCM, a frame is a collection of RCM signals. The properties of a frame described by NS include an identifier, a priority, a frame type, a transmission type, a sender node ID, a list of receiver nodes IDs, period or minimum inter-arrival time, deadline, whether a frame is IN or OUT frame, real-time requirements, etc. The transmission type of a frame can be periodic, event or mixed (transmitted periodically as well as on arrival of an event).

RCM clearly distinguishes the data flow from the control flow. The components inside a single node communicate with each other by using data and control signals separately. However, if a component on one node communicates with a component on another node via a network then the signals are packed into frames. The frames are then transmitted over the network. Here, some questions arise regarding the network communication. How are the signals packed into the frames? How are the signals encoded into the frames at the sender node? How are the signals decoded from the frames and sent to the respective SWCs at the receiver node? All the rules concerning the answers to these questions are specified in the Signal Mapping. The signal mapping is a unique object for each protocol for network communication and is an integral part of the protocol-dependent part of NS. The Signal Mapping describes the length of each signal in a frame, the type of signal encoding in a frame (e.g., signed or unsigned two's complement), maximum age of a signal guaranteed by the sender, etc.

40.4.2 Modeling Objects for Network Communication in a Node

There is a need to introduce special objects for modeling network communication in a node. These objects should be able to hide the software architecture in a node from the network protocol and vice versa. If SWCs located on different nodes intend to communicate with each other, they should only communicate (using the existing intra-node communication in RCM) with the special objects on their own node. Hence, the new objects should encapsulate and abstract the communication protocols and configuration in a node. These special objects along with NS should be able to facilitate the inter-node communication in a distributed embedded application.

To support the abstraction of the implementation of network communications in a node (processor), we will explore different options to develop new object types in the next subsection. After selecting the most suitable object types (based on the selection criteria to be discussed in the next section), we will add them in RCM. This will enable RCM to support state-of-the-practice development processes of distributed embedded systems where communication rules are defined early in the development process. The proposed extension of the model will also allow model-based and component-based development of new nodes that are deployed in the legacy systems that use predefined rules for network communication.

40.4.3 Options for Modeling Network Communication in a Node

Option 1: *reuse of existing intra-node communication*. The first option is to reuse the model of existing intra-node communication in RCM for modeling of inter-node communication (network communication). The intra-node communication in RCM is modeled by means of connectors as shown in Fig. 40.1. A connector in RCM links out-ports of one SWC with in-ports of the other in the same node. RCM allows port-to-port connections for data and trigger flow separately. In a nutshell, the idea is to use the same connectors for modeling both inter-node and intra-node communications by attaching, for example, a Boolean Specifier to it. If the connector is used for intra-node communication, then it is assigned a value "0". On the other hand, if an

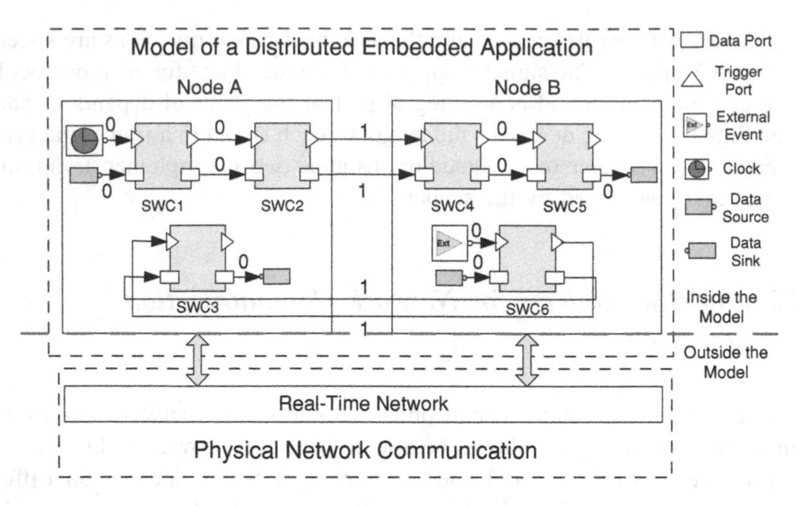

Fig. 40.2 Option1: model of a distributed real-time and embedded application

SWC on one node communicates with an SWC on another node then the specifier is assigned value equal to "1". We assume that a tool will automagically generate the run-time architecture for all communications in the modeled application. Moreover, the tool will perform the deployment of the distributed embedded application.

Figure 40.2 shows a distributed real-time application modeled with this technique. The application consists of two nodes which are connected to each other via a real-time network. There are three SWCs in each node. There are two distributed transactions (Event Chains) in the system. By distributed transaction, we mean that SWCs are in a sequence (chain) and have one single triggering ancestor (e.g., a clock, interrupt, external or internal event, etc.). In Fig. 40.2, the first distributed transaction is composed of SWC1, SWC2, SWC4 and SWC5 whereas the second is composed of only two SWCs, i.e., SWC6 and SWC3. The first distributed transaction is activated (triggered) by a clock while the second is activated by an external event.

The value of Boolean Specifier associated with each connector is "0" in case of intra-node communication, for example, a connector between SWC1 and SWC2. Similarly, the value of Boolean Specifier is equal to "1" in case of network communication, for example, a connector between SWC2 and SWC4. Although, the communication between SWC2 and SWC4 takes place via the network, a designer models the system as indicated by the upper portion of Fig. 40.2. The deployment and synthesis tools along with the run-time support will be responsible for generating intra-node and inter-node communications.

Option 2: *out- and in- SWCs for out and in frames*. The second option for modeling network communication in RCM is to introduce special purpose software circuits i.e. out software circuit (OSWC) and in software circuit (ISWC) for each frame that is to be sent or received by a node, connected to a network, respectively.

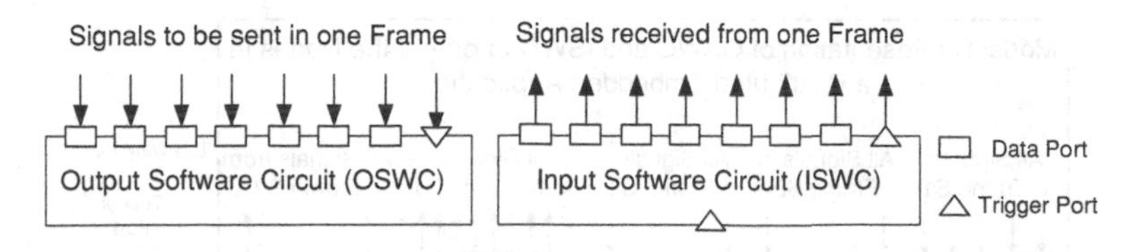

Fig. 40.3 Graphical illustration of OSWC and ISWC

Out Software Circuit (*OSWC*). it is a software circuit which denotes the data that leaves the model. An OSWC is associated with a LAN object. In RCM, LAN is an object to represent the connection between two or more nodes in the system. Formally, a LAN is defined by its name, a list of the connected nodes and NS. There is exactly one OSWC in a node for every outgoing frame on the network. Each OSWC describes all the signals that can be sent in a particular frame. As discussed earlier, a frame constains zero or more signals. An OSWC has only one trigger in-port and at least one data in-port. Each data in-port is associated with one signal in NS. Therefore, the number of data in-ports may vary depending upon the number of signals to be packed in the frame. An OSWC has no data and trigger output-ports. The OSWC component uses protocol specific rules, specified in the protocol-specific part of NS, while maping signals to frames and encoding data in the frames. In this way, OSWC provides a clear abstraction to the SWCs that send signals to one of its data in-ports. Thus, SWCs are kept unaware of the protocol-specific details such as signal-to-frame mapping, data-type encoding and transmission patterns of frames.

In Software Circuit (*ISWC*). it is a software circuit which denotes the data that enters the model. An ISWC is associated with a LAN object defined in RCM. There is exactly one ISWC component in a node for every frame received from the network. Each ISWC describes all the signals that are contained in a particular received frame. An ISWC has one unconditional trigger out-port. An unconditional trigger port produces a trigger signal every time the SWC is executed. There is at least one data out-port in the ISWC component. Each data out-port is associated with one signal in NS of the LAN object. Therefore, the number of data out-ports may vary depending upon the number of signals contained in the received frame. An ISWC has no data out-ports. There is one trigger in-port in every ISWC component which can be trig-gered when a frame arrives at a node. Figure 40.3 graphical illustrates OSWC and ISWC.

Consider an example of a node in a distributed embedded application modeled with OSWC and ISWC as shown in Fig. 40.4. Although, CAN is used for real-time network communication, this modeling approach can be applied to any real-time network protocol. Note that the figure is divided into two halves: the upper half represents the model of a node whereas the lower half depicts the physical communication including CAN controller and CAN network. There are two grey boxes outside the model called CAN SEND and CAN RECEIVE that are placed just below the sets of OSWCs and ISWCs, respectively. These gray boxed are specific for each network protocol. The frames that leave the model are denoted by

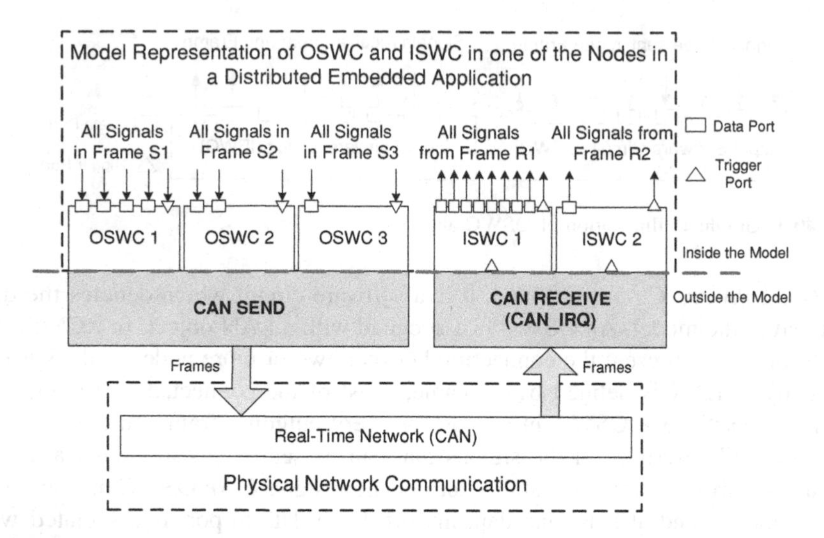

Fig. 40.4 Option2: model of a distributed real-time and embedded application

S e.g., *S1*, *S2* and *S3*. Similarly all the frames that enter the model are denoted by *R* e.g., *R1* and *R2* as shown in Fig. 40.4.

All the signals sent in frame *S1* are provided at the data in-ports of OSWC1. These signals are mapped and encoded into *S1* by OSWC1 according to the protocol-specific information available in NS. Once the frame is ready, it leaves the model as it is sent to the grey box CAN SEND. In this example, this grey box represents a CAN controller in the node which is responsible for the physical transmission of this frame. When a frame arrives at the receiving node, it is transferred by the physical network drivers to a grey box CAN RECEIVE that produces an interrupt. The frame enters the model and is transferred to the destination ISWC which extracts the signals, decodes the data and encodes it to RCM data-type. The data is placed on the data out-port of ISWC connected to the data in-port of the destination SWC and the corresponding trigger out-port is triggered (the tracing information is provided in the NS).

Option 3: *network interface components for each node*. The third option is to introduce two special purpose modeling component types in RCM: the network input interface (NII) and the network output interface (NOI) as shown in Fig. 40.5.

Network input interface (*NII*) *component*: it is the model representation of incoming signals from the network. Each node connected to a network will have one NII. Each NII contains one data-port and one trigger-port for each signal in every frame. When a frame arrives at the node, the physical bus driver and protocol-specific implementation of NII extract the signals (zero or more signals per frame) and encode their data in the RCM data-type. When the signal(s) is delivered, the data is placed on the data-port which is connected to data in-port of the destination SWC (the tracing information is provided in NS), and the corresponding trigger-port is triggered.

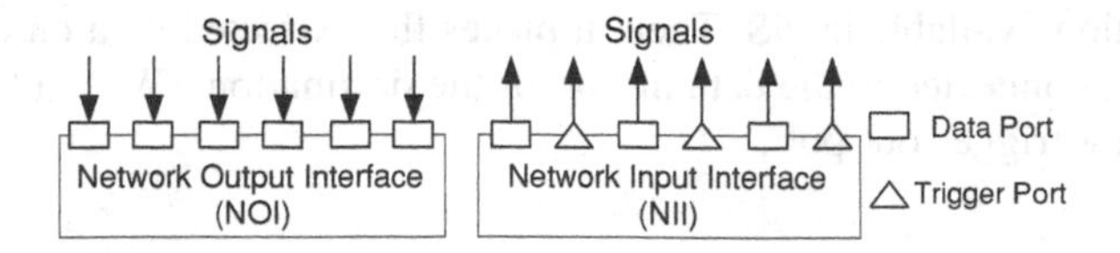

Fig. 40.5 Graphical illustration of NOI and NII Components

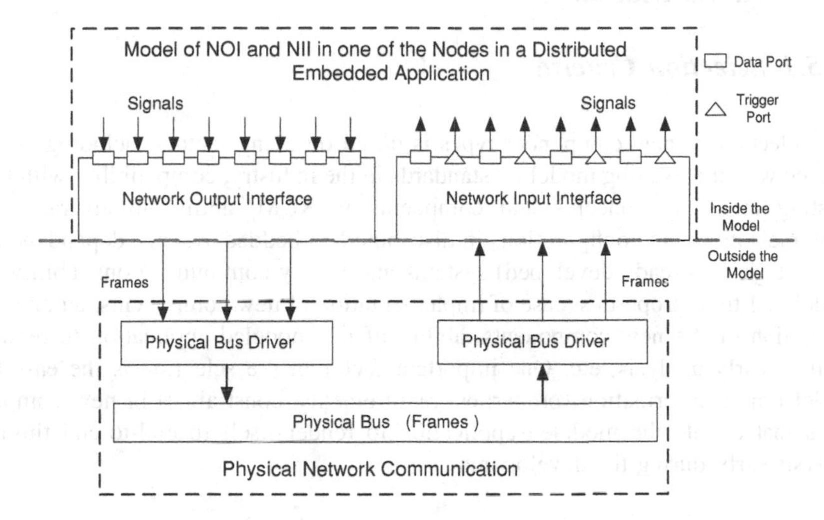

Fig. 40.6 Option3: model of a distributed real-time and embedded application

Network output interface (NOI) component: it is the model representation of the outgoing signals on a network. Each node connected to a network will have one NOI component. The major difference from the NII is that the NOI component does not have any trigger ports. Conceptually, the NOI has an implicit trigger port for each data port–however, to decrease the burden for the modeler, these ports are omitted from the model. The NII uses protocol-specific rules on how to map signals to frames and encode data in the frames specified in the protocol-specific part of NS. The NII also uses protocol specific rules to decide when to send each frame. Thus, the SWCs that use the NII are kept unaware about details such as signal-to-frame mapping, data-type encoding, and transmission patterns (common transmission patterns are: periodic, on-change, on-change with minimum distance between messages, etc.).

Consider the model representation of NOI and NII in one of the nodes connected to a real-time network in a distributed embedded application as shown in Fig. 40.6. The figure is divided into two halves: the upper half represents the model of a node in RCM whereas the lower half depicts the physical network. All SWCs in a node, requiring remote communication, send signals to the data in-ports of the NOI component. The NOI maps these signals into frames and sends the frames according to the protocol-specific rules defined in NS. When a frame arrives at a node, NII decodes the frame and extracts signals out of it according to the information available in NS. Then, it places the extracted data on data output-port which is connected to the data in-port of the destination SWC and triggers the corresponding trigger out-port.

40.5 Components for Modeling Real-Time Network Communication

40.5.1 Selection Criteria

The selection of new component types is based on many factors including compliance with the existing modeling standards in the industry, compatibility with the existing modeling concepts and compopents in RCM, ability to abstract the implementation and configurations in distributed embedded sysems, dependencies due to legacy (already developed) systems and legacy communications, ability to model real-time properties, ease of implementation of new components, automatic generation of the new components, ability of the modeled application to render itself to early analysis, etc. One important factor in the selection is the ease to model timing information (properties, requirements, constraints) in new components that enable the modeled application to render itself to end-to-end timing analysis early during the developemnt.

40.5.2 Components Selection

Although, the modeling approach in option 1 appears to be very simple and easy to model, it may not be practical in an industrial setup because of the requirements of modeling legacy systems and legacy communications, deployment of newly developed nodes in the existing systems and requirement for early analysis of the modeled application. Moreover, this approach requires very complex tools capable of automagically deploying and synthesizing the modeled system and generating the run-time implementation of all communications in the modeled application. This approach is not very expressive to perform the response-time analysis which is one of the most important requirements during the development of real-time systems.

The modeling approach in option 2 appears to be the most suitable candidate for the development of special components. One reason is that OSWC and ISWC component types are consistent with the existing modeling elements in RCM. These components are very similar to regular SWCs and hence, it is very easy to add them in the component model. The run-time implementation of these components can be easily generated. Another advantage is the generation of OSWC and ISWC automatically from NS. Moreover, it complies with the existing modeling standards and network communication protocols used in the automotive industry. For any network protocol such as CANopen, HCAN, MilCAN, Flexray, etc., all the modeling objects will remain the same except the protocol-dependent part of NS. This modeling approach may remain applicable if an application modeled with RCM uses the communication mechanism of AUTOSAR. In this case the grey boxes in Fig. 40.4 will represent the AUTOSAR run-time environment. A node developed with this approach can be easily deployed in a legacy distributed embedded system. In this case, same OSWC and ISWC components can be used and the rules of communication defined by the legacy system are encapsulated in NS. Since the network communication is modeled at frame level in a node, this approach provides an effective means to specify and trace the timing information for early timing analysis during the development.

The third modeling approach is a very general approach to model real-time network communication. It can be used for any type of real-time network protocol. Similar to the second approach, the protocol specific information is specified in NS while the implementation of NOI and NII components remains the same for all protocols including the rules of communication in legacy systems. NOI and NII can also be automatically generated from NS. This approach is very complex to implement mainly because these components are very different from the existing SWCs in RCM. Moreover, if these components are not implemented carefully then there is a risk of getting redundant rules for sending frames over the network. For example, assume CAN network in Fig. 40.6. The NOI sends a frame according to the protocol-specific information in NS where as the frame is physically transmitted by the CAN controller according to the CAN protocol. Although this approach facilitates to completely specify the timing related information, it is very difficult to extract the tracing information of event chains from the modeled application. Therefore an application modeled with this approach does not easily render itself to end-to-end timing analysis as compared to the second modeling approach. In future, this approach may be added in RCM because of being concise and general.

Based on the above discussion, we select the second modeling approach. Hence, we add special-purpose components OSWC and ISWC along with NS, in RCM to support modeling of real-time network communication.

40.5.3 Automatic Generation of the Selected Components

Both OSWC and ISWC components can be automatically generated from NS by a network configuration tool. The input to this tool is the protocol-specific information about the network communication and the tracing information of tasks in all the distributed transactions (event-based and periodic chains) present in the application. This information is made available from the configuration files that correspond to NS. The output of this tool is the automatically generated OSWC and ISWC components for each node in the network. This tool also carries out mapping from NS to components and vice versa. All OSWC and ISWC components are translated into a set of SWCs to execute the protocol at run-time.

40.6 Conclusion

We explored various options to develop special purpose component types in the rubus component model (RCM). The purpose of the special components is to model network communication in distributed real-time and embedded systems. RCM is an industrial component model for the development of resource-constrained real-time and embedded systems. We introduced the capability of modeling real-time network communication in RCM. We investigated three approaches for modeling of real-time network communication. We presented the selection criteria to select the most suitable modeling technique from the three candidates. Accordingly, we added the following objects in RCM: NS (model representation of physical network); and output and input software circuit (OSWC and ISWC) component types. These components make the communications capabilities of a node in distributed embedded system very explicit, but efficiently hide the implementation or protocol details.

The criteria for the selection of new components is based on many factors, such as, compliance with the industrial modeling standards, compatibility with the existing modeling concepts and objects in RCM, ability to abstract the implementation and configurations in distributed embedded sysems, capability of modeling legacy (already developed) systems and legacy communications, ability to model and specify timing related information (properties, requirements, constraints, etc.), ease of implementation and automatic generation of new components, ability of the modeled application to render itself to early timing analysis, etc. With the introduction of new modeling capabilities in RCM, it can be considered a suitable choice for the industrial development of distributed real-time and embedded systems. There are many reasons to support this proposition such as: ability to model and specify real-time requirements, properties and constraints during all the stages of system development; small run-time footprint (timing and memory overhead); implementation of state-of-the-art research results; ability to model real-time communication (both intra-node and inter-node); strong support for development and analysis tools, etc.

In future, we will extract an end-to-end timing model from the modeled application to perform the holistic response-time analysis. We will also demonstrate how the periodic and event chains can be traced in a distributed transaction modeled with RCM. We also plan to automatically generate the implementation of OSWC and ISWC from configuration files of other specialized protocols for real-time communication such as CANopen, HCAN, MilCAN, subsets of J1939, etc. Furthermore, we also plan to build a validator in an industrial setup with the extended RCM.

Acknowledgments This work is supported by Swedish Knowledge Foundation (KKS) within the project EEMDEF. The authors would like to thank the industrial partners Arcticus Systems Sweden and Hägglunds BAE Systems for cooperation.

References

1. Broy M (2005) Automotive software and systems engineering. Formal methods and models for co-design, MEMCODE'05
2. Charette RN (2009) This car runs on code. IEEE Spectrum 46(2). http://spectrum.ieee.org/green-tech/advanced-cars/this-carruns-on-code
3. Henzinger TA, Sifakis J (2006) The embedded systems design challenge. Proceedings of the 14th international symposium on formal methods (FM), pp 1–15
4. Crnkovic I, Larsson M (2002) Building reliable component-based software systems. Artech House, Norwood
5. Hänninen K, Mäki-Turja J, Nolin M, Lindberg M, Lundbäck J, Lundbäck K-L (2008) The rubus component model for resource constrained real-time systems. 3rd IEEE international symposium on industrial embedded systems
6. Arcticus systems. http://www.arcticus-systems.com
7. Microsoft, Distributed component object model (DCOM). http://msdn.microsoft.com/enus/library/Aa286561
8. OMG. Common object request broker architecture (CORBA), Version 3.1, January 2008. http://www.omg.org/spec/CORBA/3.1
9. DeMichiel L (2002) Sun microsystems, Enterprise JavaBeans specification, Version 2.1. Sun microsystems
10. AUTOSAR Techincal overview, Version 2.2.2. AUTOSAR—AUTomotive Open System ARchitecture, Release 3.1. The AUTOSAR Consortium, August 2008. http://www.autosar.org
11. Heinecke H et al (2006) AUTOSAR—Current results and preparations for exploitation. In: Proceedings of the 7th euroforum conference, ser. EUROFORUM'06
12. TIMMO Methodology, Version 2, TIMMO (TIMing MOdel), Deliverable 7, October 2009. The TIMMO Consortium
13. TADL: Timing augmented description language, Version 2, TIMMO (TIMing MOdel), Deliverable 6, October 2009. The TIMMO Consortium
14. The UML profile for MARTE, January 2010. http://www.omgmarte.org/
15. EAST-ADL domain model specification, Deliverable D4.1.1. http://www.atesst.org/home/liblocal/docs/ATESST2-D4.1.1
16. TIMMO-2-USE. http://www.itea2.org/projects/step/2
17. Sentilles S, Vulgarakis A, Bures T, Carlson J, Crnkovic I (2008) A component model for control-intensive distributed embedded systems. In: Proceedings of the 11th international symposium on component based software engineering, pp 310–317
18. Carlson J, Feljan J, Mäki-Turja J, Sjödin M Deployment modeling and synthesis in a component model for distributed embedded systems. 36th EUROMICRO Conference on Software Engineering and Advanced Applications, pp 74–82
19. Ke X, Sierszecki K, Angelov C (2007) COMDES-II: a component-based framework for generative development of distributed real-time control systems. 13th IEEE international conference on embedded and real-time computing systems and applications, RTCSA, pp 199–208
20. Mäki-Turja J, Hänninen K, Nolin M (2007) Efficient development of real-time systems using hybrid scheduling. 9th real-time in Sweden (RTiS'07), pp 157–163
21. Guo Y, Sierszecki K, Angelov C (2008) COMDES development toolset. 5th international workshop on formal aspects of component software FACS 08, Spain

22. Catalog of specialized CORBA specifications. OMG group. http://www.omg.org/technology/documents/
23. Lankes S, Jabs A, Bernmerl T (2003) Integration of a CAN-based connection-oriented communication model into real-time CORBA. Parallel and distributed processing symposium
24. Finocchiaro R, Lankes S, Jabs A (2004) Design of a real-time CORBA event service customised for the CAN bus. In: Proceedings of the 18th international parallel and distributed processing symposium
25. GmbH RB (1991) CAN specification Version 2.0. September
26. CANopen high-level protocol for CAN-bus, Ver. 3.0. NIKHEF, Amsterdam, March 2000
27. Westerlund J (2009) Hägglunds controller area network. Network implementation specification. BAE Systems, Hägglunds, Sweden
28. MilCAN (CAN for Military Land Systems domain). http://www.milcan.org/

Chapter 41
A Low-Latency Virtual-Channel Router with Optimized Pipeline Stages for On-Chip Network

Shanshan Ren, Zhiying Wang, Mingche Lai and Hongyi Lu

Abstract Network-on-Chip (NoC), as the next generation interconnection technology on chip, facilitates a high-bandwidth communication and improves the performance of communication observably. Since the NoC performance closely relies on the intra-router latency, a low-latency virtual-channel router with optimized pipeline stage is proposed in this study. By utilizing look-ahead routing algorithm and speculation switch allocation strategy, the pipeline depth is shortened to two stages. To further shorten the critical path of designs, this study also proposes a three-way parallel arbitration mechanism when designing low-latency virtual-channel allocator and speculative switch allocator. The experiment results show the allocators present the good scalability to different virtual channel and physical port numbers and the pipeline stage delay is reduced by 33% compared with the traditional one. Then, the proposed router can provide throughput increase and latency decrease with 19 and 25%, respectively, compared to the traditional router.

Keywords Network on chip · Virtual-channel allocator · Speculation switch allocator

S. Ren (✉) · Z. Wang · M. Lai · H. Lu
School of Computer, National University of Defense Technology, Changsha, China
e-mail: shanshanren333@163.com

Z. Wang
e-mail: zhiyingwang@nudt.edu.cn

M. Lai
e-mail: mingchelai@nudt.edu.cn

H. Lu
e-mail: hongyilu@nudt.edu.cn

J. J. Park et al. (eds.), *Proceedings of the International Conference on Human-centric Computing 2011 and Embedded and Multimedia Computing 2011*, Lecture Notes in Electrical Engineering 102,
DOI: 10.1007/978-94-007-2105-0_41,

41.1 Introduction

With the rapid development of semiconductor technology, many transistors are allowed to be integrated on a single chip. Thus the number of cores on chip is increased and the on-chip interconnect is becoming a critical bottleneck of chip design. But the traditional on-chip communication structures, such as buses, cannot satisfy the requirement of multi-core chip designs any longer [1]. Fortunately, referring to the conception of the internet field, system designers proposed a new interconnection technology, such as network-on-chip (NoC) [2], to support on-chip interconnect. The high performance, low latency and good scalability of NoC well meet the challenge of requirement for multi-core chip design.

Since NoC has been proposed, many researchers have done lots of works in this field. Some of the work is to reduce the intra-router latency, which is extremely critical for the network latency. Till now, several methods have been proposed to solve this problem. Totally speaking, because the intra-router latency is directly related to the router pipeline depth, many researchers are used to reduce the pipeline to fewer stages as possible. Speculation is a recent way to perform different pipeline stages of flit transfer in parallel [3, 4]. Look-ahead routing algorithm [5, 6] attempts to reduce pipeline depth by perform routing computation ahead. The approaches of Guess transmission [7] and advancing routing computation [8] were also proposed, hoping to reduce pipeline depth to the single stage; but always incurred low guess accurateness, complex hardware design and many other limitations, which made them hard to reach their goals.

Compared to the previous works, the contribution of this study is to propose a low-latency virtual-channel router with optimized pipeline stages. The design idea comes from the pipeline knowledge [9], which tells that there are two ways to reduce total delay of pipeline: shortening pipeline depth and shortening stage delay. We firstly shorten the pipeline depth to two stages by utilizing look-ahead routing algorithm and speculation switch allocation strategy. Then in order to shorten the delay of pipeline stage, we propose a three-way parallel arbitration mechanism when designing low-latency virtual-channel allocator and speculative switch allocators. The experimental results show that the proposed model reduces the stage delay by 33% averagely when compared to traditional ones and presents the good scalability to different channel and input numbers. Then, the proposed router provide throughput and latency is, respectively, improved by 19 and 25%.

The rest of the paper is organized as follows. Section 41.2 gives the router pipeline and presents the architecture of the proposed router. In Sect. 41.3 we analyze the delay of the traditional models. In order to reduce the stage delay, we propose a low-latency virtual-channel allocator and a low-latency speculative switch allocator in Sect. 41.4. Section 41.5 presents our experiment results of our design and Sect. 41.6 concludes the paper.

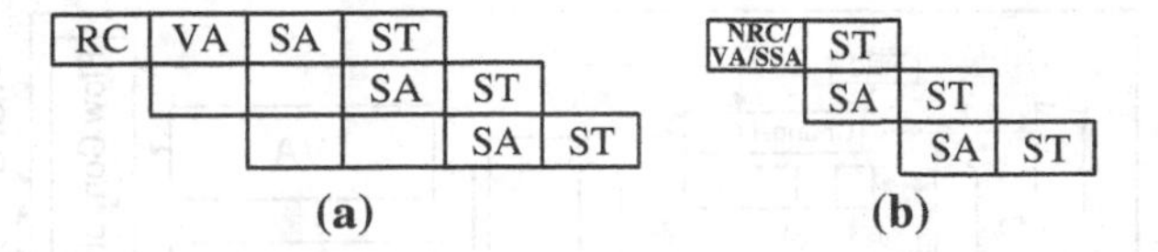

Fig. 41.1 Comparison of different pipelines for on-chip router **a** pipeline of traditional router **b** pipeline of proposed router

41.2 Architecture of Router

41.2.1 Router Pipeline

In traditional virtual-channel router, flits of a packet must proceed through four router functions in order before they leave a router, including routing computation (RC), virtual-channel allocation (VA), switch allocation (SA) and switch traversal (ST), shown in Fig. 41.1a. There are dependencies between these functions, e.g., a flit must do virtual-channel allocation after routing computation, as the result of the later will tell the head flit which virtual channels to request for. In order to improve performance, these functions usually work in a pipeline which has four stages. If we plan to shorten pipeline depth in order to reduce intra-router latency, we must remove these dependencies between them. Fortunately, there are several ways proposed by researchers to remove these dependencies. We choose look-ahead routing algorithm and speculation switch allocation strategy, as both of them do not lengthening the delay of each stage and are not complex for hardware design.

Look-ahead routing algorithm means that current router's output port is computed by the upstream router and sent to the current router in one field of the head flit. Thus, the head flit does not need to wait routing computation and can directly request for virtual-channel allocation. Look-ahead routing algorithm makes RC for the next output port and VA can be performed in parallel. In order to distinguish RC for the next output port from traditional routing computation, the former is called NRC.

Speculation switch allocation strategy is to remove the dependency between virtual-channel allocation and switch allocation. A flit sends a request for switch allocation assuming it will succeed in its virtual-channel allocation; thus SA and VA can be performed in the same cycle. In order to tell SA done in the same cycle with VA from SA done after VA, the former is called SSA.

Thanks to look-ahead routing algorithm and speculation switch allocation strategy a flit can perform NRC, VA and SSA in the same cycle and the pipeline depth reduces from four to two stages, as shown in Fig. 41.1b.

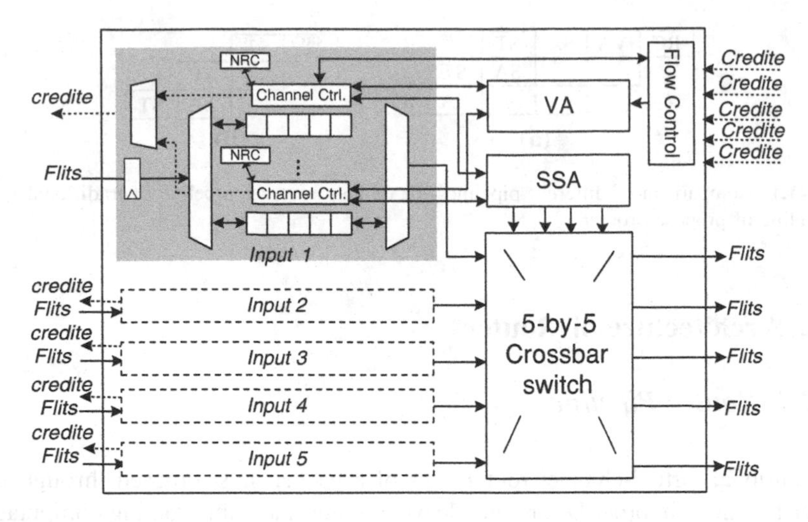

Fig. 41.2 Diagram of proposed router block

41.2.2 Block Diagram

Figure 41.2 shows the block diagram of the low-latency router designed in this paper, which is based on 2D-mesh topology and dimension-order routing algorithm. As illustrated in Fig. 41.2, there are five input ports, a virtual-channel allocator, a speculation switch allocator, a crossbar switch and a flow control. Input ports are to buffer packets, perform look-ahead routing computation, send requests to virtual-channel allocator and speculative switch allocator and transmit flow-control information to upstream router. In order to make effective use of buffer storage, each input ports of router are divided into several virtual-channels. Virtual-channel allocator and speculation switch allocator receive requests from all input virtual-channels and arbitrate which one to win. The crossbar switch, which has five inputs and five outputs, is responsible to transmit flits from its input port to its corresponding output port. And flow control is to compute the number of free buffers in each input virtual-channel downstream, based on flow-control information from input virtual-channels downstream and the number of flits sent from the current router to the downstream router.

Consider a two-flit packet, one head flit and one tail flit, traversing the proposed router. When the head flit arrives, it is buffered in the appropriate virtual-channel according to its virtual-channel identifier field. As it is a head flit, its destination field is sent to the NRC logic, which returns the output port of the next router. At this point, according to the output port computed by the upstream router, the virtual-channel control sends virtual-channel request to virtual-channel allocator and speculative switch request to speculative switch allocator. If the two requests both win, the head flit will traverse crossbar switch and leave for the next router. If virtual-channel request fails, the virtual-channel control will send the two requests

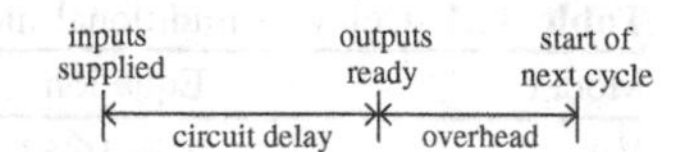

Fig. 41.3 Delay of the general on-chip router model

again. If only speculative switch request fails, it will just send switch allocation request. As to the tail flit, it will only send requests to the speculative switch allocator as it use the same virtual-channel with the head flit.

41.3 Analyze Stage Delay

Firstly, let us introduce logical effect theory [10], which will be used to compute delay of each stage. Logical effort theory proposes a method for fast back-of-the-envelope estimates of delay in a CMOS circuit using a technology-independent unit, τ. It gives us equations to compute total circuit delay along a path, one of which is shown in (1). From this equation we can know, total circuit delay is the sum of the effort delay and the parasitic delay of gates along the path. The effort delay is the product of the logical effort required to perform a logic function and the electrical effort required to drive an electrical load, while the parasitic delay is the intrinsic delay of a gate due to its own internal capacitance.

$$T_{\text{total}} = T_{\text{effort}} + T_{\text{parasitic}} \tag{41.1}$$

In order to compute the delay of each model by using logical effort theory, Ref. [11] prescribes the delay of each model is the sum of circuit delay and overhead, shown by Fig. 41.3. Circuit delay spans from when inputs are presented to the module to when the outputs needed by the next model are stable. Overhead is the additional delay required before the next set of inputs can be presented to the module. Take arbiter for example, circuit delay of it is from when requests are sent to it to when the grant signals are stable and overhead of it is the time needed to update priorities after latest arbitration.

With logical effort theory and this model, it is easy to get the parameterized equations of the delay of traditional models for our proposed router, shown in Table 41.1, in which P stands for the number of input ports, V stands for the number of the virtual channels and W stands for the data width of physical ports. And in order to analyze, we take the value of parameters into those equations. In our proposed router, we assume there are five 32-bit wide input ports and output ports and each input port has four virtual-channels, i.e., $P = 5$, $V = 4$, $W = 32$.

As delay of each model is provided, we can analyze the delay of each stage now. In the first stage, there are three models performed: routing logic for NRC, virtual-channel allocator for VA and speculative switch allocator for SSA. Since dimension-order routing only does several compares, routing logic will incur less delay than an arbiter and need not be took in mind. Thus the delay of the first stage is the delay of virtual-channel allocator, which is bigger than that of speculative

Table 41.1 Delay of traditional models

Model	Equation	Delay(τ)
VA	$T = 16.5\log_4 V + 21.5\log_4 PV + 25 + \frac{5}{6}$	89
SSA	$T = 31\log_4 V + 5\log_4 P + 16.5\log_4 P + 17.5$	74
ST	$T = 15\log_4 P + 5\log_4 W - 5$	25

switch allocator. In the second stage there is only crossbar switch for ST, thus the delay of the second stage is the delay of crossbar switch. Because the delay of pipeline stage is the biggest delay of each stage, which is 89τ, it is obvious that if we want to shorten the delay of pipeline stage we should to shorten the delay of virtual-channel allocator and speculative switch allocator.

41.4 Low-latency Allocators

41.4.1 Virtual-channel Allocator

In order to shorten the delay of virtual-channel allocator, we propose a low-latency virtual-channel allocator, which is illustrated in Fig. 41.4. Here, *P* stands for the number of physical input ports or output ports, which are the same, and *V* stands for the number of virtual channels. If the result of routing algorithm is any virtual-channel of an output port, the traditional virtual-channel allocator has two arbitration stages which are performed in order, but the proposed one has three arbitration ways which are performed in parallel. The first way needs *P V*: one arbiters for each input port to handle which virtual-channel sends request to each output port; the second way needs a *P*: one arbiter for each output port to choose an input port; the third way needs a *V*: one arbiter for each output port to decide which output virtual-channel to allocate to input virtual-channels. In order to ensure the output virtual-channel already held by packets are not to allocated to others, the inputs of the third arbitration way are the outputs of R–S triggers which latch the status of each output virtual-channel. After two 2-AND gates deal with the results of three arbitration ways, each input virtual channel can get an output virtual-channel or no output virtual-channel.

Here is an example to explain how this virtual-channel allocator works. Assume the routing result of virtual-channel 1 of input port 1 is any virtual-channel of output port 3 and in the next cycle only virtual-channel 1 of input port 1 sends requests to virtual-channels of output port 3. In the first arbitration way, virtual-channel 1 of input port 1 wins in the third arbiter of input port 1; in the second arbitration way, input port 1 wins in the arbiter of output port 3 and in the third arbitration way output virtual-channel 1 is available and wins in the arbiter of output port 3. After the results of the ways pass the two 2-AND gates, virtual-channel 1 of input port 1 is allocated virtual-channel 1 of output port 3.

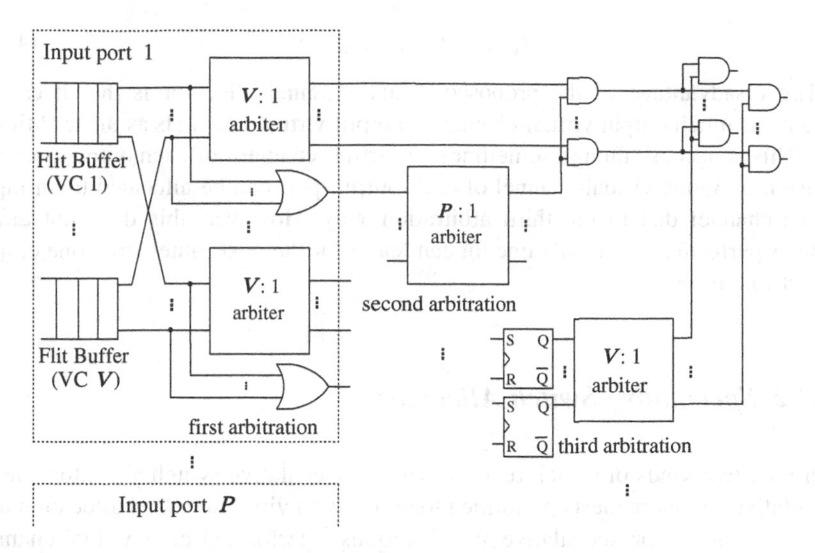

Fig. 41.4 Low latency virtual-channel allocator

The following is to derive the equations of the circuit delay of the allocator. As the results of the first way and the second way meet the result of the third way after a 2-AND gate, then the circuit delay of the first way and the second way should add the circuit delay of a 2-AND gate. The circuit delay of three arbitration ways is derived by (41.2)–(41.4). Then the circuit delay of the allocator is the biggest circuit delay of three ways plus the circuit delay of a 2-AND gate, derived by (41.5).

$$T_{\text{circuit-1}} = 16.5 \log_4 V + 10\frac{1}{12} \tag{41.2}$$

$$T_{\text{circuit-2}} = 16.5 \log_4 P + \frac{5}{3}V + 10\frac{5}{12} \tag{41.3}$$

$$T_{\text{circuit-3}} = 16.5 \log_4 V + 6\frac{3}{4} \tag{41.4}$$

$$T_{\text{circuit-total}} = \max(T_{\text{circuit-1}}, T_{\text{circuit-2}}, T_{\text{circuit-3}}) + \frac{10}{3} \tag{41.5}$$

In proposed virtual-channel allocator, there are two circuits which need to update before the next set of inputs can be presented to the module: the matrix arbiter and the R–S trigger. The updating of the R–S trigger can proceed in parallel with the updating of matrix arbiter and the former takes less time than the later. Thus the overhead of the virtual-channel allocator is the overhead of a matrix arbiter and the value is nine, which is derived in [11]. Then the total delay of the virtual-channel allocator is the sum of the circuit delay and the overhead, derived in (41.6).

$$T_{\text{total}} = T_{\text{circuit-total}} + 9 \tag{41.6}$$

The disadvantage of the proposed virtual-channel allocator is that it cannot allocate as much output virtual-channels to input virtual-channels as the traditional one. This is because though sometimes all virtual-channels of each output port can be used, only one virtual-channel of each output port can be allocated to an input virtual-channel due to the third arbitration way. However, this does not affect router's performance, as only one flit can leave for the next router from one output port at one time.

41.4.2 Speculative Switch Allocator

There are two kinds of switch requests sent to speculative switch allocator: one is speculative switch requests performed together with virtual-channel allocation and the other one is non-speculative switch requests performed after virtual-channel allocation. As can be seen, a speculative switch request which wins in switch allocation will not result in transfer of a flit if the virtual-channel request fails; while a non-speculative switch request which wins will result in transfer of a flit because the switch allocation is done after the virtual-channel allocation. In order not to influence the throughput, non-speculative switch requests should have higher priority than speculative switch requests. Thus, the detailed design idea is: non-speculative switch requests and speculative switch requests are separately arbitrated in different blocks; if one output port has non-speculative switch requests and speculative switch requests, speculative switch requests will be invalidated; if a non-speculative switch request and a speculative switch request of the same input port both wins in the different blocks the speculative switch request will be invalidated.

Figure 41.5 shows the proposed low-latency speculation switch allocator. Here, P stands for the number of physical input ports or output ports, which are the same, and V stands for the number of virtual channels. There are two blocks in this allocator: one block to deal with non-speculative switch requests and the other to deal with speculative switch requests. Both of them have three arbitration ways which are performed in parallel. In the block for non-speculative switch requests, there is a P V: one arbiter for each input port in the first arbitration way to decide which virtual-channel can use the input port to transmit a flit, followed by P V: one arbiters for each input port in the second arbitration way to decide which virtual-channel of each input port sends request to each output port, and a P:one arbiter for each output port in the third arbitration way to handle which input port wins access to the output port. Then the results of three arbitration ways are sent to a 3-AND gate to get the allocation result. As to the block for speculative switch requests, the first and second arbitration ways are almost the same as that of the block for non-speculative. Their difference is the inputs of the third arbitration way. In the block for non-speculative switch requests, before the third arbitration way, there are

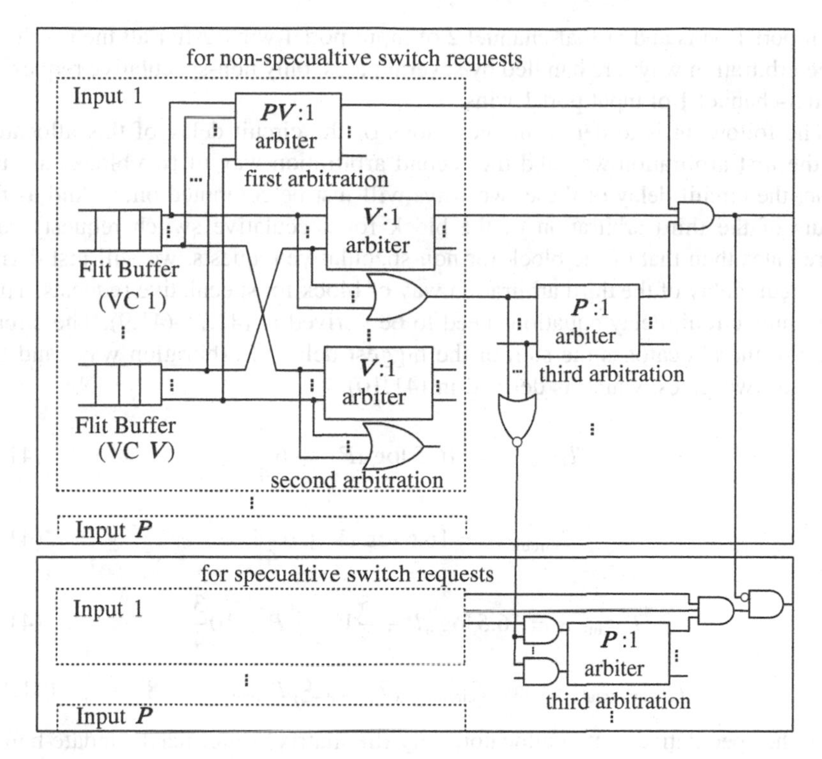

Fig. 41.5 Low latency speculative switch allocator

several V-NOR gates to detect whether there is any non-speculative switch request for each output port and send the result to 2-AND gates in the block for speculative switch requests. And in the block for speculative switch requests, before the third arbitration way, there are several 2-AND gates to invalidate speculative switch requests if there is any non-speculative switch requests for the same output port. Then the results of the three arbitration ways of the block for speculative switch requests are sent to a 3-AND and a 2-AND gates to get the allocation result of requests, the 2-AND gate of which is to avoid that requests of the same input port wins at the same time.

Here is an example to explain how speculation switch allocator works. Assume virtual-channel 1 of input port 1 sends non-speculative switch request to output port 1, virtual-channel 2 of input port 1 sends speculative switch request to output port 2, virtual channel 3 of input port 1 sends non-speculative request to output port 3 and virtual-channel 1 of input port 2 sends speculative switch request to output port 1. At the first arbitration way, as input port 1 has two virtual-channels sending non-speculative switch request, we assume virtual-channel 3 fails in the first arbitration way. After the second arbitration way, the four requests win separately. Before the third arbitration way the speculative switch request of output 1 is invalidated by the non-speculative switch request. After the third arbitration way, virtual-channel 1 of input port 1 wins and virtual-channel 2 of input port 1 wins. After all the results of three arbitration ways are handled by several gates, only non-speculative request of virtual-channel 1 of input port 1 wins.

The following is to derive the equations of the circuit delay of this allocator. As the first arbitration way and the second arbitration way of two blocks are the same, the circuit delay of these two ways will just be computed once. And as the inputs of the third arbitration of the block for speculative switch requests pass more gates than that of the block for non-speculative requests, we will just derive the circuit delay of the third arbitration way of block for speculative requests. Thus only three circuit delay equations need to be derived in (41.7)–(41.9). The circuit delay of the allocator is the sum of the biggest delay of arbitration ways and the delay of two gates, which is derived in (41.10).

$$T_{\text{circuit}-1} = 16.5 \log_4 PV + 6\frac{3}{4} \tag{41.7}$$

$$T_{\text{circuit}-2} = 16.5 \log_4 V + 6\frac{3}{4} \tag{41.8}$$

$$T_{\text{circuit}-3} = 16.5 \log_4 P + \frac{5}{3}V + \frac{5}{3}P + 10\frac{3}{4} \tag{41.9}$$

$$T_{\text{circuit}-\text{total}} = \max(T_{\text{circuit}-1}, T_{\text{circuit}-2}, T_{\text{circuit}-3}) + 8 \tag{41.10}$$

In the speculative switch allocator, only the matrix arbiter needs update before the next set of inputs can be presented to the module, thus the overhead of this allocator is 9τ. The equation of the delay of this allocator is derived by (41.11).

$$T_{\text{total}} = T_{\text{circuit}-\text{total}} + 9 \tag{41.11}$$

41.5 Experiments

41.5.1 Delay Analysis

Synopsys prime time is always used to simulate the delay of different design. Here we use it to simulate the delay of the proposed allocators in a 0.18 μm CMOS technology and in this process technology, a τ = 18ps. For our proposed router, which has five input ports and five output ports and each physical port has four virtual channels, the delay of traditional virtual-channel allocator is 87.9 τ and the delay of proposed virtual-channel allocator is 45.2 τ, which reduces 49%. And for the proposed router, the delay of traditional speculative switch allocator is 74 τ and the delay of proposed speculative switch allocator is 59.4 τ, which reduces 20%. In the proposed router with proposed virtual-channel allocator and speculative switch allocator, delay of first pipeline stage is the delay of speculative switch allocator,

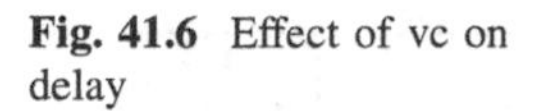
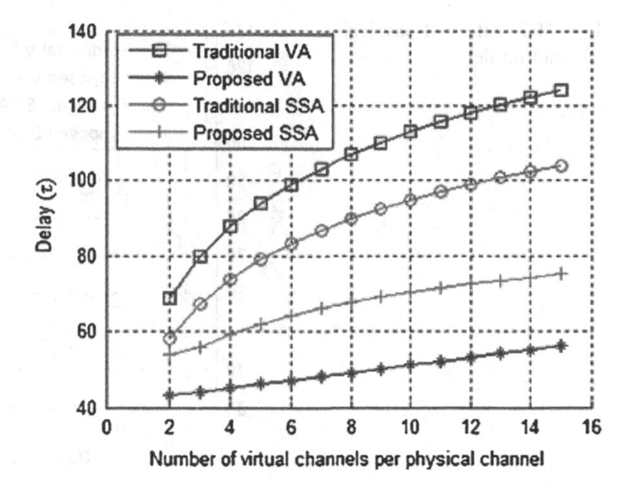

Fig. 41.6 Effect of vc on delay

which is bigger than virtual-channel allocator and the value is 59.4 τ. As the delay of second pipeline stage is smaller than that of first pipeline stage, the delay of pipeline stage of proposed router is 59.4 τ. Thus, the intra-router latency of proposed router is reduced by 33% when compared to the traditional router, the intra-router latency of which is 87.9 τ.

In order to investigate whether our proposed allocators has the good scalability to the different numbers of virtual-channels per physical ports and physical ports, we use synopsys prime time to simulate the delay of traditional allocators and proposed allocators with different numbers of virtual-channels per physical ports and physical ports. Figure 41.6 is the delay of traditional allocators and proposed allocators when the number of physical ports is five and the number of virtual-channels per physical ports changes from 2 to 15, while Fig. 41.7 is the delay of traditional allocators and proposed allocators when the number of virtual-channels per physical ports is four and the number of physical ports changes from 2 to 15. From these two figures, we can see the delay of our proposed allocators is lower than traditional allocators when the number of virtual-channels per physical ports and the number of physical ports change. Thus, our proposed allocators can apply to many different routers, i.e., our proposed allocators are strongly robust.

41.5.2 Performance Analysis

Throughput and latency [12] are two performance parameters of network which are often considered. Throughput is the rate of transmitting packets under a traffic pattern, which is the total number of packets received by all of the routers in the network in a unit time. Latency is the total time used by a packet from its first bit leaving the source router to its last bit reaching the target router. Here, we use

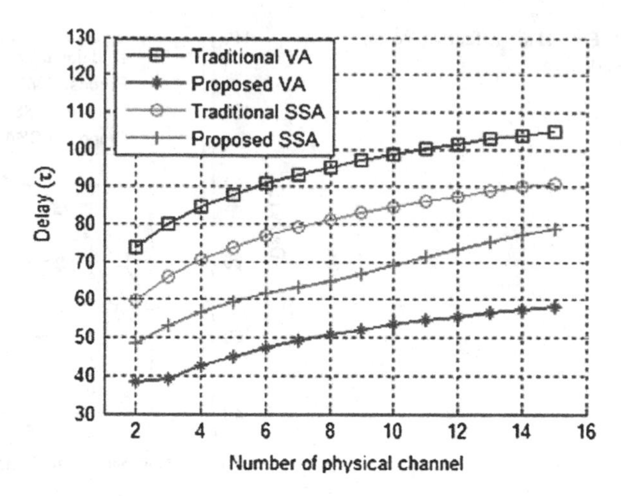

Fig. 41.7 Effect of physical channel on delay

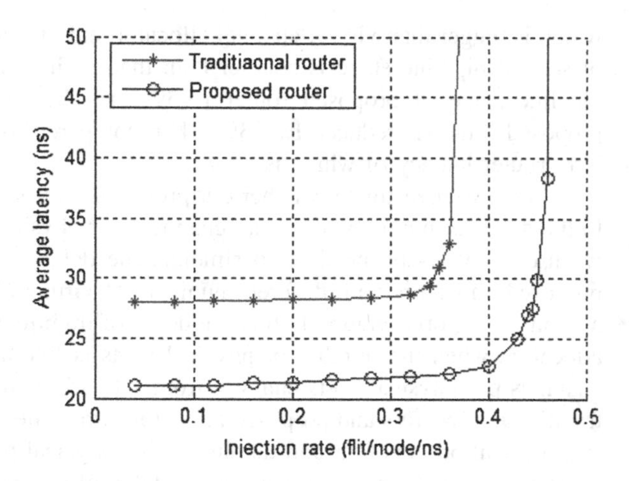

Fig. 41.8 Effect on average latency delay

BookSim monitor to analyze throughput and average latency of network. And we choose the random pattern as the traffic pattern to inject flits into network.

In order to indicate the advantage of our proposed router, we wrote two router programs: one is traditional router with look-ahead routing algorithm and speculation switch allocation strategy and the other is our proposed router. Figure 41.8 illustrates the effect on average latency of two routers when the injection rate changes. As can be seen from Fig. 41.8, network latency of proposed router is lower than that of traditional router by 25%, respectively. It can also be seen from Fig. 41.8 that when injection rate is small network latency does not change much and when injection rate is large network latency changes fast. This is because when injection rate is too large routers cannot handle all the packets and the network becomes saturated.

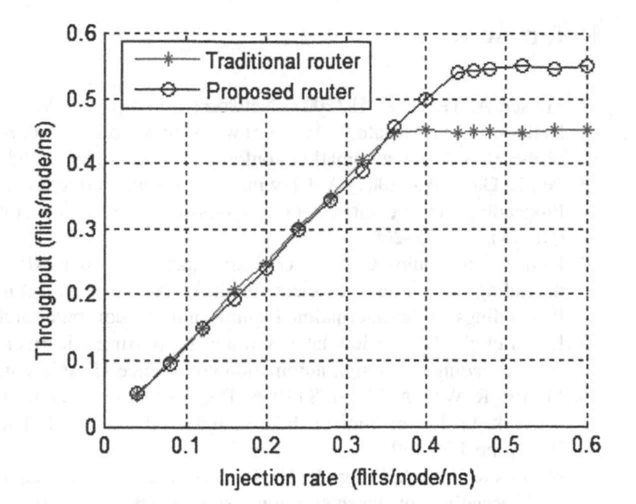

Fig. 41.9 Effect on throughput

Figure 41.9 shows the effect on network throughput of two routers when the injection rate changes. Throughput of two routers is the same when injection rate is small and throughput of proposed router is higher than that of traditional router by 19% when injection rate is large. The reason is that when injection rate is small network is not saturated and routers can send packets received and network throughput is proportional to injection rate; but when injection rate increase, the traditional router cannot send the flits accepted in time due to its intra-delay is higher than proposed router and then the network with the traditional router get saturated firstly.

41.6 Conclusions

In this paper, we proposed a low-latency virtual-channel router with optimized pipeline stage by shortening the pipeline depth and reducing the delay of pipeline stage. We apply look-ahead routing algorithm and speculation switch allocation strategy to shorten the pipeline stages to two stages. And in order to shorten the delay of critical path we propose a three-way parallel arbitration mechanism for designing a low-latency virtual-channel allocator and speculative switch allocator. The experiment results show the pipeline stage delay is reduced by 33% when compared with the traditional one and the proposed allocators presents the good scalability to different virtual channel and physical port numbers. Then, the proposed router can provide throughput increase and latency decrease with 19 and 25%, respectively, when compared to the traditional router. Future study will be concentrated on router of which pipeline is one stage in order to get lower latency.

References

1. Jantsch A, Tenhunen H (2003) Networks on chip. Academic Press, Norwell
2. Dally WJ (2001) Route packets, not wires: on-chip interconnection networks. In: Proceedings of the 38th design automation conference, Las Vegas, pp 684–689
3. Peh L, Dally WJ (2001) A delay model and speculative architecture of pipelined routers. In: Proceedings of the international symposium on high-performance computer architecutre (HPCA), pp 255–266
4. Kim J, Nicopoulos C, Park D, Narayanan V, Yousif MS, Das CR (2006) A gracefully degrading and energy-efficient modular router architecture for on-chip networks. In: Proceedings of the international symposium on computer architecture (ISCA), pp 14–15
5. Kim J et al (2005) A low latency router supporting adaptively for on-chip interconnects [J]. In: Proceedings of design automation conference (DAC). Anaheim, CA, pp 559–564
6. Mullins R, West A, Moore S (2006) The design and implementation of a low-latency on-chip network. In: Proceedings of the asia and south pacific design automation conference (ASP-DAC), pp 164–169
7. Mullins R, West A, Moore S (2004) Low-latency virtual-channel routers for on-chip network. In: Proceedings of the international symposium on computer architecture (ISCA), pp 188–197
8. Kumar A, Kundu P, Singh AP, Peh L, Jha NK (2007) A 4.6Tbits/s 3.6 GHz Single-cycle NoC router with a novel switch allocator in 65 nm CMOS. In: Proceedings of international conference on computer design (ICCD), pp 63–70
9. Hennessy JL, Patterson DA (2002) Computer architecture: a quantitative approach, 3rd edn. Morgan Kaufmann, San Francisco
10. Sutherland IE, Sproull RF, Harris D (1999) Logical effort: designing fast CMOS circuits. Morgan Kaufman, San Francisco
11. Peh L (2001) Flow control and micro-architectural mechanisms for extending the performance of interconnection networks. PhD thesis, Stanford University
12. Dally WJ, Towles B (2004) Principles and practices of interconnection networks. Morgan Kaufmann, San Francisco

Chapter 42
Towards A Web Service Based HCI Migration Framework

Yao Shen, Minjie Wang and Minyi Guo

Abstract Proliferation of mobile devices and advance in pervasive computing are leading to an urgent demand on smart human-computer interaction (HCI). In this paper, we focus on a key issue of smart HCI that is multimodal and multi-platform interaction migration. We present requirements for such smart interaction process by using simple scenarios. With these insights, we propose a web service based HCI migration framework to support open pervasive space full of dynamically emerging and disappearing devices. It integrates user preferences and context-awareness into interaction process. Our framework resides on each devices, composing a peer-to-peer structure and avoiding costly centralized services. We also illustrate our design choices for usability concerns and the migration process of our framework.

42.1 Introduction

People are no longer satisfied with sitting in front of personal computers and communicating with digital world through mouse and keyboard. Instead, people are now surrounded by various kinds of devices which connect to digital world in different ways. Besides the usage of traditional devices like projectors, the prevalent emergence of mobile devices has further provided new tunnels between physical world and digital world, thus increasing chances for people to interact with multiple devices. Unfortunately, current tools are insufficient to meet diverse

Y. Shen (✉) · M. Wang · M. Guo
Department of Computer Science and Engineering, Shanghai Jiao Tong University, 200240 Shanghai, China
e-mail: shen_yao@cs.sjtu.edu.cn

J. J. Park et al. (eds.), *Proceedings of the International Conference on Human-centric Computing 2011 and Embedded and Multimedia Computing 2011*, Lecture Notes in Electrical Engineering 102, DOI: 10.1007/978-94-007-2105-0_42,

requirements of users. As a result, the lack of convenient and flexible interaction approaches between human and multiple devices makes users more willing to use a single device rather than to migrate their tasks between several facilities [1]. Such shortage becomes even more prominent when interactions need to cater customized needs of people. Besides these challenges, ubiquitous computing (Mark Weiser, 1991 [2]) has put forward a new issue for human-computer interaction (HCI) to concern context-awareness and seamless feature. In ubiquitous computing, more sensors or probers will be deployed to make devices react properly to physical environment change rather than merely human behaviors. Therefore, how to integrate these context-driven information into interaction is an immerse problem for HCI realm to tackle.

42.2 Pervasive HCI Scenarios and Solutions

We conceive two possible scenarios for multimodal and multi-platform HCI under pervasive computing circumstance.

Scenario One: suppose a user is taking a trip with his camera. He takes photos of landscape and would like to share them to social network. Before uploading, he wants to revise these photos and selects some of these for sharing. He then unlocks his mobile phone and the photos just taken are displayed on his mobile phone. He remarks some photos and uploads them to FaceBook. The same time uploading photos, he finds a new portrait of his friend. To obtain a larger view, he steps near a public computer and the web page is automatically displayed on the computer without any login requirements. In this scenario, interaction is migrated among camera, mobile phone and public computer.

Scenario Two: suppose a dad is reading a business document on his personal computer. The document is displayed like traditional pdf on the screen. After a while, his son, a ten-year-old child, would like to use his dad's computer and continue reading the story about Peter Pan unfinished last night. After the son takes the seat and opens the story, the computer automatically reads the story for him aloud and displays story word by word according to reading speed. In this scenario, interaction is migrated from visual form to audio form according to different user contexts.

The two simple scenarios above illustrate following requirements for HCI under pervasive environment:

- *one user, many devices*: interaction needs to be migrated across *multi-platforms* according to context-awareness like physical positions of users and devices. The whole interaction system must be able to dynamically detect nearby devices and establish appropriate connection with them. Moreover, interactions may distribute among several devices.
- *many users, one device*: interaction needs to be migrated across *multimodals* according to user preferences and contexts.

Previous works have designed several solutions to these requirements. For two scenarios given above, migration among different devices (camera, mobile phone to public computer) and interaction modes (from visual display to audio form) is essential to a convenient and comfortable HCI experiences with multi-platform environment. One traditional migration technique is performed on the *process level*. Such service records process information on the source platform and create similar functionality on the target one. Nevertheless, the diverse hardware structure and operating system foundations make process migration hardly practical in current applications. Another research direction focuses on a higher level of migration—*task level*. The technique, *Task Migration* [3–5], tries to enable users to seamlessly transition a task across devices using specific technologies. The task model proposed in [3] casts a light on task definition and decomposition, theoretically establishing the methodology for task migration. However, task migration still confines itself by the capability of each task. For instance, users can never switch interaction mode since the logic presentations of the task may not contain such conversion. Moreover, task migration usually relies on centralized service which is costly and inefficient.

In this paper, we propose a new migration framework across multi-platform devices, namely *web service based migration framework*. Our objective is to support interaction migration under the environment composed of dynamically emerging and disappearing devices. Our migration framework goes beyond task level. More than merely focusing on process or task logic, we also integrate user preferences and environment context information into interaction migration to adjust platforms, contents and behaviors of interactions. Moreover, our method is based on device side web services, thus diminishing the role of centralized services.

Our main contributions are as follows:

- New framework supporting multimodal and distributed interactions across multi-platform environment.
- Web service infrastructure, which differs from centralized services, supporting dynamically emerging and disappearing devices. Our service is also able to provide open pervasive space rather than local smart area.
- Adaptive and flexible interaction methodology concerning user preferences and environment context-awareness.

42.3 Related Works

In recent years, an increasing interest has emerged in HCI with the guidance of Mark Weiser, who suggested the concept of ubiquitous computing in 1991 [2]. Predicted by Mark, computer technologies will weave themselves into fabric of everyday life until they are indistinguishable from it. Therefore, HCI technology needs to be improved to adapt to future ubiquitous environment. Antti [6] observed the urgent needs for flexible HCI technologies of workers who stepping around multiple interaction devices. Besides, the proliferation of mobile devices like PDAs and smart phones diversifies the interaction platforms and thus arises a hot topic of multi-platform interaction solutions.

Previous works have designed several framework for multi-platform collaborative web browsing. A good example of collaborative web browsing can be found in [7], where the authors designed WebSplitter, a framework based on XML to split web pages and deliver appropriate partial views to different users. Bandelloni [8], adopting similar ideas, implemented web pages migration between PC and PDA by recording system states and creating pages with XHTML. These researches provide access for users to browsing web pages on various types of devices. However, the framework is specific for collaborative web browsing, not suitable for general application interfaces.

The main obstacle to design a framework for general application user interfaces is the heterogeneity of different platforms. With various limitations and functionalities, it is hard to decide the suitable UI elements regarding the target platforms. A model-based approach developed in [3] tried to define application tasks and modeled temporal operators between tasks. This concurtasktrees (CTT) model divides the whole interaction process into tasks and describes the logic structure among them, which makes it clear the responsibilities of target platform user interfaces. Paternò and Santoro made great contributions within this realm and offered detailed support for applying the model [9, 10]. Another important issue is on automatic generation of user interfaces. Unfortunately, automatic generation is not a general solution due to the varying factors that have to be considered within design process, while semi-automatic mechanism is more general and flexible [11].

Based on CTT model and XML-based language, TERESA [4] supports transformation from task models to abstract user interfaces and then to user interfaces on specific platforms. The principle is to extract abstract description of the user interface on source platform by task model, and then to design a specific user interface by the environment on target platform. TERESA is able to recognize suitable platforms for each task sets and describe the user interfaces using an XML-based language, which is finally parsed along with platform-dependent information to generate user interfaces. TERESA tools perform rapidly in deploying applications that can be migrated over multiple devices, but it addresses only the diversity of multiple platforms while neglects the task semantics of users. With these insight, Bandelloni [12] integrated TERESA to develop a service to support runtime migration for web applications. However, the application migration service designed by Bandelloni needs centralized server and device registration, which is costly and unadaptable to newly-coming devices. Moreover, due to their basis on task model which is constrained by the task functionality, they are unable to handle multimodal interaction such as visual to audio.

Our approach differs from TERESA and Bandelloni's method in three ways. Firstly, we focus on device side web service infrastructure, which avoids centralized structure, thus alleviating the burden of server and reduce the cost. Secondly, our framework supports dynamic device environment and integrates

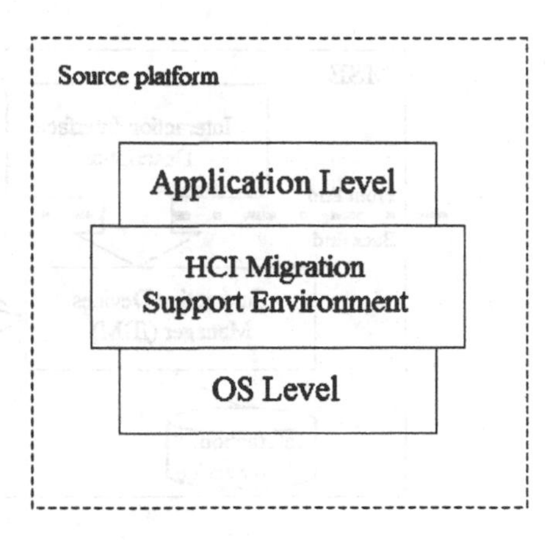

Fig. 42.1 Framework position between application level and OS level

user requirements and context-awareness. Thirdly, our framework enables interaction to be migrated across different modes.

42.4 Interaction Migration Framework

42.4.1 HCI Migration Support Environment

Traditional implementation of application user interface migration relies on centralized service, which performs costly and awkwardly facing with newly-emerging devices. Static device registration is incompetent in open pervasive surroundings since, contrary to local smart zone, we can never know what devices will be involved into interaction area. Today, however, users tend to prefer services that can support dynamic update of devices information, otherwise, they would instead use one device to finish the work regardless of its inconvenience [1]. Therefore, our design of the framework for interaction migration aims at the *hot-swapping* of interaction devices. As is shown in Fig. 42.1, the *HCI migration support environment*(MSE) lies between application level and OS level, acting as a middleware, hiding platform-dependent details from users and providing interaction migration APIs for upper applications. For each platform within interaction zone, MSE defines the boundary of application and makes application free from burdensome affairs of multi-platform interaction creations such as communication with other MSEs and so on. Application merely needs to padding interaction requirements and tasks, and then leave the details to MSE. With each device represented as an *Interaction web service*(discussed in Sect. 42.4.4), the whole connection structure resembles a peer-to-peer framework and thus avoiding centralized services. With the absence of centralized servers, our framework supports dynamically distributed devices and reduces the cost brought by large servers.

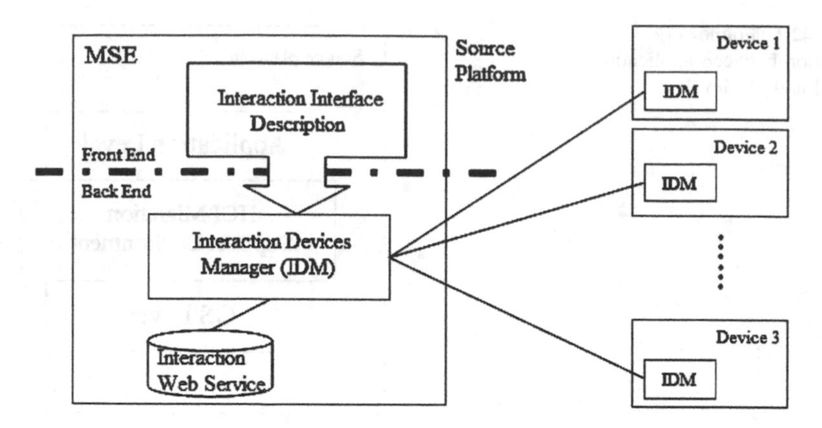

Fig. 42.2 Framework structure of HCI migration support environment (MSE)

Our framework can be further divided into two ends (shown in Fig. 42.2):

- *Front end*: which is platform-independent and responsible for handling user preferences and context-awareness
- *Back end*: which is platform-dependent and in charge of interaction migration and communication with other devices

Front end formalizes user preferences, context information and application constraints as *interaction requirements description* (IRD). Back end then utilizes these descriptions as input for *interaction devices manager* (IDM), which fetches devices information from network and exerts semantic matching to select appropriate devices for migration. Interaction Devices Manager also maintains platform-dependent information and acts as interaction device to respond to requests from other platforms. Devices that provide interaction are represented as *interaction web service*, and is provided for semantic matching. After selecting target devices, interactions are performed through networks. Sections 42.4.2–42.4.4 will discuss these components in detail.

42.4.2 Interaction Requirements Description

Human-Computer Interaction involves both human behaviors and application manners. With this consideration, we need to formalize a description, namely IRD, to include both user requirements and application response.

Figure 42.3 demonstrates the composition of an application in general: *program logic*, *interaction elements* and *interaction constraints. Program logic* assures the correct reaction of application to certain interaction input. Therefore, program logic can be separated from interaction framework, which makes our framework independent of application functionality. Our IRD includes three parts:

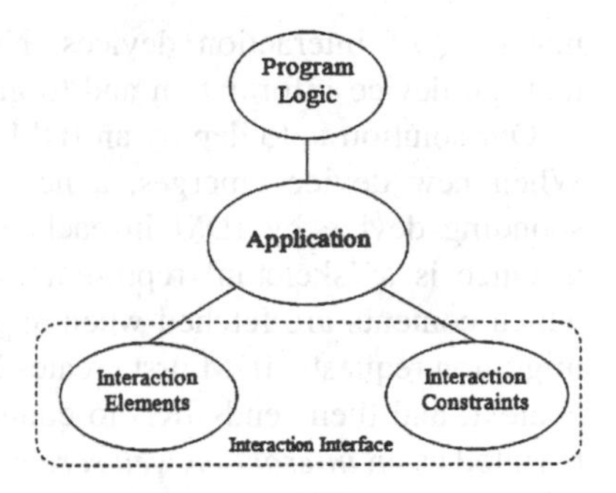

Fig. 42.3 General compositions of applications

1. *Definition of interaction elements*: for graphical user interface (GUI), interaction elements include buttons, links and other interface elements. In our condition, which is more general, the definition covers all input and output items during interaction and the descriptions of their names, types and functionality.
2. *Interaction constraints*: there are two sources of constraints. Requirements on interaction devices proposed by users and context-driven constraints are formalized as constraint descriptions used to select proper interaction devices. For example, physical position relationship between user and devices is taken into consideration. The second kind of constraints is involved in application interaction process. For example, if application interface contains audial material, constraints should involve "Support for audio output". When conflict occurs between user-defined and application-dependent constraints, we grant user-defined constraints higher privilege since interaction need to more willingly satisfy users' feelings.
3. *Interaction display moment*: defined by program logic, the moment for interaction display assures the continuity of application process. When reaching the moment for interaction display, application sends interaction requests through HCI migration support environment (MSE). All the elements displayed at the same display moment composes the distributed HCI interface.

42.4.3 Interaction Devices Manager

Heterogenous device structures always tax multi-platform interaction since it is hard to design universally compatible interface and it is costly to maintain several copies for multi-platform support. Another crucial problem for interaction migration is the matching of appropriate platforms. An intuitive idea is to analyze the migration requests from application users and fetch the most fit device service according to device description. Fortunately, the interaction requirements description discussed in Sect. 42.4.2 has combined user requests, physical contexts and application interaction elements, contributing to the accurate and swift matching of interaction devices. Nevertheless, there still remain problems to manage device information and to give proper matching mechanism.

Our solution is to deploy an IDM to take in charge of each connected devices. When new device emerges, a new instance is created to represent the corresponding device by IDM in each devices within the network connection. The instance is a 'skeleton' representation of devices to reduce storage burden, of which contents are fetched when application requests migration. When receiving migration requests, IDM first creates IRD based on application, user behaviors and context, and then sends IRD to connected devices. Each remote devices are represented as an *interaction web service* (further discussed in Sect. 42.4.4). Based on this concept, IDM on each remote device maintains an *interaction web service description* (IWSD) including interaction-related information such as display resolution, window size and supports for other interaction modes. Semantic analysis is then carried out to calculate the matching degree between its own IWSD and the provided IRD. Semantic similarity measures the ability of the devices to fulfill the constraints of this interaction. All matching degrees are returned to source platform for final decision. For successfully matched device (the one with the highest similarity), MSE announces these devices and locks the relevant resources for this device to avoid being occupied by other interaction requests (further illustrated in Sect. 42.5). It is entirely possible to match multiple devices (devices of same kinds), and MSE on source platform will decide according to interaction semantics whether to distribute parts or the total process of interaction to those devices. Once interaction devices are decided, migration requests will be sent to device terminals and connection will be established.

42.4.4 Interaction Web Service

The best feature of our framework structure is the absence of centralized server, replaced by HCI Migration Support Environment on each devices. Such framework is similar to the well-known peer-to-peer structure. Those devices which receive interaction requests from MSE on source platform are represented as *interaction web services*(IWS) for application users. For each interaction web service, IDM takes charge of maintaining the IWSD of the device (by static configuration or auto-generation), which is then used to compare with received IRD from source platform. The matching degrees are returned to source platform and MSE on source platform selects devices with the most fit Web Service description to IRD and decides target platforms (Fig. 42.4). The key consideration of such design is that we distribute the semantic matching process to remote devices, in a parallel-computing form, which largely relieves the data processing burden for source platform. Traditional semantic technologies can be applied to implement this modular such as *web service description language - semantics* (WSDL-S) [13, 14]. Another advantage for this framework is that it is independent of communication protocols which reduces the difficulty for implementation.

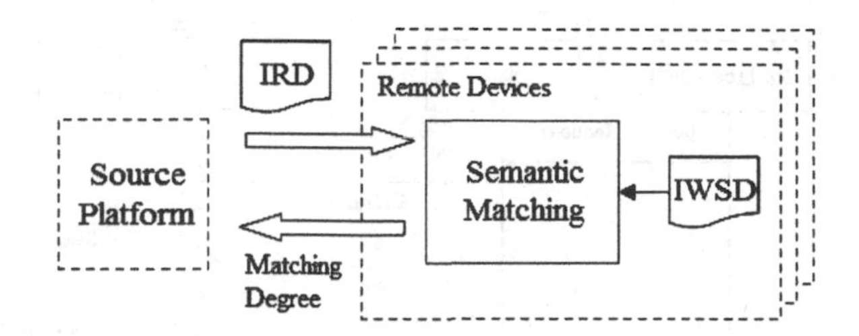

Fig. 42.4 Semantic matching

42.5 Runtime Migration Demonstration

Supporting multi-platform interaction migration means integrating platform information and creating appropriate interaction under correct circumstances. Our migration service resides on each device side, composing a peer-to-peer structure, which avoids static registration and supports dynamic device awareness. Such design, though heavy for small device as mentioned in [12], indicates its potentiality especially with the development of data processing capacity of embedded devices. To affirm practicality, our framework is designed to meet three main requirements, dynamic device awareness, user preference and usability criterion. To reach usability criterion and relieve the burden of data processing in client side, we follow two principles in our designation:

- *Less data storage on each devices*. Each device only stores its own platform information. Devices awareness is accomplished by using negotiation among devices. Based on this mechanism, device descriptions are not fetched by other devices, thus saving storage.
- *Less data process on each devices*. Task model tools like TERESA may surpass the process ability for current small devices. Therefore, we recommend semantic analysis among web service descriptions for device selection. Our distributed semantic matching process also follows this principle to ease the pressure on client side.

With these design choices, our migration process consists of these main steps (the sequence of output interaction migration is demonstrated in Fig. 42.5):

1. *Request for Interaction Migration*. Application requests for interaction migration through API provided by MSE. MSE receives Interaction Elements from application and combines with user preference to formalize Interaction Constraints. Requests can also specify the moment for interaction migration (if not specified, interaction will immediately validate). User preference can be obtained by manual input or by physical environment parameters such as position relationship between user and devices detected by GPS, wireless network or static configurations. After extracting interaction requirements description (IRD) from requests, MSE send interaction message to devices in vicinity, asking for device information.

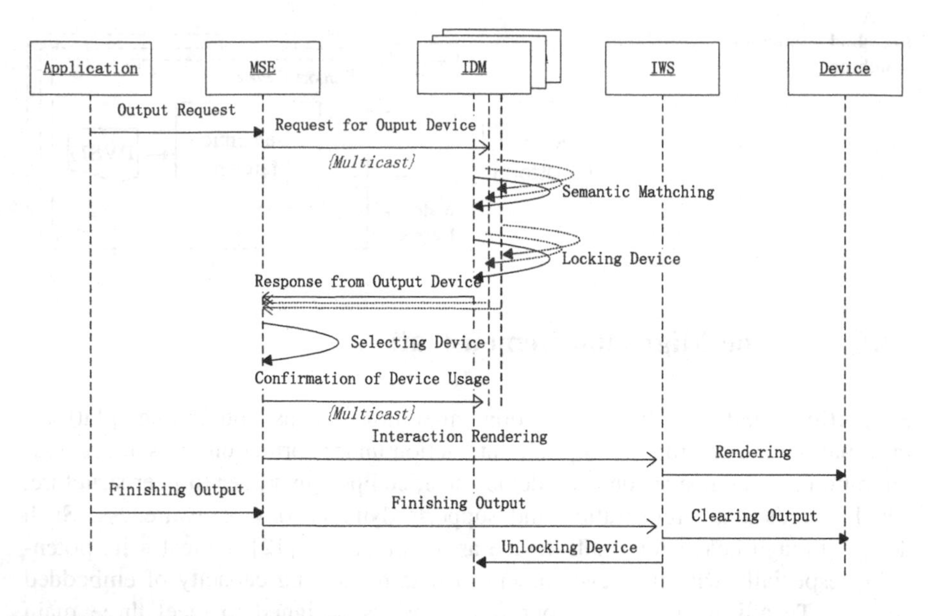

Fig. 42.5 The migration process

2. *Interaction Devices Selection*. Device information is represented by web description language (such as WSDL-S). IDM is in charge of the semantics of these Web Services. After received IRD from source platform, all IDMs calculate semantic matching degree according to IRD, lock themselves and return the results. MSE collects these semantic matching degrees from devices and selects the highest ones from them. MSE then multicasts the matching results to these devices to establish interaction connection. MSE also announces other devices not chosen to release their locks.
3. *Interaction Rendering*. When MSE determines the interaction devices within the networks, remote devices represented by matched web service will receive messages from source platform. Interaction contents are rendered in the messages for remote devices to validate by calling device drives. Interaction requirements description will also be transmitted to remote devices through network. For graphical user interface, remote devices can deploy auto-generation tools to create GUI based on interaction elements given. Moreover, callback instances will also be transmitted to remote devices for interaction feedback.
4. *Interaction Termination*. In our framework design, the display moment for interaction elements is determined by program logic. For interaction output process, application invokes Interaction termination API provided by MSE to instructs remote devices to clear output. For interaction input process, interaction finishes once input results are returned from remote devices to source platform. Resources locks on remote devices are released after interaction termination.

For input interaction migration, the whole process is similar except that input results will return to application at the end of the process. We can also divide the migration process according to physical positions. The application and MSE objects to the left part of the sequence graph reside on source platform, which requests for migration. The IDMs, IWS and device objects to the right part of the sequence graph reside on target platforms (or remote devices), which receive requests and provide interaction.

42.6 Conclusion

In this paper, we theoretically propose a new interaction framework, namely web service based HCI migration framework, which concerns multi-platform and multimodal interaction. We demonstrate requirements for such interaction using simple scenarios and present the potentiality of our framework to meet these requirements. Our framework structure avoids centralized servers, thus able to support interactions with dynamically emerging and disappearing devices. Moreover, we integrate concerns for user requirements and context-awareness into our framework, making our interaction migration more flexible and adaptive to physical environment. Our future work will continue to solve implementation problems and focus on proper semantic matching techniques. More work is needed to build an effective context-aware interaction system.

Acknowledgments This work was supported by the National Natural Science Foundation of China under Grant 61003219 and the Doctoral Fund of Ministry of Education of China under Grant 20100073120022.

References

1. Pyla P, Tungare M, Holman J, Pérez-Quiñones M (2009) Continuous user interfaces for seamless task migration. In: Julie Jacko (ed) Human-computer interaction. Ambient, ubiquitous and intelligent interaction, Lecture notes in computer science, vol 5612. Springer Berlin/Heidelberg, pp 77–85
2. Weiser M (1991) The computer for the 21st century. Sci Am 265(3):94–104
3. Paterno F (1999) Model-based design and evaluation of interactive applications. Appl Comput. Springer-Verlag
4. Mori G, Paternò F, Santoro C (2003) Tool support for designing nomadic applications. In: Proceedings of the 8th international conference on intelligent user interfaces, IUI '03, ACM, New York, USA, pp 141–148
5. Paternò F, Santoro C (2002) One model, many interfaces. In: Christophe Kolski, Jean Vanderdonckt (eds), CADUI, Kluwer, Dordrecht, pp 143–154
6. Oulasvirta A, Sumari L (2007) Mobile kits and laptop trays: managing multiple devices in mobile information work. In: Mary Beth Rosson, David J. Gilmore (eds) CHI, ACM, pp 1127–1136

7. Han R, Perret V, Naghshineh M (2000) Websplitter: a unified XML framework for multi-device collaborative web browsing. In: Proceedings of ACM CSCW'00 conference on computer-supported cooperative work, mobility, pp 221–230
8. Bandelloni R, Paternò F (2003) Platform awareness in dynamic web user interfaces migration. In: Human-computer interaction with mobile devices and services, Lecture notes in computer science, vol 2795. Springer Berlin/Heidelberg, pp 440–445
9. Paterno F, Mori G, Galiberti R (2001) CTTE: an environment for analysis and development of task models of cooperative applications. In: Proceedings of ACM CHI 2001 conference on human factors in computing systems, demonstrations: design tools, vol 2. pp 21–22
10. Mori G, Paternò F, Santoro C (2002) CTTE Support for developing and analyzing task models for interactive system design. IEEE Trans Softw Eng 28(8):797–813
11. Puerta AR, Eisenstein J (1999) Towards a general computational framework for model-based interface development systems. In: IUI, pp 171–178
12. Bandelloni R Paternò F (2004) Flexible interface migration. In: Jean Vanderdonckt, Nuno Jardim Nunes, Charles Rich (eds) IUI, ACM, pp 148–155
13. Chinnici R, Moreau JJ, Ryman A, Weerawarana S (2007) Web services description language (WSDL) version 2.0 part 1: core language. World wide web consortium, recommendation REC-wsdl20-20070626, June 2007
14. Akkiraju R et al (2005) Web service semantics–wsdl-s. W3C member submission, November 2005

Part XI
EMC 2011 Session 4: Multimedia Computing and Intelligent Services

Chapter 43
Fast Clustering of Radar Reflectivity Data on GPUs

Wei Zhou, Hong An, Hongping Yang, Gu Liu, Mu Xu and Xiaoqiang Li

Abstract In short-term weather analysis, we use clustering algorithm as a fundamental tool to analyze and display the radar reflectivity data. Different from ordinary parallel k-means clustering algorithms using compute unified device architecture, in our clustering of radar reflectivity data, we face the dataset of large scale and the high dimension of texture feature vector we used as clustering space. Therefore, the memory access latency becomes a new bottleneck in our application of short-term weather analysis which requests real time. We propose a novel parallel k-means method on graphics processing units which utilizes on-chip registers and shared memory to cut the dependency of off-chip global memory. The experimental results show that our optimization approach can achieve 40× performance improvement compared to the serial code. It sharply reduces the algorithm's running time and makes it satisfy the request of real time in applications of short-term weather analysis.

Keywords Clustering algorithm · Real time · Short-term weather forecast · GPU · CUDA

W. Zhou (✉) · H. An · G. Liu · M. Xu · X. Li
Department of Computer Science and Technology,
University of Science and Technology of China,
230027 Hefei, China
e-mail: greatzv@mail.ustc.edu.cn

H. Yang
China Meteorological Administration, Meteorological Observation Centre,
100081 Beijing, China

W. Zhou
Hefei New Star Research Institute of Applied Technology,
230031 Hefei, China

J. J. Park et al. (eds.), *Proceedings of the International Conference on Human-centric Computing 2011 and Embedded and Multimedia Computing 2011*, Lecture Notes in Electrical Engineering 102,
DOI: 10.1007/978-94-007-2105-0_43,

43.1 Introduction

There are many existing research efforts conducted in the radar data analysis for numerical weather prediction [1]. Some of them take the clustering of radar reflectivity data as a fundamental tool to depict the storm structure. The large volume and high updating frequency ($\sim$every 5 min) of the radar reflectivity data poses a strict requirement of real timing on the data processing.

Lakshmanan [2] conducted a comparing of several segmentation algorithms used to analyze weather images, and also proposed a nested segmentation method using k-means clustering [3]. However, the execution time required by k-means algorithm is significantly long and scales up with the size and dimensionality of the dataset. Such feature hinders its application into the short-term weather analysis, which emphasizes the real-timing a lot.

On the other hand, as a specialized single-chip massively parallel architecture, graphics processing units (GPUs) are becoming increasingly attractive for general purpose parallel computation beyond their traditional uses in graphics rendering. For the parallel codes that required costly computational clusters to achieve reasonable speedup, they could be ported to GPUs to gain the equivalent performance. The NIVDIA compute unified device architecture (CUDA) makes the programming easier.

GPU also brings convenience and lower cost. It has shown throughput and performance per dollar that is higher than traditional CPU by orders of magnitude. Consequently, parallel clustering algorithms for the traditional computational clusters could be ported to desktop machines or laptops now. Thus, GPU are providing weather researchers great convenience to conduct fast weather analysis and forecast.

The intrinsic feature of clustering algorithm makes it suitable to be implemented as multithreaded SIMD program in CUDA. Many researches aimed at porting the clustering algorithm to CUDA-enabled GPU [4–7]; however, most of them only port part of clustering works to the GPU, not the whole process of clustering algorithm including k centroids re-calculation. Therefore, the rich memory hierarchy of GPU has not been fully utilized yet. Meanwhile, the bottleneck in our real time system actually derives from the long memory access latency.

In this paper, we propose a novel method using on-chip registers and shared memory to reduce the global memory latency, and in turn fully utilize the feature of many-core architecture and the rich memory hierarchy to optimize the whole process of the clustering of radar reflectivity data. The result shows that our approach reduces the algorithm's execution time within 4 s with 1 million points input, and thus enable its satisfaction of the real time requirement of short-term weather analysis. Moreover, our approach could be applied in the cases of multimedia computing in which clustering is adopted in the process of images.

The paper is organized as follows. Section 43.2 presents the related works of clustering algorithm using CUDA. Section 43.3 presents the clustering of radar reflectivity data based on CPU. Section 43.4 gives the details of our fast clustering of radar reflectivity data. It shows the strategy of parallelizing and the complexity of our algorithm. The experiments of our approach are reported in Sect. 43.5. Finally, conclusions are drawn in Sect. 43.6.

43.2 Related Work

As the clustering algorithm is widely used in many areas, many researches have been made to port the clustering algorithm onto GPU [4–7].

Farivar et al. [4] parallelized the k-means algorithm and achieved a satisfying speed up. But the algorithm only dealt with 1D dataset and only parallelized the first phase of algorithm.

Chen et al. [7] propose a mean shift clustering algorithm and accelerate the k means clustering using CUDA. But after each of iteration, they use a single thread to solve the means updating and threads synchronization which reduce the performance. The GPU's rich memory hierarchy was not well utilized in this step.

Walters et al. [6] used clustering in liver segmentation. They ignored the threads synchronization between block and made the difference ratio in control. Achieve great speed up at the cost of sacrificing the precision.

Most of the work only parallelized the partial steps of clustering [4, 5, 7]. The rich memory hierarchy on GPU was seldom fully utilized to combine with the algorithm improvement. In order to meet the request of real time, our fast clustering of radar reflectivity data adopts a new method and achieves higher speed up.

43.3 Clustering of Radar Reflectivity Data

Clustering algorithm is widely used in automated weather analysis area. Lakshmanan [3] proposed a method to segment radar reflectivity data using texture. In the implementation of this method, k-means clustering algorithm was adopted to cluster the radar reflectivity data.

The k-means algorithm is one of the most popular clustering methods. In our implementation, a set of n data points $R = \{r_1, r_2, \ldots, r_n\}$ in a d-dimensional space is given, where each data point represents 1 km*1 km and the d-dimensional space is composed by the texture feature vectors extracted from radar composition reflectivity. The task of k-means is to partition R into k clusters $S = \{S_1, S_2, \ldots, S_k\}$ such that each cluster has maximal similarity as defined by a cost function related to weather analysis [8].

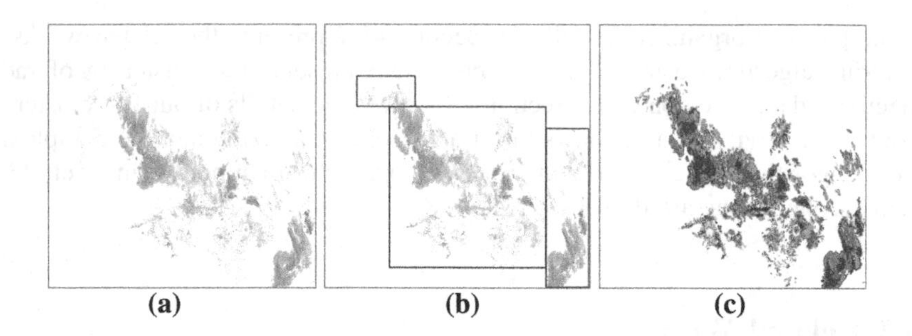

Fig. 43.1 **a** The distribution of radar reflectivity data. **b** The value of the area outside the rectangle is bellow 5dbz, thus, we can ignore it in our weather analysis, and sharply reduce the computation. **c** The result after clustering

In our clustering of radar data, we preprocess the input data by labeling the areas requiring computation. Then, we adopt k-means algorithm iteratively to partition the dataset into k clusters. It first divides up the measurement space into k equal intervals and each data point was initially assigned to the interval in which its reflectivity data value lies, and so the initial k centroids are calculated. Then, the algorithm iterates as follows: (1) Compute the cost function between each pair of data point and the centroid, and then assign each data point to the centroid with the minimum cost function. The cost function incorporates two measures, the euclidean distance and contiguity measure [8]; (2) Recalculate new centroids by taking the means of all the data points in each cluster. The iteration terminates when the changes in the centroids are less than some threshold or the number of iterations reaches a predefined threshold. The whole process is shown in Algorithm 1.

We point out that there's an interesting distribution of radar reflectivity. The points with the same value are always in the same area. We only need to compute the area we care about. The areas where reflectivity data value below a threshold can be ignored in our meteorology application. So we preprocess the input data and label the areas needing computation before performing k-means. The areas needing computation are only the rectangle areas in Fig 43.1b. the feature of the reflectivity distribution sharply reduces the computational requirement.

Our clustering of radar reflectivity data includes the following steps

1. Preprocess the input data.
2. Partition the input data using k-means with the cost function.

Algorithm1 Clustering of radar data based on CPU

```
//Preprocess(input data): labeling the area needing
computation;
//stable: show weather it reaches stable;
//max_iter: the maximum number of iterations;
// E(ri, sj): the cost value between ri and sj;
// E_max: the temporal minimum cost value;
//classes: stores the minimum centroid ID for data
point
//means: stores the centroid for each cluster
1. Preprocess ( input data)
2. Initial partition the data and generate k centroids;
3. while !stable && iter <= max_iter
4.   for ri in R
5.      for sj in S
6.           Compute E(ri, sj);
7.           if  E(ri,sj)<E_min
8.                E_min ← E(ri, sj)
9.                classes[i] ← j ;
10.           end of if
11.       end of for
12.    end of for
13.  for i from 1 to n
14.     ++count[classes[i]];
15.      for j from 1 to d
16.         centros_temp[classes[i]][j] +=data[i][j];
17.      end of for
18.  end of for
19.  for i from 1 to k
20.     for i from 1 to d
21.        centros[i][j] = centros_temp[i][j]/count[i];
22.     end of for
23.  end of for
24.  if reach stable state or threshold
25.     stable ← true;
26.  end of if
27.  iter++;
28. end of while
```

43.4 Design of Fast Clustering of Radar Reflectivity Data on GPUs

The computational complexity of algorithm 1 is O [n + m (nk + n + k)]. Lines 1–2 preprocess the data once and get complexity O (n); Lines 3–28 contain a series of 2-phase steps. Lines 3–12 compute the cost function and assign each point to the cluster whose centroid has minimum cost function values with it. This phase has a complexity of O (nk); Lines 13–23 recalculate the centroids and has a complexity O [(n + k) d], where d is dimension of the data point.

Many works have been done to parallelize the first phase: data points assignment. But after parallelizing the first phase, the problem we faced is that the second phase recalculating the centroids becomes the new bottleneck for our short-term weather forecast application which has strict real time requirements. To see this, observe that with p cores we can accelerate the first phase to O (nk/p), so the ratio between the two is O [nk/p(n + k)] $\approx$ O (k/p)(for n $\gg$ k) when d = 1. That means when k is a few orders of magnitude larger than the amount of cores, the algorithm is bound by the first phase and not by the second. For example, in the Farivar's [4] implementation, there were about k = 4,000 clusters and 32 streaming multiprocessors and the performance is bound by the first phase only. But in our clustering of radar reflectivity data, k is 3–16 [1], and there are 16 or 30 multiprocessors (p = 16 or 30). So the second phase becomes the new bottleneck for us. We show it with the experimental results in Sect. 43.5. part A. The high-dimensional of dataset causes larger memory access and makes things worse. Therefore, we adopt a new method including parallelizing the second phase on GPUs utilizing the shared memory and registers to reduce the memory access latency.

The problems in the first phase are the large scale dataset and the high dimension of the data point which causes long memory latency. The on-chip register resource must be utilized skillfully to decrease the reading latency. The strategy is simply as follow: (1) keep the multiprocessors busy; (2) keep register usage high and reduce the usage of local memory; besides, coalesced access to global memory also decreases the reading latency.

We discuss specific design decisions to accelerate k-means for the CUDA architecture in the following two subsections. Section 43.4.1 introduces parallel algorithm for assignment of data points. Section 43.4.2 illustrates our novel parallel algorithm for k centroids recalculation.

43.4.1 Data Points Assignment

The CPU-based algorithm of assignment of data points is shown in algorithm 1 lines 3–12. There are two strategies to parallel the algorithm. The first is the centroid-oriented, in which the cost function value from each centroid to all data points are calculated and then, each point get its own centroid with minimum cost function value. Another is the data points-oriented. It dispatching one data point to one thread and then each thread calculates the cost function from one data point to all the centroids, and maintains the minimum cost function value and the corresponding centroid. The former strategy has disadvantage that each point which is stored in off-chip global memory has to be read several times and causes long latency. Another disadvantage in our clustering of radar data is that our k is small (k = 3–16), resulting in making the number of threads too small for GPU scheduler to hide the latency well in this strategy. Therefore, the latter strategy is adopted in our application. The parallel algorithm of data point assignment is shown in Algorithm 2. Lines 1–2 show the design of block size and grid; line 3–5 calculate the position of the corresponding data points for each thread in global memory; line 6 loads the data point into the register; line 7–13 compute the cost function and maintain the minimum one.

Algorithm2 Data points assignment based on GPU

```
//the input data scale: 1204*1024;
// Block: the dimension of Block;
//Grid: the dimension of Grid;
//i, j: line number of the input data and row number of
it
//Gdata: the address of the corresponding data point
//E(r_i,s_j): the cost function between r_i and s_j
//classes: stores the minimum centroid ID for data
points
1.  Dim3 Block (blocksize);
2.  Dim3 Grid (1024/blocksize , 1024);
3.  i ← blockIdx.y;
4.  j ← blockIdx.x*blocksize + threadIdx.x;
5.  Gdata ← i*1024+j;
6.  Load Gdata to the on-chip register
7.  for s_j in S
8.     Compute E(r_i, s_j);
9.     if  E(r_i,s_j)<E_min
10.           E_min ← E(r_i, s_j)
11.           classes[i] ← j ;
12.       end of if
13.end of for
```

Algorithm 2 only has one loop instead of two loops in Algorithm 1. The loop for n data point has been dispatched to n threads. If the number of processing elements were equal to the number of data points, this pass could be finish in one step. However, the number of processing elements is limited in our GPU and with p cores we can accelerate the first phase to O (nk/p).

The key point of achieving high efficiency is reducing the global memory access latency utilizing the on-chip register and hiding the latency with GPU scheduler. We accomplish as follows.

To fully utilize the on-chip register, firstly, we load the data points into the on-chip registers and ensure that reading data points from global memory happens only once when calculating the cost function in the thread. Reading from the register is much faster than reading from global memory which largely reduces the latency. Secondly, we adjust the block size and optimize the kernel program to fully use the register and reduce the usage of local memory which stored in global memory. Because of the total number of registers in stream multiprocessor is limited, the kernel program have to be adjusted properly and the block size have to be adjusted to utilize the SM's limited registers resources. Our experiments in Sect. 43.5 show that a block size of 128 results better performance than block size of 32, 64 and 256. Besides, coalesced access to the global memory also decreases the reading latency. In our design of the thread block and grid, the access of global memory can be coalesced well.

Hiding the latency is mainly done by GPU scheduler automatically. The multiprocessor SIMT unit creates, manages, schedules, and executes threads in groups of 32 parallel threads called *warps*. So the block size should be a multiple of 32. The number of blocks in our application is large enough to be scheduled.

43.4.2 K Centroids Recalculation

In order to achieve the new centroids, the points have to be added in variable centros_temp in Algorithm 1, line 16. The second phase of k centroids recalculation has a relatively low computational complexity of O (nd) and is difficult to be fully parallelized due to the write conflict.

Though it has a low computational complexity, it is a time consuming processing in our clustering of radar reflectivity data due to the long memory access latency. In order to compute the new centroids on GPU, we have to accumulate all data points in k variables in centros_tmp array. It should be done in atomic operations. Thus, the process was turned into a serial process in fact. What makes things worse, because of the variables of centros_tmp should be shared by all threads in all the blocks, they have to be stored in global memory suffering long global memory access latency.

We give a new method to solve the problem of the serial accumulation process and the long latency of global memory. Our method includes two steps as follows. Figure 43.2 shows the two steps.

Firstly, we use "divide and conquer" strategy to turn the serial process into a parallel one. We divide the dataset and accumulate the partial sum of each sub dataset in different multiprocessor simultaneously. Each part of sum would be dispatched to one block. The algorithm is shown in algorithm 3. In line 1, we use shared memory instead of global memory to store the variables of centroid_temp because of the shared memory can be shared in one block. This reduces the latency caused by atomic operations on global memory. With p cores we can get the complexity of this step to O (n/p) when $d = 1$.

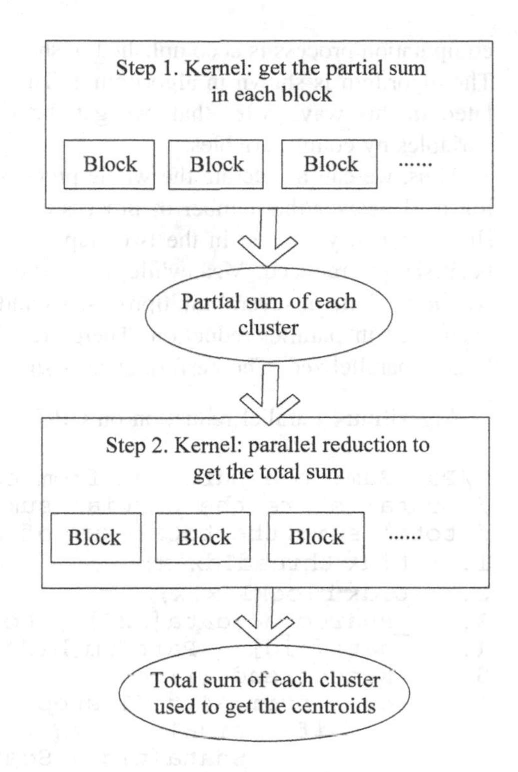

Fig. 43.2 Our method of k centroids recalculation

Algorithm3 Data points assignment based on GPU

```
// centros_temp: partial sum in each block stored in
shared memory;
//count: partial count in each block;
//Gdata: the address of the corresponding data point;
//label: the minimum centroid ID for current point
stored;
1.    _shared_  centros_temp[];
2.    _shared_  count[];
3.    i ← blockIdx.y;
4.    j ← blockIdx.x*blocksize + threadIdx.x;
5     Rdata ← data[i*1024+j];
6.    Rlabel ← classes[i*1024+j];
7.    atomicAdd(count[Rlabel]);
8.    for j from 1 to d
9.         atomicAdd(centros_temp[Rlabel][j] , Rdata[j])
10.   end of for
11.   GpartSum←centro_temp;
12.   GpartCount←count;
```

Secondly, we accumulate all the partial sums using parallel reduction algorithm which has a complexity of O (log n). We make the partial sums be accessed from global memory only once and accessed coalesced (algorithm 4, lines 4). The computation process is accomplished in shared memory (algorithm 4, lines 6–13). The algorithm is shown in algorithm 4. The variable of count can also be calculated in this way. After that, we get the centroids by dividing the total sum variables by count variables.

Thus, we can accelerate the whole process of the second phase to O (n/p + k log n). However the number of process elements is limited. In fact, because the shared memory is used in the two step of centroids calculation, the latency has been sharply reduced. Meanwhile, the serial process of accumulation is parallelized to be done in several multiprocessors and the accumulation of partial sums is calculated in parallel reduction. Therefore, the whole process of centroids was largely parallelized. The performance is shown in our experiments in Sect. 43.5.

Algorithm4 Parallel reduction on GPU

```
//PartSum: the part sum from each block;
//Sdata: store the partial sum in shared memory;
//total_sum: the total sum of all  parts of sum;
1.    tid←threadIdx.x;
2.    bid←blockIdx.x;
3.    _shared_  Sdata[128] , total_sum ;
4.    Sdata[tid]  = PartSum[bid*128+tid]
5.    _syncthreads();
6.    for i from 64 to 1 step i=i/2
7.         if   tid<1
8.              Sdata[tid]= Sdata [tid]+ Sdata [tid+i];
9.         end of if
10.       _syncthread();
11. end of for
12. if  tid = 0
13.       atomic(total_sum, s_data[0]);
14. end of  if
```

43.5 Experiments

We have implemented our fast clustering of radar data using CUDA version 2.3. Our experiments were conducted on a PC with an NVIDIA Geforce GTX275 and an Intel(R) Core(TM) Q8200 CPU. GTX 275 has 30 multiprocessors, and performs at 1.40 GHz. The memory of GPU is 1 GB with the peak bandwidth of 127 GB/s. The CPU has four cores running at 2.33 GHz. The main memory is 4 GB with the peak bandwidth of 5.6 GB/s. Our data set consist of a d-dimensional texture feature vector extract from radar reflectivity data. There are several million data points to be clustered, and the number of clusters k is smaller than 32 according to our application demands.

Table 43.1 The proportion in CPU based algorithm

Input radar data	The first phase (s)	The second phase (s)	Proportion of the second phase (%)
CREF_20100717_010	152.203	3.201	2.06
CREF_20100717_020	152.882	3.093	1.98
CREF_20100717_030	153.293	3.162	2.02

Table 43.2 The proportion after parallelizing the first phase only

Input radar data	The first phase (s)	The second phase (s)	Proportion of the second phase
CREF_20100717_010	2.204	3.201	59.2%
CREF_20100717_020	2.187	3.093	58.5%
CREF_20100717_030	2.377	3.162	57.1%

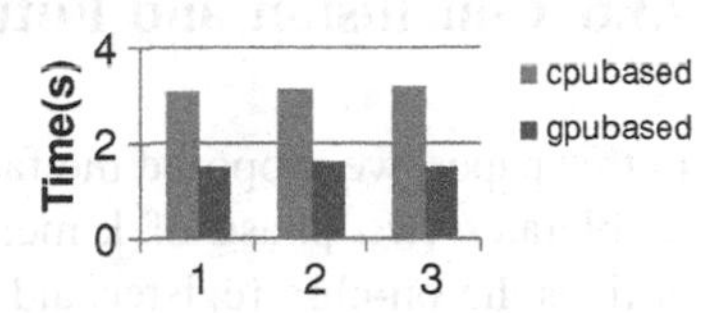

Fig. 43.3 The speed up of the second phase using our method

43.5.1 Time Consuming Analysis

The radar reflectivity data generated by multiple radars at different times in one day was used as our input data: CREF_20100717_010, CREF_20100717_020, and CREF_20100717_030.

We show the time consuming proportion of the second phase in the algorithm based on CPU in Table 43.1. We show the proportion of the second phase after only parallelizing the first phase in Table 43.2. It shows that in the CPU based algorithm, the first phase is the bottleneck to accelerate. But after parallelizing the first phase only, the second phase of k centroids recalculation becomes the new bottleneck in our real time system which takes more than 57% of the total consuming time. And the proportion doesn't change with the scale of input data.

43.5.2 Speed up of Fast Clustering of Radar Reflectivity Data

Figure 43.3 present the speedups gained by using our method on GPU for the second phase. It has a 2X speed improvement over the serial one.

The data sets with 1 , 2 , 3 million points were created. Figure 43.4 shows the speed up of our fast clustering method for the whole process. We experienced a 36 $\sim$ 40X speed improvement over a baseline application running on host machine. The speed up almost doesn't change with the input data scale.

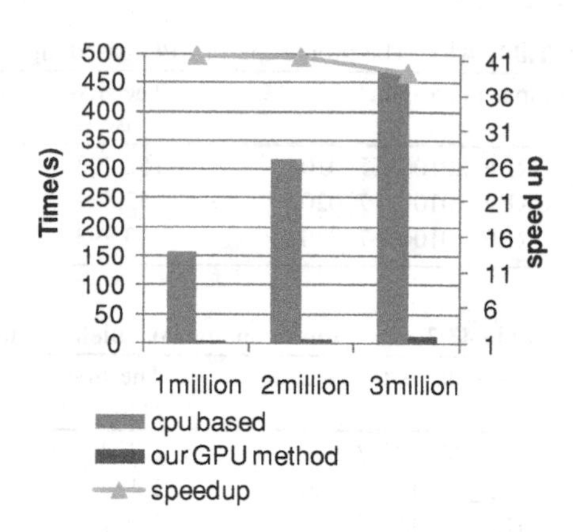

Fig. 43.4 The speed up of our parallel clustering method of whole process

43.6 Conclusion and Future Work

In this paper, we proposed the fast clustering of radar reflectivity data algorithm. It accelerates two phase of k-means clustering algorithm. The first phase mainly utilizes the on-chip register and adjusts the execution configuration to reduce the global memory latency and hide the latency with scheduler. The second phase adopts a novel algorithm. It firstly accumulates the partial sums of centroids in shared memory of different blocks in parallel. And then, it uses parallel reduction to get the total sum of partial sums and get the centroids eventually. In this way, our clustering algorithm show over a 40× performance improvement. It meets the request of real time in application of short-time weather analysis and forecast.

Acknowledgment This work is supported financially by the National Basic Research Program of China under contract 2011CB302501, the National Natural Science Foundation of China grants 60633040 and 60970023, the National Hi-tech Research and Development Program of China under contracts 2009AA01Z106, the National Science & Technology Major Projects 2009ZX01036-001-002 and 2011ZX01028-001-002-3.

References

1. Yang HP, Zhang J, Langston C (2009) Synchronization of radar observations with multi-scale storm tracking. J Adv atmos sci 26
2. Lakshmanan V, Rabin R, DeBrunner V (2003) Multiscale storm identification and forecast. J Atmos Res 67–68:367–380
3. Lakshmanan V, Rabin R, DeBrunner V (2001) Segmenting radar reflectivity data using texture. 30th international conference on radar meteorology, Zurich, Switzerland
4. Farivar R, Rebolledo D, Chan E, Campbell R (2008) A parallel implementation of K-means clustering on GPUs. International conference on parallel and distributed processing techniques and applications (PDPTA 2008), pp. 340–345
5. Miao Q, Fu ZL, Zhao XH (2009) A new approach for color character extraction based on parallel clustering. International conference on computational intelligence and software engineering, IEEE Computer Society

6. Walters JP, Balu V, Kompalli S (2009) Evaluating the use of GPUs in liver image segmentation and HMMER database searches. IEEE International Parallel And Distributed Processing Symposium, IEEE Computer Society
7. Chen J., Wu XJ, Cai R (2010) Parallel processing for accelerated mean shift algorithm with GPU. J Comput-Aided Des Comput Gr 3
8. Wang GL (2007) The Development on a multiscale identifying algorithm for heavy rainfall and methods of extracting evolvement information. Ph.D Thesis, Nanjing University of Information Science and Technology

Chapter 44
A Compiler Framework for Translating Standard C into Optimized CUDA Code

Qi Zhu, Li Shen, Xinbiao Gan and Zhiying Wang

Abstract Recent years have seen a trend in using graphic processing units (GPU) as accelerators for general-purpose computing. However, there are still challenges for developing applications on GPUs. In this work, we present a compiler framework for translating standard C applications into CUDA-based GPGPU applications. The framework transforms C applications to suit programming model of CUDA and optimizes GPU memory accesses according to memory hierarchy of CUDA. Experimental results from benchmark suites show that our compiler framework would produce efficient CUDA-based applications.

Keyword CUDA compiler framework

44.1 Introduction

Multicore architecture has become the necessary approach for improving processor performance in accordance with Moore's Law. However, as core number increases and the chip heterogeneity expands, the rapid performance improvement would not impose the flexibility of multicore resources management and the simplicity of application programming. On the contrary, the complexity of multicore structure makes it difficult to utilize a large number of chip resources efficiently.

Recently, graphic processing units (GPU) has been widely used in high performance computing applications such as biomedical, financial analysis, physical simulation and database processing due to powerful computing capability of GPU.

Q. Zhu (✉) · L. Shen · X. Gan · Z. Wang
School of Computer, National University of Defense Technology, Changsha, China
e-mail: zhudbd@live.cn

J. J. Park et al. (eds.), *Proceedings of the International Conference on Human-centric Computing 2011 and Embedded and Multimedia Computing 2011*, Lecture Notes in Electrical Engineering 102,
DOI: 10.1007/978-94-007-2105-0_44, © Springer Science+Business Media B.V. 2011

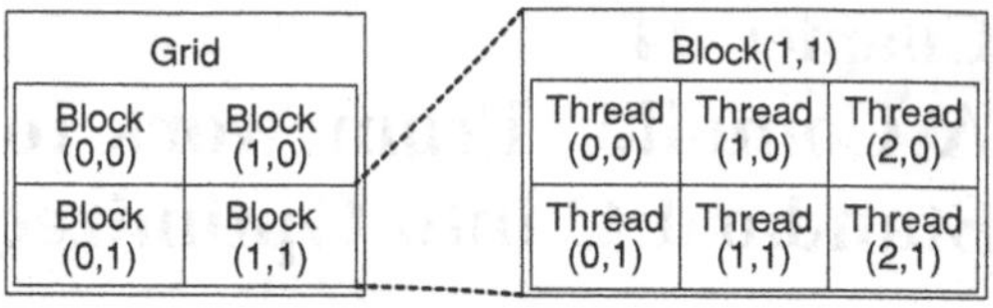

Fig. 44.1 Thread hierarchy for CUDA

Table 44.1 Properties of multi-level storage structure

Memory cell	Position	Size	Latency	Access
Global	off-chip	1 GB	300–500 cycles	read/write
Shared	on-chip	16 KB/SM	about one cycle	read/write
Constant	off-chip	8 KB/SM	about one cycle	read only
Texture	off-chip	1 GB	100 cycles	read only

More specifically, the GPU is especially well-suited to the program which can be executed on many data elements in parallel. In 3D rendering, large sets of pixels and vertices are mapped to parallel threads. Similarly, image and media processing applications such as post-processing of rendered images, video encoding and decoding, image scaling, stereo vision, and pattern recognition can map image blocks and pixels to parallel processing threads.

Compute unified device architecture (CUDA) developed by NVIDIA is a high-performance programming model for CUDA-enabled GPU. A hierarchy of thread groups, shared memories, and barrier synchronization are three critical abstractions in CUDA. And CUDA C is extended from ANSI C.

There are also three abstractions for CUDA threads in Fig. 44.1. A grid is a set of thread blocks that executes a kernel function. Each block is composed of hundreds of threads. Threads within one block can share data through shared memory and can be synchronized at a barrier.

In practice, CUDA threads may access multiple memory spaces, including Global memory, Shared Memory, Constant Memory and Texture memory, as detailed in Table 44.1.

44.2 Related Works

How to obtain the best performance from CUDA architecture is a great challenge for programmers. In order to provide a familiar and efficient programming model, Seyong Lee proposed a framework composed of optimizer and translator [1], which could translate OMP programs into efficient CUDA programs. Yi Yang and Muthu Manikandan Baskaran [2, 3] both presented an automatic CUDA code optimization process. Yixun Liu [4] described a cross-input adaptive framework for GPU programs optimization. Eddy Z. Zhang [5] reorganized data and threads dynamically to reduce the overhead introduced by sequence operations. Efficiency

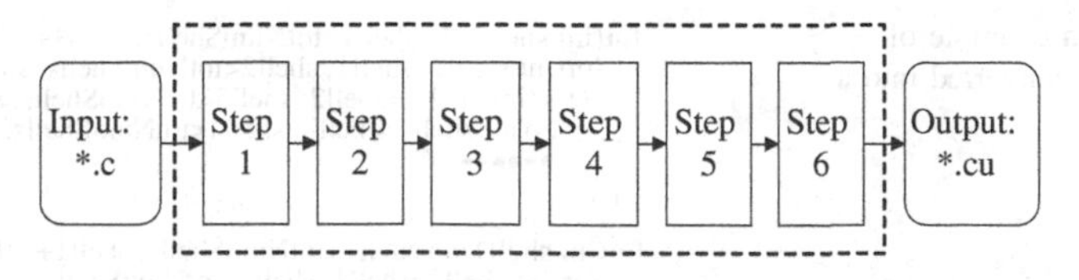

Fig. 44.2 The overall structure of the framework

and utilization were proposed by Shane Ryoo [6] to find the optimum set of parameters of GPU applications.

Differently, we proposed a novel compiler framework for translating standard *C* into efficient CUDA code, in which programmers could use standard *C* to develop GPGPU applications directly and efficiently. And we also proposed a set of general approaches to optimize GPU programs. Our purpose is to simplify parallel programming, leaving the tedious syntactic representation and complex performance tuning to our compiler framework.

44.3 Compiler Framework

The proposed compiler framework is shown in Fig. 44.2.

The input to our compiler is a *.c file and the output is a *.cu file which is supported by a CUDA compiler, in which there are six phases, including (1) identifying kernel candidate which should be suitable to execute on GPU, (2) transforming subscript for adapting to CUDA model well and further optimization, (3) generating Single-thread code, in which best kernel candidate would be converted to kernel and generate single-thread code, (4) handling Data Dependence for minimizing performance penalties, (5) merging threads hierarchically, and data layout optimization.

In this section, we highlight implementation of subscript transformation, handling data dependence, merging threads hierarchically and storage layout optimization.

44.3.1 Subscript Transformation

First, we discuss data which are not related to loop indices in loop body. Nested sequential loops can be converted to single sequential loop easily, and data can be treated as controlled by single loop indices.

Fig. 44.3 An example of nested loops converted into a single loop

```
for(int shell1=0; shell1<totNumShells; shell1++){
   for(int shell2=shell1; shell2<totNumShells; shell2++){
      for(int shell3=shell2; shell3<totNumShells; shell3++){
       for(int shell4=shell3; shell4<totNumShells; shell4++){
       ••••••
 }}}}
                    ⇩
for(int shell1=0; shell1<totNumShells; shell1++){
   for(int shell2=shell1; shell2<totNumShells; shell2++){
      for(int shell3=shell2; shell3<totNumShells; shell3++){
       for(int shell4=shell3; shell4<totNumShells; shell4++){
            data_distribute[numElements].x=shell1;
            data_distribute[numElements].y=shell2;
            data_distribute[numElements].z=shell3;
            data_distribute[numElements].w=shell4;
            numElements++;
}}}}
                    ⇩
for(int i=0;i<numElements;i++)
  ••••••
}
```

If data in the loop is related to loop indices, we should focus on how to map data which are controlled by loop indices to the thread structure of CUDA. If the loop is a single sequential loop, we can merge the neighboring threads at will, and data in the loop body can be mapped to the thread structure of CUDA naturally.

If iterative operation is a case of nested loops, and loop indices are variables, the mathematical relationship between the loop indices and threaded may be very complex. Leveraging subscript transformation, our compiler restructures nested loops to a single loop. Supposing α, β, χ, and δ are the indices of nested loops, ξ is the index of a single loop, Φ and Γ represent the relationship between indices of loops and ψ which is a variable satisfying

$$\psi = \Phi(\alpha, \beta, \chi, \delta) \tag{44.1}$$

$$\psi = \Gamma(\xi) \tag{44.2}$$

The purpose of subscript transformation is to find another relationship H, satisfying

$$\xi = H(\alpha, \beta, \chi, \delta) \tag{44.3}$$

$$\Phi = \Gamma^{\circ} H. \tag{44.4}$$

Our compiler unrolls the nested loops and records the indices in a one-dimensional array (see Fig. 44.3). Algorithm 1 describes the process in more detail.

Algorithm 1 *Subscript transformation.*

Input: A set of control parameter I = {I_1, I_2,... I_n}. Assuming i_x is i_{max} the current value of I_x, i_{xmax} is the maximum of I_x. Array F[] is the redirection of parameters.

1. numelement = 0
2. for (flag=1 to *n*)
3. Calculating the initial value of i_{flag} and $i_{flagmax}$
4. end for

Fig. 44.4 The original abstract code

```
for(i=4; i<100; i++){
S1:     a[i]=b[i-2]+1;
S2:     c[i]=a[i]+f[i];
S3:     b[i]=a[i-1]+2;
S4:     q=b[i-1]+1;
}
```

5. while (i_n<=i_{nmax}) do
6. for (flag=1 to *n*)
7. F[numelement].f_{flag} = i_{flag}
8. end for
9. numelement++
10. i_1++
11. for (flag=1 to *n*)
12. if(i_{flag}>$i_{flagmax}$)
13. Recalculating the value of i_{flag} and $i_{flagmax}$
14. i_{flag+1}++
15. end if
16. end for
17. end while

    Output: F[]

The original data structure and the algorithm of loop are reserved, making this approach more general. Furthermore, subscript transformation does not change data location, and does not copy data to other free memory. So, there is no additional overhead of transferring.

At the same time of converting nested loops to single loop, our compiler also reschedules threads, concentrating the locations of threads which are sharing data, and facilitating the operations of merging threads hierarchically.

Usually, in quasi–kernel, we can find the presence of loop-carried dependences, or the presence of shared variables. The former cannot be eliminated by our compiler. Sometimes the overhead of the latter presence is very small, and it does not worth for parallel optimization. If we try to eliminate the presence, we may find the overall performance of output code is unsatisfying. However, when the overhead of the presence cannot be ignored, data dependence should be handled for optimization.

44.3.2 Handling Data Dependence

Data dependence can be classified as flow-dependence, anti-dependence and output-dependence. An abstract example is shown in Fig. 44.4.

In Fig. 44.4, some dependence relationships exist in the *for* loop. It's obviously that S2 depends on S1, S1 depends on S3, S3 depends on S1, S3 depends on S4, and S4 output-depends on S4. Accordingly, the relationships can be illustrated by a loop dependence graph of *G*, as shown in Fig. 44.5a, b represents the agglomerative graph of *G*.

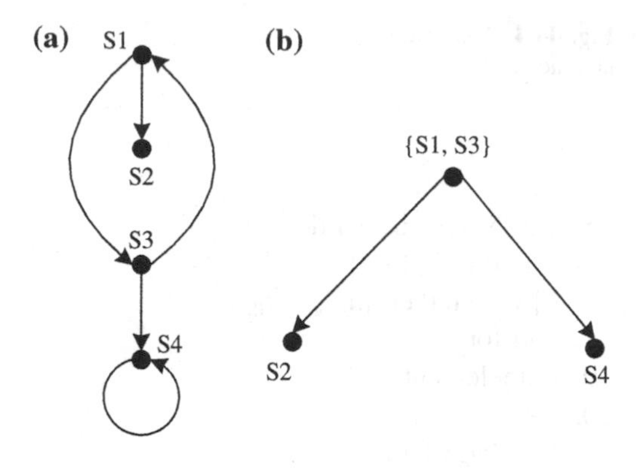

Fig. 44.5 Data dependence graph **a** Dependence graph of G **b** Agglomerative graph of G

```
      for(i=4; i<100; i++){
S1:     a[i]=b[i-2]+1;
S3:     b[i]=a[i-1]+2;
      }

      for(i=4; i<100; i++){
S2:     c[i]=a[i]+f[i];
      }

      for(i=4; i<100; i++){
S4:     q=b[i-1]+1;
      }
```

Fig. 44.6 The divided abstract code

```
      for(i=4; i<100; i++){
S4:       q[i-4]=b[i-1]+1;
      }
```

Fig. 44.7 Privatization of shared variable

According to the agglomerative graph of G, The original abstract codes can be divided into three sets which are {S1, S3}, {S2} and {S4}. The divided abstract code is shown in Fig. 44.6.

Analyzing the three sets respectively, we can find out that {S1, S3} cannot be parallelized because of data dependence, so it must be executed sequentially. {S2} which is without data dependence can be parallelized. {S4} contains mutually exclusive accesses to the shared variable q, which can be partially parallelized after being processed. According to the analysis, {S1, S3} must be firstly executed, followed by {S2} and {S4} which can be executed simultaneously.

In {S4}, our compiler converts the centralized access to q into the distributed access to copies of q, reducing the access collision region of q. Then, a reduction operation of the copies is carried out to ensure correctness. An extreme case is the privatization of shared variable. As shown in Fig. 44.6, q is shared by every iteration of S4. Privatizing q is to arrange each iteration a separate copy of q, shown in Fig. 44.7. After privatization, {S4} could be fully parallelized.

Reducing the access collision region of shared variable can effectively improve the performance of memory accessing. First, all threads which access a shared variable will certainly cause conflicts. To guarantee correctness of the execution, these threads must be executed in sequence. If certain copies of the shared variable

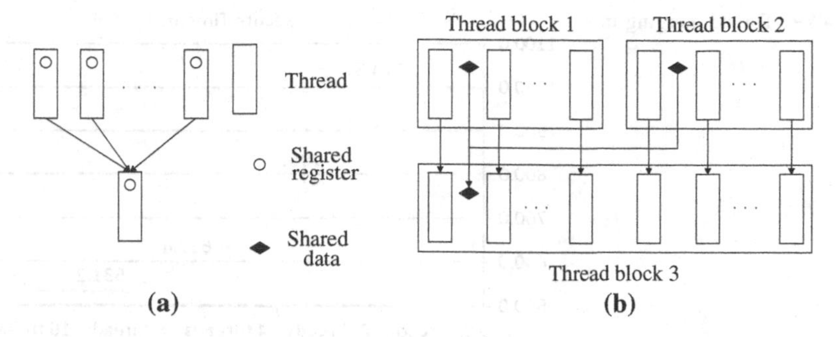

Fig. 44.8 Schematic diagram of merging threads hierarchically [2] **a** Thread merging **b** Thread block merging

exist, parallel access to this variable can be implemented by arranging different copies for these conflicts-cause threads. Theoretically, the access overhead will decrease to 1/16 of original, if 16 copies exist. When concern with memory access latency, the access to shared variables will be slow down heavily by the high latency of global memory, as the global shared data must be placed in global memory. If conflicts occur only in block, access to the copies can fully take advantage of the high-speed shared memory of CUDA.

In the application of *Cp*, a single thread producing eight results consumes time least. The method of merging eight single threads into one would achieve expected performance. Whenever the number of combined threads increases or decreases, the performance of *Cp* will become worse.

Thread block merging is to merge multiple thread blocks into one. Shared data sets exit among the original blocks. According to CUDA programming manual [13], the number of threads contained within a thread block should be a multiple of 32 which can be better adapted to the underlying hardware of CUDA. In the actual processing, our compiler merges neighboring blocks which share certain data sets to form one block to meet these requirements (Fig. 44.8).

Thread block merging does not change the total number of threads, but it increases the number of threads within thread block. Since the number of registers assigned to a thread block is certain, such merging will reduce the number of registers used by a single-threaded, impacting the overall performance (Fig. 44.9).

44.3.3 Storage Layout Optimization

The final step of our compiler framework is storage layout optimization.

The texture memory space resides in video memory and is cached in texture cache, so a texture fetch costs one memory read from device memory only on a cache miss, otherwise it just costs one read from texture cache. The delay of texture memory addressing calculation has been better hidden, for random data

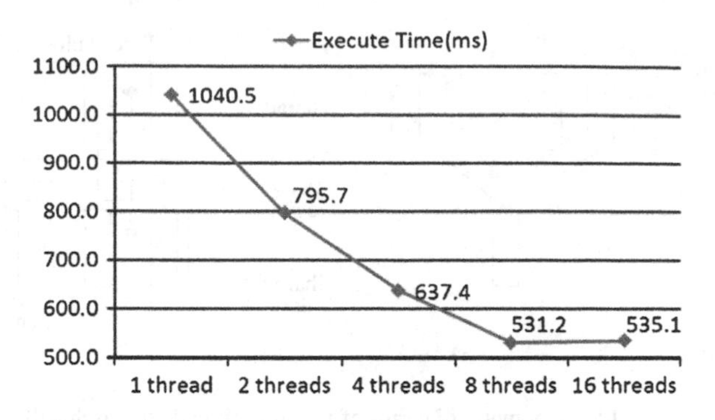

Fig. 44.9 Thread merging in *Cp*

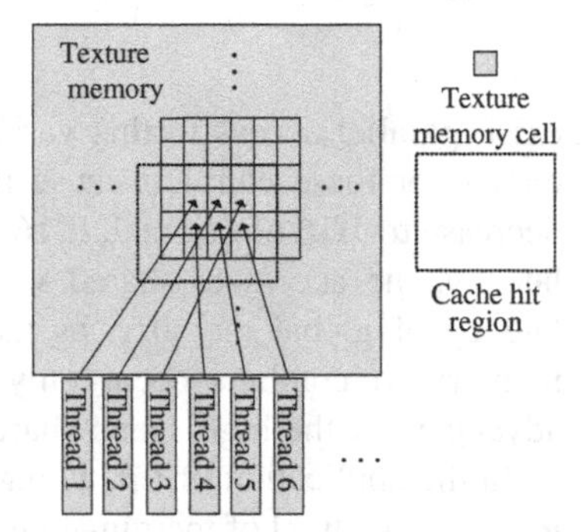

Fig. 44.10 Spatial locality of texture cache

accessing, performance of memory accessing is improved. The overhead of data binding from global memory to texture memory is very trivial.

If the memory access does not follow the access patterns that global memory access must respect to get good performance, higher bandwidth can be achieved providing that there is locality in the texture fetches.

In the CUDA Best Practices Guide3.0 [13], performance of texture memory access and global memory access are compared. Test shows that the access bandwidth of texture memory is much higher than that of global memory. On GTX280 platform, even compared with coalesced access bandwidth of global memory, the access bandwidth of texture memory still shows its inherent superiority.

As discussed above, our compiler mainly binds data in video memory to texture memory in this step.

Only after writing back to global memory, can output data of a kernel be transferred through PCI-E to memory. Therefore, each time output data being transferred from video memory, the global memory must be accessed at least once. The underlying hardware of texture memory is optimized for read-only data accessing. Writing data to texture memory cell is time-consuming. So the latency of accessing global memory for data outputting cannot be eliminated by accessing texture memory.

Table 44.2 Description of benchmark [12]

	Description
Cp	A benchmark for computing the columbic potential at each grid point over one plane in a 3D grid in which point charges have been randomly distributed.
Mri-Fhd	Computation of a matrix FHd used in a 3D magnetic resonance image reconstruction algorithm in non-Cartesian space. The vector *d* represents the scan data. The vector FH represents the scan trajectory.
Mri-*Q*	Computation of a matrix *Q*, representing the scanner configuration, used in a 3D magnetic resonance image reconstruction algorithm in non-Cartesian space.
Pns	A benchmark for petri-net simulation. This benchmark implements a generic algorithm for Petri net simulation.
Rpes	A benchmark which implements a Rys polynomial equation solver.
Sad	This benchmark computes SADs for pairs of blocks, where an SAD is one metric for how closely two images match.
Tpacf	The two point angular correlation function (TPACF) describes the angular distribution of a set of points. This benchmark computes the angular correlation function for a data set of astronomical bodies.

Our compiler framework binds read-only data to the texture memory, optimizing memory access with little additional overhead of binding. To increase the cache hit rate and speed up the texture memory accessing by leveraging spatial locality of texture cache, threads are organized in accordance with the memory space which is accessed by them (shown in Fig. 44.10).

44.4 Evaluation and Results

44.4.1 Experimental Environment And Benchmark

The configuration of experimental environment is described as follows.

1. Inter Core2 Quad 2.33GHz CPU, 4GB of memory, Microsoft Visual Studio 2005
2. 2 GeForce GTX280, 1GB memory, GeForce GT9600, 512MB memory, CUDA toolkit and SDK2.0, NVIDIA Driver for Microsoft Windows XP (177.98)

In the paper, Parboil [12], developed by UIUC, is used as the benchmark to evaluate the performance of our compiler framework. Compared with traditional benchmarks, Parboil consists of programs which are more suitable for GPU platform performance evaluation. It includes applications of *Cp, Mri-Fhd, Mri-Q, Pns, Rpes, Sad* and *Tpacf*. Details of these applications are shown in Table 44.2.

44.4.2 Experiment Results

Parboil contains a *C* code version and a hand-optimized CUDA code version (HO code) of each application. The results of experimental running on the compiler

Table 44.3 Performance comparison before and after transformation (*Rpes*)

	Io(ms)	Copy(ms)	Gpu(ms)	Compute(ms)	Tot(ms)
Original code	58.4	18,063.6	216,061.8	2,038.9	236,222.7
Converted code	61.0	11.6	710.3	54.5	837.4

Table 44.4 Performance comparison before and after privatization (*Tpacf*)

	Io(ms)	Copy(ms)	Gpu(ms)	Compute(ms)	Tot(ms)
Original code	1,077.5	26.3	10,570.1	3.8	11,677.7
Converted code	1,074.4	23.8	1,488.7	3.9	2,590.8

framework are shown in Tables 44.3, 44.4, and 44.5 *Io* represents the time of transferring data from disk to memory. *Copy* represents the time of transfer data from memory to video memory. *Gpu* represents computing time on the device (GPU), while *Compute* represents computing time on the host (CPU).*Tot* represents the total run time of a program.

As shown in Table 44.3, the speedup of the converted code over the original code is about 282*X*. In the subscript transformation, the subscript of input data of kernel is converted to the form we expected, dramatically reducing the amount of data being transferred from memory to video memory. Furthermore, our compiler framework is able to map data to the CUDA thread structure naturally, which is the main reason of reduction in computing time both on device and host.

The performance comparison before and after shared variable privatization of *Tpacf* is shown in Table 44.4. By reducing the access collision region of shared data, the computing performance on device achieves a speedup of 7.1*X*. And the overall performance of *Tpacf* obtains a 4.5 times improvement.

As shown in Table 44.5a, HO code achieves consistently better performance than C code. The HO code of *Cp* shows the most significant speedup, as much as 1,071.8. The speedup of *Sad*'s HO code over the C code is only 1.16, which is the minimum. The performance bottleneck of *Sad* is data transfer but not calculation, which is the reason that the acceleration is not obvious. In Table 44.5b, c, two kind experimental results are contained. One is the results of codes (pre-SL code) which are not optimized by storage layout optimization, and the other is the results of the final codes. Both of them have been executed on GTX280 and GT9600, respectively, as shown in Table 44.5b, c.

We can find out the performances of final code improve over that of pre-SL code varies from 1.1 to 24.2 for different applications. The performance of application optimized by storage layout shows an expected speedup, as much as 12.1 on averages.

In Table 44.6, Speedup 1 represents the ratio of the running time of *C* code to the running time of HO code, Speedup 2 represents the ratio of the running time of *C* code to the running time of pre-SL code, Speedup 3 represents the ratio of the running time of C code to the running time of final code, and Speedup 4 represents the ratio of the running time of HO code to the running time of final code.

Table 44.5 Description of benchmark

(a) Test of parboil (GTX280)

	C code			HO code				
	Io (ms)	Compute (ms)	Tot (ms)	Io (ms)	Copy (ms)	Gpu (ms)	Compute (ms)	Tot (ms)
Cp	47.0	199,203.0	199,250.0	25.9	3.4	149.1	7.5	185.9
Mri-Fhd	156.0	119,344.0	119,500.0	113.5	5.5	86.4	1.3	206.7
Mri-Q	109.0	83,422.0	83,531.0	127.7	7.5	82.9	0.4	218.5
Pns	16.0	122,031.0	122,047.0	1.3	13.7	1,412.4	3.6	1,431.0
Rpes	125.0	289,453.0	289,578.0	48.0	27.1	224.0	255.4	554.5
Sad	1,891.0	328.0	2,219.0	1,898.0	5.4	0.8	6.6	1,910.8
Tpacf	5,100.0	153,978.0	159,078.0	3,123.4	9.1	989.1	15.3	4,136.9

(b) Test of compiled Code (GTX280)

	Pre-SL code					Final code				
	Io (ms)	Copy (ms)	Gpu (ms)	Compute (ms)	Tot (ms)	Io (ms)	Copy (ms)	Gpu (ms)	Compute (ms)	Tot (ms)
Cp	19.4	2.8	158.2	5.4	185.8	18.9	5.2	168.7	6.1	198.9
Mri-Fhd	97.5	11.3	161.0	0.4	270.2	99.4	16.3	84.4	0.4	200.5
Mri-Q	95.8	8.1	117.9	1.2	223.0	97.2	9.7	79.2	1.3	187.4
Pns	1.7	3,085.4	2,537.2	4,053.8	7,603.1	1.8	3,088.2	1,642.6	4,049.0	7,581.1
Rpes	61.2	18.9	950.7	53.1	1,083.9	61.0	11.6	710.3	54.5	837.4
Sad	1,731.9	11.3	17.9	6.6	1,767.7	1,727.0	12.0	21.3	6.3	1,766.6
Tpacf	1,069.8	24.7	5,515.1	3.4	6,613.0	1,074.4	23.8	1,488.7	3.9	2,590.8

(c) Test of compiled code (GT9600)

	Pre-SL code					Final code				
	Io (ms)	Copy (ms)	Gpu (ms)	Compute (ms)	Tot (ms)	Io (ms)	Copy (ms)	Gpu (ms)	Compute (ms)	Tot (ms)
Cp	36.5	15.3	3,872.9	5.0	3,929.7	36.2	5.2	486.6	5.0	533.0
Mri-Fhd	11.3	20.5	8,547.8	0.4	8,580.0	11.6	7.2	335.2	0.4	354.4

(continued)

Table 44.5 (continued)

(c) Test of compiled code (GT9600)

	Pre-SL code					Final code				
	Io (ms)	Copy (ms)	Gpu (ms)	Compute (ms)	Tot (ms)	Io (ms)	Copy (ms)	Gpu (ms)	Compute (ms)	Tot (ms)
Mri-Q	11.8	10.5	7,204.7	1.1	7,228.1	11.5	8.1	313.3	1.1	334.0
Pns	0.7	3,269.6	46,246.5	383.9	49,742.8	0.7	3,186.0	4,292.2	379.6	7,678.1
Rpes	29.6	18.6	6,585.5	43.3	6,677.0	29.1	14.8	3,655.5	42.5	3,741.9
Sad	1,043.5	15.1	174.0	6.0	1,238.6	1,043.3	14.0	87.7	5.4	1,150.4
Tpacf	760.9	52.6	95,897.3	2.2	96,713.0	751.5	38.1	3,572.0	1.8	4,363.4

Table 44.6 Comparison of test time

	Speedup 1	Speedup 2	Speedup 3	Speedup 4
Cp	1,071.81	1,072.39	1,001.76	0.93
Mri-Fhd	578.13	442.26	596.01	1.03
Mri-Q	382.29	374.58	445.74	1.17
Pns	85.29	16.07	16.11	0.19
Rpes	522.23	267.16	345.81	0.66
Sad	1.16	1.26	1.26	1.08
Tpacf	38.45	24.06	61.40	1.60

Analyzing Speedup 4 in Table 44.6, it can be noticed that the performance of the final codes of *Cp*, *Mri-Fhd*, *Mri-Q*, *Sad* and *Tpacf* approximate the performance of the HO codes. We find the HO codes of *Pns* and *Rpes* are different with their respective *C* codes in the algorithm level. And our compiler framework is unable to complete the algorithm level optimization. Therefore, the performance of the final codes of *Pns* and *Rpes* is dissatisfying.

44.5 Conclusion

In the paper, we introduce a compiler framework from *C* to CUDA architecture. Some novel and effective methods are used to ensure optimal performance of CUDA code, including subscript transformation, handling data dependence and storage layout optimization. Experimental results show that codes produced by our compiler framework achieve a satisfied speedup of performance comparing with the hand-optimized codes in Parboil.

In future, we will focus on the transformation from *C* to another parallel platform to make our compiler architecture more universal. In addition, the development of performance analysis tools for the framework will be carried out to assist programmers use our framework more efficiently, and shorten developing span of GPGPU applications.

Acknowledgment This work is supported by the National basic research program of China (973 Program) under grant No.2007CB310901.

References

1. Lee S, Min S-J, Eigenmann R (2009) OpenMP to GPGPU—a compiler framework for automatic translation and optimization. In: 14th ACM SIGPLAN Symposium on Principles and Practice of Parallel Programming, ACM Press, Raleigh, pp 101–110
2. Yang Y, Xiang P, Kong J, Zhou H (2010) A GPGPU compiler for memory optimization and parallelism management. In: 2010 ACM SIGPLAN conference on programming language design and implementation, ACM Press, Toronto, pp 86–97

3. Baskaran MM, Bondhugula U, Krishnamoorthy S, Ramanujam J, Rountev A, Sadayappan P (2008) A compiler framework for optimization of affine loopnests for GPGPUs. In: 22nd annual international conference on supercomputing, ACM Press, Island of Kos, pp 225–234
4. Liu Y, Zhang E, Shen X (2009) A cross-input adaptive framework for GPU programs optimization. In: 23rd IEEE international symposium on parallel and distributed processing, Rome, pp 1–10
5. Zhang EZ, Jiang Y, Guo Z, Shen X (2010) Streamlining GPU applications on the fly thread divergence elimination through runtime thread-data remapping. In: 24th international conference on supercomputing, ACM Press, Tsukuba, pp 115–126
6. Ryoo S, Rodrigues CI, Stone SS, Baghsorkhi SS, Ueng S-Z, Stratton JA, Hwu WW (2008) Program optimization space pruning for a multithreaded GPU. In: Sixth international symposium on code generation and optimization, ACM Press, Boston, pp 195–204
7. Hong S, Kim H (2009) An analytical model for a GPU architecture with memory-level and thread-level parallelism awareness. In: 36th international symposium on computer architecture, ACM Press, Austin, pp 152–163
8. Huowang C, Chunlin L, Qingpin T, Yue L (2007) Principles of programming language compilation, 3rd edn. Beijing: national defence industrial press (In Chinese) pp 329–352
9. Ryoo S, Rodrigues CI, Baghsorkhi SS, Stone SS, Kirk DB, Hwu WW (2008) Optimization principles and application performance evaluation of a multithreaded GPU using CUDA. In: 13th ACM SIGPLAN symposium on principles and practice of parallel programming, ACM Press, Salt Lake City, pp 73–82
10. Volkov V, Demmel JW (2008), Benchmarking GPUs to tune dense linear algebra. In: ACM/IEEE conference on high performance computing, IEEE/ACM Press, Austin, pp 1–11
11. Bakhoda A, Yuan , Fung WWL, Wong H, Aamodt TM (2009) Analyzing CUDA Workloads Using a Detailed GPU Simulator. In: IEEE international symposium on performance analysis of systems and software IEEE press, Boston, pp 163–174
12. CUDA benchmark suite (2010) http://www.crhc.uiuc.edu/impact/cudabench.html.
13. CUDA 3.0 (2010) http://developer.nvidia.com/object/cuda_3_0_downloads.html. (2010)
14. http://code.google.com/p/gpgpucompiler.
15. Xinbiao G, Li S, Zhiying W (2010) Parallelizing full search algorithm for motion estimation using CUDA. In: Wu E (eds.). Journal of computer-aided design and computer graphics 22(3):457–460, Science Press
16. Xinbiao G, Li S, Zhiying W (2010) A method for encryption based on elliptic curve cryptography with exclusive OR operation. In: Wenquan Tao (eds.) Journal of Xi'an Jiaotong University 44(12):28–31, Editorial Office of Journal of Xi'an Jiaotong University.
17. Hong S, Kim SK, Oguntebi T, Olukotun K (2011) Accelerating CUDA graph algorithms at maximum warp. In: 16th ACM SIGPLAN symposium on principles and practice of parallel programming ACM Press, San Antonio, pp 267–276
18. Zhang EZ, Jiang Y, Shen X (2010) Does cache sharing on modern CMP matter to the performance of contemporary multithreaded programs? In: 15th ACM SIGPLAN symposium on principles and practice of parallel programming ACM Press, Bangalore, pp 203–212
19. Lee SI, Johnson TA, Eigenmann R (2003) Cetus —an extensible compiler infrastructure for source-to-source transformation. In: languages and compilers for parallel computing, 16th International Workshop, Springer Press, College Station, pp 539–553
20. Baskaran MM, Bondhugula U, Krishnamoorthy S, Ramanujam J, Rountev A, Sadayappan P (2008) Automatic data movement and computation mapping for multi-level parallel architectures with explicitly managed memories. In: 13th ACM SIGPLAN symposium on principles and practice of parallel programming, ACM Press, Salt Lake City, pp 1–10

Chapter 45
A Dynamic Balancing Algorithm for VOD Based on Cloud Computing

Shu Chang, Zhang Xingming, Tang Lianzhang and Yang Yubao

Abstract Large-scale and high-quality video-on-demand (VOD) technology has been a hot topic to researchers. In recent years, the development of cloud computing theory and virtualization cluster technology also provides a new solution of thoughts to the construction of large-scale VOD system, especially in the load balancing strategy. This article take examples from the traditional distributed VOD system, designed and constructed a distributed VOD model based on cloud computing, and highlighted the popularity of a content-based dynamic load balancing control algorithm. The algorithm sacrifice requests for cold contents to ensure the service quality of hot contents, at the meanwhile, the use of virtual dynamic resource scheduling technology that significantly increased the utilization of physical resources and overall service capabilities. Finally, results of comparison tests in the simulated environment and nearly three months of use in production environment show that this algorithm can make the greatest possible use of physical resources to ensure the service quality of hot video contents, and laid the foundation of the further research on VOD service based on cloud computing.

Keywords Distributed system · Cloud computing · Virtualization · Video on demand · Load balancing

S. Chang (✉) · Z. Xingming
School of Computer Science and Engineering,
South China University of Technology, Guangzhou, China
e-mail: shuchangok@gmai.com

Z. Xingming
e-mail: cszxm@scut.edu.cn

T. Lianzhang · Y. Yubao
Network Center, Guangzhou university, Guangzhou, China
e-mail: lzhtang@gzhu.edu.cn

Y. Yubao
e-mail: xiaobaoyang@gzhu.edu.cn

J. J. Park et al. (eds.), *Proceedings of the International Conference on Human-centric Computing 2011 and Embedded and Multimedia Computing 2011*, Lecture Notes in Electrical Engineering 102,
DOI: 10.1007/978-94-007-2105-0_45, © Springer Science+Business Media B.V. 2011

45.1 Introduction

At present, video-on-demand (VOD) service is one of the largest internet service. According to a study from comscore [1] company, by January 2010, 77.8% of US. Internet users often use online video service. The total number of the month's visits has reached as high as 173 million, and the annual growth rate of the video users maintains above 45%. Meanwhile, well-known online video services websites, such as YouTube, FIM, Yahoo, and so on, at least have hundreds of thousands of daily visits. With the development of third generation, VOD services based on mobile terminals becomes more and more popular. Therefore, to achieve VOD services in a variety of bandwidths becomes a hot research topic.

With the development of internet video services, multi-user, concurrent flow and high quality are the major development trends of VOD system. In current VOD system, research of this trend of VOD has concentrated on the video codec and high-performance dedicated server and increase network bandwidth more. However, it's little research of the system structure of the system. In recent years, the development of cloud computing theory [2–4] and virtualization cluster technology [5, 6] also provides a new solution of thoughts to the construction of large-scale VOD system. The author is meticulous in-depth analysis and comparison of the current popular VOD system design concept, realization, deployment approach and its performance advantages and disadvantages, and presents a design of distributed VOD system based on cloud computing (DCC-VOD system), which can be widely applied to various large and medium network environments. The system consists of dynamic balance server, distributed video server and VOD client. Compared with traditional centralized VOD system, it distributes each server module to deploy. At the same time using hardware virtualization, dynamic resource allocation, dynamic migration, distributed switching technologies, makes the different modules of the server running independently and stability, with no mutual interference, and good scalability. It greatly improves the reliability and stability of the system, and achieves high cost-effective to meet the current needs of the variety of medium-sized and even the large-scale-sized network environment of VOD.

This paper describes the architecture of distributed VOD system based on cloud computing, and also introduces the components of the core technology, focusing on the implementation of the load balancing server studies for distributed system.

45.2 Distributed VOD System Based on Cloud Computing

45.2.1 System Architecture

Figure 45.1 shows the overall architecture of DCC-VOD system, which contains three layers:

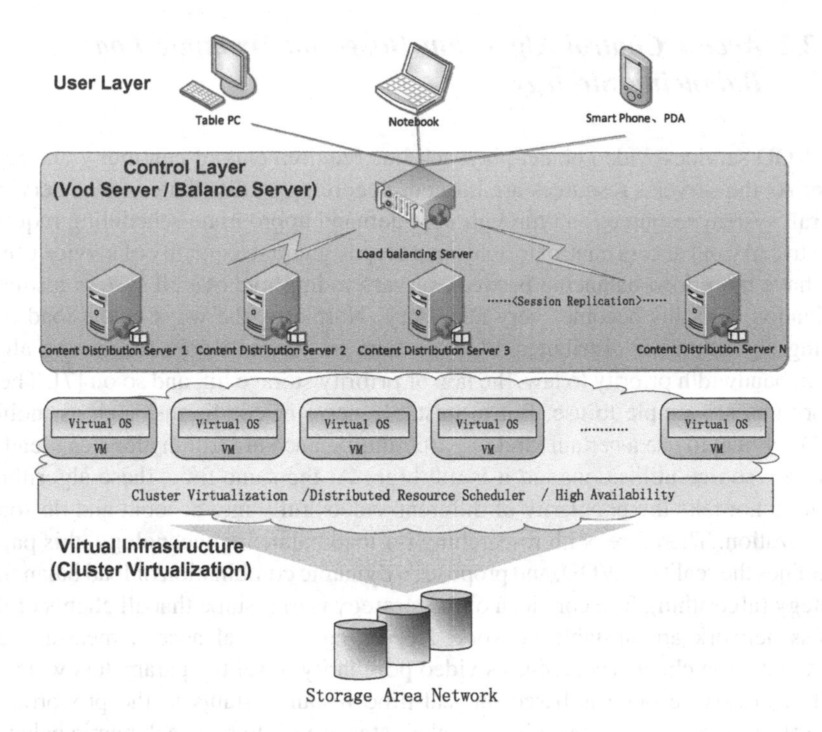

Fig. 45.1 DCC-VOD system architecture

User layer. In this layer, end users can access VOD server through the terminal equipment, such as table PC, notebook, smartphone, PDA and so on.

Control layer. This layer is mainly composed of the load balancing servers and content distribution server clusters. Based on cyclostationarity, the load balancing algorithm uses the balance of the dynamically terminal server connection requests to a relatively down loading distribution server nodes to be balanced.

Virtual infrastructure (*cluster virtualization*). Using virtualization technology of cloud computing, creating a large number of virtual machines in the limited physical environment, and achieving high availability virtual-cloud cluster based on SAN, distributed resource scheduling, dynamic migration technology provide high efficiency and stable foundation support for distributed VOD services.

In the figure above, content distribution server cluster is formed by a number of virtual machine-based servers which are linked by the high-speed virtual network. With storage systems (disk arrays) through high-speed SAN network, it provides video streaming services. Because the system is shared memory, content distribution server itself does not store any files. And it will not be streaming media server as local storage of videos are in different popularity, resulting in different content distribution server load imbalance.

45.2.2 Access Control Algorithm Based on Dynamic Load Balancing Strategy

For VOD services, video object playback has requirements of continuity and real-time. As the server's resources are limited, according to distributed video services overall system resources and through user demand appropriate scheduling request, how to carry out access control to achieve not only guarantee quality of service users, and have more load balancing between servers to improve overall system resource utilization, and this becomes very necessary. Normally, the ways of the load balancing algorithm for distributed VOD system are Round-Robin scheduling algorithm, bandwidth priority to law, the law of priority storage hit, and so on [7]. These algorithms are simple to use. But in unstable network conditions, such as mobile VOD service, to use a certain kind of particular balance algorithm alone can lead to service resource utilization, and it is not high. At the same time, these algorithms often do not take the popularity of different videos fully into account and do some optimization. Therefore, with researching on load balancing technology, this paper combines the reality of VOD, and proposes a dynamic combination of load balancing strategy (algorithm). The core idea of the strategy is to assume that all clients of the access network are unstable network. According to initial access, measurement bandwidth, the client type, requests video popularity to set the parameters weights. And each service node is based on real-time resource status to the proportional control strategy, to keep the entire virtual cluster service to be in a dynamic balance.

As video services system, the main bottlenecks are the server's disk resources to read and write bandwidth, network exit bandwidth (network card bandwidth), CPU's processing capability. Therefore, if we want that all users access requests can be fulfilled, when need to access to new users, you must consider the worst situation of balance control strategy. According to available system resources, user's performance requirements and the relationship between the number of users can access the main line, and we focus on the following indicators based access control model design:

- User client network bandwidth *Tp* (mainly used to distinguish between PC and mobile phones mobile client access control);
- In the time frame (e.g. 2 months), the number of on-demand video files requests is *H*;
- According to research, we know that the number of on-demand video file obey Zipf distribution model (Zipf-like) [8], which follows the 80/20 principle to access the content, that is 20% content and it will share 80%. Therefore, we are here to demand systems in the popularity of video files, or on demand number within the range of time as an important parameter, so as to achieve as much as possible to satisfy the need of requesting access to resources on hot spots.
- Bandwidth requirements of a video is *B*;
- Depending on the quality of on-demand content, we set different bandwidth needs, such as the 1,080 p file needs over 256 kb bandwidth, 720 p file needs over 128 kb bandwidth, and so on;

- The server node has access to the number of users is N;
- The maximum bandwidth of the network server node is W;
- The server node network available bandwidth is A;

The strategies are as follows:

"System load test" module of the balanced server will real-time detect each service node network bandwidth, the number of users has access and collect information used for access control.

The node i available bandwidth:

$$A_i = W_i - \sum_{i=1}^{n} B_i \tag{45.1}$$

When a new access request arrives, according to target request video file and client network bandwidth, controlling module will calculate the reserved bandwidth B_0 first:

$$B_0 = \mathrm{Min}\left(\frac{\text{mpegPackets}}{\text{PCR delta}} \times \ln K, T_p\right) \tag{45.2}$$

PCRdelta. The difference between the K neighbored PCR (program clock reference).

mpegPackets. The number of packet (MPEG packet) between the K neighbored PCR.

When the server node responds to a new remote request, the bandwidth is required to meet minimum reserve relations $\exists(B_0 \le A_i)$. So then, the system needs the node topology, one by one node for the new flow requests to get into admission of control test. If there is enough available bandwidth, comprehensive consideration of the number of users, access and the popularity of the requested video, we can determine the request to allow access and balance the load to the appropriate node, to reject the request or to join the waiting queue. The Judgment principles are as follows:

$$\begin{cases} \text{Acceptable and balanced to node i} & \exists(\mathrm{B}_0 \le \mathrm{A}_i) \\ \text{Pending connections, and the waiting queue} & \mathrm{H} \ge \frac{\sum H_i}{R^{i-\alpha}} \\ \text{Refused access request} & \text{others} \end{cases} \tag{45.3}$$

Because according to Zipf's law, a video file of the requested frequency and inversely proportional to its rank R are estimated to be approximately $\sum H_i/R^{1-\alpha}$, $\sum H_i$ is the sum of the number of visits of all the files. In general VOD services, the tilt index number α is usually taken 0.269 α (Different types of content will be different from the size of α). Therefore, in practical applications, within a certain video file, when the cycle time is requested more than or equal to runs estimated value, We believe that the video in the cycle is very popular, and the system gives the priority to ensure that the video was the requested service. When the new connection allows access to the module under the balanced schedule to designated server, the client can get data of the server to send requests and data sent by the server for subsequent modules to deal with.

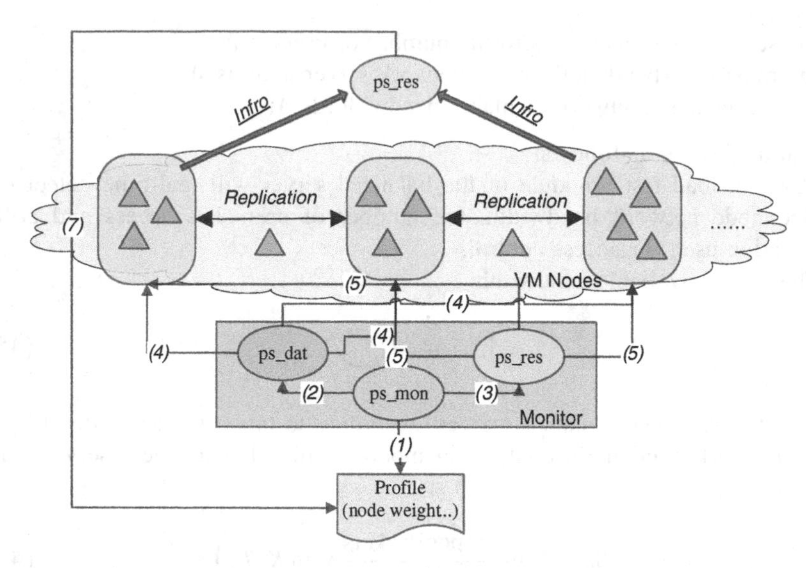

Fig. 45.2 Virtual cluster resource scheduling process

45.2.3 Virtual Cluster Resource Scheduling

Virtualization technology [9] is the key delivery technology in cloud computing. In a cloud, virtualization refers primarily to platform virtualization or the abstraction of physical IT resources from the users and applications using them. Virtualization allows servers, storage devices, and other hardware to be treated as a pool of resources rather than discrete systems, so that these resources can be allocated on demand.

Taking into account the characteristics of VOD services, we will lease mechanism to achieve control based on the node resources to achieve the rental of private slots [10]. Assessment of benefits and resource utilization, and to consider the operational phase of the application may be free of private slots, therefore, can take advantage of idle slots to help complete other tasks. At the same time, make use of circulation (lender/borrower) information between VM nodes, the task allocation of resources, taking into account the weight of the load balancing and node-effective use of influencing factors such as policy server adjust the resources, While taking advantage of the process based on process migration mechanism, according to the load status of resources and resource pool of free slots changes, the physical system to dynamically adjust the load balancing.

As shown in Fig. 45.2:

1. Monitor starting the process: *ps_mon*. first read the configuration information from configuration files, such as CPU, memory, VM node weights, etc.;
2. Monitor starts the child process: *ps_dat*;
3. Monitor starts the child process: *ps_res*;
4. VM nodes of real-time data synchronization/backup by *ps_dat* process;

5. Lists of all physical resources management and maintenance by *ps_res* process;
6. The global monitoring process *ps_cal* collecting operation and maintenance information for each node, and according to algorithm idle resources (idle slots), then dynamically adjust the weight with service configuration information, and write into the configuration file.

Algorithm is as follows:

$$f(x,i) = a - \sum_{i=1}^{n} xy_i - \omega(x) \tag{45.4}$$

$$f(x) = f(x,i) + c - \sum_{i=1}^{n} y_i x - \omega(x) \tag{45.5}$$

$$f(x,r) = \begin{cases} (s-\omega(x)) - (s-\omega(x))(1-r), & x \in \text{left node} \\ (s-\omega(x)), & x \in \text{right node} \end{cases}$$

$$f(x,c) = \sum_{i=1}^{n} xy_i + f(x,r) \tag{45.6}$$

$$f(x,u) = f(x) + f(x,c) + g(m) \tag{45.7}$$

Global process *ps_cal* use $f(x,i)$ computing applications (x)'s free private slots, y_i represents the application which has borrowed resources from x. xy_i represents the slot number that x lent to y_i; a representative of the configured value of the private slot; (x,i) represents other factors such as resources, the loss of crash number of slots; c represents the number of shared pool free slots; $f(x)$ represents the slot number x lent to y_i; s representative of the total number of slots shared pool; r represents the default weight (percentage); $g(m)$ represents the recovered slot number. If $f(x,i)$ is greater than or equal to the number of slots required, then call the system process of x, deployment schedule process to complete the task; If $f(x,i)$ is less than the required slots, we must use $f(x)$ to expand resources for search, calculate whether there are more free resources. If $f(x)$ is less than the required number of slots, we use $f(x,c)$ function calculates the number of recovery slots. $f(x,u)$ be the result of the final number of resources, and $f(x,u)$ will be adjust with the dynamic allocation of resources.

45.3 Experiment and Analysis

45.3.1 Test Environment

System test environment topology as shown in Fig. 45.3.

Based on this framework we will implement multi-group experiments. The same configuration of two physical machines, one for virtualization, virtual machines were constructed seven, one of which as a load balancing server.

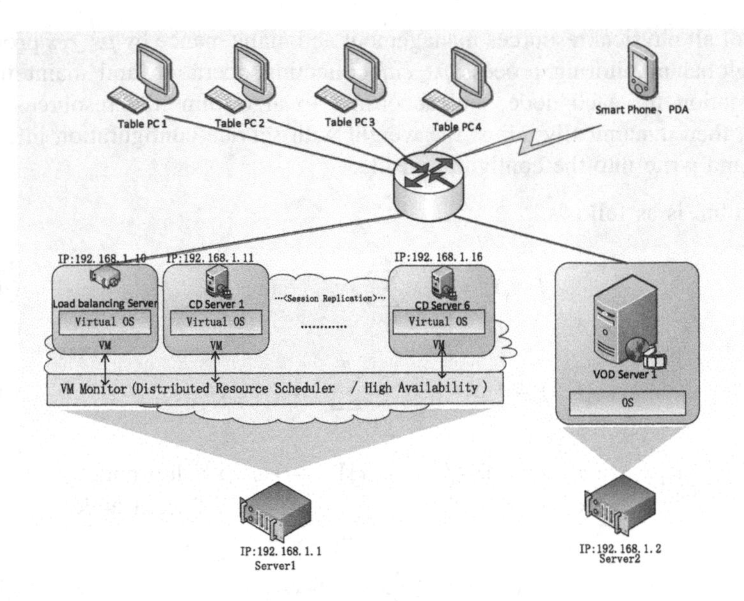

Fig. 45.3 DCC-VOD system test environment topology

By using load balancing algorithm, the load balancing server is distributed terminal connection request to the relatively lower load balanced server nodes up to achieve a balanced distribution purposes. The other six units, respectively, as CD_Server1-CD_Server6, and configure the VM monitor the resources used weighted priority scheduling policy to guarantee the heavy weight services. The physical machine will install operating system based entirely and deployment of VOD business applications directly. then, LAN access using several sets of clients to simulate stress tests, and usability testing by mobile phone.

Streaming media content related to testing the following Table 45.1.

45.3.2 Test Model

During the test, we first launch a large number of client server performance testing in ethernet, and were using the PC in the internet, and mobile in the network of GPRS online satisfaction of the client's subjective test. It takes a lot of streaming services, streaming server used to test the performance of the session, these sessions according to access streaming service streaming media content distribution and flow of service requests the distribution of time intervals, the space should be defined separately in different models (request content model) and time model (request time interval model).

- Content access model
  PC analog terminal of the requested streaming media content distribution services to the server will affect the performance shown on the access to streaming

Table 45.1 Test video file parameters

Group	File name	File size (MB)	Play time	Codec	Height	Width	Video Bit rate (kbps)	Frame rate (fps)	Audio bit rate (kbps)	Initial runs(H)
PC VOD	Normal.avi	42	5:00	AAC/MPEG4	480	656	1,045	30	126	600
	HDFile.avi	76.5	5:01	AAC/H.264	576	1,024	1,833	32	256	3,000
Mobilephone VOD	MoFile.3 gp	1.8	5:02	AMR/H.264	176	240	41	6	10	1,000

media streaming media file in the distribution of the source database can be divided into centralized distribution, uniform distribution and Zipf distribution. Centralized distribution is the simplest condition, that all access is concentrated in a streaming media source. Uniform means to access all of the simulated clients were evenly distributed in the entire streaming media source. The distribution of these two is the ideal model, not the truth, we cannot simulate the actual situation of the request for content distribution, and therefore, according to the analysis of Sect. 45.2.2, we will be here closer to the actual situation with Zipf distribution model to test. Access control policy based on Eq. 45.3.

- Time access model

  PC simulation of the terminal when the server access, will be the performance of the server performance has a great relationship. Session start in the timeline according to the distribution of interval can be designed uniform, Poisson distribution model for two time model.

  Uniform distribution is the number of clients at a fixed time interval, one after the flow of service requests on the server. This of course is an ideal model, a relatively realistic model is the Poisson distribution. This is a statistics and probability in common to the discrete probability distribution, the formula is:

$$P\{\xi = k\} = \frac{l^k}{k!} e^{-l}, \quad k = 0, 1, \ldots \tag{45.8}$$

λ can be set to request rate, K is the number of simulated clients.

Video server in the actual running process, the user's demand by the load command, the load simulator by here we submit a request to the video server. Command in the production of on-demand load simulator process, will be based on defined "content access model and time access model" to determine the two random parameters, as the number of programs on demand and on-demand request interval.

45.3.3 Test Results and Analysis

In analyzing the test results before the first state a few definitions:

Effective connections. in the load connection, a new on-demand to the normal connection, and play to maintain smooth, clear the cache does not appear, and no notice to stop, jitter and other phenomena;

CD node. That is based on the VOD CD Server virtual machine number;

Service response time. each valid connection request from launch to end the use of video content delivery time. The time depends on network bandwidth, client buffer size, as well as server load balancer scheduling policy, etc. (the client did not take the case of cache strategy, in theory, the time should be approximately equal to the file playback time, if a play length on the description in the process of emergence of the phenomenon of delay or queue). The less time that the higher the quality of on-demand service and response;

1. Effective connections—deploy and balance policy analysis

Table 45.2 Effective connections and system resource utilization test results

Group \ Effective Connections \ CD Node	1	2	3	4	5	6
(Group1) Access control algorithm based on dynamic load balancing strategy & Virtualization clusters deployment	**300**	**600**	**840**	**1050**	**1260**	**1500**
(Group2) Balancing Strategy ordinary nodes polling method & Virtualization clusters deployment	**288**	**576**	**806**	**1052**	**1250**	**1490**
(Group3) A single physical server deployment	**598**	**-**	**-**	**-**	**-**	**-**

	Group1	**Group2**	**Group3**
VM/physical Server Effective Connections Curve	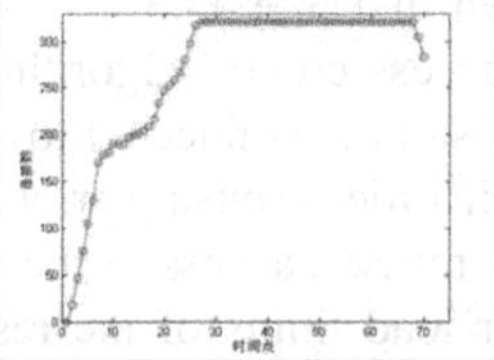	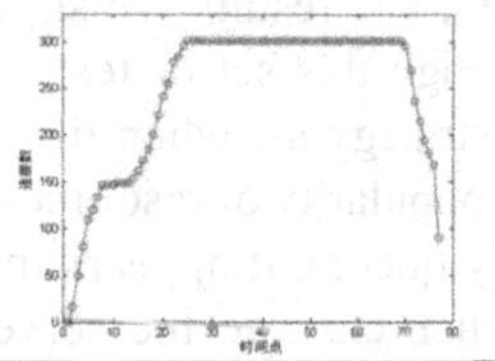	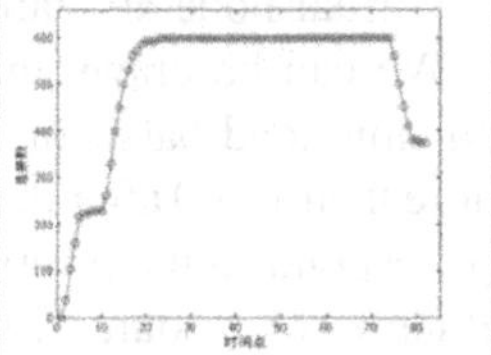
VM CPU Utilization	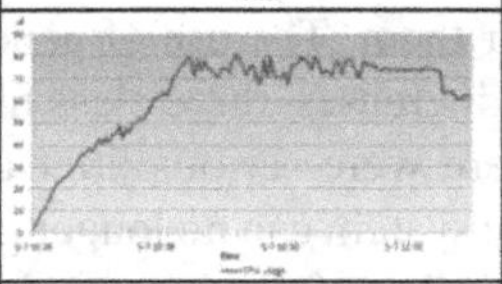	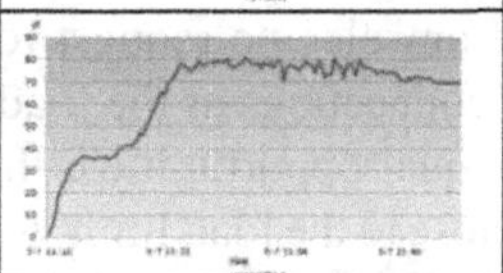	—
Physical Server CPU Utilization	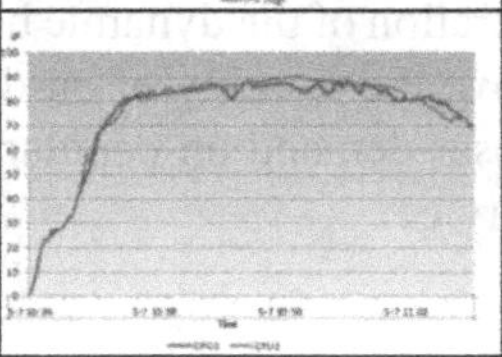	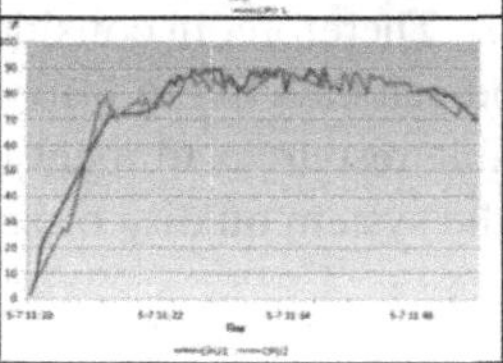	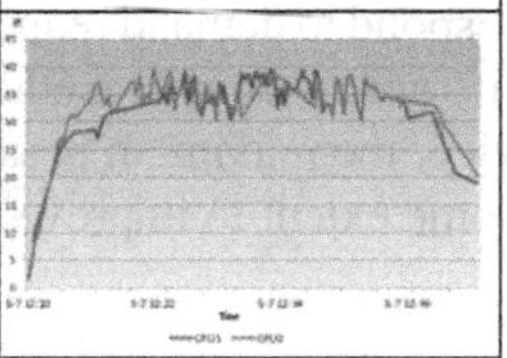

To test the effective load balance the number of connections in different algorithms and different deployment of the difference between the way we use simulated load generator, based on time access model to a single video file (MoFile.3 gp) to initiate a effective connections continuous connection requests and analog video stream data received, sustained about 35 min after the start phasing out the connection request, during which time the payload of each node records the number of connections, at the same time the whole process, use the resource monitoring application to monitor all VM virtual server and physical server node performance information. Test results are as follows:

This set of tests (Table 45.2) can be drawn by the load balancing algorithm on a simple and effective load different connections, and little effect on the utilization of physical resources; but cloud-based virtualization clusters on the effective deployment of the load connected to upgrade (increased by nearly three times), and server resource utilization (increased twofold) increase is a very important significance.

2. Service response time—balance policy analysis

To test the popularity of video content on demand in different service request response time in a different distinction between equilibrium strategy, we use simulated load generator, according to Zipf distribution model, the heat of the video was launched to different connection requests and receiving analog video stream data increasing the number of connections until the server is almost unable to respond to new requests began to stop after the connection request, during which real-time service record of each an effective linkage system response time and then demand the payload connection. System uses virtualization deployment, six virtual node all started, test results are shown in Table 45.3.

We can be drawn through this set of tests, access control algorithm based on dynamic load balancing strategy as, when the system is connected to the payload more than 1,000 after the popularity of resources in order to ensure priority response (or respond to the priority queue), it appears the refused access to the undesirable situation of resources, when close to the server load limits of the resources on-demand total rejection of unwelcome request for them. The use of general polling algorithm, the number of connections as the load increases, three class files requests for resources are nearly equal opportunities, there were serious jitter and refused access to the case, when close to the load limit, the system can no longer guarantee to respond to demand request. Therefore, in consideration of the dynamic heat demand file access control method can effectively improve the stability of on-demand service, the maximum possible resources to meet user demand on popular request, to some extent to improve the system quality of service.

45.4 Applied Analysis

Based on the above analysis of experimental data, we use "access control algorithm based on dynamic load balancing strategy and virtualization clusters" to deployed a distributed VOD system based on cloud computing in the school network center. Provide services to students and teachers. The specific structure shown in Fig. 45.3, the virtualization deployment model.

While running, we collected periodically for "effective connections, physical server CPU utilization and physical server memory utilization". At the same time to conduct subjective tests for VOD effects. After nearly three months of data collection and analysis, statistical conclusions can be obtained as follows (Table 45.4).

Table 45.4 indicates that in practical application environment, with the increase of effective connections, when physical server CPU and memory average utilization below 85%, compressive capacity of the system were in line with the results of simulated environmental testing. But more than 85%, especially more than 90% CPU usage, although the system allows on-demand requests access any course, but the actual performance results are not acceptable, there a serious distortion and jitter phenomenon. This is our next step for improvement.

Table 45.3 Effective connections and system resource utilization test results

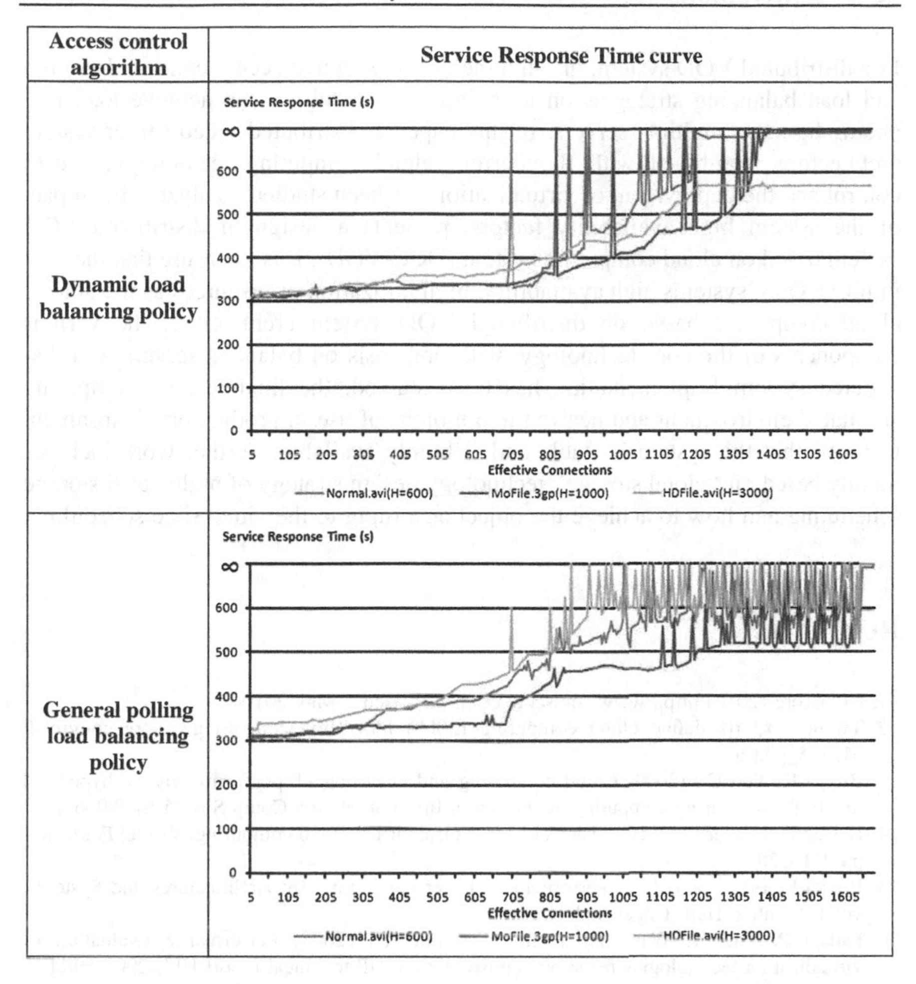

Access control algorithm	Service Response Time curve
Dynamic load balancing policy	(chart)
General polling load balancing policy	(chart)

Table 45.4 VOD Service Statistical Conclusions

Effective connections	Physical server CPU utilization (%)	Physical server memory utilization	Subjective test results		
			Is buffering	Average buffer time (s)	Play fluency
0–500	<55	<50	No	<1	Very smooth
500–800	55–70	50–75	No	<1	Smooth
800–1,000	70–85	75–90	Yes	<2	Intermittent
>1,000	85–95	>90	Yes	>5	Unacceptable

45.5 Conclusions

In a distributed VOD system, through the use of effective access control algorithm and load balancing strategies on user request scheduling, can achieve load balancing between multiple servers. In this paper, a distributed video server system architecture, combined with the current cloud computing technology, access control and the deployment of virtualization has been studied, analyzed the impact of the system high availability factors, presents a design of distributed VOD system based on cloud computing system (DCC-VOD). It is to ensure that the user intuitive QoS, systems high availability, high utilization of resources as the goal of cloud computing based on distributed VOD system architecture, the various components of the core technology, with emphasis on balanced servers in a distributed system Implementation has been studied, the final test by comparing simulated environment and nearly three months of use in production environment to prove that the system is stable and is highly available. Further work includes mainly based on "cloud storage" technology design strategy of multi-level storage scheduling and how to achieve the object according to the video slice scheduling.

References

1. Comscore (2010) http://www.comscore.com. Accessed 2 May 2010
2. Twenty experts define cloud computing (2009) http://cloudcomputing.sys-con.com/read/612375_p.htm
3. Buyya R, Yeo CS (2009) Cloud computing and emerging IT platforms: vision, hype, and reality for delivering computing as the 5th utility. Future Gen Comp Sys 25(6):599–616
4. Barrett D, Kipper G (2010) Visions of the future: virtual cloud computing. Virtual Forensics, pp 211–220
5. Buyya R (Ed) (1999) High performance cluster computing. In: Architectures and Systems, vol 1, Prentice Hall, Upper Saddle River.
6. Padala P, Zhu X, Wang Z, Singhal S, Shin KG (2008) Performance evaluation of virtualization technologies for server consolidation, HP techincal report HPL-2007-59R1
7. Waraich SS (2008) Classification of dynamic load balancing strategies in a network of workstations. Fifth international conference on information technology: New Generations, USA
8. Zornig P (2010) Statistical simulation and the distribution of distances between identical elements in a random sequence. Comput Statist Data Anal 54(10):2317–2327, 1 October 2010.
9. Virtualization Technology (2010) http://www.kernelthread.com/publications/virtualization/ Accessed 5 May 2010
10. Cloud computing with Linux [EB/OL]. http://www.ibm.com/developerworks/linux/library/l-cloud-computing/2010.10

Chapter 46
High-Quality Sound Device Virtualization in Xen Para-Virtualized Environment

Kan Hu, Wenbo Zhou, Hai Jin, Zhiyuan Shao and Huacai Chen

Abstract I/O virtualization plays an important role in client virtualization, and audio performance directly impacts user experience. Solutions for I/O virtualization includes full device emulation, split driver model and direct I/O. Full device emulation supports sound device virtualization but suffers in heavy load environment. Split driver model has better performance but does not support sound devices. Direct I/O has the best I/O performance which is close to native, but it sacrifices fault isolation and device transparency. In order to improve user audio experience in virtual machine, a method is proposed to implement high-quality sound device virtualization based on split driver model. Built on top of Xen, the frontend and backend drivers of sound device are both provided. The test results show that sound device virtualization based on split driver model enhances user audio experience in Xen virtual environment.

Keywords Xen · I/O virtualization · Split driver model · Sound device driver

46.1 Introduction

With development of hardware, virtualization technologies are popularly used [1, 6]. While the technologies used for virtualizing CPU and memory resources have very low overhead [1, 3, 14], I/O intensive applications still suffer much in

K. Hu · W. Zhou · H. Jin (✉) · Z. Shao · H. Chen
Services Computing Technology and System Lab,
Cluster and Grid Computing Lab, School of Computer Science and Technology,
Huazhong University of Science and Technology, Wuhan, China
e-mail: hjin@hust.edu.cn

J. J. Park et al. (eds.), *Proceedings of the International Conference on Human-centric Computing 2011 and Embedded and Multimedia Computing 2011*, Lecture Notes in Electrical Engineering 102,
DOI: 10.1007/978-94-007-2105-0_46,

virtual environment, which has an direct impact on user experience in virtual machines.

Existing solutions for I/O virtualization include full device emulation, split driver model and direct I/O. In full device emulation, such as the QEMU device model in Xen [2–5], the privileged instructions are trapped into emulator. Full device emulation does not need to modify native drivers, but incurs significant performance overhead. Split driver model is also known as the para-virtualized (PV for short) device driver [13]. In this model, the unprivileged domain executes the frontend driver, which intercepts I/O requests and passes them to the backend driver running in the privileged domain. This model significantly reduces the cost of I/O device virtualization; however it does not support sound device by now. Direct I/O has very close to native I/O device performance [10, 11, 15]. However, it needs hardware support, increases guest image complexity, reduces guest portability, and complicates live migration between systems with different devices [12].

Though sound devices can be virtualized by full device emulation, it can not meet the high-quality requirement, especially in heavy load environment. In order to enhance user audio experience in virtual environment, we implement the sound device virtualization based on split driver model.

The remainder of this paper is organized as follows. Section 46.2 gives a brief survey on the related work for I/O virtualization. Section 46.3 describes the key design issues of our approach of sound device virtualization. Performance evaluation is shown in Sect. 46.4. Conclusions are summarized in Sect. 46.5.

46.2 Related Works

Researchers provide optimizations on existing I/O virtualization methods. Based on split driver model, moving the device driver to driver domain can provide a safe execution environment [5, 8]. Optimization to improve the performance of virtual device (such as network cards) includes new network interface, new implementation of I/O channel [11], etc. Paul Willmann et al. [15] propose hardware and software mechanisms to enable concurrent network access from guests to reduce overhead. Jiuxing Liu et al. [7] propose VMM-bypass I/O, significantly improve I/O and communication performance for guests. Though most these optimizations focus on virtual network/block devices, they will benefit sound device virtualization because they have many similarities.

To improve the audio quality in Xen virtualization environment, in our previous research we have tried dynamically promoting the priority of audio guests (and the driver domain) in the Credit scheduler [16]. However, scheduler improvement cannot simplify the complexity of the device model, and somewhat impact other domains' performance. So in this paper we use a more direct and reasonable solution.

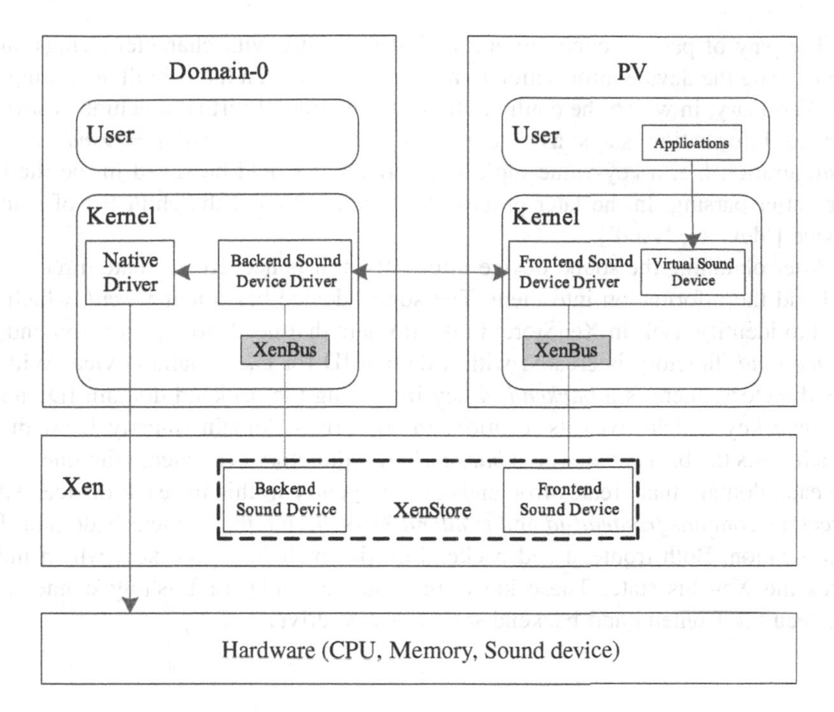

Fig. 46.1 System framework of sound device virtulization

46.3 Key Design Issues

Figure 46.1 is the system framework of our high-quality sound device virtualization. The frontend driver provides a virtual sound device to applications, intercepts the audio requests via the virtual sound device and passes them to the backend driver. The backend driver receives the audio requests and delivers them to the native sound device driver.

46.3.1 Register Frontend and Backend Sound Devices

When operating system boots, it should find devices and then register to the PCI bus. When sound device driver registered, it will match with sound devices and begin to initialize. In Xen environment, devices will be enumerated by XenStore at boot time, and the PCI bus is simulated by the XenBus. In XenStore, configuration information of domains is saved under corresponding directories. When a domain boots, all the directories in XenStore will be traversed to find out configured devices, and then registered to XenBus. So the first thing of registering the frontend and backend sound devices is to populate the XenStore with the frontend and backend sound devices.

The way of parsing configuration is firstly dealing with character strings, and then saving the device information to a dictionary. Next, a list is built according to the dictionary, in which the configuration information of virtual machine is saved. Device information stays as a child list. If we set *soundhw* = *vsnd* in the configuration file, a key-value tuple *soundhw*:*vsnd* would be saved in the dictionary after parsing. In the later processing, it needs to add the child list of sound device ['device', 'vsnd'].

After obtaining the sound device information, it is needed to create directories and add the information into them. The sound device has a name *vsnd*, which is used to identify itself in XenStore. In the domain that needs to register frontend, a *device/vsnd* directory is created with a device ID for each sound device. Within this directory, there is a *backend-id* key indicating the backend domain ID and a *backend* key which gives its location. In the driver domain (mostly Domain-0) which hosts the backend driver, a *backend/vsnd* directory is created, with one entry for each domain that creates frontends and one entry in this for each device. This directory contains *frontend-id* and *frontend* keys, giving the frontend's domain ID and location. Both frontend and backend entries include a *state* key, which indicates the XenBus state. These keys are required while establishing connection between the frontend and backend sound device driver.

46.3.2 Establishing Communication for Data Transmission

It is required to pass audio requests and responses between the frontend and backend. In Xen environment, the frontend communicates with the backend via device I/O rings, event channel, and grant tables. I/O rings provide a message-passing abstraction. The frontend places a request in the ring, whereas the backend removes it and inserts a response. Event channel provides the mechanism for asynchronous notifications between domains. Grant table provides a way for guests' share memory communication. Memory can only be shared at a page granularity, and shared pages are identified by grant references. When frontend transfers data to backend, it needs allocate grant references to the pages containing the data, and passes them to the backend via I/O rings. The backend maps the pages into its own address space with the offered grant references. To establish a connection between the frontend and the backend, these mechanisms should be initialized at first.

The frontend and backend are registered and matched with the sound devices on XenBus, and then begin to initialize. To set up the I/O ring, the frontend gets a free page, initializes the device I/O rings on this page, allocates a grant reference, and finally writes the grant reference into XenStore. To set up event channel, the frontend allocates an unbound event channel and passes it to the backend via XenStore. The connection process of the backend sound device driver is symmetrical to the frontend's operation. It first allocates some pages in its virtual

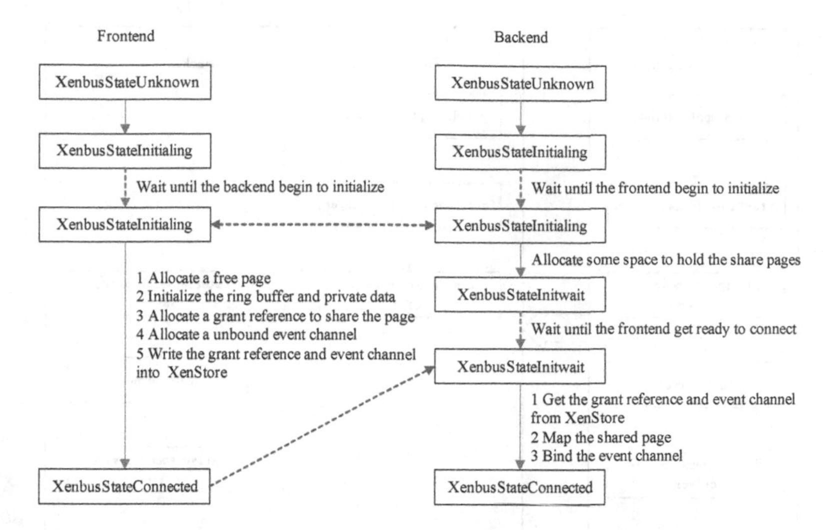

Fig. 46.2 The connection establishing between frontend and backend

address space to hold the shared pages for transferring the audio data, then enters *XenbusStateInitWait* state. While the XenBus state of the frontend is changed to *XenbusStateConnected*, the backend gets the grant reference and event channel from XenStore, then maps the share page and binds the event channel. Finally, the backend sets its XenBus state as *XenbusStateConnected* to indicate that the device is connected and ready to transfer data. Figure 46.2 gives the process of establishing connections between frontend and backend sound driver.

46.3.3 Creating a Virtual Sound Device

The frontend sound device driver needs to create a virtual sound device with which some interfaces can be provided to the applications. ALSA is the default sound driver architectures of Linux. In ALSA, a sound device is composed of a card record and some additional components. The card record is the headquarters of the sound device and manages all the additional components of the sound device, and users operate the sound device via the components on the sound device. For our para-virtualized sound device driver, the virtual sound device is created based on ALSA. The creation needs three steps. The first step is creating a card instance and the second one is allocating a PCM instance. A PCM instance consists of PCM playback streams and captures streams, each stream corresponds to a device file, via which application reads/writes the audio data from/to the sound device. After the PCM component is assigned, the last step is registering the card instance. After that, all the device files attached to the card are created and access to the device files is enabled.

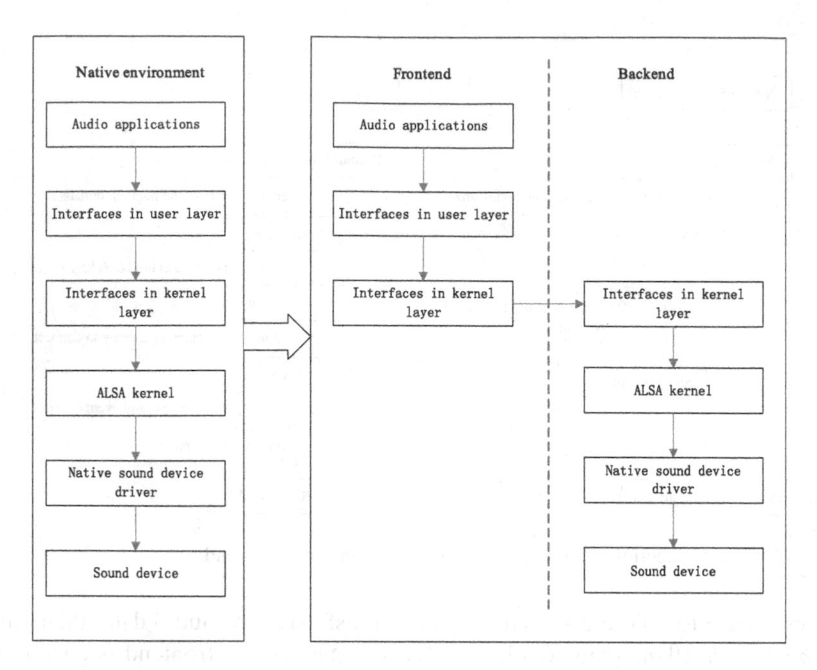

Fig. 46.3 The handling process of the audio requests

46.3.4 Processing the Audio Requests

The frontend driver intercepts I/O requests and passes them to the backend, and the backend driver delivers the received requests to the native driver to process I/O.

ALSA provides some interfaces to user-space to operate sound devices. The interfaces are read/write operations on some device files in*/dev/snd* directory, which are created when sound devices registered. In kernel, corresponding interfaces are file-operation functions. The audio applications open a device at first, and then write data into them to play audio data (or read data from them to capture). When user operates sound devices, audio requests are sent to corresponding kernel interfaces and delivered to the device via ALSA kernel driver.

For para-virtualized sound devices, both frontend driver and backend present an abstraction to high-level as device files. In order to intercept the audio requests, the frontend implements a file-operation function set. With these functions, after the device files received the audio requests from applications, they are sent to the backend rather than ALSA kernel driver. In order to deliver the audio requests to native sound device driver, the backend operates the device files in kernel layer. Figure 46.3 gives the handling process of the audio request.

In native environment, when application operates the device files, the operations will be sent to the ALSA kernel driver and then access the sound device. In split driver model, the operation will be sent to the backend. Taking the *open* operation as an example, while an application opens the device file, the frontend

Table 46.1 CPU usage by domain-0 and audio guest

	Domain-0 (%)	Audio guest (%)
HVM	4.9	4.5
PV	2.0	0.5

sends an *open* request to the backend via the I/O ring and waits. The request includes the operation type (i.e., *open*) and arguments. Then the backend opens the device file and sends a response. The response includes the operation type, the opened file descriptor, the result of the operation, etc. After that, the frontend returns the result to the application. The backend operates the device files in kernel layer is very similar to native. Besides *open*, the most important operations are *write* and *read*. We use a *batching* way to reduce the cost of data transmission: the frontend keeps the audio data from the applications, and then transmits audio data to the backend only when data has accumulated to a threshold. In the implementation of *mmap* operation, the pages are immediately mapped rather than later in page fault exception handing. This is also used to reduce the cost of grant operations and context-switching between domains.

46.4 Evaluation

Our para-virtualized sound device is implemented on Xen 3.4.2 and Linux-2.6.18-xen. The testbed has an Intel Core2 Duo E5300 processor, 2.60 GHz, with 64 KB I/D L1 and 2 MB L2 Cache, 2 GB memory. In order to evaluate the audio performance, we compare our para-virtualized sound device with full virtualized sound device. In the experiments, there are four domains: Domain-0, which contains native drivers, two PV guests, to simulate the co-existing HPC workloads, and the 4th guest in which a media playing program is employed to run to play some music. Full emulated sound device is used for HVM audio guest, and for PV audio guest, we use our para-virtualized sound device. We configure the audio guests with one virtual processor and the PV guests which simulate HPC workloads with four virtual processors.

Experiments are carried out in both light and heavy load environment. In light load environment, we keep HPC guests idle and compare the CPU usage of playing (since there is no playback jitters, lower CPU usage implies a better solution). From Table 46.1 we can see that compared with full emulated sound device, para-virtualized sound device results in lower CPU usage in both Domain-0 and the audio guest. As mentioned above, the para-virtualized sound device is implemented on the high-level interfaces, so full emulated sound device requires more I/O requests and responses between domains. In addition, full virtualization has a more complex process in handling I/O requests, so it incurs more performance overhead.

In heavy load environment where there are 2 PV guests simulating HPC workloads, we compare the underrun time during playing back. The *estimate time*

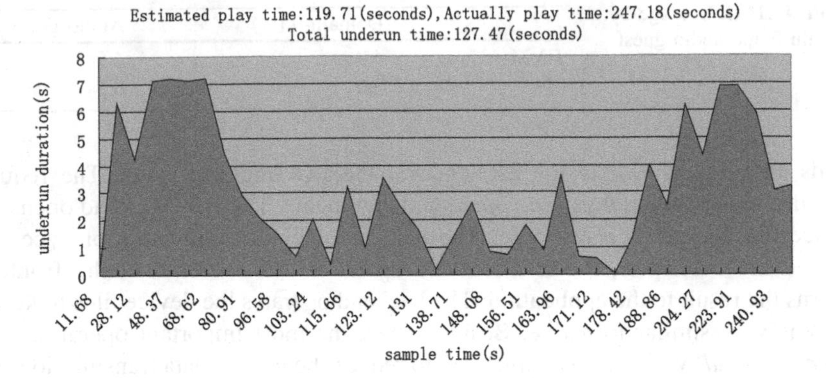

Fig. 46.4 Underrun situation in HVM

Table 46.2 The total underrun time in heavy load environment

	One	Two	Three	Four	Five	Six	Seven	Eight	Nine	Ten
HVM	130.5	110.7	95.4	115.2	124.7	82.1	85.0	103.0	117.6	87.6
PV	0.21	0.4	0.03	0.03	0.04	0.06	0.04	0.06	0.03	0.41

to play audio data is calculated with file data size, sample rate, sample bits, and number of channels. The gap between the actual playing time and estimate time is underrun time. Figure 46.4 shows that there are serious jitters in HVM audio guest because the accumulated underrun time for a 120-second file is more than 100 s. Underrun time in PV audio guest is too little to plot in figures, so we repeat the experiment for ten times and record the accumulated underrun time in Table 46.2. We can see that when playing audio, there are nearly no underruns in PV (less than 1 s) solution which leads to fluently playback.

46.5 Conclusion

In this paper, we design and implement sound device virtualization based on split driver model. By implementing frontend/backend sound device drivers under ALSA architecture, we construct a solution of high-quality sound device virtualization. The frontend sound device driver provides abstract interfaces to applications and intercepts audio requests from applications, while the backend sound device driver delivers the audio requests from frontend to native sound device driver. From the experiments results, the para-virtualized sound device can provide far better audio experience than full device emulation.

Acknowledgment The work is supported by National 973 Basic research program of China under grant No.2007CB310900, China National Natural Science Foundation under grant No.60903022 and No.61003007.

References

1. Adams K, Agesen O (2006) A comparison of software and hardware techniques for x86 virtualization. In: Proceedings of the Conference on Architectural Support for Programming Languages and Operating Systems (ASPLOS'06) Oct 2006, pp 2–13
2. Bellard F (2005) QEMU, A fast and portable dynamic translator. In: Proceedings of the USENIX Annual Technical Conference (USENIX'05), USA, 10–15 April 2005, pp 41–46
3. Barham P, Dragovic B, Fraser K, Hand S, Harris T, Ho A, Neugebauer R, Pratt I, Warfield A (2003) Xen and the art of virtualization. Proceedings of the 9th ACM symposium on operating systems principles (SOSP'03) The Sagamore, Bolton Landing, Lake George, New York, USA, 19–22 October 2003, pp 164–177
4. Dong Y, Li S, Mallick A, Nakajima J, Tian K, Xu X, Yang F, Yu W (2006) Extending Xen with Intel virtualization technology. Intel Technol J 10(3)
5. Fraser K, Hand S, Neugebauer R, Pratt I, Warfield A, Williamson M (2004) Safe hardware access with the Xen virtual machine monitor. In: Proceedings of the workshop on operating system and architectural support for the on demand IT Infrastructure (OASIS'04), Oct 2004, pp 3–7
6. Figueiredo R, Dinda P, Fortes J (2005) Resource virtualization renaissance. IEEE Comput, May 2005, pp 28–31
7. Liu J, Huang W, Abali B, Panda DK (2006) High performance VMM-bypass I/O in virtual machines. In: Proceedings of the USENIX Annual Technical Conference (USENIX'06), June 2006, pp 29–42
8. Levasseur J, Uhlig V, Stoess J, Gotz S (2004) Unmodified device driver reuse and improved system dependability via virtual machines. In: Proceedings of the operating systems design and implementation (OSDI'04), pp 17–30
9. Menon A, Cox AL (2006) Optimizing network virtualization. In: Xen. Proceedings of the USENIX Annual Technical Conference (USENIX'06) June 2006, pp 15–28
10. Mansley K, Law G, Riddoch D, Barzini G, Turton N, Pope S (2007) Getting 10 Gb/s from Xen: safe and fast device access from unprivileged domains. In: Proceedings of the Euro-Par 2007 workshops parallel processing
11. Rixner S (2008) Network virtualization: breaking the performance barrier. ACM Queue, January/February
12. Ram KK, Santos JR, Tuner Y, Cox AL, Rixner S (2009) Achieving 10 Gb/s using safe and transparent network interface virtualization. In: Proceedings of the 2009 ACM SIGPLAN/ SIGOPS International Conference on virtual execution environments, March 2009
13. Whitaker A, Shaw M, Gribble SD (2002) Scale and performance in the Denali isolation kernel. In: Proceedings of the operating systems design and implementation (OSDI'02), Dec 2002, pp 195–209
14. Waldspurger C (2002) Memory resource management in Vmware ESXServer. In: Proceedings of the Operating Systems Design and Implementation (OSDI'02), Dec 2002
15. Willmanm P, Shafer J, Carr D, Menon A, Rixner S, Cox AL, Zwaenepoel W (2007) Concurrent direct network access for virtual machine monitors. In: Proceedings of the International Symposium on high-performance computer architecture(HPCA'07), Feb 2007
16. Chen H, Jin H, Hu K, Yuan M (2010) Adaptive audio-aware scheduling Xen Virtual Environment. In: Proceedings of 2010 ACS/IEEE International Conference on computer systems and applications (AICCSA'10). Hammamet, Tunisia: IEEE Computer Society, 2010, pp 1–8

Chapter 47
Emotional Communication Efficacy, Group Cohesion, and Group Performance

Linying Dong and Bharat Shah

Abstract Despite the recent advances in affective computing, research has yet studied how users' emotion communication efficacy affects group cohesion and performance. Through a survey of 113 undergraduate students (34 groups) working in groups to complete a group project, the study reveals the significant and positive impact of emotion communication efficacy on social cohesion, confirms the salient influence of social and task cohesion on group performance. The insightful findings of the research have significant implications to researchers and practitioners on affective computing and emotion communication.

Keywords Affective computing · Emotion communication efficacy · Group cohesion · Group performance

47.1 Introduction

Affective computing has become popular since 1990 [1]. Researchers on affective computing devote to designing technology applications and features that recognize and support human emotions [2]. One of the examples is emoticons incorporated in communication technologies (e.g., electronic mail) to allow users to express emotions. Those communication technologies have quickly become the most commonly used form of communication [3].

L. Dong (✉) · B. Shah
Ted Rogers School of Information Technology Management,
Ryerson University, Toronto, Canada
e-mail: ldong@ryerson.ca
B. Shah
e-mail: bshah@ryerson.ca

J. J. Park et al. (eds.), *Proceedings of the International Conference on Human-centric Computing 2011 and Embedded and Multimedia Computing 2011*, Lecture Notes in Electrical Engineering 102,
DOI: 10.1007/978-94-007-2105-0_47,

While users are not shy away from expressing their emotions in communication technologies [4], effectively conveying emotions through those communication technologies has proven to be challenging due to the lack of nonverbal cues (e.g., head nods, smile, eye contact, tone of voice), simultaneous exchange of information and with its low social presence, and simultaneous exchange of information [3, p.325]. Emotions, however, constitute a daily experience of people who are at work. Emotions help clarify reasons, affiliate with people, and make decisions. Individual's ability to utilize emotions affects individual and organizational effectiveness [e.g., 5]. As organizations have increasingly adopted communication technologies, effective communication skills over have become essential. Current research, however, has not adequately addressed the subject [6]. With the backdrop, the research intends to address the research question "What is the impact of emotion communication efficacy on group cohesion and group performance?". We focus on the group environment that it has been widely employed in various industries to achieve a variety of tasks.

47.2 Conceptual Background

47.2.1 Emotion Communication Efficacy

Self-efficacy is denoted as "people's judgments of their capabilities to organize and execute courses of action required to attain designated types of performances" [7]. Self-efficacy has its theoretical roots in Bandura's (1977) social cognitive theory, which posits that by watching others perform a behaviour, an individual's perception of his or her ability to perform the action affects his or her decision to perform the behaviour. Based on a detailed review of studies on self-efficacy, Marakas and colleagues stress the difference between computer self-efficacy and task-specific self-efficacy [8]. Recognizing this distinction, our focus is on self-efficacy that is related to one's capability to effectively communicate emotions using communication technologies. And we call this self efficacy as emotion communication efficacy (ECE).

47.2.2 Group Cohesion

Group cohesion is an important concept in the small group literature and has attracted research for decades. Cohesiveness, "the resultant of all the forces acting on the members to remain in the group" [9, p. 274], reflects "the tendency for a group to stick together and remain united" (Carron et al. 1998, p. 213). While early conceptualization defines group cohesion as a unitary construct [e.g., 9], recent research effort has offered a strong support for decomposing the construct into two dimensions—task cohesion ("an attraction to the group because of a liking for or commitment to the group" [10, p. 388]) and social cohesion (SC) ("an attraction to the group because of satisfactory relationships and friendships with other members of the group") [11, p. 164].

The two dimensions acknowledge that there are different forces/needs drawing people to join a group. In other words, task cohesion (TC) is based on the potential of a group to attain goals which cannot be achieved by an individual and SC is based on affiliating with members in a group, which is independent of "the actual activities involved" [12, p. 228]. Empirically, TC has been discovered to strongly affect role uncertainty, absenteeism, and ultimately individual performance, while SC has less strong or no effect [10].

47.2.3 Research Hypotheses

When working in a group environment, individuals tend to encounter three types of conflicts: relationship conflict, process conflict, and task conflict, all of which are detrimental to group performance. Emotions are an important element of these conflicts [13]. Affectionate communicate helps create and reinforce individual relationships under the work group context [14]. Contrary to common beliefs, high performance groups are not immune to any type of group conflict; instead, a group performs well when disagreements and disputes are resolved through openly discussing emotions (e.g., anger, disappointment) [13]. Ineffective emotion communication via communication technologies could induce dysfunctional team behavior and hinder group members' willingness to cooperate with each other; in contrast, effective email communication could reduce conflicts, enhance group cohesion, and improve task performance. Accordingly, we hypothesize that:

Hypothesis 1a Individual emotion communication efficacy is positively and significantly related to TC.

Hypothesis 1b Individual emotion communication efficacy is positively and significantly related to SC.

Hypothesis 1c Individual emotion communication efficacy is positively and significantly related to group performance.

The theoretical rationale for the relationship between group cohesion and performance is that members of cohesive groups tend to be motivated to advance group's objectives [e.g., 15]. Empirically, group cohesion has been found to be positively and significantly associated with group performance [16]. Cohesive groups engender effectiveness in combat situations and group cohesion impacts effectiveness of global virtual teams [17]. A number of meta-analyses have shown cohesion as a significant determinant of group performance [e.g., 18, 19]. As a result, we propose that.

Hypothesis 2a Social cohesion is positively and significantly related to group performance.

Hypothesis 2b Task cohesion is positively and significantly related to group performance.

47.3 Research Methodology

47.3.1 Data Collection

The subjects for this study were recruited from students enrolled in three undergraduate courses offered by a School of Technology Management at an urban university in one of the largest cities in Canada. One of the mandatory requirements of those courses was that students formed a group of 3–6 at the beginning of a semester and worked together throughout the entire semester to complete an assigned group project. All students were equipped with a notebook with access to the campus-wide wireless network, and were avid users of social media and information technologies. At the end of the semester, an online survey was sent to the students to collect empirical data for this study.

To ensure that the survey was user friendly and easy to navigate, we applied the techniques recommended by Dillman [20] in web survey design. Several tests were run to check format consistency, instruction clarity, navigation, length, survey output, security, and privacy protection. The web survey was available for 5 weeks, and resulted in 113 responses. Of the 113 respondents, 90.3% (102) were male, and 9.7% (11) were female (reflecting the demographics of the program in which the students were enrolled), which made up to 34 groups. The minimum size of a group was three, and maximum was six, with the average of group size being 4.54. 60% (63.2%) of respondents indicated that email communication accounted for at least 50% of their communication on the group project. Email was used to allocate group work (71.7%), discussion the group project (69.9%), discussion project plan (60.2%), and communicate with the course instructor about the group project (65.5%).

47.3.2 Construct Measures

Emotion communication efficacy (*ECE*): Marakas and his colleagues [21] suggest that self-efficacy (SE) questions "must be constructed such that the subject is focused only on his or her ability within that specific task context if any interpretable results are to be obtained". They state that "the predicative capability of an SE estimate is strongest and most accurate when determined by specific domain-linked measures rather than with general measures". As a result, we tailored the measures to emotion communication efficacy under the email communication context, and developed three items (e.g., "I am able to express my emotions [e.g., anger, happiness] without being misunderstood").

Group cohesion: Group environment questionnaire, developed by Carron and his colleagues, is a widely applied measurement instrument and believed to be a conceptually and psychometrically sound measuring instrument reflecting the multidimensional nature of group cohesion [22]. Because the instrument for group cohesion is developed under the sport team context, researchers are advised to carefully select the instrument measurements when applying the instrument to a different group context [15]. Since the subjects investigated in this study were undergraduate students who were required to form groups to complete a group project, measures for task and SC were adjusted to the context. For example, the term "team" was changed to "group," and ome measurement items were deleted as they strongly reflected the sports context (e.g., "I like the style of play on this team"). In the end, four items were chosen for SC (e.g., "Some of my best friends are in the group") and four items for TC (e.g., "I like how we work together in the group on the project").

47.3.2.1 Group Performance

Researchers take interests in various aspects of group performance (GP) including group work efficiency, innovativeness, and work excellence [e.g., 23], team's adherence to schedules and budgets and ability to resolve conflicts [e.g., 24], and productivity, cooperation, and adequacy of problem identification [25]. In this research, we used the marks given by the course instructor to each group to measure each group's performance.

47.3.2.2 Control Variable

The research results could be affected by the extent of familiarity among group members. By controlling familiarity, we re-ran the structural model, and the results remained the same. Familiarity did not exert significant influence on group performance.

Common error biases: There is a potential for common method biases for all self-reported data [26]. To reduce common error biases, we enforced a procedural remedy by assessing group performance from a different source. In addition, a Harmon one-factor test was performed on the measures for all constructs in the model, and results showed that four factors were present and the most covariance explained by one factor was 38%, suggesting that common method biases would not likely contaminate the results of the study [27].

Table 47.1 Descriptive statistics, composite reliability, AVE, and construct intercollelations

Variable	M	SD	Composite reliability	One	Two	Three	Four
Group performance	79.18	6.30	1	1	–	–	–
Emotion communication efficacy	4.96	1.30	0.889	0.020	0.854*	–	–
Social cohesion	3.91	1.15	0.839	0.316	0.188	0.755	–
Task cohesion	4.91	0.93	0.873	0.263	0.066	0.631	0.796

47.3.3 Data Aggregation

Since group cohesion (task and SC) is a group-level construct, within-group agreement and between-group variance was examined before aggregating individual level data for group cohesion to the group level [28]. The r_{wg} results indicated that all groups exhibited high levels of within-group agreement (r_{wg} indices >7) in terms of TC (average $r_{wg} = 0.91$) and SC (average $r_{wg} = 0.93$). As a result, the individual level data for task and SC were aggregated to the group level ($N = 34$ groups).

47.3.4 Measurement Reliability and Validity

Composite reliability was applied to assess internal consistency reliability. The examination of internal consistency reliability indicates that the composite reliability for all constructs is higher than the .seven threshold (see Table 47.1), suggesting that all construct measures are reliable.

Convergent validity was examined based on average variance extracted (AVE) score for each construct should be above 0.50 [29]. As shown in Table 47.1, the square roots of AVEs are above 0.50, satisfying convergent construct validity. To evaluate discriminant validity, the square root of AVE for each construct must be larger than its correlation with other constructs. Our analysis shows that these two criteria are met, indicating satisfactory discriminant validity (Table 47.1).

47.3.5 Data Analysis

Partial least square, a rigorous structural equation modeling analytical tool, was employed to run the proposed model. The results indicate that the model explains 11.2 percent of the variance in group performance. Emotion communication efficacy significantly and positively affects SC. In particularly, the path coefficient from emotion communication efficacy to SC is 0.188 ($p < 0.001$). Both task and SC significantly affect group performance

($\beta = 0.099$, $p < 0.02$ and $\beta = 0.267$, $p < 0.001$). Emotion communication efficacy explains 0.4% of variance in task and 3.5% of variance in SC.

47.4 Conclusions and Future Directions

Affective computing helps support human emotion communication through advanced computer systems. As emotion is an important part of organization communication, effective communication of emotions via advanced computer technologies becomes essential. This study is among early research exploring the impact of emotion communication efficacy on group cohesion and group performance. Through the survey of 113 students who were working in groups and using emails to complete their group projects, we have identified the significant impact of emotion communication efficacy on SC and its indirect influence on group performance. Further, although both TC and SC are significant predictors of group performance, SC has been found to exert a heavier influence on group performance than does TC.

The research offers insightful findings that are useful for future studies. First, it confirms the important of emotion communication efficacy on SC. The significant and positive influence of emotion communication efficacy on SC suggests that communicating emotions (talking jokes, skilful in expressing positive/negative feelings) via emails helps mitigate/resolve group conflicts (e.g., disagreement on work allocation, different opinions on project quality, coordination of group work), reduce tensions, and promote a collaborative, trusting and positive environment. In contrast, ineffective communication of emotions (e.g., complaints, blaming others for mistakes, ignoring others' feelings) would cause frictions among group members, ignite group conflicts, consequently eroding cohesion among group members.

Emotion communication efficacy, however, does not affect group performance directly. Rather, emotion communication efficacy takes effect by influencing SC. In other words, by enhancing SC, emotion communication efficacy indirectly helps groups improve their performance.

Our research findings indicate that emotion communication efficacy is not a significant predictor of TC. A possible explanation could be that TC is concerned with completion of tasks that an individual could not finish alone, not affiliating with group members. Therefore, emotion communication is not the focus. However, further analysis on the relationship between emotion communication efficacy and TC.

In addition, our research has confirmed that both social and TC significantly and positively affect group performance. Interestingly, SC exerts a heavier influence on group performance than does TC. While previous studies suggest that the influence of social and TC depends on outcome variables being selected, our study used the marks for group project, which is an objective assessment of group performance.

Therefore, the heavier influence of SC suggests that when both the high level of TC and SC presents, SC affects group performance more.

The findings of the research have significant implications to practitioners as organizations often rely on cross-functional and virtual teams to generate new ideas, produce new innovations, and design and implement new information systems. First and foremost, it highlights the importance of emotion communication efficacy in improving group performance. Organizations need to understand the significant impact of emotion communication and help employees improve their skills in communicating emotions via emails in order to help groups be productive. Second, vendors specializing in affective computing could develop new technologies checking negative emotions being expressed and advising users of positive ways of expressing negative emotions. This would help minimize various conflicts encountered by group members and ultimately improve group performance.

The findings of the research offer directions for future research. First, the proposed model explains only 11.2% of variances in group performance and 4% of variances in task and SC. Other factors may affect these variables along with emotion communication efficacy. Therefore, future research could explore the impact of salient predecessors (at the organizational, group, and individual level) on group cohesion and performance and examine the role of emotion communication efficacy in the influence.

Second, future research could explore the extent to which affective computing such as bimodal emotion recognition [30] could help improve individual emotion communication efficacy and how advanced affective computing technologies could be applied to support teams (e.g. virtual teams) to freely communicate emotions without raising confusions.

Lastly, will individual emotion communication efficacy change over time? The extant research has indicated that group cohesion will evolve over time [31] and that virtual teams become known to each other and perform better over time. Will group members improve their skills in emotion communication as time goes by and if so, how will the improvement affect group cohesion and performance?

In conclusion, despite the recent advances in affective computing, research has yet studied how users' emotion communication efficacy affects group cohesion and performance. The study based on a survey of 113 undergraduate students working in groups to complete a group project reveals the significant and positive impact of emotion communication efficacy on SC, confirms the salient influence of social and TC on group performance. The insightful findings of the research have significant implications to researchers and practitioners on affective computing and emotion communication.

References

1. Nissan E (2009) Computational models of the emotions: from models of the emotions of the individual to modelling the emerging irrational behaviour of crowds. AI Soc 4:403–414
2. Bunt A, Conati C, McGrenere J (2009) Mixed-initiative interface personalization as a case study in usable AI. AI Magazine 4:58–64
3. Walther JB, D'Addario KP (2001) The impacts of emoticons on message interpretations in computer-mediated communication. Soc Sci Comp Rev 3:324–347
4. Sloan A (1999) Review of affective computing. AI Maga 1:127–133
5. Weisinger H (1998) Book emotional intelligence at work: the untapped edge for success. Jossey-Bass, San Francisco
6. Bryon K (2008) Carrying too heavy a load? The communication and miscommunication of emotion by email. Acad Manag Rev 2:309–327
7. Bandura A (1986) Book social foundations of thought and action: a social cognitive theory. Prentice Hall, Englewood Cliffs, p 391
8. Marakas GM, Johnson RD, Clay PF (2007) The evolving nature of the computer self-efficacy construct: an empirical investigation of measurement construction validity, reliability and stability over time. J AIS 1:16–46
9. Festinger L (1950) Informal social communication. Psychol Rev, pp 271–282
10. Zaccaro SJ (1991) Nonequivalent associations between forms of cohesiveness and group-related outcomes: evidence for multidimensionality. J Soc Psychol 3:387–399
11. Festinger L, Schachter S, Back K (1950) Book social pressures in informal groups. Stanford University Press, CA
12. Tziner A (1982a) Differential effects of Group cohesiveness types: a clarifying overview. Soc Behav Pers 2:227–239
13. Jehn KA, Mannix EA (2001) The dynamic nature of conflict: a longitudinal study of intragroup conflict and group performance. Acad Manag J 2:238–251
14. Chatman JA, Flynn FJ (2001) The influence of demographic heterogeneity on the emergence and concequences of cooperative norms in work teams. Acad Manag J 5:956–974
15. Carron AV, Brawley LR (2000) Cohesion: conceptual and measurement issues. Small Gr Res 1:71–88
16. Schwarz A, Schwarz C (2007) The role of latent beliefs and group cohesion in predicting group decision support systems success. Small Gr Res 1:195–229
17. Kayworth TR, Leidner DE (2001/2002) Leadership effectiveness in global virtual teams. J Manag Inf Sys 3:7–40
18. Mullen B, Cooper C (1994) The relationship between group cohesiveness and performance: an integration. Psychol Bull 2210–2227
19. Oliver LW, Harman J, Hoover E, Hayes SM, Pandhi NA (1999) A quantitative integration of the military cohesion literature. Milit Psychol 1:57–83
20. Dillman DA (2000) Book mail and internet surveys: the tailored Design method. Wiley, New York
21. Marakas GM, Yi MY, Johnson RD (1998) The multilevel and multifaceted character of computer self-efficacy: toward clarification of the construct and an integrative framework for research. Inf Sys Res 2:126–163
22. Carron AV, Widmeyer WN, Brawley LR (1985) The development of an instrument to assess cohesion in sport teams: the group environment questionnaire. J Sport Psychol, 244–266
23. Wang S-s (2008) Task knowledge overlap and knoweldge variety: the role of advice network structures and impact on group effectiveness. J Organ Behav 5:591–614
24. Ancona DG, Caldwell DF (1992) Demography and design: predictors of new product team performance. Organ Sci 321–341
25. Gibson CB (1999) Do they do what they believe they can? Group efficacy and group effectiveness across tasks and cultures. Acad Manag J 2:138–142

26. Podsakoff PM, MacKenzie SB, Lee J-Y, Podsakoff N (2003) Common method biases in behavioral research: a critical review of the literature and recommended remedies. J Appl Psychol 5:879–903
27. Liang H, Saraf N, Hu Q, Xue Y (2007) Assimilation of enterprise systems: the effect of institutional pressures and the mediating role of top management. MIS Q 1:1–29
28. Hofmann DA, Griffin MA, Gavin MB (2000) The application of hierarchical linear modeling to organizational research. In: Klein KJ, Kozlowski SW (eds.): Multilevel theory, research, and methods in organizations, Jossey-Bass, San Francisco, pp 467–511
29. Fornell C, Larcker DF (1981) Evaluating Structural Equation Models with Unobservable Variables and Measurement Error. Journal of Marketing Research 1:39–50
30. Huang L, Xin L, Zhao L, Tao J (2007) Combining audio and video by dominance in bimodal emotion recognition. In: Hutchison D, Kanade T, Kittler J (eds) Affective computing and intelligent interaction: second international conference, Lisbon, Portugal
31. Chang A, Bordia P (2001) A multidimensional approach to the group cohesion-group performance relationship. Small Gr Res 4:379–405

Part XII
PDC 2011

Chapter 48
A Robust Audio Aggregation Watermark Based on Vector Quantization

Juan Li, Rangding Wang and Jie Zhu

Abstract In this paper, vector quantization is applied to protect the audio aggregation's copyright and a robust watermarking algorithm which takes audio aggregation as carrier has proposed. The quantizer's input vectors are composed by the audio aggregation's DCT coefficients, the input vectors are divided into two clusters according to the parity of codeword index, then distributing information '0' and '1' to the two clusters respectively. According to the watermark information and the genus bits of clusters, modify the middle-frequency part of DCT coefficients to embed the watermark. When detected, the watermark can be recovered only by classified the middle-frequency part of DCT coefficients to which clusters without original audio aggregation. The experimental results show the algorithm has good imperceptibility, and has stronger robustness when resist to various attacks such as lossy compression, normalization and so on.

Keywords Vector quantization · Audio aggregation · Robust watermark

J. Li (✉) · R. Wang · J. Zhu
CKC Software Lab, Ningbo University, Ningbo, China
e-mail: li.juan317@163.com

R. Wang
e-mail: wangrangding@nbu.edu.cn

J. Zhu
e-mail: zhujiefirst@gmail.com

J. J. Park et al. (eds.), *Proceedings of the International Conference on Human-centric Computing 2011 and Embedded and Multimedia Computing 2011*, Lecture Notes in Electrical Engineering 102,
DOI: 10.1007/978-94-007-2105-0_48,

48.1 Introduction

In recent years, with the rapid development of Internet technology, digital audio information is vulnerable to be edited, copied and distributed maliciously. How to protect copyright and the publisher's legitimate rights and interests effectively has become increasingly important. Digital watermarking technology as the most effective means to achieve copyright protection has become a hot area research in information security industry.

Vector quantization is an efficient lossy compression technique which has been widely used in image compression and speech coding for its high compression ratio and simple decoding [1–4]. In recent years, with in-depth study of vector quantization, digital watermarking technology based on vector quantization has been widely carried out. Literature [5] divided the codebook into several parts, then modified the indexes by finding substitutional codewords according to the corresponding relationship between watermarking bits and subcode books. An algorithm has been proposed in literature [6], which embeds the watermark (binary image) into the public key according to the correlation of neighboring VQ indexes. A novel VQ-based digital image watermarking algorithm is proposed in literature [7]. The associated dual codebooks composed of basic codebook and novel extended codebook with signs. The algorithm is robust against the channel error, thus causing loss of watermarked indices. Above algorithms are all proposed for digital image. With the development of study, this technology embed secret information in cover medias such as images, audios and videos in order to protect the intellectual property of them, with the hope of no perception. Literature [8] proposed an audio watermark algorithm based on the codeword which marked with '0' and '1'. An audio watermarking scheme based on VQ codeword pairing is proposed in literature [9]. Watermarks are embedded by modulating the middle-frequency coefficients after MDCT transform according to the watermarking sequence and the genus bits of subcodebooks. These methods are still single watermarking technology based on single audio for the carrier.

This paper presents a robust watermarking algorithm based on vector quantization for audio aggregation. We divide the input vectors into two clusters according to the parity of codeword index and allocate the genus bits '0' and '1' respectively. The embedding operation is conducted by modulating the middle-frequency coefficients after DCT transform according to the watermarking sequence and the genus bits of clusters. Choose the optimal codeword in codebook to replace when embedding. Control the repeated times of watermark embedding to ensure the algorithm's imperceptibility, while embedding operations are repeated continuously to ensure the robustness of the algorithm. For comprising the robustness and imperceptibility, selecting the middle-frequency part of DCT coefficients to modify. Experiments show that the algorithm is simple and effective, while is better than the algorithm proposed in literature [8] and literature [9] in terms of imperceptibility. The algorithm reaches a very good balance between robustness and imperceptibility.

48.2 Vector Quantization

Vector quantization is a new coding method to quantify which becomes popular in image, speech coding in recent years. The signal to be coded is first divided into vectors. Vectors can be various types of data, such as a short speech waveform or a small piece of the image. Speech coding parameters (LPC coefficients) and the image transform coefficients (DCT coefficients) can be also expressed, so it is widely used in image retrieval and speech recognition [10]. Vector quantizer Q is defined as a map of Rk in k-dimensional Euclidean space to a in finite set C containing N points and the size of Q is N, Q: $R^k \rightarrow C$, where $C = \{y_0, y_1, \ldots, y_{N-1}\}$, $y_i \in R^k$, $i \in \{0, 1, \ldots, N-1\}$. We denote the set C as codebook, and the size is N. N elements in the codebook are called codeword or vector, which are also the vectors in R^k.

48.3 Audio Aggregation Watermarking Algorithm Based on Vector Quantization

Since the input vectors of vector quantization can be expressed various types of data, such as a short speech waveform or a small piece of the image, so it can also be the characteristic coefficients of audio aggregation. In this paper, we take audio aggregation which contains n audio works as carrier, so the input vectors are produced from the audio aggregation. Suppose x_i $(1 \leq i \leq n)$ is the ith audio in the audio aggregation which contains M samples. Then segment the original audio signal into several frames. Therefore, the number of frames is $\mathbf{H}_{(H \geq (M_1 * M_2)/2)}$ and the beyond part doesn't involve during the embedding process. Select the middle-frequency coefficients as the input vectors after conducting DCT transform on every frames.

48.3.1 Watermark Embedding

According to the property of codebook to embed the watermark without making any changes on the codebook is the basic idea of this algorithm. When embedding, we will choose the corresponding codeword in codebook to replace. The imperceptibility will be improved because we keep the codebook same to ensure selecting the optimal codeword to replace. The embedding flow chart is shown in Fig. 48.1.

Specific process is as follows:

Step 1: The input vectors are provided by the audio aggregation in the process of quantify. Then a training algorithm such as LBG [11, 12] is used to generate a codebook $C_1 = \{C_{(1,0)}, C_{(1,1)}, \ldots, C_{(1,N_1-1)}\}$.

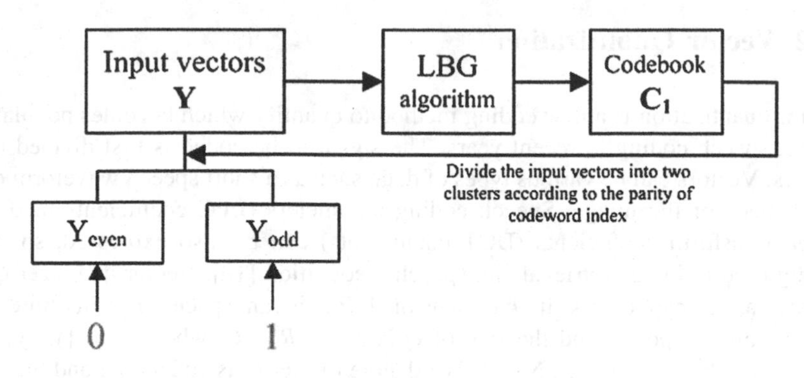

Fig. 48.1 The embedding flow chart

Step 2: Divide the codebook into two parts $C_{1,\text{even}}$, $C_{1,\text{odd}}$ according to the parity of codeword index, promptly $C_{1,\text{even}} = \{C_{(1,0)}, C_{(1,2)}, \ldots\}, C_{1,\text{odd}} = \{C_{(1,1)}, C_{(1,3)}, \ldots\}$.

Step 3: The input vectors will be divided into two clusters Y_{even}, Y_{odd} when finding the corresponding vector in inputvectors according to the codeword in $C_{1,\text{even}}$, $C_{1,\text{odd}}$, promptly $Y = Y_{\text{even}} \cup Y_{\text{odd}}, Y_{\text{even}} \cap Y_{\text{odd}} = \phi$

Step 4: If the embedding watermark bit $V_r(k)$ is '0', select the vector in Y_{even} randomly as embedding position; If the embedding watermark bit $V_r(k)$ is '1', select the vector in Y_{odd} randomly as embedding position; promptly

If $V_r(k) = 0$

then $y(k) \in Y_{\text{even}}$, $y(k) = C_{(1,k)}$, $C_{(1,k)} \in C_{(1,\text{even})}$;

If $V_r(k) = 1$

then $y(k) \in Y_{\text{odd}}$, $y(k) = C_{(1,k)}, C_{(1,k)} \in C_{(1,\text{odd})}$;

Step 5: Replace all the middle-frequency coefficients with the corresponding codeword and IDCT transform is employed to obtain the watermarked audio signal.

Step 6: After completing embedding, the codebook $\boldsymbol{C_1}$ as secret key has to be transmitted.

In literature [5] and literature [9], they firstly divide the original codebook into some subcodebooks using different algorithm. Then find the match codeword according to the watermarking sequence and the genus bits of subcodebooks to complete embedding operation. However, the match codeword does not necessarily the optimal codeword in the codebook which has a great influence on perceived quality of audio. From the above analysis, the algorithm in this paper has overcome this shortcoming because it can ensure the embedding operations completed by selecting the optimal codeword in codebook to replace. Therefore, the watermarked audio aggregation has maintained good perceived quality.

48.3.2 Watermark Extracting

Watermark extraction is simple. Firstly, obtain the input vectors from the audio aggregation. Then find a codeword in codebook $\boldsymbol{C_1}$ for input vector Y'_{k} which has the smallest Euclidean distance to it. It can be seen from the previous section, the extracted watermark bit $V'_{\mathrm{r}}(k)$ is '0' if the index of codeword $C'_{(1,k)}$ is even. Otherwise, the extracted watermark bit $V'_{\mathrm{r}}(k)$ is '1', as shown in following equation:

$$V_r'(k) = \begin{cases} 0, & C'_{(1,\mathrm{K})} \in C_{1,(\mathrm{even})} \\ 1, & C'_{(1,\mathrm{K})} \in C_{1,(\mathrm{odd})} \end{cases} \tag{48.1}$$

48.4 Experimental Results and Analysis

In our experiment, five types of music including blues, classical, country, folk, pop, are used to compose an original audio aggregation. Two audio works are selected from each kind of music, each audio is a 16-bit mono audio work in the WAVE format sampled at 44,100 Hz. So an audio aggregation has $\boldsymbol{n} = 10$ audio works. The original watermark is the binary image of size 36*36 bits, shown in Fig. 48.2. The size of codebook $\boldsymbol{C_1}$ is 128, every 32 samples in each frame has been put into DCT transform.

Compute the correlation coefficient (*NC*) between the original watermark and the extracted watermark, and observe the similarity between them to evaluate the vulnerability of algorithm, *NC* defined as follows:

$$NC\left(W, W'\right) = \frac{\sum_{i=1}^{M_1} \sum_{j=1}^{M_2} w(i,j) w'(i,j)}{\sqrt{\sum_{i=1}^{M_1} \sum_{j=1}^{M_2} w^2(i,j)} \sqrt{\sum_{i=1}^{M_1} \sum_{j=1}^{M_2} w'^2(i,j)}} \tag{48.2}$$

where W is the original watermark, and W' is the extracted watermark, and i, j are indices of the binary watermark image. If *NC* is close to 10,000, then the similarity between W and W' is very high. If *NC* is close to zero, then the similarity between W and W' is very low.

Fig. 48.3 is the extracted watermark from the detected audio aggregation which is not suffered from any attacks. The $NC = 1$ which means the watermark can be recovered correctly if the audio aggregation without any attacks.

48.4.1 Imperceptibility Analysis

Usually, we hope people cannot aware of the embedded watermark which means the higher imperceptibility, the better perceived quality of audio. For watermark

Fig. 48.2 Original watermark

Fig. 48.3 The extracted watermark

Table 48.1 SNR comparison

	Method [8]	Method [9]	Proposed method
SNR (dB)			
Blues1	48.6646	54.9435	59.1157
Blues2	49.4357	54.6325	59.6987
Classical1	53.0564	59.4060	64.1502
Classical2	53.8945	59.9234	64.9845
Country1	52.7573	59.0893	63.7738
Country2	48.5172	53.1214	58.4420
Folk1	48.2166	53.1487	63.8287
Folk2	47.8077	53.6981	63.7953
Jazz1	61.8862	61.6092	74.8480
Jazz2	56.0414	61.3351	69.2576

can be seen as noise adding to signal, we can use the signal to noise rate (SNR) to evaluate perceptual quality of the watermark embedded audio. SNR is defined as (48.3):

$$SNR = -10\log\left[\frac{\sum_i |f_i' - f_i|^2}{\sum_i |f_i|^2}\right] \tag{48.3}$$

where f_i is the original audio signal and f_i' is the audio signal contained watermark.

Compare the proposed algorithm with the methods proposed in literature [8] and literature [9] under the same conditions. In literature [8], the original codebook with size 128 has been extended to the labeled codebook with size 256 by using the conventional LBG algorithm. After generating the labeled codebook, we can use it to perform the watermark embedding. In literature [9], the embedding operation is conducted by modulating the middle-frequency coefficients after MDCT transform according to the watermarking sequence and the genus bits of subcodebooks. The subcodebooks are formed by codeword pairing. The SNR values of the three methods are shown in Table 48.1:

According to the results in Table 48.1, the value of SNR of the algorithm in this paper is larger than methods in literature [8] and literature [9], it means the imperceptibility of the audio aggregation that is got by the method in this paper is better.

Table 48.2 The watermark detection results for stirmark for audio attacking

	Method [8]	Method [9]	Proposed method
Robustness comparison NC ($L = 9$)			
No attack	0.9921	1.0000	1.0000
Addbrumm_100	0.9915	1.0000	1.0000
AddBrumm_1100	0.9791	0.9988	1.0000
AddSinus	0.7176	0.9258	0.8902
Addnoise_100	0.6386	0.8071	0.8661
Crompressor	0.9890	0.9994	0.9970
Echo	0.5294	0.6765	0.5735
FFT_real_reverse	0.9915	1.0000	1.0000
Flippsample	0.7687	0.8874	0.8619
LSBZero	0.9915	1.0000	1.0000
Normalize	0.9921	1.0000	1.0000
RC_HighPass	0.6947	0.8897	0.8929
RC_lowpass	0.5228	0.7141	0.6742
Smooth	0.4523	0.5907	0.6736
Zerocross	0.6310	0.7710	0.8251

48.4.2 Attack Testing

48.4.2.1 Traditional Attack

We make use of stirmark for audio 2.0 to attack the watermarked audio aggregation, default parameters are accepted in our experiments and [13] gives the description of all attacks in this system. Watermark image extracted from different attacks and its NC values are shown in Table 48.2. All audio works in the audio aggregation will be attacked when evaluating the robustness. It's easy to know the robustness will improve if increasing the capacity of watermark embedding, but at the same time, it also easy to be perceived and will certainly be more prone to attack. Therefore, we take the repeated times of embedding operation as 9.

Table 48.2 summarizes the proposed watermark detection results comparing with scheme [8] and scheme [9] against various attacks. According to the Table 48.2, the three algorithms are all robust to some common signal processing attacks including noise addition, lossy compression and so on. But the ability of resisting to attacks like flippsampple, echo and low pass filtering is poor. It can be seen from the data shown in Table 48.2, the proposed method can achieve better robustness.

48.4.2.2 Special Attacks

To further prove the robustness of the proposed algorithm, we must consider the special nature of audio aggregation. So the audio aggregation will suffer deleting

Table 48.3 The watermark detection results for substituting audio attack

	Method [8]	Method [9]	Proposed method
Substituting audio attack (NC)			
Substitute 1 audio	0.9909	0.9993	0.9993
Substitute 2 audio	0.9799	0.9335	0.9725
Substitute 3 audio	0.7073	0.8137	0.9120
Substitute 4 audio	0.6497	0.5878	0.8452

Table 48.4 The watermark detection results for deleting audio attack

	Method [8]	Method [9]	Proposed method
Deleting audio attack (NC)			
Delete 1 audio	0.9921	0.9993	0.9993
Delete 2 audio	0.9821	0.9944	0.9397
Delete 3 audio	0.9103	0.9609	0.8720
Delete 4 audio	0.6851	0.7198	0.8022

audio attack and substituting audio attack. Substituting audio attack means one or more audio works replaced by some non-correlated audio works, then extracting watermark from the replaced audio aggregation. The experimental results are shown in Table 48.3. Deleting audio attack means deleting one or more audio works in the audio aggregation, then extracting the watermark from the audio aggregation which is suffered from the above attack. The experimental results are shown in Table 48.4.

According to Tables 48.3, 48.4 three methods have shown certain robustness for substituting audio attack and deleting audio attack. The resistance of the proposed algorithm is slightly better.

48.5 Conclusion

This paper presents an audio aggregation watermarking algorithm based on vector quantization. the main features of the algorithm are shown: (1) The input vectors construct by the DCT coefficients of audio aggregation, (2) When embedding, it can ensure the operations completed by selecting the optimal codeword in codebook to replace without make any changes on codebook, such as codebook partition, (3) Watermark embedding and extraction in the algorithm is simple and effective, In addition, the watermark can be extracted without the help from the original audio aggregation. The experimental results show that the proposed watermarking algorithm has good performance on imperceptibility and robustness.

Acknowledgments This work is supported by the National Natural Science Foundation of China (60873220), Zhejiang Natural Science Foundation of China (Y108022, Z1090622, Y1090285), Zhejiang Science and Technology Preferred Projects of China (2010C11025), Zhejiang province education department key project of china (ZD2009012), Ningbo science and technology preferred projects of china (2009B10003), Ningbo Key Service Professional education project of china (2010A610115), Ningbo natural science foundation (2009A610085) and Ningbo university foundation (XYL10002, XK1087). In addition, our programs are supported by High school special fields construction of computer science and technology (TS10860), Zhejiang province excellent course project (2007), Zhejiang province key teaching material construction project (ZJB2009074).

References

1. Lu ZM, Pan JS, Sun SH (2000) Image coding based on classified side-match vector quantization. IEICE Trans Inf Sys 83(12):2189–2192
2. Chang CC, Chou YC, Hsieh YP (2009) Search-order coding method with indicator-elimination property. Syst Softw 82:516–525
3. Koishida K, Tokuda K, Masuko T, Kobayashi T (2001) Vector quantization of speech spectral parameters using statistics of static and dynamic features. IEICE Trans Inf Syst 84(10):1427–1434
4. Sun Z, Hou Z, Wang C (2010) Image coding algorithm based on classified side-match vector quantization in DCT domain. Shuju Caiji Yu Chuli/J Data Acquis Process 25(6):772–776
5. Lu ZM, Sun SH (2000) Digital image watermarking technique based on vector quantisation. Electron Lett 36(4):303–305
6. Huang HC, Wang FH, Pan JS (2001) Efficient and robust watermarking algorithm with vector quantization[J]. Electron Lett 37(13):826–828
7. Anping W, Jian T, Li G (2008) New vector quantization-based watermarking algorithm[J]. J Data Acquis Process 23(2):146–152
8. Liu J-X, Lu Z-M, Pan J-S (2008) A robust audio watermarking algorithm based on dct and vector quantization. 8th International Conference on intelligent systems design and applications, ISDA, pp. 541–544.
9. Zhou Y, Wang R, Yan D (2010) An audio watermarking scheme based on VQ codebook pairing. Image and signal processing (CISP) 3rd International Congress. 8:4011–4015.
10. Gersho A, Gray RM (1992) Vector quantization and signal compression. Kluwer Academic Publishers, MA
11. Anderberg MR (1973) Cluster Analysis for Application. Academic Press, London
12. Linde Y, Buzo A, Gray RM (1980) An algorithm for vector quantizer design. IEEE Trans Commun 28(1):84–95
13. Jana D, Christian K (2006) Audio benchmarking tools and steganalysis. Revision 1.1.Online:http://www.ecrypt.eu.org/documents/D.WVL.10-1.1.pdf

Chapter 49
A Service Giving a Case-Based Instruction of Bioinformatics Workflow Running on High Performance Computer for Engineering Design and Education

Feng Lu, Hui Liu, Li Wang and Yanhong Zhou

Abstract The role of bioinformatics workflows to support the Life Sciences has become fundamental for the comprehensive analysis of large amount of biological data concerning different bioinformatics tools and databases. In this paper, we design and develop a web service giving a case-based instruction of bioinformatics workflow running on high performance computer for engineers and educators. The service is designed for interactive usage and the possibility to learn from the case in a straightforward way. It will help to integrate bioinformatics software tools to address questions in life science by providing the designer, student, or educator specific questions in bioinformatics workflow design and hands-on experience. It is publicly accessible at http://bioinfo.hust.edu.cn:8080/bio-estexplore/.

Keywords Bioinformatics workflow · EST analysis · High performance computing · Case-based instruction · Web-based service

F. Lu (✉) · H. Liu
School of Computer Science and Technology,
Huazhong University of Science and Technology, Wuhan, China
e-mail: lufeng@hust.edu.cn

H. Liu
e-mail: huiliu.mail@gmail.com

L. Wang · Y. Zhou
School of Life Science and Technology, Huazhong University Science and Technology,
Wuhan, China
e-mail: hust08wangli@gmail.com

Y. Zhou
e-mail: yhzhou@hust.edu.cn

J. J. Park et al. (eds.), *Proceedings of the International Conference on Human-centric Computing 2011 and Embedded and Multimedia Computing 2011*, Lecture Notes in Electrical Engineering 102,
DOI: 10.1007/978-94-007-2105-0_49,

49.1 Introduction

Nowadays, the role of bioinformatics workflows to support the Life Sciences has become fundamental for the comprehensive analysis of large amount of biological data concerning different bioinformatics tools. Researchers in molecular biology always analyse genetic data using bioinformatics tools in their daily work [1]. Advances in ecoinformatics are addressing issues regarding data access and integration by adopting semantic approaches and building scientific workflow systems to assist researchers in locating and documenting their data and analyses [2]. The scientific workflow systems typically support multiple analytical frame-works and components and have been successfully used in a variety of disciplines, including ecology, the geosciences, molecular biology, and other areas where data access, modeling, and visualization are complex and multistaged [3–8].

However, it is relatively difficult for engineers to establish workflows for bioinformatics applications. Learning any workflow programming environment will still put some cognitive burden on the biologist [9]. Questions arise like: How do I build a workflow for comprehensive data analysis using appropriate different software tools? How do I reduce execution time of the specific steps required for a comprehensive analysis using distributed processes and parallelized software? Teaching designer, student, or educator to answer these questions means to produce individuals with a truly cross-disciplinary set of skills who are able to integrate bioinformatics software applications to address questions in life science research programs. Being an effective training method, it is useful to construct meaningful biological cases for this workflow-establishing purpose.

Here, we report on a web service giving a case-based instruction. As a comprehensive workflow system for EST data management and analysis concerning different bioinformatics tools, databases and large-scale computing technology, the famous ESTExplorer [4] is chosen as the case of workflow. According to the ESTExplorer, we deploy a particular bioinformatics workflow on a 8-node Linux cluster (44 GHz CPUs, 20 GB RAM, 1.6 TB HD). Then, in accordance with the dynamic and source code-based way of the ADP [1], a web service for our workflow is constructed. Our service is designed for interactive usage and the possibility to learn from the case in a straightforward way. This service is regularly used in our local bioinformatics curricula at Huazhong University of Science and Technology (HUST), which implies that it will remain up-to-date in the foreseeable future.

49.2 Related Works

The major funding agencies in many countries have programs to support increased bioinformatics education and training. In USA, the training programs in bioinformatics and computational biology can be supported by the National Institutes of

Health (NIH), the National Science Foundation (NSF) and the National Institute of General Medical Sciences (NIGMS) [10]. In UK, the agencies include the Biotechnology and Biological Sciences Research Council (BBSRC), the Engineering and Physical Sciences Research Council (EPSRC), the Medical Research Council (MRC), and Wellcome Trust [11]. The German Research Foundation (DFG) (Deutsche Forschungsgemeinschaft) started a 5 years training initiative in 2000 [12]. In the first 2 years, the DFG spent around 5 million EUR per year [12]. In Australia, together with co-investment by the States and the Commonwealth Scientific and Industrial Research Organization (CSIRO), the National Collaborative Research Infrastructure Strategy (NCRIS) will allocate more than AUD 1 billion from early 2007 through mid 2011 [13, 14]. In Malaysia, investments include various governmental research funds from the Ministry of Science, Technology and Innovation (MOSTI) and the Ministry of Higher Education Malaysia (MOHE) [15]. These investments can be taken as an indicator that bioinformatics training is still a bottleneck in our field's process of becoming an established scientific discipline.

For bioinformaticians, it is a common pattern of their work to use computational tools and data sources via the Internet. Hence, web services are commonly used in bioinformatics engineering training. Currently, there are many bioinformatics training services provided online in the form of e-learning. One of the typical services supplies various bioinformatics training resources. These services include eLearning facility of the European Molecular Biology Network (EMBnet, http://elearning.embnet.org/), EBI's eLearning courses (http://www.ebi.ac.uk/training/elearningcentral/course/category.php?id=1) ensuring that users can learn to make the most of the EMBL-EBI's data resources and tools, and ExPASy's e-Proxemis (http://e-proxemis.expasy.org/), the first Bioinformatics learning portal for Proteomics with an case-based training methodology and frequently enriched/updated content. The other typical service offers interactive trainings in a dynamic and source code-based way of dealing with bioinformatics problems, such as Algebraic Dynamic Programming in Bioinformatics (ADP, http://bibiserv.techfak.uni-bielefeld.de/dpcourse/) [1]. This service allows the transfer of theoretical background knowledge to hands-on experience. The last typical service is web-based bioinformatics applications integrating a variety of common bioinformatics tools and resources for teaching and other training programs. The famous example is Biomanager [16], developed by University of Sydney, which integrates over 280 bioinformatic programs and sequence databases from a range of different software packages.

However, these existing online bioinformatics training services need further development as bioinformatics in biology and biomedicine is in its infancy [17]. Most of the above services tend to emphasize training in "tool use", or cook-book style instructions for specific tools, out of accord with the training goal—answering fundamental questions of Life Sciences with aid of bioinformatics workflows.

49.3 A Service Giving a Case-Based Instruction of Typical Bioinformatics Workflow

Our goal is to develop a case-based instruction service that is effective in training the ability of building a workflow for analysis large amount of data in high performance computer. Therefore, the reference case using in this service should be carefully selected. Then, the workflow has been implemented on a cluster, using public software integrated by Perl scripts, with Linux as default operating system and the OSCAR distribution that provides the tools and the software packages for cluster management and parallel job executions.

49.3.1 Choose the Reference Case of Workflow

An appropriate workflow is chosen as the reference case. Firstly, the workflow should be concerned with significant biological meaning in life science, large amount of molecular biological data, and several bioinformatics databases and software tools available for the processing in workflow. Secondly, the main process of the workflow should be designed to serialize and to control the parallel execution of the different steps required for the analysis and to parse into database the collected results. Finally, the workflow should have examples which can be used to verify the correctness of the workflow procedures.

The suitable reference case could be found among the expressed sequence tag (EST) processing systems. EST sequences are generated by single-pass 5′ or 3′ DNA sequencing of clones randomly picked from cDNA libraries [18]. The analysis of EST data can enable gene discovery, complement genome annotation, aid gene structure identification, establish the viability of alternative transcripts, guide single nucleotide polymorphism (SNP) characterization and facilitate proteomic exploration [19]. EST processing system always comprises a suite of software tools, manipulates huge EST datasets, and uses a 'distributed control approach' in which the most appropriate bioinformatics tools are implemented over different dedicated processors. Furthermore, useful information resulting from each single step of the processing workflow always integrates into a relational database and can be analysed by Structured Query Language (SQL). Therefore, one of the EST processing systems should be selected as our reference case.

Several EST processing systems, such as ParPEST [3], ESTExplorer [4], ESTAnnotator [20], TIGCL [21], ESTminer [22], EGassembler [23], ESTpass [24], and PESTAS [25], have been developed. Among them, ESTExplorer is a comprehensive workflow system for EST data management and analysis, runs on a Linux cluster and is freely available for the academic community at http://estex plorer.biolinfo.org. The detailed processing steps, analysis programs and datasets of the ESTExplorer workflow can be received and understood according to the

Contig1

Hit ID	Definition	E-value	% Identity	% Similarity	Query	Subject	Alignme Length
gi|12643508|sp|Q42435|CCS_CAPAN	Capsanthin/capsorubin synthase, chloroplast precursor >gi|468748|emb|CAA54495.1| capsanthin/capsorubin synthase [Capsicum annuum] >gi|522120|emb|CAA53759.1| capsanthin/capsorubin sythase [Capsicum annuum]	2.91232e-10	33	33	203 - 301	466 - 498	33
gi|8247354|emb|CAB92977.1|	neoxanthin synthase [Solanum tuberosum]	4.64959e-08	28	30	203 - 301	466 - 498	33
gi|10644119|gb|AAG21133.1|	chromoplast-specific lycopene beta-cyclase [Lycopersicon esculentum]	1.76684e-07	26	30	203 - 301	466 - 498	33
gi|8249885|emb|CAB93342.1|	neoxanthin synthase [Solanum lycopersicum]	1.76684e-07	26	30	203 - 301	466 - 498	33
gi|11131528|sp|Q9SEA0|CCS_CITSI	Capsanthin/capsorubin synthase, chloroplast precursor >gi|6580973|gb|AAF18389.1|AF169241_1 capsanthin/capsorubin synthase [Citrus	3.01378e-07	25	31	203 - 301	471 - 503	33

Fig. 49.1 Results of ESTExplorer deployed on our linux cluster

reference, detailed tutorial and FAQ released in their website. Several detailed examples which provide datasets of ESTs and complete analysis results from ESTExplorer have been given as references. And then, the analysis results can be easily collated at the final output stage when datasets are analysed using a workflow constructed by us.

49.3.2 Choose HPC Architectures

Availability of high performance computing (HPC) architectures, including Single Processor, SIMD, SMP, MPP, Cluster, Constellation and Grid, is becoming quite common. The massively parallel computation offered by these architectures has proven to be an important tool in a number of bioinformatics projects [26], especially a number of EST analysis systems [3, 4, 20–25], over the years. Cluster, which was born in 1994 at the NASA Goddard Space Flight Center [27], offers tremendous processing power and advantageous price-to-performance ratio to any research group or scientists. Since 2000, in TOP500 List (http://www.top500.org), which serves as a valuable tool for tracking trends in supercomputer performance and architectures, the proportion of Cluster has been rising remarkably. In November 2010, Clusters are the most popular supercomputing platform with over 80% of the top 500 supercomputers (http://www.top500.org/stats). Moreover, there are several studies [3, 4, 21] addressing the parallel execution of EST analysis programs in Cluster environments. So, the ESTExplorer workflow is deployed on a Linux cluster. The results are shown in Fig. 49.1.

Table 49.1 Software or programming environment for our cluster system

Environment component	Software or programming environment
Operation system	CentOS
Cluster middleware	OSCAR
Cluster resource management software	PBS
Database management system	MySQL5
Programming language for Web development	PHP5
Workflow control script	Perl

49.3.3 Software and Programming Environment

The components in the Cluster software and programming environment include operation system, middleware and resource management software of cluster, relational database system, parallel programming style, workflow control scripting language and web interfaces production technique. The software or programming environment for our cluster system is described in Table 49.1.

We chose the CentOS, OSCAR, and PBS, respectively, as the operating system, middleware and resource management software of the Linux cluster. MySQL is used to store our EST sequences, intermediate data of the reference workflow and annotated results. MySQL is an open-source relational database management system and widely used for bioinformatics databases, such as databases in Ensembl [28], UCSC [29], NCBI [30] and databases in EST analysis systems, such as ParPEST [3], ESTExplorer [4], ESTPass [24] and PESTAS [25]. Moreover, the constructed database can be queried through SQL calls implemented in a suitable PHP-based interface. We provide a pre-defined web-based query system to support users.

At last, Perl is chosen as the scripting language to control workflow. Since it has been extremely successful for connecting software applications together into sequence analysis workflows, converting file formats, and extracting information from the output of analysis programs and other text files [31], Perl is selected by most of the present EST analysis systems, such as ParPEST [3], TGICL [21], ESTMiner [22], EGassembler [23] and ESTPass [24]. Furthermore, cause the Bioperl is flexible enough to support applications, while maintaining an easy learning curve for novice Perl programmers, and is capable of executing analyses and processing results from programs such as BLAST, ClustalW. So, the BioPerl and Perl are both selected to control analysis workflow and operate database separately.

49.3.4 Parallel Strategy

EST analysis is generally time-consuming due to the large number of EST. Parallel computing technique is always used to enhance the performance of EST analysis workflows. For the programs in the ESTExplorer [4], the most time-consuming

Table 49.2 Correlation between Mpiblast threads and time

Number of threads (nodes)	Time required to run blast (s)
3 (1)	816.932
4 (2)	492.355
5 (3)	240.609
6 (4)	226.62
7 (5)	212.74
8 (6)	196.193
9 (7)	177.519
10 (8)	169.068

step is sequence blast, which carries out annotation at the nucleotide level, using the BLASTX program and NCBI's non-redundant protein database.

There are two different methods to parallelize the blast computing. One method is intra-program parallelism. In this method, the raw EST data and assembled contigs are divided into several subsets firstly. Then, the subsets are parallel executed by BLASTX against the nr database on respective computing node (e.g. cluster-node, CPU-core, etc.) in the same way. In this method, however, since the database can't reside in the main memory of each node generally, the effective computer speed to the parallel blast is slowed down due to the heavy disk I/O operation. Another method is intra-search parallelism. In this method, the blast database is fragmented according to some criteria and the generated data fragments are allocated to the different computing nodes. The input sequence data is then parallel aligned against different database fragmentation. Finally, the results are integrated into one single output file.

The typical examples of the latter parallel method are NCBI BLAST on SMP machines [32] and MpiBLAST. The NCBI BLAST toolkit itself provides support for multi-processor computation on SMP systems using database segmentation. Processors within such a system search on separate portions of the database, and results are collated as one. However, although NCBI has developed a parallel BLAST using the thread on SMP machines for the speedup of BLAST, the speedup is still limited due to the architectural limitations of SMP machines [32]. MpiBLAST is a freely available, open-source, parallel implementation of NCBI BLAST. Database fragmentation reduces execution latency because each node's database fragment resides in main memory, yielding a significant speedup due to the elimination of disk I/O. Furthermore, query segmentation increases execution throughput as a single multi-sequence query is now split into separate, independent subqueries and each subquery is executed in parallel. Here, MpiBlast is used to speed-up blast computing, just as in ParPEST [3].

After test we find when divide the nr database into eight sub databases, the time required by MpiBLAST varies in accordance with number of threads and nodes open at the same time. The correlation is shown in Table 49.2. The speedup ratio is shown in Fig. 49.2.

It is concluded from the curves that, when there are fewer nodes, increasing the number of nodes will have a significant increase in speed, which is approximately

Fig. 49.2 Time measured running BLAST

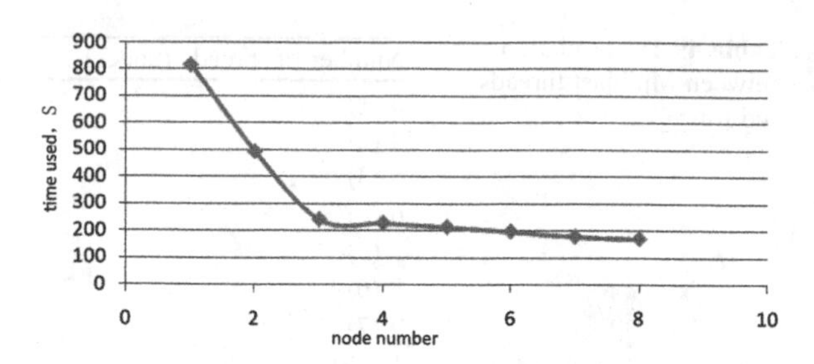

linear speed-up. But when there are many nodes, node–node interaction will take up too much memory and CPU, thus leading to further adding nodes would not increase the Blastx speed dramatically.

49.3.5 Presentation Technique

Our service website was constructed according to the method using by interactive bioinformatics course ADP at the Bielefeld University Bioinformatics Server (http://bibiserv.techfak.uni-bielefeld.de/dpcourse/). The web pages are mainly built with navigation bar and small screen-readable portions in the main window. The navigation bar gives the workflow framework in a highly abstract manner. The corresponding small portions give the essential ideas about the EST analysis workflow, implementation details and source code. Through this hands-on experience, the service is convenient for interactive usage and the possibility to guide students to conduct study in an efficient way.

49.4 Conclusion

While there are several developed comprehensive bioinformatics workflow systems which are freely available for the academic community at different website supplying processing steps, analysis programs, datasets and detailed examples online, the service giving a detailed workflow case which guides users to integrate software applications to address questions in life science on high performance environment is not available yet.

Our service is aimed at designers, students and educators who are expected to familiar with bioinformatics software application and computing technology with aid of this service website.

Overall, the website is structured into a tutorial text and interactive application pages (Fig. 49.3). While the first part of the website is initial reading, the later part mainly works by inviting the user to practice the technique by modifying existing examples, or writing their own. After an introduction to the theoretical background information, the experience about the comprehensive workflow's employment on a

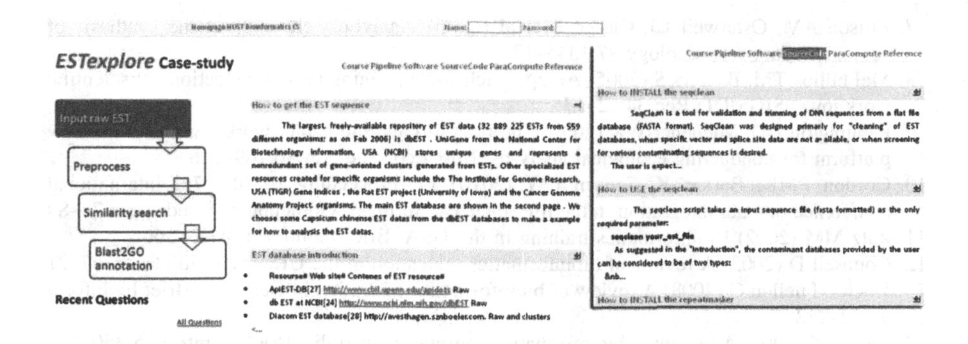

Fig. 49.3 This figure shows two of the content pages of the website

linux cluster is shown to give users an idea of the programs and operations they have to write or do. The hands-on experience includes bioinformatics software installation, parallel processing script, construction of the EST database, script of the web-based database query, procedures and methods for computing efficiency analysis and correctness verification of the workflow's employment. The source example code is shown.

All the necessary information can be arranged on the screen side-by-side to the introductory text, allowing the transfer of theoretical background knowledge to hands-on experience. Present-windows, such as code fragment and hints for experimenting, shall motivate students to modify the examples and thereby understand the topics in more detail.

Acknowledgments This research was supported by the National Key Technology R&D Program (2008BAH29B05), 2008 Innovational Experimental Area Program for Talented Person Raise Pattern by the Ministry of Education of China (DHE-Doc[2009], No.4, SN.71), the Educational Reform Program of Hubei Educational Committee, and the College Student Innovation Experiment Plan of Huazhong University of Science & Technology (2010131).

References

1. Sczyrba A, Konermann S, Giegerich R (2008) Two interactive bioinformatics courses at the Bielefeld university bioinformatics server. Brief bioinform 9:243–249
2. B J, M P, S MJ, R O et al The new bioinformatics: integrating ecological data from the gene to the biosphere, Vol 37 (Annual Reviews, 2006)
3. D'Agostino N, Aversano M, Chiusano ML (2005) ParPEST: a pipeline for EST data analysis based on parallel computing. BMC Bioinform 6(4):S9
4. Nagaraj SH, Deshpande N, Gasser RB et al (2007) ESTExplorer: an expressed sequence tag (EST) assembly and annotation platform. Nucleic Acids Res 35:W143–W147
5. Altintas I, Berkley C, Jaeger E et al (2004) Kepler: an extensible system for design and execution of scientific workflows. 16th International conference on scientific and statistical database management, proceedings. 423–424
6. Deelman E, Blythe J, Gil Y et al (2004) Pegasus: mapping scientific workflows onto the grid. Lecture Notes in Comput Science, vol 3165, pp 11–20

7. Ellison AM, Osterweil LJ, Clarke L et al (2006) Analytic webs support the synthesis of ecological data sets. Ecology 87:1345–1358
8. McPhillips TM, Bowers S (2005) An approach for pipelining nested collections in scientific workflows. SIGMOD Rec 34:12–17
9. Pennington DD, Michener WK (2005) The EcoGrid and the kepler workflow system: a new platform for conducting ecological analyses. Bull Ecol Soc Am 86:169–176
10. Gordon PMK, Barker K, Sensen CW (2010) In: Proceedings of the 7th international conference on data integration in the life sciences. Springer, Gothenburg, Sweden, pp 74–89
11. Zatz MM (2002) Bioinformatics training in the USA. Brief bioinform 3:353–360
12. Counsell D (2003) A review of bioinformatics education in the UK. Brief bioinform 4:7–21
13. Koch I, Fuellen G (2008) A review of bioinformatics education in Germany. Brief bioinform 9:232–242
14. Cattley S (2004) A review of bioinformatics degrees in Australia. Brief bioinform 5:350–354
15. Ragan MA, Littlejohn T, Ross B (2008) Genome-scale computational biology and bioinformatics in Australia. PLoS Comput Biol 4:e1000068
16. Zeti AMH, Shamsir MS, Tajul-Arifin K et al (2009) Bioinformatics in Malaysia: hope, initiative, effort, reality, and challenges. PLoS Comput Biol 5:e1000457
17. Cattley S, Arthur JW (2007) BioManager: the use of a bioinformatics web application as a teaching tool in undergraduate bioinformatics training. Brief bioinform 8:457–465
18. Pevzner P, Shamir R (2009) Computing has changed biology–biology education must catch up. Science 325:541–542
19. Adams M, Kelley J, Gocayne J et al (1991) Complementary DNA sequencing: expressed sequence tags and human genome project. Science 252:1651–1656
20. Nagaraj SH, Gasser RB, Ranganathan S (2007) A hitchhiker's guide to expressed sequence tag (EST) analysis. Brief bioinform 8:6–21
21. Hotz-Wagenblatt A, Hankeln T, Ernst P et al (2003) EST annotator: a tool for high throughput EST annotation. Nucleic Acids Res 31:3716–3719
22. Pertea G, Huang X, Liang F et al (2003) TIGR Gene Indices clustering tools (TGICL): a software system for fast clustering of large EST datasets. Bioinformatics 19:651–652
23. Nelson RT, Grant D, Shoemaker RC (2005) ESTminer: a suite of programs for gene and allele identification. Bioinformatics 21:691–693
24. Masoudi-Nejad A, Tonomura K, Kawashima S et al (2006) EGassembler: online bioinformatics service for large-scale processing, clustering and assembling ESTs and genomic DNA fragments. Nucleic Acids Res 34:W459–W462
25. Lee B, Hong T, Byun SJ et al (2007) ESTpass: a web-based server for processing and annotating expressed sequence tag (EST) sequences. Nucleic Acids Res 35:W159–W162
26. Nam S-H, Kim D-W, Jung T-S et al (2009) PESTAS: a web server for EST analysis and sequence mining. Bioinformatics 25:1846–1848
27. Pekurovsky D, Shindyalov IN, Bourne PE (2004) A case study of high-throughput biological data processing on parallel platforms. Bioinformatics 20:1940–1947
28. Hargrove WW, Hoffman FM, Sterling T (2001) The do it yourself supercomputer. Sci Am 285:72–79
29. Hubbard TJP, Aken BL, Ayling S et al (2009) Ensembl 2009. Nucleic Acids Res 37:D690–D697
30. Kuhn RM, Karolchik D, Zweig AS et al (2009) The UCSC Genome Browser Database: update 2009. Nucleic Acids Res 37:D755–D761
31. Sayers EW, Barrett T, Benson DA et al (2010) Database resources of the National Center for Biotechnology Information. Nucleic Acids Res 38:D5–D16
32. Stajich JE, Block D, Boulez K et al (2002) The Bioperl toolkit: perl modules for the life sciences. Genome Res 12:1611–1618
33. Kim H-S, Kim H-J, Han D-S (2003) In: Proceedings of the 2003 international conference on Computational science: PartIII. Springer, Melbourne, Australia, pp 213–222

Chapter 50
A Fast Broadcast Protocol for Wireless Sensor Networks

Yang Zhang and Yun Wang

Abstract Broadcast is a hot topic in wireless sensor network research. The main challenges are from network instability and complexity of communication collision. Currently, the broadcast protocols for sensor network focus on reducing the energy overhead without concerning thoroughly about the delay issue. In some real-time scenarios, these broadcast protocols are not very appropriate to some extent. This paper proposes a new kind of rapid broadcast protocol for sensor network, which introduces the concept of the umbrella topology. The simulation experiment shows that an umbrella topology structure is more efficient to solve the delay issue. Finally, we compare the protocol we proposed with the simple broadcast protocol called flooding. The result shows that the fast broadcast protocol can reduce the number of transmissions and decrease the routing delay.

Keywords Broadcast protocol · Routing delay · Wireless sensor network

50.1 Introduction

Broadcast is a basic problem in wireless sensor networks. The broadcast protocol in wireless sensor networks ensures that all the active nodes in the networks can receive the packet sent by the source node. General criteria for broadcast protocols

---

Y. Zhang (✉) · Y. Wang
School of Computer Science and Engineering,
Southeast University Key Laboratory of Computer Network and Information Integration, MOE, Nanjing, China
e-mail: zhangyang@seu.edu.cn

Y. Wang
e-mail: yunwang@seu.edu.cn

J. J. Park et al. (eds.), *Proceedings of the International Conference on Human-centric Computing 2011 and Embedded and Multimedia Computing 2011*, Lecture Notes in Electrical Engineering 102,
DOI: 10.1007/978-94-007-2105-0_50, © Springer Science+Business Media B.V. 2011

include the average network delay, the number of transmissions and the average network energy consumption.

The simplest broadcast protocol is called flooding with the characteristic of computation simplicity. However, it brings high transmission rate and high communication overhead, which at certain stages can even cause the serious broadcast storm problem [1] and reduce the network performance. Gossip flooding protocol [2] improves the flooding which is almost the same as flooding protocol except the feature of probability-based. Compared with flooding, gossip flooding enjoys higher performance, but it still brings high energy consumption and high communication overhead.

The broadcast protocol based on the global information can achieve the approximate optimal solution in theory [3], which is based on computing the Minimum Connected Domination Set (MCDS) [4]. However, it requires that every node has good computation capability and knows the global topology information, which is impractical in wireless sensor network environment. Article [5] proposes approximation algorithms. However, this method is not scalable and adaptive. The algorithm becomes too complex when the network environment changes.

The distributed broadcast protocols require neither the global information nor high computation capability. Scalable Broadcast Algorithm (SBA) [6] and Ad Hoc Broadcast Protocol (AHBP) [7] are both distributed broadcast protocols and based on the two-hop neighbor information. Therefore, the number of transmission is small. However, they need the "hello" message to maintain the two-hop neighbor information, which brings high communication overhead and energy overhead.

Distance-based broadcast protocols [8] require each node knows its geographic location information. Broadcast Protocol for Sensor Networks (BPS) [9] is a distance-based protocol. In BPS, when the node receives the packet, it can decide whether to transmit it or not based on the location information. This protocol can maximize the distance of each hop. Thus, it reduces the number of transmissions and decreases the energy overhead. In the meanwhile, it brings high network delay. Moreover, this method may cause the network disconnected when the density of nodes is low. BPS uses the idea that the hexagonal lattice arrangement is more efficient than any circle arrangement [10]. Therefore, each node receiving packet should compute the distance between itself and the vertices of the hexagons called strategy locations. However, the number of effective nodes is less than six in practice. There are only three effective nodes in the optimal scenario, including one upstream node.

The broadcast protocols for wireless sensor network should consider thoroughly with many relative issues because of instability of wireless sensor network, irregularity of the deployment of nodes, and the complexity of collision. Currently, the broadcast protocols for sensor network focus on reducing the energy overhead, the communication overhead and the computation overhead, without consideration about the real-time requirement in the sensor network. More transmissions can bring high energy overhead. The energy conservation protocols always bring high network delay such as BPS, while low-delay protocols always consume high energy such as SBA and AHBP. These protocols are not applicable in the real-time environment. This paper proposes a new distributed, umbrella topology based, scalable and self-adaptive broadcast protocol for wireless sensor network, which meets network delay requirement as well as reduces energy consumption.

50.2 Fast Broadcast Protocol

50.2.1 System Model

It assumes that each node knows its geographic location with the usage of GPS system in current research situation. The communication radius of each node is R. The scale of network may increase, and it will be stable for a period of time after a change. Although there might be possibility that few nodes make errors, the majority ones can work well. Each broadcast packet has a unique broadcast ID, which is the identification of broadcast packet and the broadcast ID increase automatically. The broadcast protocol is considered successful so long as 99.5% of nodes receive the packet. Otherwise, it is considered failed.

50.2.2 Fast Broadcast Protocol

The fast broadcast protocol requires the packets reach the boundary of network as soon as possible. The fundamental idea is that the broadcast packet is transmitted to the boundary of network along with the umbrella topology and spread in local. As shown in Fig. 50.1, the backbone in the umbrella topology is defined as the primary path while other branches are defined as auxiliary path.

50.2.2.1 Building up the Primary Path

The broadcast in wireless sensor network is the covering problem in essence. Kershner had proved the hexagonal lattice arrangement is most efficient arrangement in the network area [10]. BPS computes six strategy points in each hop [9]. However, there are only three effective nodes in the optimal scenario, including one upstream node. As shown in Fig. 50.2, the effective nodes are V2, V4 and V6. Thus, it is unnecessary to compute the hexagonal lattice points. Instead, each node only computes the distance from last hop to decide whether it is in the primary path.

In our broadcast protocol, when a node receives a packet, it will decide whether it is in the primary path or not. If yes, it will transmit the packet. Otherwise, it will spread the packet locally. As shown in Fig. 50.3, we define the primary path is in the scope of 60° of the extended line in this paper. Node S transmits the broadcast packet and node T receives the packet and transmits it to node U and node V.

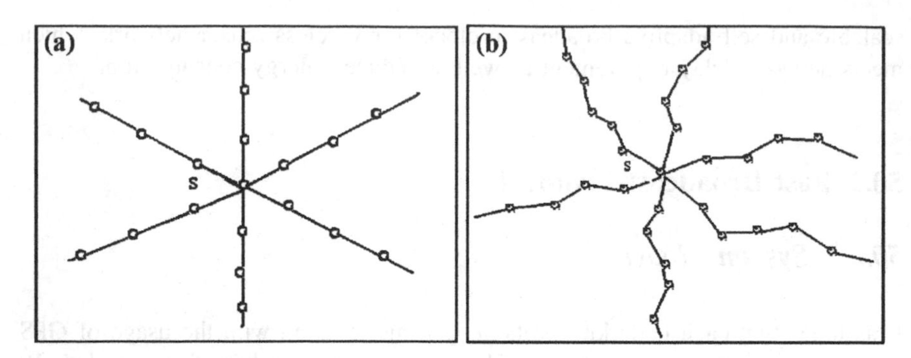

Fig. 50.1 Umbrella topology. **a** Possible umbrella topology in ideal scenario. **b** Possible umbrella topology in practice

Fig. 50.2 The hexagonal lattice strategy points (3 nodes transmit)

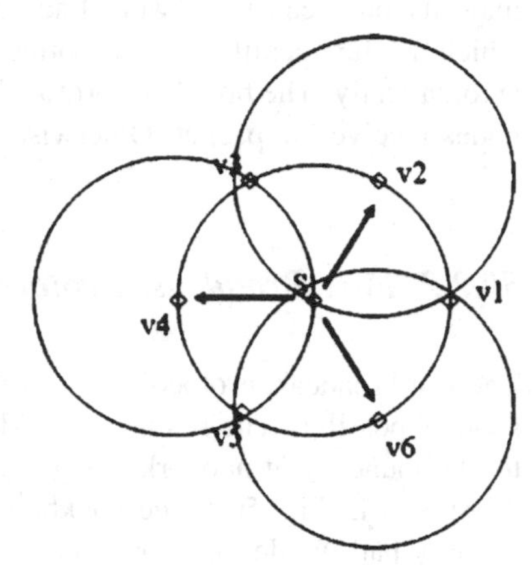

Fig. 50.3 Transmission strategy based on primary path

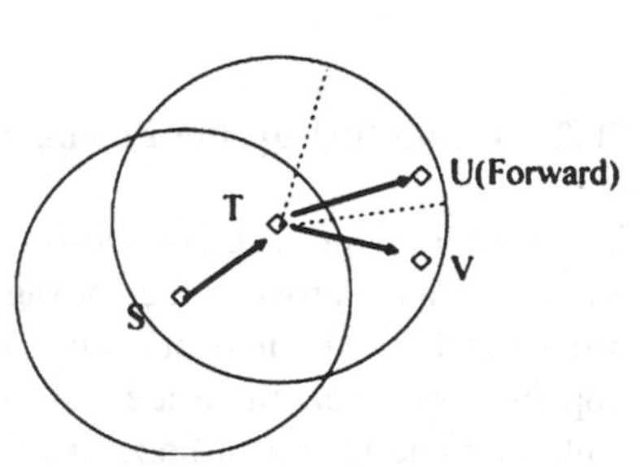

When U and V receive the packet, they should decide whether they are in the primary path or not. Since node U is in scope of 60° of the extended line from S to T, U is in the primary path and it may transmit the packet. V is not in the primary path, thus it will spread the packet locally. The two nodes should compute the distance between themselves and the boundary of the communication radius of T. The distance is denoted as d_u, d_v. They compute their own waiting time based on (50.1) and (50.2), which are denoted as $delay_u$ and $delay_v$.

Table. 50.1 Routing table

Source Node ID	Forwarding
1	TRUE
5	TRUE
10	FALSE
25	FALSE
99	TRUE

$$\text{delay}_u = a * CONSTANT\_DELAY * \frac{d_u}{R}, \quad 0<a<1. \tag{50.1}$$

$$\text{delay}_v = CONSTANT\_DELAY * \frac{d_v}{R}. \tag{50.2}$$

In (50.1) and (50.2), delay_u and delay_v are the waiting time of U and V, which are also the waiting time of the primary path and the auxiliary path. *CONSTANT_DELAY* is a predefined delay parameter in the protocol. a is predefined waiting time ratio between the primary path and auxiliary path. Equation (50.1) and (50.2) indicate that longer distance d_i means longer waiting time and node in the primary path means shorter waiting time.

Then they start the timer respectively. When timer is expired, if they do not receive the packet transmitted by other nodes, they will transmit their own packet. Otherwise, it is implied that some more appropriate node has transmitted the packet, thus, they should not transmit the packet. Since the waiting time is short when the node is in the primary path, the nodes in the primary path will have higher priority to transmit the packet. Therefore, the nodes in the primary path will transmit the packet earlier. This process formulates the primary path.

50.2.2.2 Protocol Description

In order to transmit the packet rapidly, the nodes should maintain the routing table dynamically, which includes the source node ID and the forwarding flag. It is shown in Table 50.1.

The fast broadcast protocol is divided into two phases: the route establishment phase and the routing phase. The route establishment phase is triggered by an arbitrary node. The trigger node sends the broadcast packet with the route establishment type and its own ID_i. Every node in the sensor network receives this kind of packet and follows the protocol of building up the primary path as mentioned above, establishing the routing table of source node ID_i. Repeat these steps until all the nodes have triggered the route establishment phase. The routing phase is available after the route establishment phase. When the node receives the packet, it will look up into its own routing table. If the source node is in the routing table and the transmission flag is true, the node will transmit the packet. Otherwise, it will do nothing. Later, the node will trigger another route establishment phase and maintain new routing table.

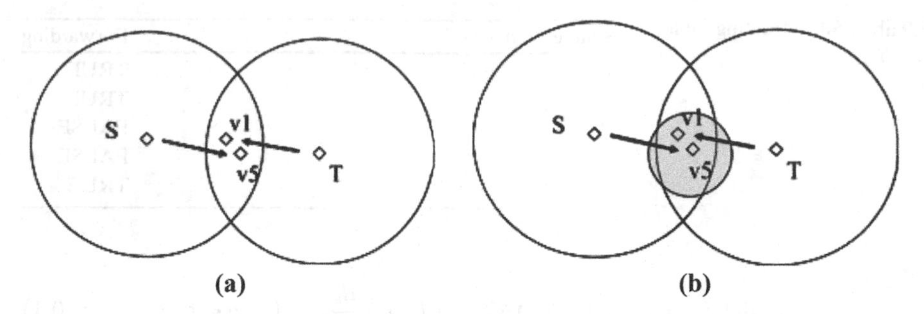

Fig. 50.4 Adjacent nodes transmission. **a** No reject area scenario. **b** Reject area scenario (gray area is reject area)

The adjacent nodes may be the candidate node at different times. As illustrated in Fig. 50.4a, node S sends the packet out at time T1 then V1 and V5 receive the packet. After the computation phase, V5 decides to transmit the packet. However, when node T transmits the packet at time T2, both V1 and V5 receive it again. At this time, V1 has longer distance from T than V5. Therefore, V1 is the candidate node and V1 transmits the packet. The transmission causes redundancy. In order to avoid this scenario, the reject area is proposed in the protocol. When a node receives a packet, it will compute whether it is in the reject area immediately based on 50.3.

$$\text{occupied} = \begin{cases} \text{true, Dist}(u, v) \ \leq \mathit{REJECT\_AREA} \\ \text{false, Dist}(u, v) > \mathit{REJECT\_AREA} \end{cases} \tag{50.3}$$

In 50.3, occupied is Boolean variable which indicates whether the node is in the reject area. *REJECT_AREA* is a predefined parameter in the system which is a metric of the distance between two nodes. If the distance between two nodes is less than *REJECT_AREA*, the node is in the reject area, which means it will not forward the packet. Otherwise, the node can be the candidate node to transmit the packet. As shown in Fig. 50.4b, when V1 receives the packet sent by V5, it will calculate that itself is in the reject area of V5 and it will not transmit the same packet.

The steps of fast broadcast protocol are shown as follows:

Step 1 Receive the packet and determine whether the broadcast ID is smaller than the current ID. If yes, it means the packet is an old broadcast packet and the node should drop the packet; otherwise, go to step 2.

Step 2 Determine the broadcast type. If the type is route establishment, go to step 3; otherwise, go to step 4.

Step 3 Route establishment phase. Compute the reject area. If the node is in the reject area, it will not transmit the packet; otherwise, determine whether it is in the primary path. If yes, start the primary transmission strategy and start the timer; otherwise, it cancels transmission. When the timer is expired, if the node does not receive the same packet, then it will transmit the packet and add the source node into the routing table; otherwise, it will cancel transmission.

Step 4 Routing phase. Check whether the source node is in the routing table. If yes, forward the packet; otherwise, do nothing.

50.3 Experiments and Analysis

50.3.1 Experiment Platform and Parameter Settings

NS2 is an integrated network simulation platform. In this study, the experiment runs on NS2 simulation platform. Both the network scale and experiment parameters are tested. We make two scenarios: the sensor network covers 1,200 × 1,200 m^2 and 2,500 × 2,500 m^2. The communication radius of nodes R is 250 m. The primary path is in the scope of 60° of the extended line. The radius of reject area is 0.7 × R. The number of nodes is 200, 300, ..., 1,000. The waiting time ratio between primary path and auxiliary path is 0.1, 0.2, 0.3, 0.5 and 1, representing the priority of node in the primary path.

50.3.2 Formulation of Umbrella Topology

In the scenario of 2,500 × 2,500 m^2 network areas, we compare the different waiting time in the primary path, which depends on a in 50.1. The two experiments with different waiting time of primary path can both achieve the broadcast target. Moreover, the result shows the bigger difference between primary path and auxiliary path is, the clearer the umbrella topology is. As shown in Fig. 50.5a, the red line is backbone of the umbrella topology. The waiting timer of primary path is much faster than any other timer. Therefore, the backbone is clear. As illustrated in Fig. 50.5b, since wait timer of primary path is a little faster than any other timer, the umbrella topology is not clear. The experiment also shows that in these two scenarios, transmissions quantity is almost the same.

This experiment shows that the waiting time in the primary path can influence the formulation of umbrella topology while there is no effect on the number of transmissions. Because the umbrella topology has well topology control character, the waiting time in the primary path should be set much shorter to formulate better umbrella topology.

50.3.3 Effect of Reject Area

As illustrated in Fig. 50.6a, the node number of four curves is 300, 500, 700 and 900 and the radius of reject area which is defined as *REJECT_AREA* in (50.3) increases from 0.4 × R to 0.9 × R. It shows that the routing delay decreases when the radius of reject area grows. In Fig. 50.6b, the node number of the curves is 200,

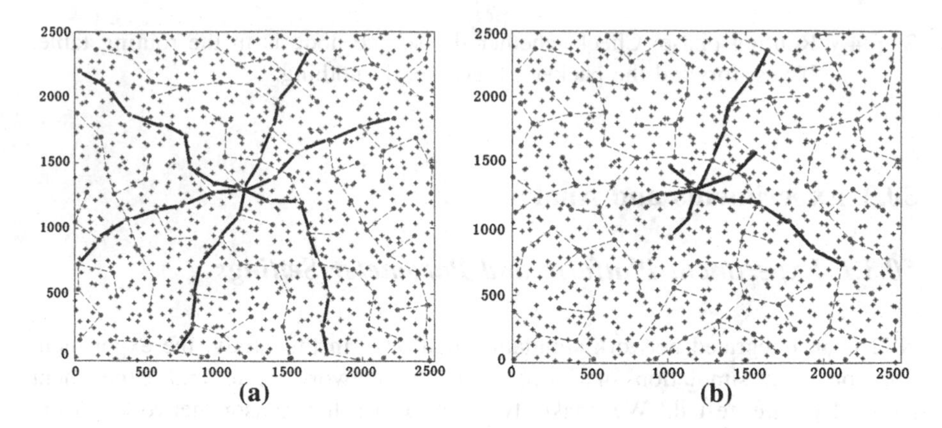

Fig. 50.5 Umbrella topology comparison. **a** Much shorter waiting time in the primary path. **b** A little shorter waiting time in the primary path

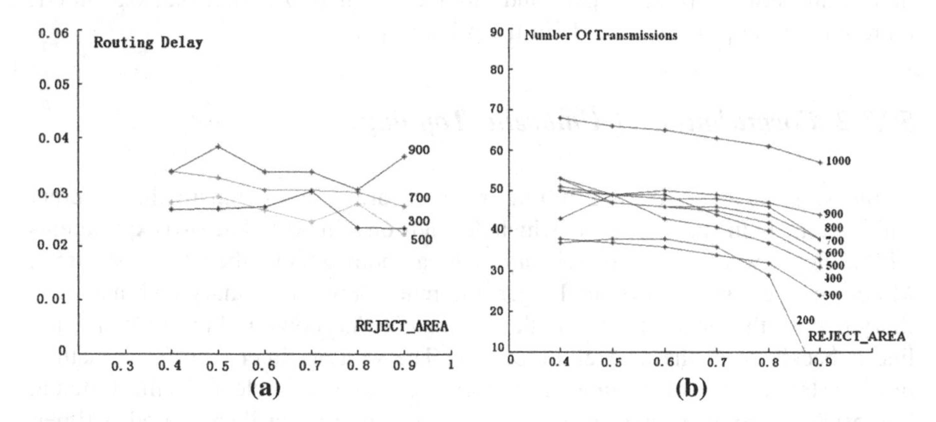

Fig. 50.6 Influence of reject area. **a** Comparison of routing delay. **b** Comparison of transmission number

300, ..., 1,000. It shows that the number of transmissions is significantly reduced. Therefore, the energy consumption is low. However, in the long reject radius scenario, the broadcast may fail since the network is disconnected. It can be seen that the longer radius of reject area, the smaller transmission number and the lower routing delay. In the meanwhile, the longer radius of reject area may cause network disconnection in a low density network scenario.

50.3.4 Effect of Waiting Time

As shown in Fig. 50.7a, the routing delay decreases with the increase of predefined waiting time which is defined as *CONSTANT_DELAY* in (50.1). In Fig. 50.7b, in

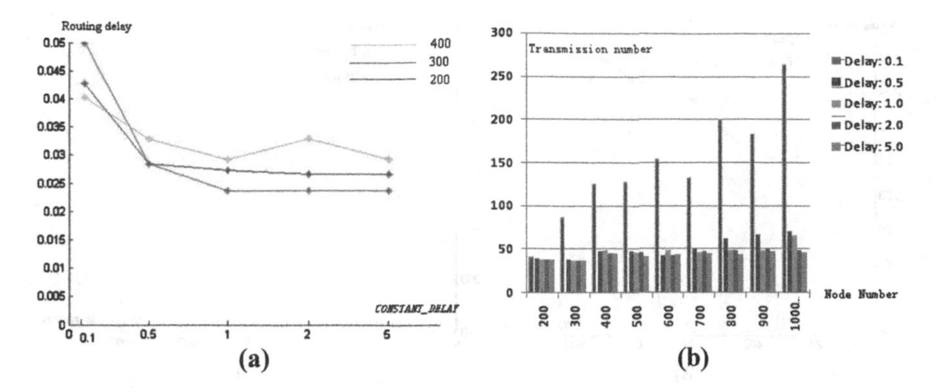

Fig. 50.7 Influence of predefined waiting time. **a** Comparison of routing delay. **b** Comparison of transmission mumber

the scenario that waiting time is predefined as 0.1 s, the transmission number is much more than that in other scenarios. The overall trend is going down. This is because longer waiting time can tell the minor difference between the different nodes' waiting time.

It can be seen that either the transmission number or the routing delay will not decrease when the waiting time is long enough. That means the waiting time should be chosen appropriately in practice. Shorter waiting time may cause higher transmission rate and higher routing delay whereas longer waiting time may bring more waste.

50.3.5 Comparing with Flooding

Flooding is a classic broadcast protocol which is the basis of broadcast. It can be implemented in any environment fairly. Therefore, flooding is compared with our protocol. As shown in Fig. 50.8, MFlood is the curve of flooding and ZFlood is the curve of the fast broadcast protocol proposed in this paper. The experiments run in the scenarios of different number of nodes. The number of nodes is 200, 300, …, 1,000 and the radius of reject area is 0.5 × R. As illustrated in Fig. 50.8a, the routing delay of MFlood is increasing with the increase of nodes number. In the meanwhile, the routing delay of ZFlood does not increase significantly. The routing delay of MFlood is 5–6 times as much as ZFlood which improves the network performance nearly 70%.

As shown in Fig. 50.8b, the curve of the flooding transmission number grows linearly while the number of transmissions of ZFlood is near a constant. In the scenario of 1,000 nodes, the flooding transmission number is 20 times as much as ZFlood. Since high transmission rate can bring high communication overhead, it can consume large amounts of energy. Thus low transmission number can save energy. In this scenario, 95% energy will be saved.

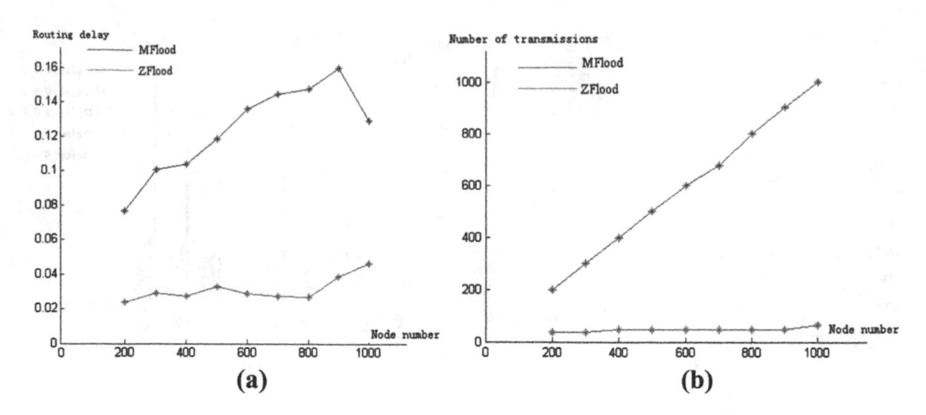

Fig. 50.8 Compare with Flooding. **a** Comparison of routing delay. **b** Number of transmission

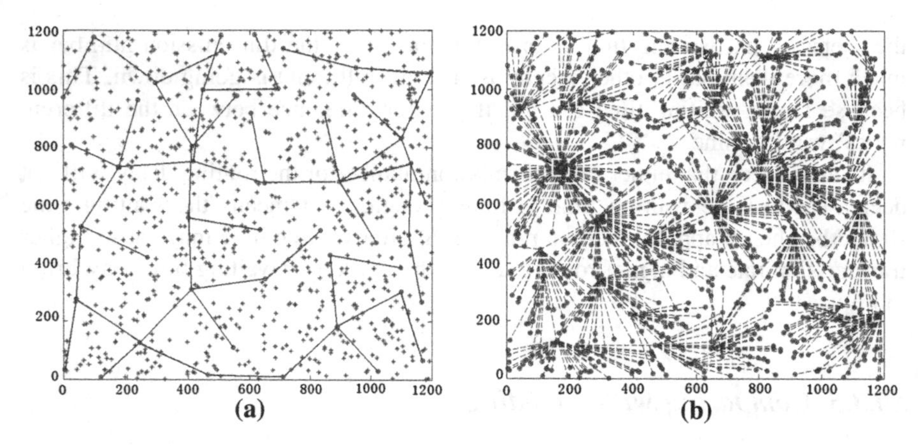

Fig. 50.9 Transmissions in the scenario of 1,000 nodes. **a** Transmission topology of ZFlood. **b** Transmission topology of flooding

Figure. 50.9a is the ZFlood transmission topology in the scenario of 1,000 nodes and Fig. 50.9b is the flooding transmission topology in the scenario of 1,000 nodes. It shows that the number of transmissions of flooding is much more than ZFlood. The proposed broadcast protocol improves the performance remarkably.

50.4 Conclusion

In this research, a fast, distributed, distance-based, scalable and self-adaptive broadcast protocol is proposed, which diminishes the network delay and improves the network performance. A new concept of umbrella topology is introduced which makes the packet reach the boundary of the network in a short time and builds a backbone of the umbrella topology. The protocol is based on distance and each hop is almost farthest and most effective, which is also used to decrease the network delay. Energy is saved because of low transmission rate. Each node can determine whether it could transmit the packet by itself. In addition, the protocol need not maintain the neighbor information. Thus, "hello" message is not necessary in the protocol. The communication overhead is low and the energy consumption is low. The sensor node is unnecessary to have great computing capability because the protocol is a distributed algorithm, which is helpful in the engineering.

Different waiting time between the primary path and the auxiliary path can affect the formulation of the umbrella topology which is useful in control area. The transmission number and the routing delay decrease with the increase of reject radius in the experiments. The predefined waiting time can also influence the performance of the protocol. Therefore, appropriate radius of reject area and the predefined waiting time can benefit the network performance, which is flexible in practice. The experiments show the protocol proposed is more effective than flooding and it improves the network performance remarkably.

In the real scenario, the maintenance of backbone is important. Therefore, further study can focus on maintaining the backbone and reducing the quantity of transmissions further in order to improve the network performance. In the wireless sensor network environment, the problem of the collisions in routing phase is also a serious and important issue. Therefore, it is necessary to avoid the collision further in the routing phase so that the network stability is improved more.

Acknowledgment The research work is partially supported by Aeronautics Foundation Program under Grant No. 20091969022 and Natural Science Foundation Program of China under grant No. 60973122.

References

1. Ni SY, Tseng YC, Chen YS, Sheu JP (1999) The broadcast storm problem in a mobile ad hoc network. In: 5th annual ACM/IEEE international conference on mobile computing and networking, New York
2. Haas Z, Halpern J, Li L (2002) Gossip based ad hoc routing. In: 21st annual joint conference of IEEE computer and communications societies
3. Alon N, Bar-Noy A, Linial N, Pelg D (1991) A lower bound for radio broadcast. J Comput Syst Sci 43:290–298
4. Sudipto G, Samir K (1996) Approximation algorithms for connected dominating sets. In: 4th annual European symposium on algorithms, pp 179–193
5. Wu J, Gao M, Stojmenovic I (2001) On calculating power-aware connected dominating sets for efficient routing in ad hoc wireless networks. In: International conference on parallel processing
6. Peng W, Lu X (2000) On the reduction of broadcast redundancy in mobile ad hoc networks. In: 1st annual workshop on mobile and ad hoc networking and computing

7. Peng W, Lu X (2001) AHBP: an efficient broadcast protocol for mobile ad hoc networks. J Comput Sci Technol 16(2):114–125
8. Giordono S, Stojmenovic I (2004) Position based routing algorithms for ad hoc network: a taxonomy. In: Cheng X, Huang X, Du DZ (eds) Ad hoc wireless networking. Kluwer, Boston, pp 103–136
9. Durresi A, Paruchuri VK, Iyengar SS, Kannan R (2008) Optimized broadcast protocol for sensor networks. IEEE Trans Comput 54(8):1013–1024
10. Kershner R (1939) The number of circles covering a set. Am J Math 61(3):665–671

Chapter 51
Survey of Load Balancing Strategies on Heterogeneous Parallel System

Wenliang Wang, Weifeng Liu and Zhiguang Wang

Abstract Load balancing is a key problem in parallel computing. Under the environment of increasing heterogeneous parallel systems, load balancing strategies facing new challenges. This paper is a survey of the developments of load balancing strategies on heterogeneous parallel system. First, we compare the differences between homogeneous and heterogeneous systems. Second, we review the status of research on related load balancing strategies and analyze existing load balancing algorithms. Finally, we discuss future challenges for load balancing in heterogeneous parallel computing.

Keywords Load balancing · Parallel computing · Heterogeneous system

W. Wang (✉) · Z. Wang
College of Geophysics and Information Engineering,
China University of Petroleum, Beijing, China
e-mail: wang_wenliang@sina.cn

Z. Wang
e-mail: wzg0202@cup.edu.cn

W. Liu
Sinopec Exploration and Production Research Institute, Beijing,
People's Republic of China
e-mail: liuwf.syky@sinopec.com

W. Liu
Sinopec Key Laboratory of Multi Components Seismic Technology, Beijing, China

J. J. Park et al. (eds.), *Proceedings of the International Conference on Human-centric Computing 2011 and Embedded and Multimedia Computing 2011*, Lecture Notes in Electrical Engineering 102,
DOI: 10.1007/978-94-007-2105-0_51,

51.1 Introduction

In recent years, hardware heterogeneity arises in many computing environments. Through a lot of experimental data, heterogeneous architectures have been proven cost-effective in designing future high performance computers. An emerging trend that is now beginning to receive more attention is of increasingly heterogeneous computing platforms.

Load balancing is one of the most essential problems in parallel computing. A series of related works have been used to solving problems which decompose computation into many tasks, and finishing all tasks in the shortest time. While load balancing for homogeneous parallel computing systems has been well studied for more than a decade, load balancing for heterogeneous parallel systems is a relatively new subject of investigation with a less-explored landscape. In heterogeneous computing environment, balancing of computational load needs special attention and the goal is the same: to minimize idle processor time and to lower the overhead of inter-process interaction. This is done by distributing the work such that no processor is waiting for the completion of another. However, there are many problems with uneven and unpredictable computation and communication caused load imbalance and poor efficiency. These bottlenecks are more prominent due to the heterogeneity of different hardware architectures.

In this paper, we place the focus on the load balancing strategies on heterogeneous parallel system and review existing approaches appearing in the literature. These strategies evolve from different research communities, aim to solve different problems. The remaining sections of the paper are organized as follows. In Sect. 51.2, we compare the differences between homogeneous and heterogeneous systems. In Sect. 51.3, we discuss and analyze static and dynamic load balancing strategies on heterogeneous parallel system. In Sect. 51.4, we discuss future challenges for load balancing in heterogeneous parallel computing.

51.2 Homogeneous and Heterogeneous System

Generally, a homogeneous parallel system is constructed by many interconnected equal computational elements; a heterogeneous parallel system is constructed by diverse computational elements of different performances, and each computational element consists of diverse chips of different architectures. For example, the overall computing pool is constructed by new multi-core clusters and early massively parallel processing (MPP) systems, or a single computing node consists of multi-CPU and multi-GPU.

Here we give two directly comparable figures. Figure 51.1 shows an example of homogeneous parallel system consists of n nodes, each node contains equal number of CPUs. Figure 51.2 shows an example of heterogeneous parallel system

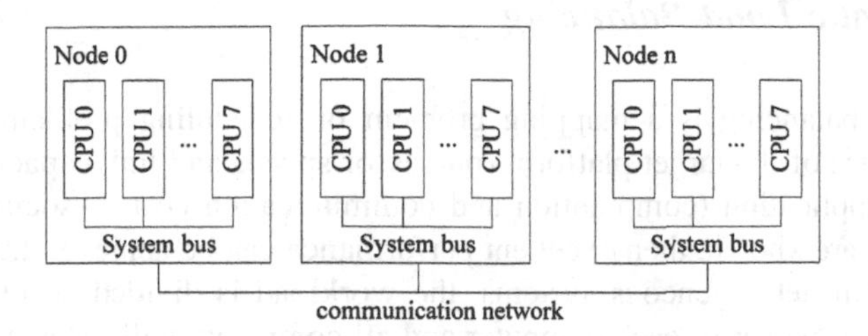

Fig. 51.1 An example of homogeneous parallel system

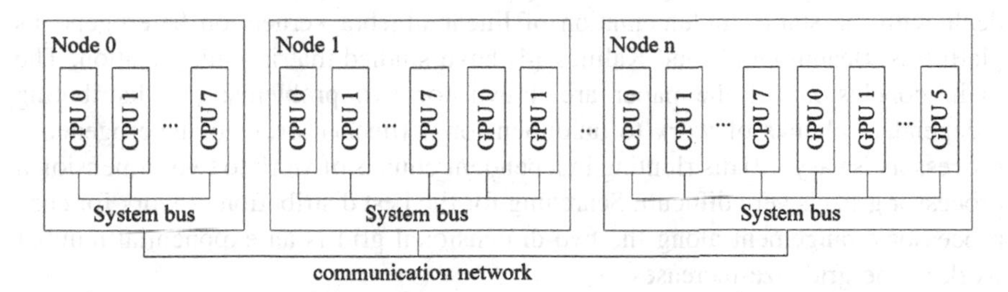

Fig. 51.2 An example of heterogeneous parallel system

also consists of n nodes, but each node contains unequal number of CPUs and GPU accelerators.

51.3 Load Balancing Strategies

The load balancing of an application has a direct impact on the speedup to be achieved as well as in the hardware computational power of parallel system. Load balancing can be basically classified into two categories: static and dynamic. Static load balancing distributes tasks among processors before the execution of the algorithms. These policies need fairly accurate prediction of the resource utilization of each task at compile time, which is generally very difficult in most of the applications. Dynamic load balancing, on the other hand, adjusts tasks which are unpredictable during the computation. It can vary workload and improve more efficiency on heterogeneous environment.

For the known workload problems, such as matrix multiplication, adopting static load balancing strategies, a "predictive" load balancing considering the calculation power of processors can be obtained; however, many real problems have a variable or dynamic workload depending on the data. In these cases, it is necessary to adjust data or processes allocation dynamically while the application is being executed. Obviously dynamic load balancing that can vary workloads according to the compute resource's capabilities could utilize the heterogeneity more effectively.

51.3.1 Static Load Balancing

Static load balancing is a mapping problem or scheduling problem [1]. If the characteristics of the target platform (processor speeds and link capacities) and of the target application (computation and communication costs associated to each data chunk) are known, then excellent performance can be achieved through static strategies. On heterogeneous systems, the workload is divided according to the computational power of each computer and all computers will receive a workload proportional to the capacity and load situation.

The static load balancing is developed by Graham [2]. Several authors have dealt with the static implementation of linear algebra kernels on heterogeneous platforms. Beaumont [3] and Kalinov [4] have studied matrix multiplication. The main conclusions of the paper are drawn for two problems: (1) distributing independent chunks of work to unidimensional (linear) arrays of heterogeneous processors is easy; (2) distributing independent chunks of work to two-dimensional processor grids is very difficult. Searching for the best distribution of work for each processor arrangement along the two-dimensional grid is an exponential number work as the grid size increases.

Static partitioning schemes to map a two-dimensional data matrix onto heterogeneous resources have been investigated by Crandall [5] and Kaddoura [6]. The two papers are all based on Recursive Bisection algorithm. Kaddoura improve the algorithm using two variations. Crandall proposed a method for block data decomposition on the heterogeneous workstation network. But heterogeneous block partitioning cannot benefit from the existing grid size. Furthermore, relaxing the geometrical constraints induced by two-dimensional grids leads to irregular partitioning that allow for a good load-balancing but are much more difficult to implement.

51.3.2 Dynamic Load Balancing

Dynamic load balancing has become a focused area of research from past ten or more years. As many problems existing, such as uneven and unpredictable computation and communication, the tasks are assigned to processors during execution of the program. A real dynamic load balancing technique should use observed computational and communication performance to predict the time a task would complete on a given component and the time needed to query the component [7]. Several dynamic load-balancing algorithms are proposed in the recent past.

Willebeek-LeMair [8] compared five dynamic load balancing strategies. The sender/receiver initiated diffusion (SID/RID) [9] strategies are asynchronous schemes which only use near-neighbor information. The hierarchical balancing method (HBM) organizes the system into a hierarchy of subsystems within which balancing is performed independently. The gradient model (GM) [10] employs a gradient map of the proximities of underloaded processors in the system to guide the migration of tasks between overloaded and underloaded processors. The dimension exchange method (DEM) requires a synchronization phase prior to load balancing and the balances iteratively. Moreover, another researcher Watts [11] concluded that the diffusion algorithm is well suited for calculating the work transfer matrix for the algorithm tends to transfer less work than DEM.

Diffusion algorithm was developed by Cybenko [9] and several variances of Diffusion algorithms like demand driven model [8] dimensional exchange method [9] were proposed. Lingen [12] presented a diffusion algorithm for heterogeneous parallel computers. Elsasser [13] extended these schemes for computational environments that are heterogeneous both with respect to the processing performances and communication speeds. Watts [14] described the Diffusion algorithm that supports heterogeneous parallel computers. Unfortunately, this algorithm is less effective than original Diffusion algorithm, even if the target computer is homogeneous. Other experimental data show that Lingen's algorithm faster than Watts' algorithm on heterogeneous computers.

Work-stealing is a technique that implements load balancing. Effectively, each thread maintains its own pool of tasks. The owner thread stores and takes items from the task pool. Typically, when there are no more tasks in the pool (the owner thread has nothing more to do), to keep busy, the thread can steal work items from other threads. A regular work stealing strategy is not adapted to cope with the heterogeneity. Specially, Hermann [15] investigated the efficiency of affinity guided work stealing in heterogeneous environment, where both CPUs and GPUs are involved. They use a priority guided work stealing to favor the execution of low weight partitions on CPUs and large weight ones on GPUs.

Maheshwari [16] present a priority-based dynamic load balancing algorithm for heterogeneous computing environment. The base case local scheduler in each machine uses the round-robin process scheduling policy. The algorithm determines the task precedence graph of the parallel jobs dynamically at run-time and assigns appropriate priorities to the processes to resolve the dependencies. This paper has analyzed two different heterogeneous parallel program models under different background load conditions.

Vignesh [17] discussed why they have taken work sharing approach instead of work stealing on heterogeneous computing environment using multi-CPU and GPU. And this paper presented two schemes which are uniform-chunk distribution scheme and non-uniform-chunk distribution scheme.

Rus [7] developed an algorithm for structured and free two dimensional quadrilateral finite element meshes based on the rearrangement of elements among respective subdomains and this algorithm enable efficiency on heterogeneous systems. By adapting the allocated work to the unknown computing power directly during the computation, it is unnecessary to know the performance characteristics of the new added computers to the existing system.

51.4 Conclusions and Future Work

While some progress has been made in load balancing strategies on heterogeneous parallel system, new algorithms and architectures require new partitioning methods. Each of them is tightly related to each other and exerts great challenges to the parallel computing. Though we review many load balancing strategies on heterogeneous parallel system, there still remain many problems due to the existence of uncertain factors. We summarize and conclude the survey with listing some important issues and research trends for load balancing strategies on heterogeneous parallel system.

First, there is no load balancing strategy that can be universally used to solve all problems. Usually, strategies are designed with certain assumptions and favor some type of biases. In this sense, it is not accurate to say "best" in the context of load balancing strategies, although some comparisons are possible. These comparisons are mostly based on some specific applications, under certain conditions, and the results may become quite different if the conditions change.

Second, new algorithms continue to generate more complex and challenging tasks, requiring more flexible load balancing strategies on existing heterogeneous environments.

Third, if some essential relationships between new heterogeneous parallel systems and well-studied homogeneous parallel systems could be established, existing effective and efficient load balancing algorithms may be applied to solve new problems on new heterogeneous architectures.

Acknowledgments Research of the first and third authors was supported by the State Key Basic Research Program (973) (No. 2007CB209602) and the National Natural Science Foundation of China (No. 60803159).

References

1. Wilkinson B, Allen M (2005) Parallel programming: techniques and applications using networked workstations and parallel computers, 2nd edn. Pearson Education, Prentice Hall
2. Graham RL (1972) Boundes on multiprocessing anomalies and packing algorithms. In: Spring joint computer conference. ACM, NY, pp 205–217
3. Beaumont O, Boudet V, Petitet A, Rastello F, Robert Y (2001) A proposal for a heterogeneous cluster ScaLAPACK (dense linear solvers). IEEE Trans Comput 50(10): 1052–1070
4. Kalinov A, Lastovetsky A (1999) Heterogeneous distribution of computations while solving linear algebra problems on networks of heterogeneous computers. In: Sloot P, Bubak M, Hoekstra A, Hertzberger B (eds) HPCN Europe 1999. LNCS, vol 1593. Springer, Heidelberg, pp 191–200
5. Crandall PE, Quinn MJ (1993) Block data decomposition for data-parallel programming on a heterogeneous workstation network. In: 2nd international symposium on high performance distributed computing. IEEE Computer Society Press, Spokane, pp 42–49

6. Kaddoura M, Ranka S, Wang A (1996) Array decomposition for nonuniform computational environments. J Parallel Distrib Comput 36:91–105
7. Rus P, Stok P, Mole N (2003) Parallel computing with load balancing on heterogeneous distributed systems. Adv Eng Softw 34(4):185–201
8. Willebeek-LeMair MH, Reeves AP (1993) Strategies for dynamic load balancing on highly parallel computers. IEEE Trans Parallel Distrib Syst 4(9):979–993
9. Cybenko G (1989) Dynamic load balancing for distributed memory multiprocessors. J Parallel Distrib Comput 7:279–301
10. Lin FCH, Keller RM (1987) The gradient model load balancing method. IEEE Trans Softw Eng 13(1):32–38
11. Watts J (1995) A practical approach to dynamic load balancing. Technical report, California Institute of Technology
12. Lingen FJ (2000) A versatile load balancing framework for parallel applications based on domain decomposition. Int J Numer Methods Eng 49(11):1431–1454
13. Elsasser R, Monien B, Preis R (2002) Diffusion schemes for load balancing on heterogeneous networks. Theory Comput Syst 35:305–320
14. Watts J, Taylor S (1998) A practical approach to dynamic load balancing. IEEE Trans Parallel Distrib Syst 9(3):235–248
15. Hermann E, Raffin B, Faure F, Gautier T, Allard J (2010) Multi-GPU and multi-CPU parallelization for interactive physics simulations. Euro Par (2):235–246
16. Maheshwari P (1996) A dynamic load balancing algorithm for a heterogeneous computing environment. In: 29th Hawaii International Conference on System Sciences. Wailea, pp 338–346
17. Vignesh TR, Ma W, Chiu D, Agrawal G (2010) Compiler and runtime support for enabling generalized reduction computations on heterogeneous parallel configurations. In: 24th ACM international conference on supercomputing. ACM, NY, pp 137–146

Chapter 52
Data Management and Application on CPSE-Bio

Xiao Cheng, Jiang Xie, Ronggui Yi, Jun Tan, Xingwang Wang, Tieqiao Wen and Wu Zhang

Abstract With the rapid development of bioinformatics, large amounts of data, algorithms and tools were accumulated. In order to effectively utilize these resources, a problem solving environment (PSE) is emphasized by many researchers. This paper proposes cloud PSE of bioinformatics (CPSE-Bio). For solving the problem of mass data and data format in CPSE-Bio, a new data management mechanism is discussed and an example of data querying and biomolecular networks constructing is given in this paper.

Keywords Could computing · Problem solving environment · Bioinformatics · Data management

---

X. Cheng · J. Xie (✉) · R. Yi · J. Tan · X. Wang · W. Zhang
School of Computer Engineering and Science, Shanghai University, Shanghai, China
e-mail: jiangx@shu.edu.cn

X. Cheng
e-mail: xchengshu@gmail.com

R. Yi
e-mail: ronggui.yi@gmail.com

J. Tan
e-mail: tanjun2525@gmail.com

X. Wang
e-mail: wangx_w@shu.edu.cn

W. Zhang
e-mail: wzhang@shu.edu.cn

J. Xie · T. Wen · W. Zhang
Institute of Systems Biology, Shanghai University, Shanghai, China
e-mail: tqwen@staff.shu.edu.cn

J. J. Park et al. (eds.), *Proceedings of the International Conference on Human-centric Computing 2011 and Embedded and Multimedia Computing 2011*, Lecture Notes in Electrical Engineering 102,
DOI: 10.1007/978-94-007-2105-0_52,

52.1 Introduction

In recent years, bioinformatics has become the hot area of various disciplines, including biology, computer science, mathematics, and so on. With the rapid development of bioinformatics, large amounts of data, algorithms and tools were accumulated [1]. It was told in a recent survey that there are 1,078 biological databases [2] and over 1,200 bioinformatics tools [3] publicly available online. Among them, is BioGRID an online interaction repository with data compiled through comprehensive efforts [4]. Moreover, people have studied various tools and technologies to solve the problems of bioinformatics, including the computing environment for bio-molecular networks [5], bio-molecular network matching (BNMatch) [6], web service management system for bioinformatics [7], cloud computing technologies for bioinformatics applications [8]. Therefore, it is necessary to develop a problem solving environment of bioinformatics (PSE-Bio) in order to make full use of these resources.

There are, in general, two problems on PSE-Bio. On the one hand, data of biological experiment are disarray. Faced with mass data, PSE-Bio cannot store and manage them easily [9]. We have to establish a data management mechanism which can support different data requirement of applications. On the other hand, it is hard to complete enormous data analysis by sequential computing [10]. Compared with the previous PSE based on grid computing technologies, PSE based on cloud computing technologies cannot only store and manage mass data easily, but also process data by parallel computing which can provide more efficient services.

This paper develops the architecture, data management mechanism and applications of CPSE-Bio. It uses Apache Hadoop, the latest cloud computing technology platform, to manage the data of CPSE-Bio based on MapReduce of cloud programming environment. Hadoop platform of cloud technologies supports Linux and Windows, and manages data through distributed file system (HDFS) [11]. Services of data querying and bio-molecular networks constructing are provided on the CPSE-Bio. Users can query data to construct bio-molecular networks through friendly interface of CPSE-Bio.

The architecture of CPSE-Bio is proposed in Sect. 52.2. Section 52.3 discusses data management, data querying and bio-molecular networks constructing in CPSE-Bio. In Sect. 52.4, we provide an example of bioinformatics application and performance comparisons are given in Sect. 52.5. Section 52.6 makes conclusions.

52.2 Architecture

CPSE-Bio is characterized by applications of bioinformatics relying on data analysis. Cloud computing technologies can manage and analyze large data sets. Figure 52.1 shows CPSE-Bio architecture based on Hadoop platform. This model,

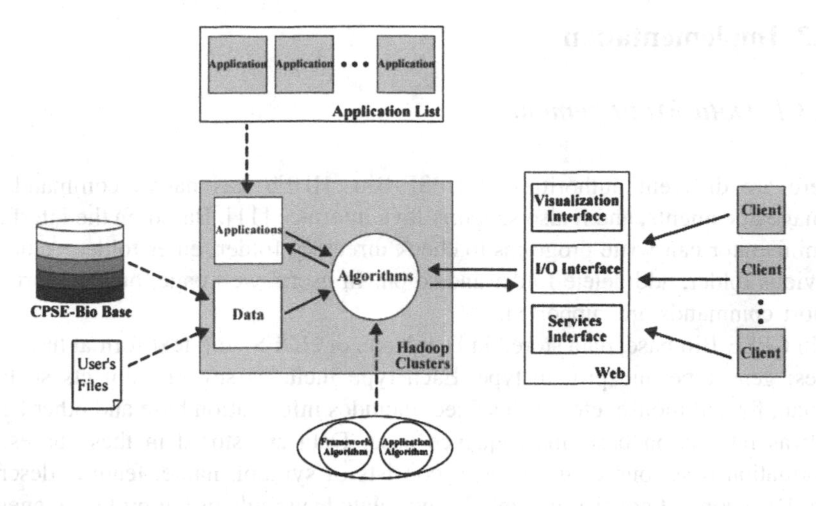

Fig. 52.1 CPSE-Bio architecture. The *black solid line* represents work relation; the *black dashed line* represents dependency relation

based on web services technology, consisting of front-end portal and background distributed clusters, is convenient for users to visit platform through the Internet. In the portal, there are three important interfaces:

- Visualization interface visualizes the results and shows to the users as graphics;
- I/O interface allows users to upload data and uses cluster to get computing results;
- Service interface provides users a large number of optional bioinformatics applications.

CPSE-Bio consists of three modules, i.e. algorithms modules, applications modules and data modules on distributed cluster. Algorithm modules contains some algorithms of CPSE-Bio, from framework algorithms such as application packaging algorithms, visualization object generation algorithms [6], file input and output and storage algorithms to concrete application algorithms such as service selection algorithms [7], bio-molecular networks match algorithms [5, 6], Meta-Pathway-Hunter algorithms [13], etc. Data querying and bio-molecular networks constructing algorithms also belong to Algorithms Modules. Applications Modules covers some types of bioinformatics analysis tools, such as Cytoscape [14], DIVID [15], Blast [16], which are packaged by Algorithm Modules to serve CPSE-Bio. Data Modules involves CPSE-Bio base and users' files and manages data through HDFS. Working process of CPSE-Bio is as follows. Users upload data or query data through Data Modules firstly. Then, users select an application from Applications Modules to compute the data. Finally, the result of application is processed by Algorithms Modules to obtain final result.

52.3 Implementation

52.3.1 Data Management

There are different authorities in CPSE-Bio. HDFS has native commands to manage documents, and it also supports Java interface [11]. Based on the interface, administrator can write programs to check directory folder, enter folder, return to previous folder, add/delete folder and so on. In users' web page, only import and export commands are supported.

In CPSE-Bio base, data stored in DataNode of HDFS with text format have two types: gene type and protein type. Each type includes several subjects such as human, fly and mouse, etc. Each subject includes information base and other bases such as interaction base and sequence base. Data are stored in these bases. In information base, one datum content consists of symbol, name, feature, description. Data content consist of symbols and related symbols of the proteins/genes in interaction base. Figure 52.2a shows the datum content. Address Mapping Tables are stored in NameNode, while each datum is stored in one text file in DataNode. The name of the text file is the symbol of the datum. Therefore, each base is composed of a large number of small files. Figure 52.2b illustrates the logical structure of the data. This structure has following advantages.

- To allocate tasks to each slave node in accordance with the size of the Map in JobTracker of MapReduce. When the sizes of files exceed the size of the Map, Hadoop will arrange tasks into line sequence. Therefore, using small files can reduce the Map and Reduce computing time. Figure 52.2c gives an example to describe the computing time of different sizes of files in the Map;
- To reduce computing amount of MapReduce program, and avoid computing repeatedly and redundantly;
- To facilitate users to query targeted information and construct bio-molecular networks, and facilitate administrators to maintain data in HDFS.

CPSE-Bio contains some application services, and each service requires different input files. Uniform data format is to help programmer to extract data correctly and reduce the complexity of code when Map functions are written. Writing corresponding MapReduce can convert original data format defined in Map to required format by application in Reduce. Library functions of Hadoop can help to output file format required by application. Parallel computing is applied to data processing. Figure 52.2d shows the flow chart of the process.

52.3.2 Data Querying

Users often need to understand some genes or protein information and functions before they begin to study bioinformatics problems. CPSE-Bio based on mass data can implement data querying lightly. Generally, with the scope of querying

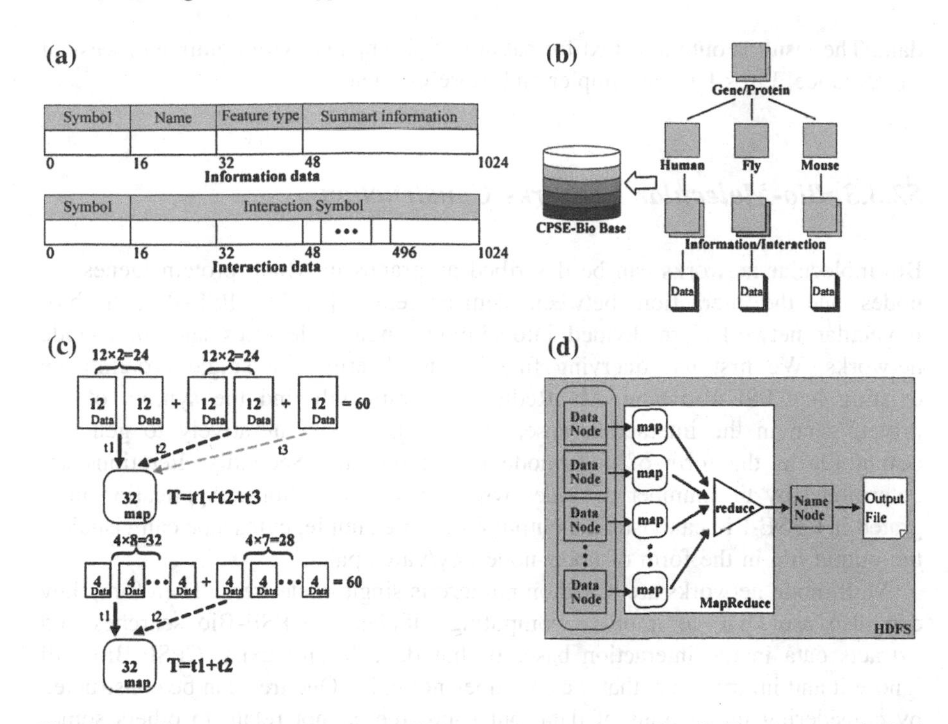

Fig. 52.2 **a** Shows the datum content. Figures represent data offset. **b** Illustrates the logical structure of the data. **c** Gives an example to describe the computing time of different sizes of files in the Map. The *black solid line* represents current task, and the computing time is t1; the *black dashed line* represents the first task in line sequence, and the computing time is t2; the *gray dashed line* represents the second task in line sequence, and the computing time is t3; T represents the total computing time; figures represents file size and compute power. **d** Shows the flow chart of MapReduce process

provided by users, MapReduce programs of CPSE-Bio can screen out corresponding data in HDFS. But MapReduce cannot implement precise querying, because it needs many steps to match data, which may lead to very high time complexity. Pig script language, developed by Apache, is a high level querying language based on MapReduce. It provides a command of checking data structure repeatedly to solve this puzzle. User can define his own functions (UDF) to achieve the desired functions [11].

In CPSE-Bio, we can use Pig to write UDF to query data. Specifically, users input the name or symbol of gene/protein and set the search scope in portal. Then information input is passed to UDF as a parameter. MapReduce searches CPSE-Bio base according to users' settings to get intermediate result stored in NameNode. And every record of the intermediate result is taken as the key, while data content as the value. (If the querying scope cannot be set, MapReduce will search the whole DataNodes of HDFS.) Pig uses relational statement to display data secondly that MapReduce generates. Then MapReduce matches information that users input with the key of intermediate result, and combines key/value of data. The result is output as text format at last. Compared with iterative process of MapReduce, Pig's UDF is simpler and more efficient.

52.3.3 Bio-Molecular Networks Construction

Bio-molecular networks can be described as graphs in which proteins/genes are nodes, and the interactions between them are edges [1]. In CPSE-Bio, the bio-molecular networks are divided into simple single-node trees and multi-node networks. We first use querying function to determine a single protein/gene existing in CPSE-Bio base. MapReduce program will find the symbol of the protein/gene in the interaction base, and compute data iteratively to generate output file as the form of node–node key/value pairs Secondly. Iterations are determined by the number of nodes which have interactions. Application integrated in CPSE-Bio can visualize output file. For example, cytoscape can visualize the output file in the form of node–node key/value pairs.

Multi-node networks are based on numerous single-node trees. Therefore, they can also use UDF to improve computing efficiency. CPSE-Bio searches and extracts data in the interaction base. If that data doesn't exist, CPSE-Bio will ignore it and inform users that the data does not exist. One tree can be constructed by considering interactions of data, but each tree cannot relate to others sometimes. It is very difficult to link data sets by MapReduce, which show not only in the adding connected paths between two trees, but also in deleting unconnected paths. Pig provides a good built-in support. Jointing, ordering and filtering statements of Pig written in UDF can integrate data sets. However, not all interactions are direct. Most nodes interact with others through one or more intermediate nodes which are often key nodes. The computational amounts of establishing networks diagrams with indirect interactions are very great. But CPSE-Bio that uses parallel computing can deal with such problems well. Specifically, pathway length (PL) is set in UDF firstly, and secondly according to the set PL, Pig begins to joint nodes. Thus the relationships between nodes are built, but the file format cannot be accepted by applications of CPSE-Bio, so MapReduce makes them to node–node key/value files as input files of application. Figure 52.3 describes a specific instance that UDF constructs bio-molecular networks.

52.4 Experiment

The current experiment presents that user submits data through CPSE-Bio, obtains the relationships of data, and establishes the bio-molecular networks, according to the computing results. Experimental procedures are as follows.

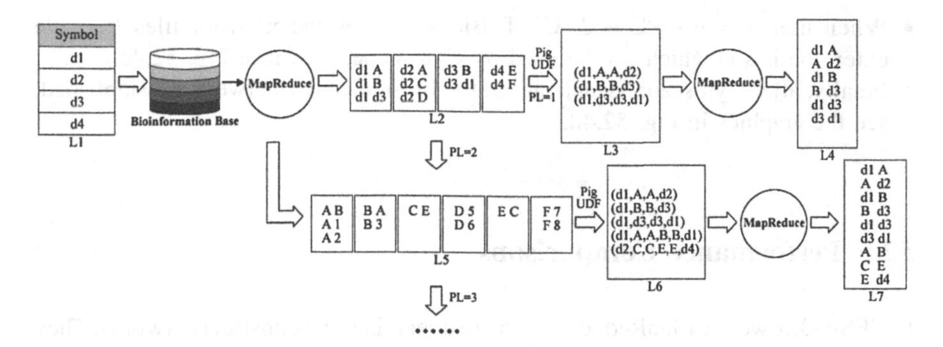

Fig. 52.3 Describes a specific instance that UDF constructs bio-molecular networks. L1 represents the file that user inputs, and MapReduce searches bioinformatics base to get L2 related with L1. When the PL is set as 1, UDF program of Pig will connect directly the data of L2 to get L3. And then MapReduce computes L3 to get final file L4 that can be identified by application. When the PL is set as 2, MapReduce will search data related with L2 in bioinformatics base to get an added new table L5. When user changes the PL, it uses the above procedures similarly

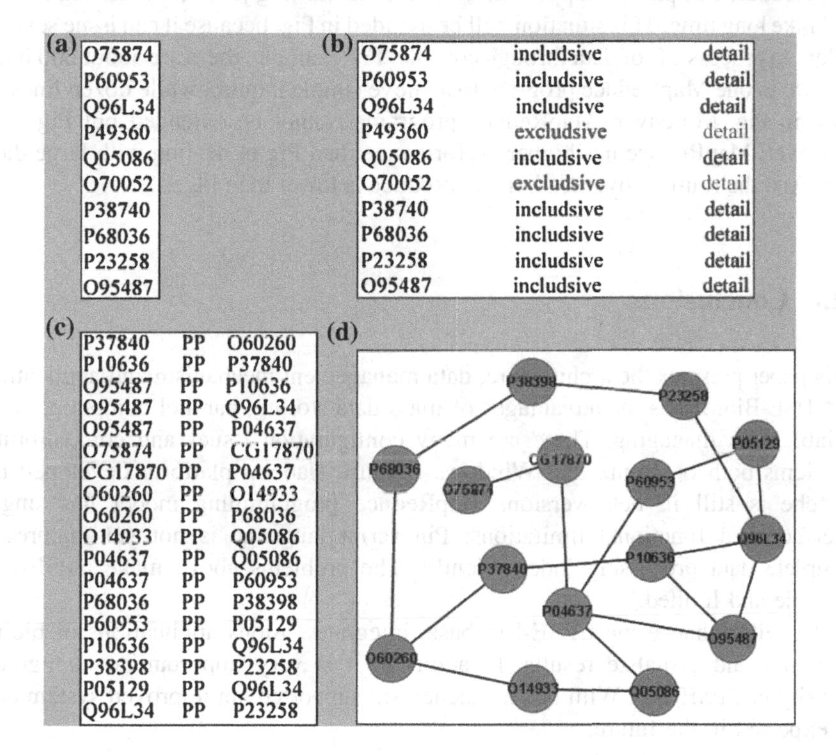

Fig. 52.4 Experiment data and result

- Users upload data (the file extension is txt) through portal, as shown in Fig. 52.4a, and sets the search scope as: gene/human/interaction.
- The data are searched in CPSE-Bio base. The result as shown in Fig. 52.4b.

- When user sets the PL = 2, CPSE-Bio generates the network files (the file extension is sif) which can be used by Cytoscape, shown in Fig. 52.4c.
- Finally, after Cytoscape visualization, the bio-molecular network is established, see the graphics in Fig. 52.4d.

52.5 Performance Comparisons

In CPSE-Bio, we use MapReduce and Pig to query data and construct networks. They have respective advantages in data processing. Through the experiment in Sect. 52.4, we compare with the performance of MapReduce and Pig. It is found that Pig has faster computing speed than MapReduce in the respect of matching data. Suppose that one file has m records and each record contains n interaction nodes. MapReduce needs to compute $m \times n$ times, but Pig only needs to compute m times. In MapReduce, compiling Mapper and Reducer, submitting jobs and indexing results will take long time. This situation will be avoided in Pig, because it can issue several orders to process TB of data through console. For example, there are about 200 lines of code in one MapReduce program to achieve simple inquire, while dozen lines of code in Pig. Moreover, MapReduce programs cannot be extended but Pig can. However, MapReduce has higher performance than Pig in dealing with large data sets. And the runtime overhead of MapReduce is lower than Pig.

52.6 Conclusions

This paper presents the architecture, data management mechanism and application of CPSE-Bio. It has the advantages of mass data storing, parallel computing, and reliable data managing. There are many configuration issues and compatibility problems both on Linux and Windows because Hadoop platform developed by Apache is still in beta version. MapReduce programming model has single interface and functional limitations. Pig script language is not yet mature to complete data processing independently. The problems above make CPSE-Bio unstable and limited.

We will enhance the CPSE-Bio base, integrate various applications of bioinformatics and visualize results. In addition, it is very important to strengthen security in CPSE-Bio. With better mechanism supported, an improved system can be expected in the future.

Acknowledgments This research is part supported by Shanghai Leading Academic Discipline Project [J50103], Ph.D. Programs Fund of Ministry of Education of China [200802800007], the National Science Foundation of China [31070954], the Science and Technology Commission of Shanghai [09JC1406600], Innovation Program of Shanghai Municipal Education Commission [11YZ03] and Innovation Fund of Shanghai University.

References

1. Sharan R, Ideker T (2006) Modeling cellular machinery through biological network comparison. Nat Biotechnol 24:427–433
2. Galperin MY (2007) The molecular biology database collection: 2008 update. Nucleic Acids Res 36:D2–D4
3. Brazas MD, Fox JA, Brow T, McMillan S, Ouellette BFF (2008) Keeping pace with the data 2008 update on the bioinformatics links directory. Nucleic Acids Res 36:D2–D4
4. Stark C, Breitkreutz BJ, Chatr-Aryamontri A, Boucher L, Oughtred R, Livstone MS, Nixon J, Van Auken K, Wang X, Shi X, Reguly T, Rust JM, Winter A, Dolinski K, Tyers M (2010) The BioGRID interaction database: 2011 update. Nucleic Acids Res 39:D698–D704
5. Xie J, Mao GY, Zhang SL, Zhang W (2010) An integrated computing environment for bio-molecular networks. J Converg Inf Technol
6. Yu L, Xie J, Cheng X, Zhang W (2010) BNMatch: a Cytoscape plugin for querying and visualizing matched similar networks. ICCCI2010
7. Xu K, Yu Q, Liu Q, Zhang J (2011) Athman Bouguettaya. Web Service management system for bioinformatics research: a case study. Springer, London
8. Jaliya E, Thilina G, Judy Q (2011) Cloud computing technologies for bioinformatics applications. IEEE Trans Parallel Distrib Syst
9. Katarzyna R, Marian B, Peter M, Vladimir G (2007) Problem solving environment for distributed interactive applications. CoreGRID Technical Report
10. DeBardeleben NA, Ligon III WB, Pandit S, Stanzione DC Jr (2002) Coven-a framework for high performance problem solving environments. In: Proceedings of the 11th IEEE international symposium on high performance distributed computing HPDC-11
11. White T (2009) Hadoop: The definitive.Yahoo!Press
12. Andréa M, Maurício TJF (2008) CloudBLAST: Combining MapReduce and Virtualization on Distributed Resources for Bioinformatics Applications. Fourth IEEE Int Conf eSci
13. Pinter R, Rokhlenko O, Yeger-Lotem E, Ziv-Ukelson M (2005) Alignment of metabolic pathways. Bioinformatics 21(16):3401–3408
14. Shannon P, Markiel A, Ozier O (2003) Cytoscape: a software environment for integrated models of bio-molecular interaction networks. Genome Res 13(11):2498–2504. http://www.cytoscape.org
15. Huang DW, Sherman BT, Lempicki RA (2009) Systematic and integrative analysis of large gene lists using DAVID Bioinformatics Resources. Nat Protoc 4(1):44–57
16. Papadopoulos JS, Agarwala R (2007) COBALT: constraint-based alignment tool for multiple protein sequences. Bioinformatics 23:1073–1079

Part XIII
ENC 2011

Chapter 53
Implementation of SOAS for SDR Processor

Liu Yang, Xiaoqiang Ni and Hengzhu Liu

Abstract Software-defined radio (SDR) processor is the core component in Wireless Communication Equipments. This paper proposes a novel implementation of optimized assembler for SDR processor, which is called SOAS (SDR processor optimized assembler). It is more convenient than legcy assembler for programmers. And it helps developing high performance applications before the compiler is ready. It can solve some optimization problems that the compiler can not solve. With the optimized SOAS, we can make good use of the high performance SDR processor.

Keywords SOAS · Assemble · SDR

53.1 Introduction

With the development of new communication technique: software-defined radio (SDR), which implements radio functions in software, the SDR become popular because it meets the development trend for better flexibility and scalability [1]. To achieve higher performance, SDR processor requires more complicated architecture than traditional processor or DSP. The corresponding development tools for SDR processor, such as compiler, assembler, linker, are harder to be developed. To make good use of SDR processor and get higher performance, compiler and optimized-compiler play a very important role.

Compiler optimization includes three levels: manual assembly optimization, auto assembly optimization and high-level language optimization. Manual

L. Yang (✉) · X. Ni · H. Liu
National University of Defense Technology, ChangSha, China
e-mail: yangliujoy@gmail.com

J. J. Park et al. (eds.), *Proceedings of the International Conference on Human-centric Computing 2011 and Embedded and Multimedia Computing 2011*, Lecture Notes in Electrical Engineering 102,
DOI: 10.1007/978-94-007-2105-0_53,

assembly optimization needs professional programmers who are very familier to the processor architecture such as pipeline structure, instruction latency and some other hardware information. It also requires that the programmers have the application background knowledge. What's more, the manual assembly optimization costs a lot of time and may cause more bugs. High-level language optimization doesn't have high requirements to programmers. The programmers don't need to know about the hardware information. But this kind of optimization is complicated to compiler design. The development of compiler is more difficult and needs longer period. Also, the effect of high-level language optimization is limited and some difficult problems can not be solved in the high-level language optimization. The auto assembly optimization combines the benefit of ease programming and less time spending. This paper gives the design and implementation of auto optimizing assembler for SDR processor, which is called SOAS (SDR processor Optimized Assembler). It is not like the normal assembler, the input of SOAS is unscheduled assembly code. It is arranged according to the logic order that programmers hope. It allows the programmers to write assembly code without knowing about the pipeline structure or registers allocation. This unscheduled assembly code is called un-optimized assembly code. SOAS transfers the un-optimized assembly code to highly-parallel assembly code. The main tasks of SOAS include: arranging instructions to fully make use of the parallelism of our target architecture; making sure that the instructions fulfill the latency requirements of our target architecture; allocating registers and functional units [2].

This paper first introduces the new architecture of our SDR processor. Then this paper describes the implementation of SOAS designed for our SDR processor and introduces a kind of assembly language for SOAS, which is called un-optimized assembly language.

53.2 New SDR Processor

Our new SDR processor architecture is based on VLIW + SIMD [3, 4]. It is newly designed and it has a new instruction set for some high performance applications. It includes unified instructions-fetching unit and instructions-dispatch unit. It includes two processing units, which are called scalar unit (SU) and vector unit (VU). The fetching and dispatch units distribute instructions for SU and VU simultaneously. SU is in charge of scalar tasks and controls the flow of the execution of VU. VU is used for vector computation and it includes several VEs. Each VE has a local register file, four accumulators and three parallel functional units. The local register file has 16 registers. All registers with the same number in all VEs form a 512bit vector. These functional units support fixed-point and floating-point operations. Although most of vector instructions are run on each VE independently, there are two kinds of operations are based on all VEs. Reduction operation adds up or compares data of the same local register in all VEs to get sum, maximum or minimum value. Shuffle operation permutes data of the same local register in all VEs by byte, half word or word. SU and VU can exchange data through a set of shared registers.

To catch up with the high throughput of VU, vector data accessing unit is used for load and store of vector data, which includes a vector memory and supports 2 load/store instructions each cycle.

53.3 Introduction of SOAS

SOAS includes two modules:

- Assembly-optimizing module: the input of this module is un-optimized assembly code and the output is normal assembly code. It is used to optimize the un-optimized assembly code. It includes four sub-modules: registers allocation sub-module, functional units' allocation sub-module, latency processing sub-module and parallelism development sub-module.
- Normal assembling module: the input of this module is normal assembly code and the output is binary code. It works like the other normal assembler. It is used to translate assembly code to binary code. It doesn't do any optimization.

The working flow of SOAS is showed as Fig. 53.1.

53.4 Un-Optimized Assembly Language

Un-optimized assembly code is the input of SOAS. It is unscheduled assembly code and is arranged according to the logic order that programmers hope. It allows the programmers to write assembly code without knowing about the pipeline structure or registers allocation. Un-optimized assembly language is a kind of programming language between high-level language and normal assembly language. It provides an efficient programming platform to the programmers that do not know much about the internal structure of our target architecture [2].

Un-optimized assembly language looks like normal assembly language. They use the same instruction set. But un-optimized assembly language has its own assembly optimizer directives and it does not need to describe some information that normal assembly language must describe. It does not need to describe the functional units where the instructions are executed on because SOAS chooses them automatically. It does not need to describe the registers that instructions use because SOAS allocates them. It does not need to describe the parallel information within instructions because SOAS analyzes their parallelism. It does not need to describe the latency and pipeline of instructions because SOAS describes the information automatically according to our architecture [5].

Un-optimized assembly language includes assembly instructions, assembly optimizer directives and comment. The format is showed as follows:

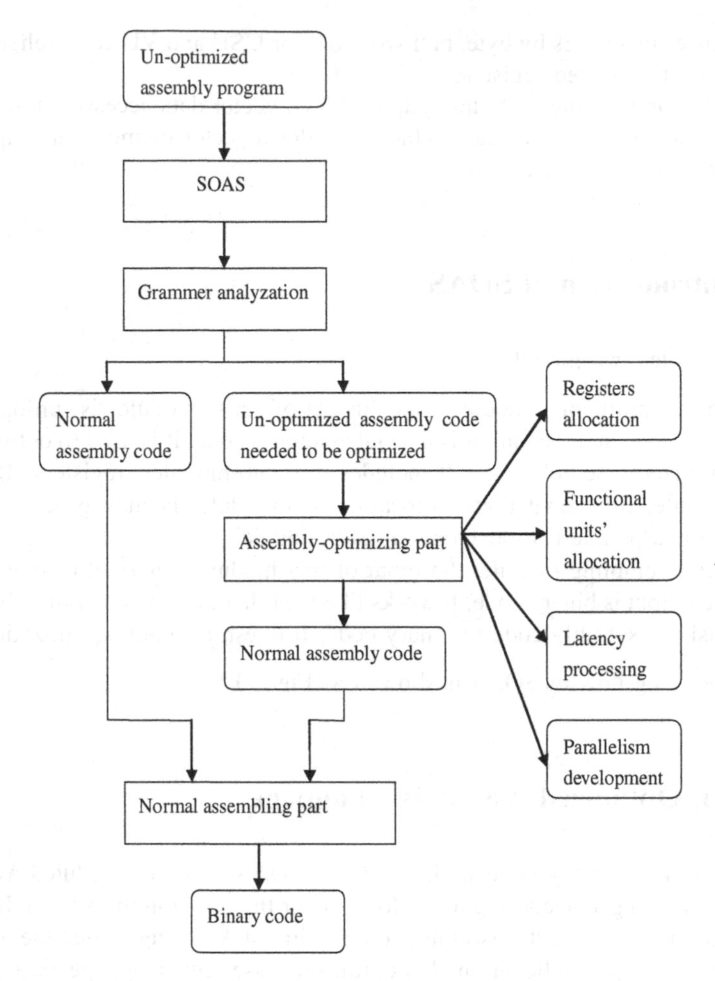

Fig. 53.1 Working flow of SOAS

[label[:]][[register]] mnemonic [unit specifier] [operand list] [;comment]

[register] represents conditional register. The value of the condition register in [ ] is used to judge this instruction is to be executed or not. The mnemonic is an instruction or SOAS directive. [unit specifier] is used to describe functional unit and it is optional. [operand list] is optional too. Not all instructions and directives need operands. The operands can be symbol, constant or expression [6, 7].

Assembly optimizer directives are used to provide the information to control the process of assembly optimizing. Un-optimized assembly program can include un-optimized assembly code and normal assembly code. We use the assembly optimizer directives to distinguish the two kinds of code and provide other information to SOAS. Such as:

Fig. 53.2 Format of un-optimized assembly language

```
                    .global       _function name
_function name      .startp       ;function parameter
                    .reg          ;register valuable name
                    ......
Loop:               .trip
                    instruction [operands]
                    ......
Label :
                    instruction [functional unit]   [operands]
                    instruction [functional unit]   [operands]
                    ......
                    . return
                    . endp
```

".startp" and ".endp" specify the code section needed to be optimized. ".startp" is put at the start of the section and ".endproc"is put at the end of section. Programmers can use them to arrange the code sections needed to be optimized.

".reg" allows programmers to set a descriptive name for each value and these values are stored in registers. When using".reg", SOAS chooses a register for the value and this register is corresponding to the chosen functional unit for the instruction which operates on the value.

".trip" is used to specify the iterative times.

The format of un-optimized assembly program is showed in Fig. 53.2. The text following "," is comment[5–7].

53.5 Assembly-Optimizing Module of SOAS

The key points of software optimizing include fulfilling various resource bound, allocating resource reasonably and making full use of the resource, putting the same work together to form loops in order to do software pipeline, making full use of the locality of instruction and data.

When optimizing our code, we always meet some difficult problems, such as memory conflict, the outer loop can not be software pipelined and so on. If we use un-optimized assembly code to describe these codes and optimize them with SOAS, good efficiency can be achieved. After we use compiler to develop and optimize the code, we can use un-optimized assembly code to rewrite the low-efficient parts and optimize them with SOAS [8].

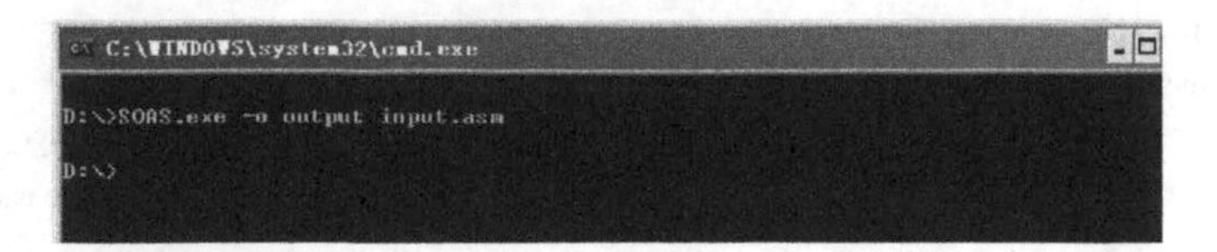

Fig. 53.3 Running SOAS in console

The SOAS assembly-optimizing module contains four sub-modules: Registers allocation module, functional units' allocation module, latency processing module and parallelism development module.

Registers allocation sub-module: registers can be used directly as user's symbols. The real allocation can be completed by SOAS. We use coloring algorithm to allocate the registers. In the allocation of registers, we consider the registers' life cycle and the conflict with the registers which are used directly by programmers. The effect of some special registers which impact the function and performance is also considered.

Functional units' allocation sub-module: In un-optimized assembly code, we can specify functional units or not. Specifying functional units allow programmers to control on which functional unit the instruction will be executed. If programmers don't specify functional units, SOAS will allocate functional units according to the instruction mnemonic and the busy-or-idle state of functional units [9, 10].

Latency processing sub-module: Based on the characteristic of the pipelines of our architecture, this module schedules instructions according to the executing cycle of related instruction in the pipelines. This module is in charge of judging latency and arranging instructions according to the latency slots. It is also in charge of doing software pipeline to arrange loop code and execute multiple iterative of a loop in parallel.

Parallelism development sub-module: this module is used to develop the parallelism among instructions. It completes this work by analyzing data dependency and resource dependency. There are multiple functional units in our architecture. In order to make full use of these functional units, the instructions need to have high parallelism to fill in the effective operation slots. This module uses linear code scheduling algorithm and loop code scheduling algorithm (mainly software pipeline algorithm), including hyperblock scheduling technique and various loop transforming techniques, such as loop titling, strip mining, loop fusion to improve the parallelism [11–13].

53.6 Using of SOAS

SOAS runs as Microsoft Windows console application and it can be ported to many other operating systems. The command line is as follows and the running interface of SOAS is as Fig. 53.3.

SOAS.exe –o outputfile inputfile

53.7 Conclusion

This paper introduces the characteristics of a SDR processor with novel architecture. It also describes the importance of compiling optimization and introduces three-level compiling optimization. Basing on it, this paper gives the implementation of SOAS and introduces the assembly language for SOAS, which is called un-optimization assembly language. SOAS can make programming more convenient than manual assembly optimization. And it can also help developing applications before the high-level language compiler is ready. What's more, it can solve some optimization problems that the compiler can not solve. With SOAS, we can make good use of high performance SDR processor.

Acknowledgment The support from the National Natural Science Foundation of China under grant 60970037 and HGJ(2009ZX01034-001-001-006)is gratefully acknowledged.

References

1. WHW Tuttlebee (2002) Software defined radio: enabling technologies. Wiley
2. Hsu C, Corretjer I, Ko M, Plishker W, Bhattacharyya SS (2007) Dataflow interchange format: language reference for DIF language. Users guide for DIF package, version 1.0. In: Technical report UMIACS-TR-2007-32, Institute for Advanced Computer Studies, University of Maryland, College Park, June 2007
3. Fisher J (1983) Very long instruction word architectures and the ELI-512. In: Proceedings of the 10th annual international symposium on computer architecture, June 1983, pp 140–150
4. Lorenz M, Wehmeyer L, Drager T (2002) Energy aware compilation for DSPs with SIMD instructions. LCTES/SCOPES '02. In: Proceedings of the joint conference on languages, compilers and tools for embedded systems, ACM Press, pp 94–101
5. TMS320C6000 optimizing compiler user's guide. Texas Instruments Inc
6. TMS320C6000 assembly language tools user's guide. Texas Instruments Inc
7. TMS320C6000 programmer's guide. Texas Instruments Inc
8. Yang Y, Xiang P, Kong J, Zhou H (2010) A GPGPU compiler for memory optimization and parallelism management. PLDI'10, 5–10 June 2010
9. Leupers R (2000) Instruction scheduling for clustered VLIW DSPs. IEEE PACT 2000, pp 291–300
10. Lapinskii SV, Jacome FM, De Veciana AG (2002) Cluster assignment for high-performance embedded VLIW processors. ACM Trans Design Autom Electron Syst 7(3):430–454
11. M. L. Software pipelining: an effective scheduling technique for VLIW machines. In: Proceedings of the ACM SIGPLAN 1988 conference on programming language design and implementation. ACM Press, 1988, pp 318–328
12. Allan VH, Jones RB, Lee RM, Allan SJ (1995) Software pipelining. ACM Comput Surv 27(3):367–432
13. Rau BR (1994) Iterative modulo scheduling: an algorithm for software pipelining loops. MICRO 27. In: Proceedings of the 27th annual international symposium on microarchitecture, ACM Press, pp 63–74
14. Ha S, Kim S, Lee C, Yi Y, Kwon S, Joo Y (2007) PeaCE: a hardware–software codesign environment for multimedia embedded systems. In: CM transactions on design automation of electronic systems, Aug 2007

Chapter 54
Design of Greenhouse Monitoring System Based on ZigBee Technology

Yong Chen

Abstract According to greenhouse need to monitor temperature and humidity real-time, ZigBee which was low-power wireless communication technology was used to design temperature and humidity monitoring system. CC2430 radio transceiver is used to send and receive data, to drives temperature and humidity sensor through the I2C bus. The temperature and humidity are sent to the monitoring platform by the ZigBee network. Test results show that the system has the advantages of fast networking, flexible nodes, good scalability and low power consamptian.

Keywords ZigBee · Monitoring system · CC2430

54.1 Introduction

Temperature-controlled greenhouses can provide an all-weather growing environment for crops and protect them effectively from inclement weathers in modern agriculture. Hence, how to monitor and control the temperature and humidity of greenhouses properly is a very important and economical measure to improve agricultural productivity and to maintain safety production [1].

Unfortunately, traditional monitoring and controlling methods of temperature and humidity commonly have the following drawbacks: (1) routine and fixed-point inspection is needed, which consumes much time and labor, (2) generic wiring is difficult, and rewiring is needed as equipments added or removed, (3) installation and maintenance costs are high, while movability is poor. Nowadays, the greenhouse monitoring and controlling systems develop towards low cost, high

---

Y. Chen (✉)
Department of Computer Science, Shaanxi University of Technology, Beijing, China
e-mail: snutchen@126.com

J. J. Park et al. (eds.), *Proceedings of the International Conference on Human-centric Computing 2011 and Embedded and Multimedia Computing 2011*, Lecture Notes in Electrical Engineering 102,
DOI: 10.1007/978-94-007-2105-0_54,

precision, high reliability, unmanned production, informatization, and adopts more and more data processing and controlling technologies, in contrast to simple data collection and displaying ones [2].

With the development of embedded and wireless communication technologies, especially recent advances of ZigBee, people can construct low-power and low-cost wireless network more easily than before, which brings great convenience to design remote greenhouse monitoring and controlling system. In this paper, we use CC2430 RF chips as the wireless transmission devices and SHT11 sensors as the collecting devices of temperature and humidity, design one effective remote greenhouse monitoring and controlling system based on the ZigBee technology [3].

54.2 General Design of System

Usually, there needs to install about 10 $\sim$ 20 monitoring and controlling points in one 1,200 m^2 (40 $\times$ 30) greenhouse. The points and the transmission distances will increase further as more than two greenhouse needing temperature monitoring simultaneously. As one rising two-way wireless communication technique, ZigBee (IEEE 802.15.4 standard) has characteristics of short space, low complicacy, low power consumption, low data rate and low cost. More specifically, it operates at 2.4 GHz ISM frequency band with transmission rate around 10 $\sim$ 250 kbps, and its communication distance can be extended from several hundred meters to several kilometers, even infinity. Hence, ZigBee is very suitable for data transmission in greenhouse monitoring and controlling systems. Figure 54.1 describes the architecture of our designed remote greenhouse monitoring and controlling systems. In each greenhouse, different type of ZigBee nodes (include ZigBee terminal nodes, ZigBee routing node and ZigBee coordination nodes) are connected in one network with treelike topology. In which, routing node is the root node, terminal nodes are the leaves, while coordination nodes are responsible for controlling and coordinating the whole network through ZigBee routing nodes in greenhouses. They are responsible for communicating and monitoring system through the serial port.

54.3 Hardware Design of System

54.3.1 Zigbee Terminal Node

ZigBee terminal node is composed of data collecting module, data processing module, data transmission module and power management module. The data collecting module is responsible for collecting necessary information for greenhouse monitoring and controlling systems through all kinds of sensors. The data processing module is responsible for processing operations, power management, as well as tasks management of the whole node. The data transmission module is

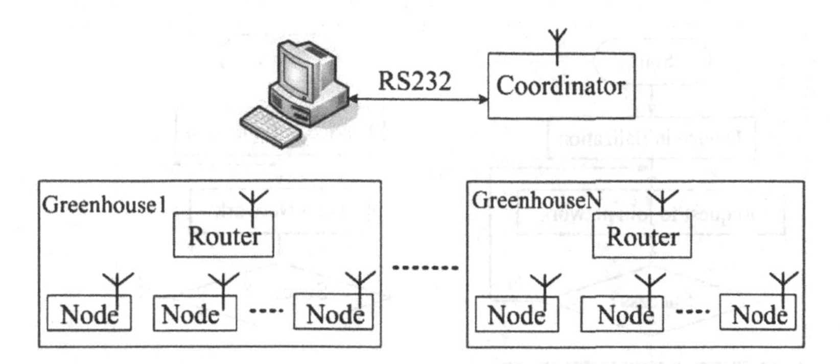

Fig. 54.1 Architecture of remote greenhouse monitoring and controlling system

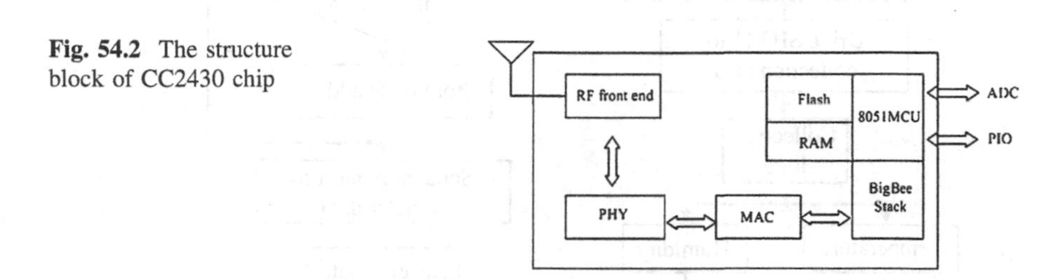

Fig. 54.2 The structure block of CC2430 chip

responsible for wireless communication with other nodes, and transmits data to routing nodes via ZigBee radio wave. The power management module is responsible for power supplies of the whole node. In the designed system, we select CC2430 chip of TI Corporation as hardware developing platform, and attach one SHT11 temperature/humidity sensor of Switzerland Sensirion Corporation to it. The structure block of CC2430 is shown in Fig. 54.2. The CC2430 chip belongs to typical system-on-chip (SoC) chips, in which are embedded one enhanced 8051 single-chip processor (high-performance and low-consumption), as well as ZigBee RF front-end, memory, and microcontroller, so that it can immediately work only needs few peripheral elements to constitute a clock circuit and a RF bias circuit. The SHT11 sensor is one new intelligent digital sensor with I^2C bus interface, on the basis of CMOSens technology, which is suitable for measuring of relative humidity, temperature, dew point and other parameters. The measuring range of temperature is about $-40 \sim 120°C$, and that of relative humidity is about $0 \sim 100\%$, which can meet the needs of greenhouse entirely [4].

54.3.2 ZigBee Routing Node

The main responsibility of ZigBee routing node is to transmit data of different area from sensor nodes to coordination nodes. Thus, the circuit is simply composed of CC2430 wireless transceiver, RF antenna, power module and crystal oscillating circuit.

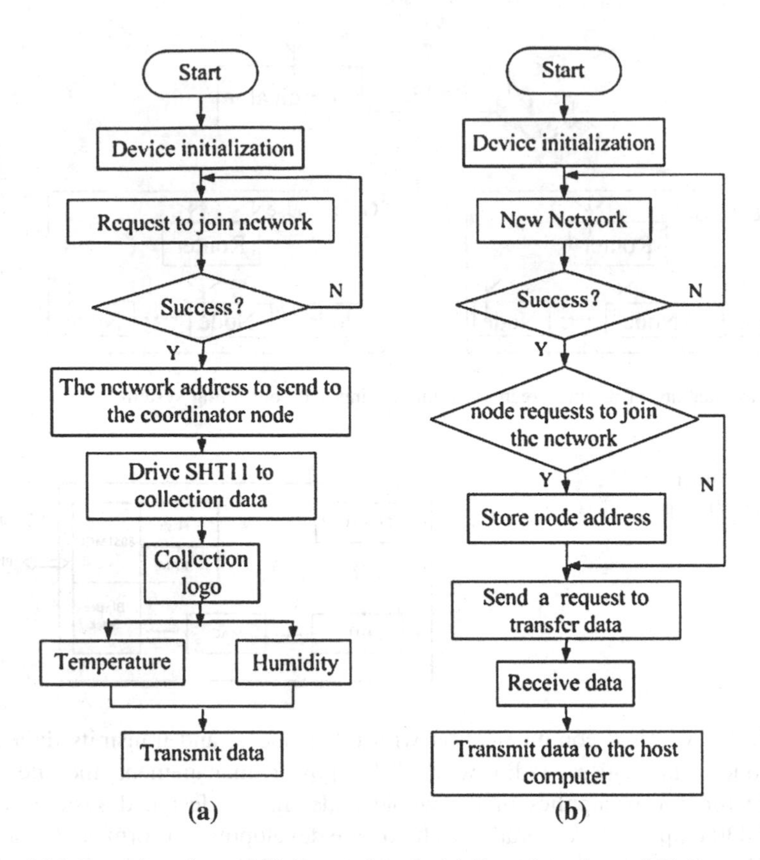

Fig. 54.3 Procedure flow of ZigBee end node (**a**) Procedure flow of ZigBee end node and coordinator node (**b**)

54.3.3 ZigBee Coordination Node

The major responsibility of ZigBee coordination node is to receive the temperature and humidity data from all ZigBee routing node. The collected data are processed, sent to the host computer through the serial port, serial port monitoring system to read and write through to show sheds a collection point or the temperature and humidity control of a point. This node constitutes of CC2430 wireless transceiver, RF antenna, power module, crystal oscillating circuit, as well as one serial port for transferring signals to PC device.

54.4 Software Design of System

Based on the standard ZigBee protocol, we use Z-Stack software development platform (in compliant with ZigBee2006 specification) to realize software component in our designed system. Relevant application programs are programmed with C language using IAR Embedded Workbench tools, to implement functions of network constructing, binding establishing, data collecting, data sending and receiving etc. The flowchart of program in terminal nodes is presented as Fig. 54.3a. The ZigBee terminal nodes initialize themselves before ask for joining the network (having treelike topology), then they call sensors to collect temperature and humidity data according known collecting signs. After that, they process the collected data before send them to ZigBee coordination nodes. Figure 54.3b gives flowchart of program in coordination nodes. Since coordination nodes need network address of each sensor, they require that all sensors must send their addresses before joining the network. Then coordinators can establish and store address table, so that the collected sensor data is in accordance with the table.

54.5 Conclusion

Combing characteristics of greenhouses in modern agriculture production, we design one remote monitoring and controlling system by utilizing the ZigBee wireless communication technology in this paper. The system has merits of good scalability, low developing costs, high performance-price ratio, convenient installation and maintenance etc., and can work for a long time with one single installation. Furthermore, our system solves difficulties of generic wiring and relocation much better than traditional monitoring systems, and very suitable for applying in environmental monitoring fields.

References

1. Zhao-hui J et al (2010) Design of general agricultural wireless monitoring system based on ZigBee. J Anhui Agric Sci 38:3419–3451
2. Qi-shen Z (2010) Temperature monitoring system of greenhouse based on CAN bus. J Anhui Agric Sci 38:4241–4243
3. Zhao-jun L (2009) Wireless transmission design of greenhouse environment monitoring system based on ARM. J Anhui Agric Sci 37:11188–11189
4. Na P, De-fu C (2010) Design of greenhouse monitoring system based on ZigBee wireless sensor networks. J Jilin Univ 28:55–60

Chapter 55
Preheating Simulation of Power Battery

Maode Li, Feng Wang and Duying Wang

Abstract When under lower temperature state such as in the winter, power battery of the electric car has the lower capacity which is easy to be decreased and even to be damaged. So it is necessary to preheat the battery before to use. To understand the preheating process, the simulation of power battery with temperature rising is simulated under the cases of forced convections. The properties of temperature rising in preheating process of the battery are discussed under different air velocities and temperatures.

Keywords Battery · Convection · Preheating · Simulation

55.1 Introduction

Because the capacity of output power of power battery under lower temperature is lower than that of under normal environment condition, the more output power of battery is needed when the electric car starts. So internal heat dissipation will be highly increased with the load. This will also cause some other effects such as function degradation of battery poles, separator film and electrolyte. With different heat effects of electrical-chemical reaction, different kinds of batteries will work well only in its self ranges of temperature, output ratio and current. During the process of charging and discharging, with the inserting or deserting of active ions, the chemical and physical properties of battery poles, separator film and electrolyte may be changed in some irreversible ways when under a condition of non

M. Li (✉) · F. Wang · D. Wang
College of Mechanical Engineering, Tongji University,
4800 Cao-an Road, 201804 Shanghai, China
e-mail: limaode@tongji.edu.cn

J. J. Park et al. (eds.), *Proceedings of the International Conference on Human-centric Computing 2011 and Embedded and Multimedia Computing 2011*, Lecture Notes in Electrical Engineering 102,
DOI: 10.1007/978-94-007-2105-0_55, © Springer Science+Business Media B.V. 2011

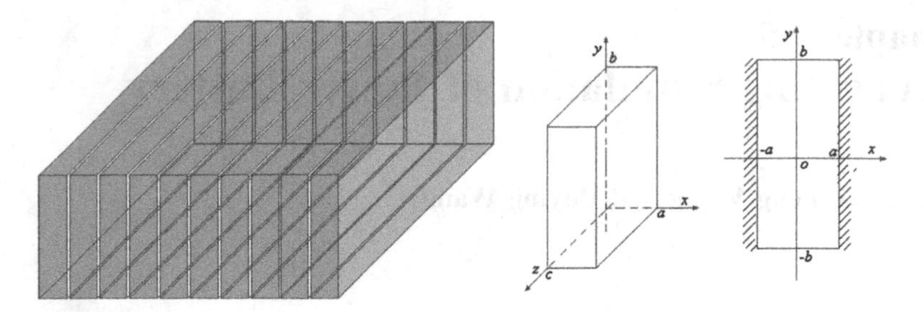

Fig. 55.1 Battery model

Table 55.1 Parameters of battery

Battery type	Mass (kg)	Size (mm × mm × mm)	Density (kg/m^3)	Special heat (J/kg.K)	Conductivity (W/m.K)
$LiFePO_4$ (36 V/10Ah)	1.5	80 × 120 × 150	1,952	1,292	0.62

proper parameter ranges. Even more, when under higher temperature condition, the separator film will tend to easy to be decomposing and with a worse insulation [1–3]. So for its safety of battery using, it is very important to ensure the proper range of temperature.

For lithium ion battery, its internal resistance will be increased obviously when in a lower temperature state, so to cause a higher inner power dissipation and lower output power. The another important factor for lithium ion battery is economical value which is always relative with usage life. So, it is necessary to preheat the power battery when in a lower temperature state [4–6].

55.2 Simulation Model

55.2.1 Basic Parameters of Battery

The lithium ion battery used to analyze and simulate is as in Fig. 55.1, and its physical parameters as in Table 55.1. In Fig. 55.2, the aluminum shell of battery is shown.

To simplify the battery block as the geometric model as in Fig. 55.1, which will be as the model to make simulation of temperature rising. The practical battery will be packed with some special materials, here the aluminum shell of battery as in Fig. 55.2 is supposed, which has not inner heat produced, but with the heat transferred. The out shape and shell construction will effect the heat transfer with environment, so it is needed to simulate together with the battery.

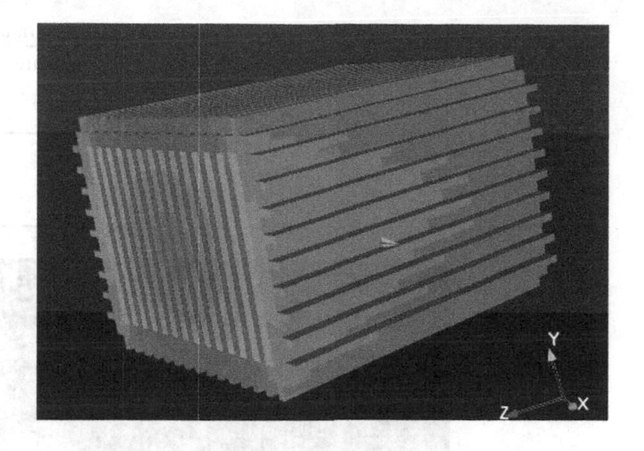

Fig. 55.2 Aluminum shell of battery

With the aluminum shell of battery as in Fig. 55.2, the heat transfer of battery will be better and practical. The preheating process will be divided into heat transfer model of aluminum shell and battery block. The initial temperature of battery block and aluminum shell is taken as −20°C, through heat conduction, to heat the temperature of battery center to be 15°C, so to keep a better thermal condition for electric chemical process.

55.2.2 Mathematical Model of Battery and Aluminum Shell

According to heat transfer theory, the basic equation of battery block can be taken as (55.1) and boundary conditions as (55.2).

$$\rho c_p \frac{\partial t}{\partial \tau} = \lambda \left(\frac{\partial^2 t}{\partial x^2} + \frac{\partial^2 t}{\partial y^2} + \frac{\partial^2 t}{\partial z^2} \right) + q \tag{55.1}$$

$$\begin{aligned} \left(\frac{\partial t_1}{\partial n} \right)_{ni} &= \left(\frac{\partial t_2}{\partial n} \right)_{ni} \\ t_{1,ni} &= t_{2,ni} \end{aligned} \tag{55.2}$$

The equation for aluminum shell is taken as (55.3) and its boundary condition as (55.4).

$$\rho c_p \frac{\partial t}{\partial \tau} = \lambda \left(\frac{\partial^2 t}{\partial x^2} + \frac{\partial^2 t}{\partial y^2} + \frac{\partial^2 t}{\partial z^2} \right) \tag{55.3}$$

$$-\lambda \left(\frac{\partial t_2}{\partial n} \right)_{nj} = \alpha (t_2 - t_f) \tag{55.4}$$

Table 55.2 Conditions of preheating with constant air temperature

Conditions	Initial temperature (°C)	Inlet temperature (°C)	Inlet velocity (m/s)	Temperature of battery center (°C)
Forced heat convection	−20	40	1	15
Al shell with fin			3	
			5	

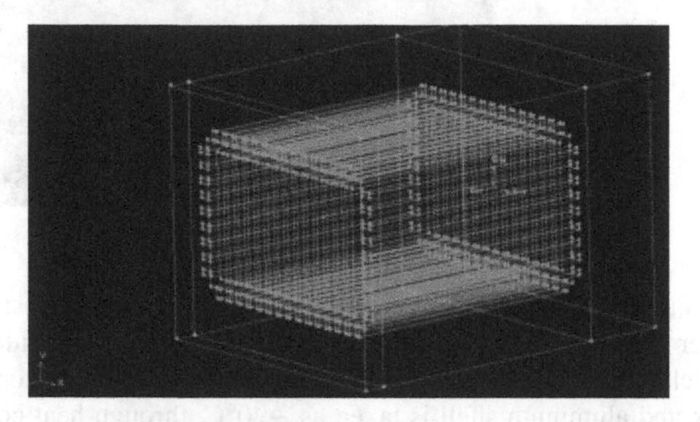

Fig. 55.3 Geometry model of battery and shell

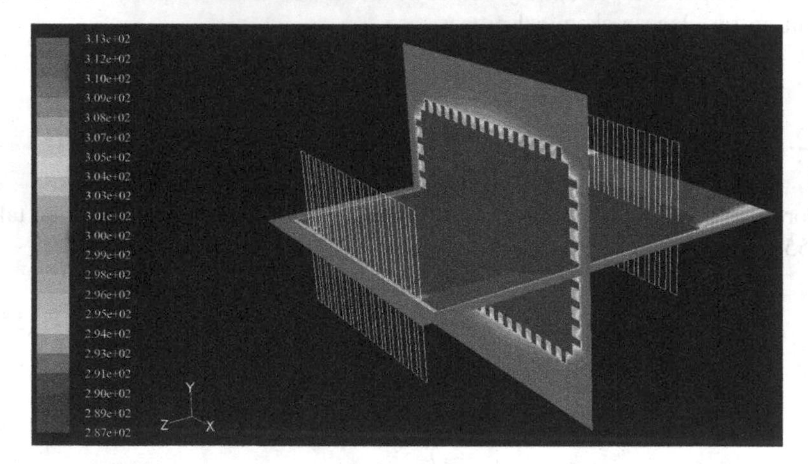

Fig. 55.4 Temperature distribution when heating with 1 m/s air velocity

55.3 Simulation Results

55.3.1 Conditions of Preheating With Constant Air Temperature

In the preheating process, it is supposed that with forced convection, initial air temperature is 40°C, with different air velocity as 1, 3 and 5 m/s, respectively, as in Table 55.2. By means of Fluent Software, some important and representative results are shown as in Figs. 55.3, 55.4, 55.5, 55.6 and 55.7.

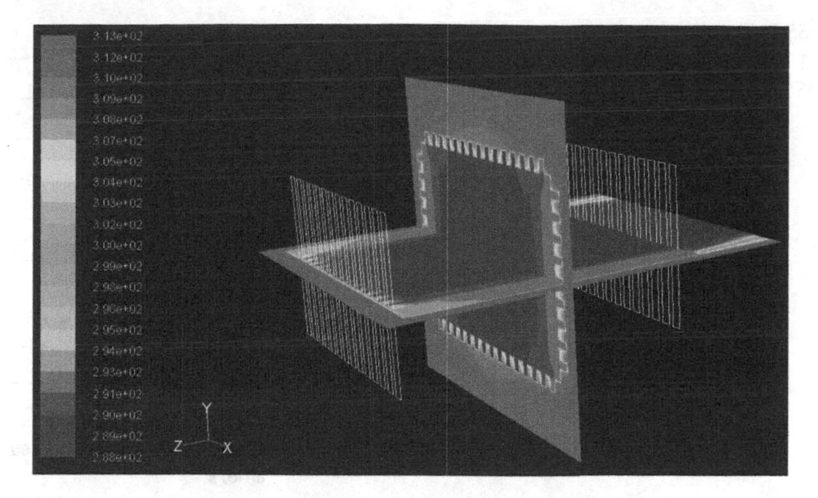

Fig. 55.5 Temperature distribution when heating with air velocity of 3 m/s

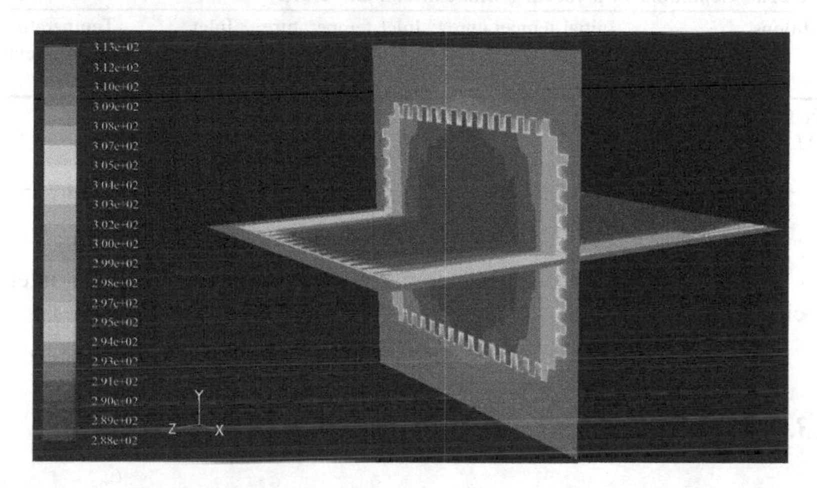

Fig. 55.6 Temperature distribution when heating with air velocity of 5 m/s

From Figs. 55.3, 55.4, 55.5, 55.6 and 55.7, it shows that, for a inlet air given temperature, the bigger the air velocity, the shorter preheating time needed, and preheating effect will be better. When inlet air temperature to 40°C and air velocity to 1 m/s, the preheating time $\Delta\tau_s$ which is heating from −20 to 15°C reaches to 2,200 s. After increasing air velocity from 1 to 3 and 5 m/s, the preheating time $\Delta\tau_s$ will be to 1,080 and 720 s, respectively. So from the viewpoint of heating effect of heating speed, to increasing the inlet air temperature will be helpful which can decrease the preheating time. But from another point of heating effect of battery characteristic, the distribution of temperature difference with bigger inlet air velocity will obviously bigger than that of smaller air velocity, which causes

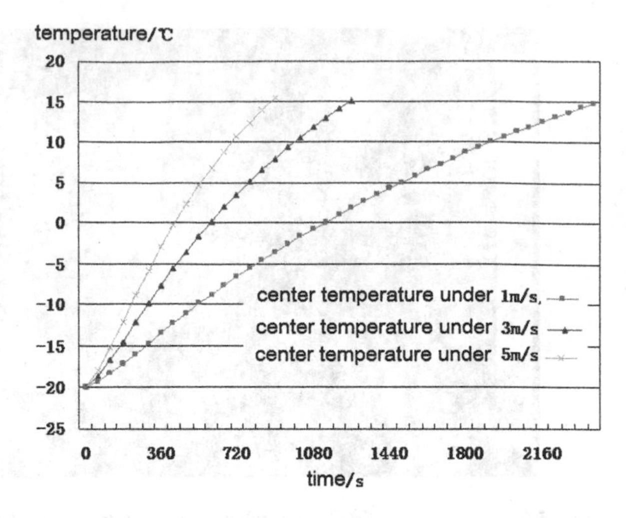

Fig. 55.7 Comparison of center temperatures under different air velocities

Table 55.3 Conditions of preheating with constant air velocity

Conditions	Initial temperature (°C)	Inlet temperature (°C)	Inlet velocity (m/s)	Temperature of battery center (°C)
Forced heat convection Al shell with fin	−20	40 50 60	4	15

inner variation of electric chemical property, and the more the bigger inlet air velocity, which is harmful to battery life.

55.3.2 Conditions of Preheating With Constant Air Velocity

In this kind of preheating process, it is supposed that with forced convection, initial air temperature is 40, 50 and 60°C respectively, with air velocity of 4 m/s, as in Table 55.3. By means of Fluent software, some important and representative results are shown as in Figs. 55.8, 55.9, 55.10, 55.11 and 55.12.

From Figs. 55.8, 55.9, 55.10, 55.11 and 55.12, it can be seen that, for a given inlet air velocity, the preheating time $\Delta\tau_s$ which is heating from −20 to 15°C can be shortened obviously by means of increasing the inlet air temperature. When inlet air temperature as 40°C and air velocity as 4 m/s, then the preheating time $\Delta\tau_s$ will be to about 1,080 s, and it will be 900 and 600 s when inlet temperature as 50 and 60°C, respectively. The inner variation of electric chemical property which is caused by temperature difference in battery block, will be improved because of the smaller difference of temperature difference in battery block.

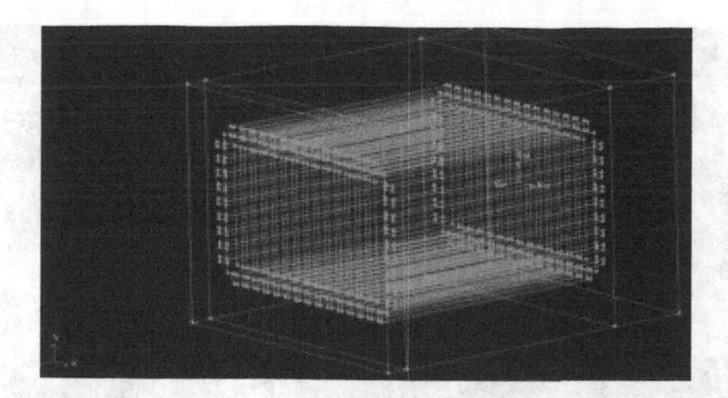

Fig. 55.8 Geometry model of battery and Al shell

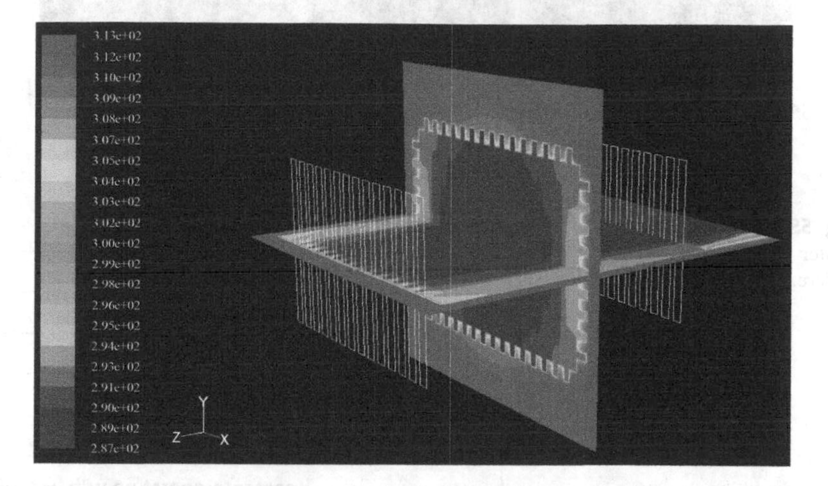

Fig. 55.9 End temperature distribution heating with air temperature 40°C

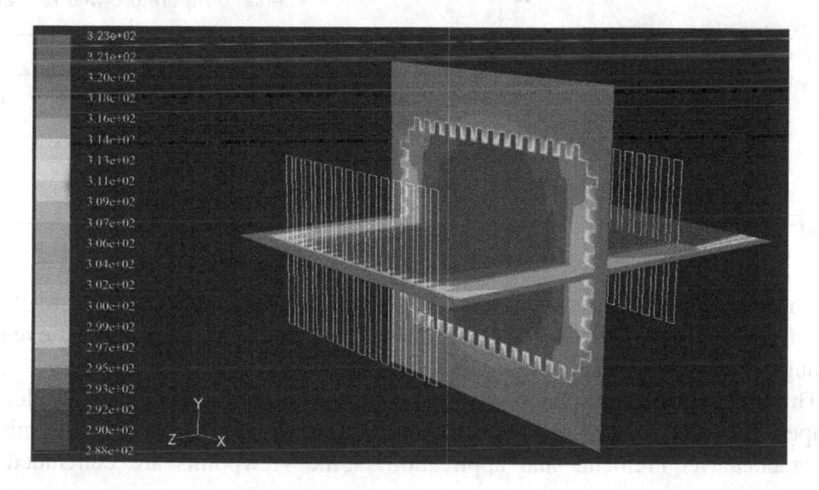

Fig. 55.10 End temperature distribution heating with air temperature 50°C

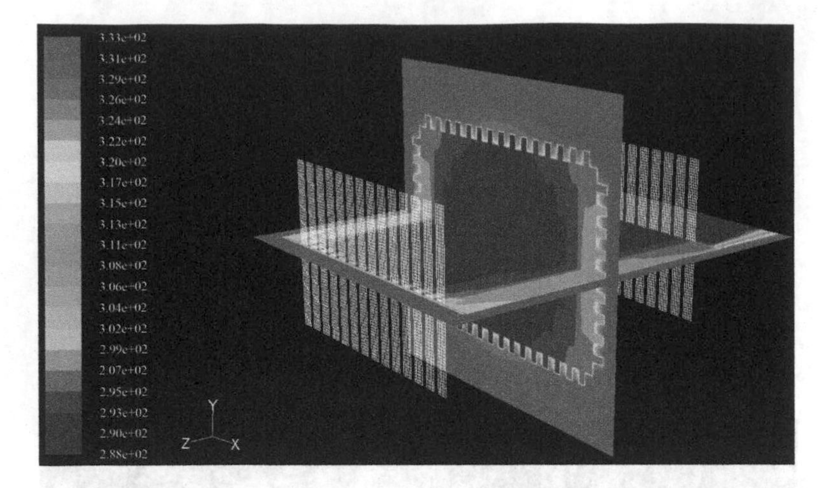

Fig. 55.11 End temperature distribution heating with air temperature 60°C

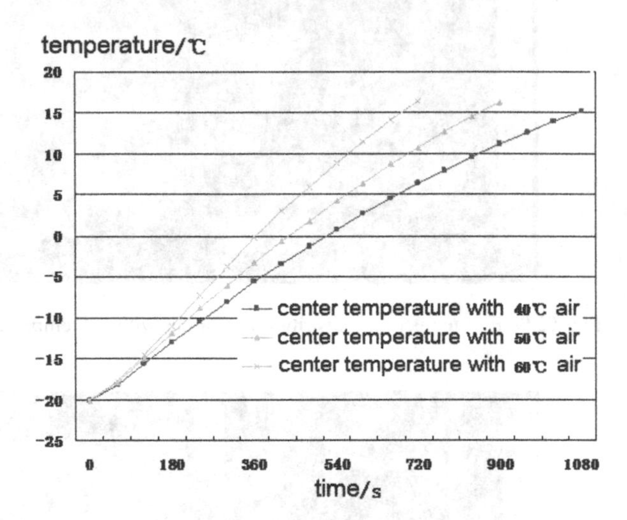

Fig. 55.12 comparison of center temperature with different air temperatures

55.4 Conclusions

When Lithium—ion power battery in a lower temperature, Power output of battery will decrease greatly because of electric chemical reaction weakened and even to damage its innerconstruction and to shorten its life.

Through the above simulation and analysis, several results of different inlet air temperature and velocity are compared about its preheating process. To combine its practical requirements and application, some viewpoints are concluded as following.

a. In the preheating process, to heat the battery with the higher temperature air is more effective, its heat time will be shortened. From the results, when inlet temperature as 50°C and velocity as 4 m/s, its time heating from −20 to 15°C is about 15 min.

b. When heating with higher velocity but with a lower temperature, although its heating speed will be fastened and preheating time shortened, but the bigger temperature difference from battery boundary to its center will cause a dis-equilibrium of electric chemical reaction in the battery, so as to damage its inner construction and to decrease its using life.

Actually, the preheating process is concerned with many other factors such as its construction and inner material. From Some reports, it is concluded that pre-heating process is useful to maintain the battery and life. The simulation and analysis made here is only the primary step, which is to be researched further for satisfactory the need of application.

References

1. Pals CR, Newman J (1995) Thermal modeling of the lithium/polymer battery. J Electrochem Soc 42:3274
2. Chen S, Wan C, Wang Y (2005) Thermal analysis of lithium-ion batteries. J Power Sour 140(1):111–124
3. Gomadam PM, Weidner JW, Dougal RA (2002) Mathematical modeling of lithium ion and nickel battery systems. J Power Sour 110(2):267–284
4. Kim GH, Pesaran A, Spotnitz RA (2007) Three-dimensional thermal abuse model for lithium-ion cells. J Power Sour 170(2):476–489
5. Wang C, Srinivasan V (2002) Computational battery dynamics (CBD) electrochemical thermal coupled modeling and multi-scale modeling. J Power Sour 110(2):364–376
6. Wu ZJ, Chen G, Lee A (2000) Prediction of internal temperature of a battery using a non-linear dynamic model. Google Patents

Chapter 56
Intelligent Guide System Based on the Internet of Things

Jintao Jiao and Hongji Lin

Abstract The application of intelligent guide system involves computer technology, wireless communication technology, positioning technology, voice technology, tourism information database technology, and how to develop high-quality service resources and accurately deliver the services to the tourists is the focus for intelligent guide development. Intelligent guide system is designed in this paper. Three-tier solution that contains cloud service center, guide sensor network and phone program is proposed; this paper also describes the solution of each tier. Guide service can be provided to tourists much better.

Keywords Cloud Service · Sensor · Web service · Internet of things

56.1 Introduction

Guide lies in the center of the travel services, which links booking, accommodation, catering, transportation, sightseeing, shopping and entertainment. So the value of other service providers can be achieved [1].

As the self-guided tour is becoming more common; the traditional guide form can not meet the needs of those tours. Driven by demand, intelligent guide system appears which has broad prospects for development. Because the system is not attached on traditional guide service, it must be used by those tourists if they want to travel by themselves.

J. Jiao (✉) · H. Lin
Department of Computer Science, Wuyi University,
Wuyi Road No. 16, Wuyishan, Fujian, China
e-mail: Jiaojintao@163.com

H. Lin
e-mail: Lhj057@163.com

J. J. Park et al. (eds.), *Proceedings of the International Conference on Human-centric Computing 2011 and Embedded and Multimedia Computing 2011*, Lecture Notes in Electrical Engineering 102,
DOI: 10.1007/978-94-007-2105-0_56,

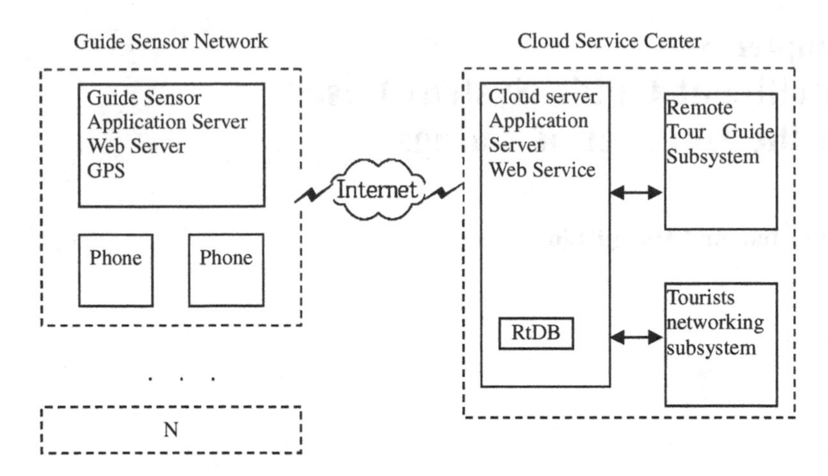

Fig. 56.1 Cloud Services Architecture

56.2 Cloud Services Architecture of Guide System

Intelligent guide use three parts—smart phone operation system, guide sensor network and cloud service center for the technology platform. Through it explain and tourist information services can be provided. Cloud services architecture is shown in Fig. 56.1.

First of all, establish tourism information, and then establish guide sensor based guide private network. Tourist use wireless communication technology to connect to the guide sensor network, and get location information through the tourist handheld Global Positioning System (GPS); Cloud service center can get the information through the wireless network deployed in the scenic. The phone program can get voice data from the guide sensor network; cloud service center can also provide other services that the tourists need.

56.3 System Design

The system mainly consists of three parts: cloud service center, guide sensor network and phone programs. The following describes how to implement the three parts.

56.3.1 Cloud Service Center

Create guide dedicated server, using Java EE technology, by creating Web Services for remote database operations. Figure 56.2 shows how the center works.

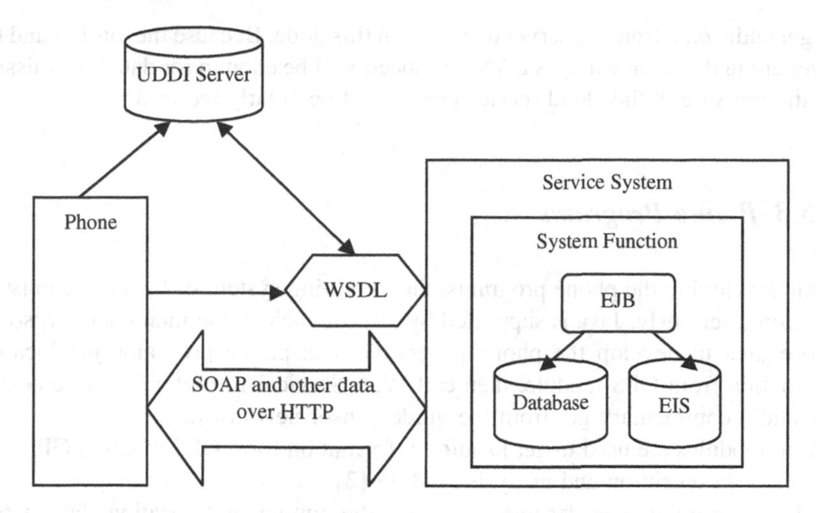

Fig. 56.2 Cloud Service Center Works

There are two major modules in cloud service center.

1. Remote tour guide subsystem. Tourist calls Web Service deployed in the service center through the mobile phone, then the cloud service center can get the location of tourists. After compared the location and the existing information, one Web server in the guide sensor network will be chosen to provide audio commentary service for the tourist, then the cloud service center will send an order to the chosen server. Also secure warming service is available for the tourist who is in danger.
2. Tourists networking subsystem. The main function of this module is to record tourist travel records and to provide download service. After analysis the records, the system can also optimize the tour route in able to provide better services.

56.3.2 Guide Sensor Network

Nowadays, the smart phones are equipped with Wi-Fi module generally, and the prices is not too high. The system use Wi-Fi phones to connect to the sensor deployed in the scenic. After perceived visitors reach the sensors work scope, the system can use the event sensor technology and identification technology to sense and collect the network data; The service center can receive sensor network data, and then provide tourists their required services.

In order to build the guide sensor network well, the system use genetic algorithms to identify the location of each sensor. Each network node contains two parts, wireless router and a server, so that when tourists connect to the wireless router they can get audio data from the server deployed in this node. Because the tourists and the server are in the same wireless LAN, the speed will be enough for data transmission and the pressure of the cloud service center will be greatly reduced.

56.3.3 Phone Programs

When developing the phone programs, the operating system of the phone must be considered seriously. Java is supported by almost each phone manufacturer, so we choose java to develop the phone programs. The phone programs get location information from GPS module, then call Web Service from cloud service center, play audio commentary get from the guide sensor network [2].

Two modules are used to get location information form GPS module: GPS data module, data reception and analysis module [3].

GPS data module is designed to store the timing and location data, while provide methods to access an update these data. Data reception and analysis module update these data by calling method SetXyz(), use these data by calling method GetXyz().

Data reception and analysis module is the core of the program. This module monitors the GPS module and gets location information from the data output by the GPS module. The core of this module is a thread which is designed as an inner class. External program can start and stop the thread by the method start() and stop(). In method start() the boolean variable named DevOn is set to true, while in method stop() DevOn is set to false. The thread continuously monitors the value of DevOn, if the value is changed to false then terminated itself. Because each information get from GPS module is ended by enter or newline symbol, the program use BufferedReader to store these information. Then the method readLine() can be used to read each line from the buffer.

56.4 System Features

Tour location and event sensing: scenic tourist groups location service, safety hazards and emergency events sensor.

Visitors statistics: tourist groups tour information networking, tourists travel records statistics.

Tour guide service: to provide automatic voice service, to provide mobile phone one button for help, one key guide service and provide tourist traveling record download service.

Phone guides and management control: cloud service center provides authorized remote management, providing early warning to prevent safety hazards and safety emergency [4], providing scenic flow management, phone emergency remote control and linkage.

56.5 Conclusion

The limitations of the system is the requirement of phone supports Wi-Fi and GPS, so only about 40% of the phone can be covered. Also tourists need to download the java program to use the service. The video service is not available for the requirement of higher bandwidth.

Technology gradually changing human habits, but it also brings the convenience of life. Because of the high mobile phone penetration rate, mobile phone and guides are combined to make mobile phones become a tour guide habits. In the tourist activities, mobile phone will play a more and more important role. I believe that there will be more tourists like the phone guide service model in the future.

References

1. Cheverst K, Davies N, Mitchell K, Friday A, Christos Efstratiou (2000) Developing a context-aware electronic tourist guide: some issues and experiences. In: Proceeding of the SIGCHI conference on human factors in computing systems, pp 17–24
2. Wakkary R, Hatala M (2007) Situated play in a tangible interface and adaptive audio museum guide. Pers Ubiquitous Comput 3(11):171–191
3. Malet JP, Maquaire O, Calais E (2002) The use of Global Positioning System techniques for the continuous monitoring of landslides: application to the Super-Sauze earthflow. Geomorphology 5(43):33–54
4. Chakravarthy A, Song KY, Feron E (2004) A GPS_based slow down warning system for automobile safety. IEEE intelligent. Vehieles symposium, pp 241–244